Advances in the Biology of Turbellarians and Related Platyhelminthes

Developments in Hydrobiology 32

Series editor
H.J. Dumont

Advances in the Biology of Turbellarians and Related Platyhelminthes

Proceedings of the Fourth International Symposium on the Turbellaria held at Fredericton, New Brunswick, Canada, August 5–10, 1984

Edited by
Seth Tyler

With the assistance of the Editorial Board:
Reinhard M. Rieger
Julian P.S. Smith, III
Michael D.B. Burt
Nathan W. Riser
Barbara M. McKinnon

Reprinted from Hydrobiologia, vol. 132 (1986)

1986 **DR W. JUNK PUBLISHERS**
a member of the KLUWER ACADEMIC PUBLISHERS GROUP
DORDRECHT / BOSTON / LANCASTER

Distributors

for the United States and Canada: Kluwer Academic Publishers, 190 Old Derby Street, Hingham, MA 02043, USA
for the UK and Ireland: Kluwer Academic Publishers, MTP Press Limited, Falcon House, Queen Square, Lancaster LA1 1RN, UK
for all other countries: Kluwer Academic Publishers Group, Distribution Center, P.O. Box 322, 3300 AH Dordrecht, The Netherlands

Library of Congress Cataloging in Publication Data

International Symposium on the Turbellaria (4th :
1984 : Fredericton, N.B.)
Advances in the biology of turbellarians and related
platyhelminths.

(Developments in hydrobiology ; 32)
"Reprinted from Hydrology vol. 131 (1985)."
Includes index.
1. Turbellaria--Congresses. 2. Platyhelminththes--
Congresses. I. Tyler, Seth. II. Rieger, Reinhard M.
III. Title. IV. Series.
QL391.P7I57 1984 595.1'23 85-23984
ISBN 90-6193-542-3

ISBN 90-6193-542-3 (this volume)
ISBN 90-6193-751-5 (series)

Cover design Max Velthuijs

Copyright

PRINTED IN THE NETHERLANDS

Preface

While the reality of the taxon Turbellaria has been called into question lately, turbellarians are nevertheless the subject of active research by a sizable group of biologists. Turbellarians are relatives of the major groups of parasitic platyhelminthes – monogeneans, digeneans, and tapeworms – and most are free-living. Because the ancestors to the major parasitic groups would be classified as turbellarians, strict application of principles of phylogenetic systematics dictates that the Turbellaria is not properly considered a separate taxon; i.e., it is, in the parlance of systematics, a paraphyletic group. The relationships of turbellarians to other invertebrates are even more problematic than their relationships to other platyhelminthes; their relatively simple morphology has been variously interpreted as quintessentially primitive – meaning a turbellarian-like ancestor would have given rise to most of the major groups of invertebrates – or as secondary simplification, meaning they would essentially be a dead-end group.

Modern research on turbellarians covers a broad spectrum. Questions of phylogenetics have inspired ultrastructural studies; the simply structured nervous systems of turbellarians make them good subjects for neurophysiology; simplicity of their tissue structure and the limited number of cell types make them good subjects of embryological and regeneration studies; they are emerging as important indicator species in ecology; and improvements in biochemical methodology have meant they are at last amenable – despite their small size – to molecular biological study.

The 45 papers in this volume are representative of current research on turbellarians. These papers were presented at the Fourth International Symposium on the Turbellaria held on the campus of the University of New Brunswick, Fredericton, Canada, in August of 1984. As with the other three symposia in this series (1970 at Chicago; 1977 at Tvaerminne, Finland; and 1980 at Diepenbeek, Belgium), invitations were sent to all biologists publishing research dealing with turbellarians; but the scope of this particular symposium was expanded with invitations also to biologists working on parasitic platyhelminth taxa. It was hoped by this to open the way to gaining a better perspective on the position of the Turbellaria in evolution and to bridge the artifical gap that now stands between studies on parasitic and studies on free-living platyhelminthes. Hence, the title of this volume departs from the tradition followed in the published proceedings of the preceeding symposia*: obviously the scope is broadened with inclusion of other platyhelminths related to the turbellarians, and, in deference to phylogenetic systematists, we use the term 'turbellarian' in its vernacular sense.

The program for the symposium encompassed 70 titles submitted for presentation by scientists from 19 countries. The task of managing the program as well as the accommodation, feasting, and entertainment of the participants was handled in a truly herculean manner by Dr. Michael D. B. Burt, Chairman of the Department of Biology at the University of New Brunswick, with help from many members of that Depart-

* Proceedings of the other symposia are published as

- Riser, N. W. & M. P. Morse (eds), Biology of the Turbellaria. McGraw-Hill, New York, 533 pp.
- Karling, T. G. & M. Meinander (eds), The Alex. Luther Centennial Symposium on the Turbellaria. Acta zool. fenn. 154: 1–207.
- Schockaert, E. R. & I. R. Ball (eds), The Biology of the Turbellaria, Developments in Hydrobiology 6. Dr. W. Junk Publishers, The Hague, Boston, London, 301 pp.

ment. Undaunted by the unexpectedly sultry weather of those August days, he directed all of the local arrangements, making the entire gathering feel royally entertained. Other members of the organizing committee, with admittedly smaller roles, were Reinhard Rieger, N. W. Riser, and Seth Tyler.

We gratefully acknowledge the University of New Brunswick for so graciously making its facilities available to the symposium and for the financial support it provided. We thank, also, for financial support, the City of Fredericton, the Natural Sciences and Engineering Research Council of Canada, and the US National Science Foundation (Systematic Biology Program, grant BSR 8401462).

For help in reviewing manuscripts, the Editorial Board thanks Ian R. Ball, Daniel R. Brooks, J. Richard Cook, Stephen R. Fegley, Roman Kenk, Marianne D. Klauser, J. Klima, Irving L. Kornfield, Charles W. Major, Edward E. Ruppert, Mary Beth Thomas, J. M. Turbeville, W. Wieser, and Bonnie G. Wood.

Seth Tyler

Contents

ONTOGENY

REPRODUCTIVE SYSTEM

REGENERATION

SYSTEMATICS AND CYTOLOGY

ECOLOGY

SYMBIOTIC RELATIONSHIPS

HISTORY

APPENDIX

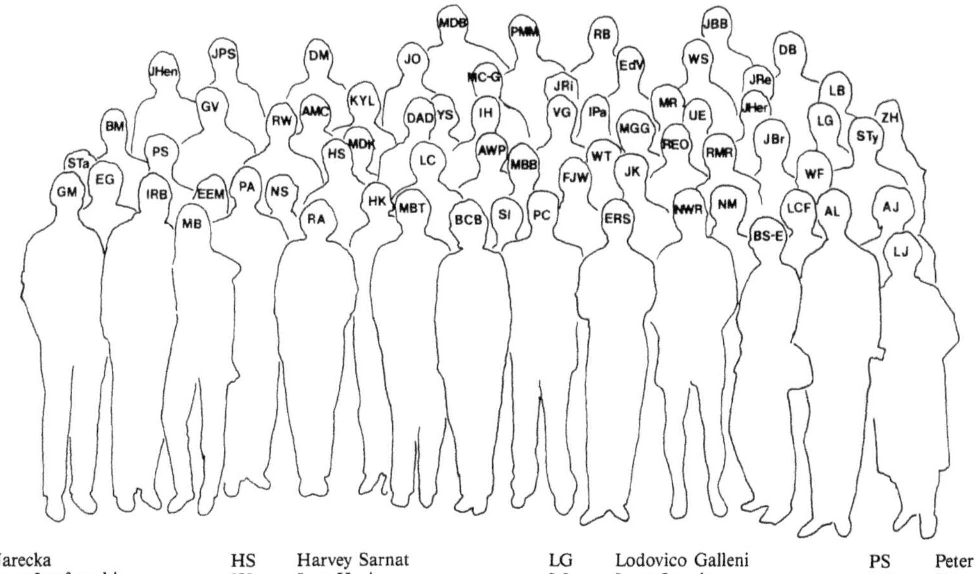

AJ	A. Jarecka
AL	Alberto Lanfranchi
AMC	Aage Møller Christensen
AWP	Alan W. Pike
BCB	Barbara C. Boyer
BM	Barbara M. MacKinnon
DAD	David A. Doe
DB	Don Beattie
DM	Daniele Murith
EdV	Elisabeth J. de Vries
EEM	Els E. Martens
EG	E. Ghirardelli
ERS	Ernest R. Schockaert
FJW	Frederick J. Wrona
GM	Göran Malmberg
GV	Gerard Van der Velde
HK	Harold Koopowitz
HS	Harvey Sarnat
IH	Isao Hori
IPa	Irmeli Palmberg
IRB	Ian R. Ball
JBB	Jay Boyd Best
JBr	Jochen Brüggemann
JHen	Jan Hendelberg
JHer	Jan Hermann
JK	Jerzy Kolasa
JO	Jean Ogren
JPS	Julian P. S. Smith, III
JRe	J. Reuter
JRi	Jean Riser
KYL	Kwang-Yang Lue
LB	Louise Bush
LCF	Lesley C. Fleming
LC	Leser Cannon
LG	Lodovico Galleni
LJ	Lena Jarecka
MBB	Mary Beverly-Burton
MB	Maria Ball
MBT	Mary Beth Thomas
MC-G	Marco Curini-Galletti
MDB	M. D. B. Burt
MDK	Marianne D. Klauser
MGG	M. Gabriella Gremigni
MR	Maria Reuter
NM	Naoya Makino
NS	Naomi Sarnat
NWR	Nathan W. Riser
PA	Peter Ax
PC	Peter Calow
PMM	Paul M. Martens
PS	Peter Schwank
RA	Renate Ax
RB	Rod Bray
REO	Robert E. Ogren
RMR	Reinhard M. Rieger
RW	Robert Wink
SI	Sachiko Ishida
STa	Sachiko Tamura
STy	Seth Tyler
UE	Ulrich Ehlers
VG	Vittorio Gremigni
WF	Witold Ferens
WS	Wolfgang Sterrer
WT	Wataru Teshirogi
YS	Yasuko Shirasawa
ZH	Zenia Heller

List of participants and contributors

Peter Ax, II. Zoologisches Institut und Museum der Universität, Berliner Strasse 28, D-3400, Göttingen, Federal Republic of Germany.

Renate Ax, II. Zoologisches Institut und Museum der Universität, Berliner Strasse 28, D-3400, Göttingen, Federal Republic of Germany.

Ian R. Ball, Instituut voor Taxonomische Zoologie, Universiteit van Amsterdam, Postbox 20125, NL-1000 HC Amsterdam, The Netherlands.

Maria Ball, Instituut voor Taxonomische Zoologie, Universiteit van Amsterdam, Postbox 20125, NL-1000 HC Amsterdam, The Netherlands.

Don Beattie, Department of Biology, University of New Brunswick, Fredericton, N.B., Canada E3B 6E1.

Celina Bedini, Istituto di Biologia Generale, Università di Pisa, Via Volta 6, I-56100 Pisa, Italy.

Jay Boyd Best, Colorado State University, Fort Collins, Colorado, USA.

Burton J. Bogitsh, Department of General Biology, Vanderbilt University, Nashville, Tennessee 37235, USA.

Barbara C. Boyer, Department of Biological Sciences, Union College, Schenectady, New York 12308, USA.

Rod Bray, British Museum of Natural History, Cromwell Road, London, England.

Jochem Brüggemann, II. Zoologisches Institut und Museum der Universität, Berliner Strasse 28, D-3400, Göttingen, Federal Republic of Germany.

M. D. B. Burt, Department of Biology, University of New Brunswick, Fredericton, N.B. E3B 6E1, Canada.

Mary Beverly-Burton, Department of Zoology, University of Guelph, Guelph, Ontario, N1G 2W1, Canada.

Louise Bush, Gray Museum, Marine Biology Laboratory, Woods Hole, Massachusetts 02543, USA.

Peter Calow, Department of Zoology, University of Sheffield, Sheffield S10 2TN, England.

Lester Cannon, Queensland Museum, Brisbane, Queensland, Australia.

Aage Møller Christensen, University of Copenhagen, Helsingør, Denmark.

Marco Curini Galletti, Istituto di Zoologia e Anatomia Comparata, Università di Pisa, Via A. Volta 4, Pisa, Italy.

Elisabeth J. de Vries, Instituut voor Taxonomische Zoologie, Universiteit van Amsterdam, Postbox 20125, NL-1000 HC Amsterdam, The Netherlands.

A. Di Marcotullio, Sez. Zoologia, Dipartimento di Biologia, Università degli Studi, I-34127, Trieste, Italy.

David A. Doe, Westfield State College, Westfield, Massachusetts 01086, USA.

I. M. Drobysheva, Laboratory of Evolutionary Morphology, Zoological Institute, Academy of Sciences, Leningrad 199034, USSR.

Ulrich Ehlers, II. Zoologisches Institut und Museum der Universität, Berliner Strasse 28, D-3400 Göttingen, Federal Republic of Germany.

Witold Ferens, Department of Biology, University of New Brunswick, Fredericton, N.B. E3B 6E1, Canada.

Lesley C. Fleming, Department of Biology, University of New Brunswick, Fredericton, N.B. E3B 6E1, Canada.

R. Franquinet, Laboratoire de Biologie Animale, Université Paris Val de Marne, Avenue Général de Gaulle, Créteil 94010, Cedex, France.

Lodovico Galleni, Istituto di Zoologia E Anatomia Comparata, Università di Pisa, Via A. Volta 4, Pisa, Italy.

E. Ghirardelli, Sez. Zoologia, Dipartimento di Biologia, Università degli Studi, I-34127, Trieste, Italy.

M. Gabriella Gremigni, Istituto di Zoologia e Anatomia Comparata, Università di Pisa, Via Volta n. 4, 56100 Pisa, Italy.

Vittorio Gremigni, Istituto di Zoologia e Anatomia Comparata, Università di Pisa, Via Volta n. 4, 56100 Pisa, Italy.

Margaretha Gustafsson, Institute of Biology, Åbo Akademi, Porthansgatan 3 – 5, SF-20500 Åbo 50, Finland.

Zênia Heller, Università do Vale do Rio dos Sinos, Praça Tiradentes, 35-CXP 275, 93.000 São Leopoldo, Brasil.

Jan Hendelberg, Department of Zoology, University of Göteborg, Box 25509, S-40031 Göteborg, Sweden S-40031.

Sherman Hendrix, Department of Biology, Gettysburg College, Gettysburg, Pennsylvania 17325, USA.

Jan Herrmann, Department of Animal Ecology, University of Lund, S-223 62 Lund, Sweden.

Isao Hori, Department of Biology, Kanazawa Medical University, Uchinada-machi, Ishikawa-ken, 920-02, Japan.

Sachiko Ishida, Department of Biology, Hirosaki University, 3 Bunkyo-cho, Hirosaki, Aomori-ken 036, Japan.

Lena Jarecka, Department of Biology, University of New Brunswick, Fredericton, N.B. E3B 6E1, Canada.

Masaharu Kawakatsu, Biological Laboratory, Fuji Women's College, Kita-16, Nishi-2, Kita-ku, Sapporo (Hokaido) 001, Japan.

Marianne D. Klauser, Department of Zoology, Murray Hall, University of Maine, Orono, Maine 04469, USA.

Jerzy Kolasa, Yale Arbovirus Research Unit, 60 College St., P.O. Box 3333, New Haven, Connecticut 06510, USA.

Harold Koopowitz, Developmental & Cell Biology, University of California, Irvine, California 92717, USA.

E. A. Kotikova, Laboratory of Evolutionary Morphology, Zoological Institute, Academy of Sciences, Leningrad 199034, USSR.

Elena B. Krichinskaya, Department of Embryology, State University of Leningrad, 199164 Leningrad, USSR.

Alberto Lanfranchi, Istituto di Biologia Generale, Università di Pisa, Via Volta 6, I-56100 Pisa, Italy.

J.-P. L'Hardy, Laboratoire de Biologie Animale, Université du Maine, B. P. 535 72017 Le Mans-Cedex, France.

Kwang-Yang Lue, Department of Biology, National Taiwan Normal University, 88 Section 5, Roosevelt Road, Taipei, Taiwan 117.

Barbara M. MacKinnon, Department of Biology, University of New Brunswick, Fredericton, N.B. E3B 6E1, Canada.

Naoya Makino, Biology Laboratory, Tokyo Medical College, 6-1-1 Shijuku, Shinjukuku, Tokyo 160, Japan.

Göran Malmberg, Department of Zoology, University of Stockholm, S-106 91 Stockholm, Sweden.

Yu. V. Mamkaev, Laboratory of Evolutionary Morphology, Zoological Institute, Academy of Sciences, Leningrad, 199034, USSR.

T. G. Markosova, Laboratory of Evolutionary Morphology, Zoological Institute, Academy of Sciences, Leningrad 199034, USSR.

I. Martelly, Laboratoire de Biologie Animale, Université Paris Val de Marne, Avenue Géneral de Gaulle, Créteil 94010, Cedex, France.

Els E. Martens, Department SBM, Limburgs Universitair Centrum, B-3610 Diepenbeek, Belgium.

Paul M. Martens, Department SBM, Limburgs Universitair Centrum, B-3610 Diepenbeek, Belgium.

J. Moraczewski, Laboratoire de Biologie Animale, Université Paris Val de Marne, Avenue Général de Gaulle, Créteil 94010, Cedex, France.

Daniele Murith, Department of Zoology, University of Guelph, Guelph, Ontario, Canada.

J. Murith, Department of Zoology, University of Guelph, Guelph, Ontario, Canada.

M. Nigro, Istituto di Zoologia e Anatomia Comparata, Universita di Pisa, Via Volta n. 4, 56100 Pisa, Italy.

Jean Ogren, Department of Biology, Wilkes College, Wilkes-Barre, Pennsylvania 18766, USA.

Robert E. Ogren, Department of Biology, Wilkes College, Wilkes-Barre, Pennsylvania 18766, USA.

Irmeli Palmberg, Institute of Biology, Åbo Akademi, Faculty of Education, Kyrkoesplanaden 12–14, 65100 Vasa 10, Finland.

F. Persi, Sez. Zoologia, Dipartimento di Biologia, Università degli Studi, I-34127, Trieste, Italy.

Alan W. Pike, Department of Zoology, University of Aberdeen, Tillydrone Ave., Aberdeen AB9 2TN, Scotland.

Ileana Puccinelli, Istituto di Zoologia e Anatomia Comparata, Università di Pisa, Via A. Volta 4, Pisa, Italy.

D. A. Read, Department of Biology, University of Southampton, Southampton, S09 3TU, England.

Maria Reuter, Institute of Biology, Åbo Akademi, Porthansgatan 3–5, SF-20500 Åbo 50, Finland.

Reinhard M. Rieger, Department of Biology, University of North Carolina, Chapel Hill, North Carolina 27514, USA. [present address: Zoologisches Institut, Universität Innsbruck, Universitätstrasse 4, A-6020 Innsbruck, Austria.]

Jean Riser, Marine Science Institute, Northeastern University, Nahant, Massachusetts 01908, USA.

Nathan W. Riser, Marine Science Institute, Northeastern University, Nahant, Massachusetts 01908, USA.

Harvey Sarnat, Alberta Children's Hospital, 1820 Richmond Road S.W., Calgary, Alberta T2T 5C7, Canada.

Naomi Sarnat, Alberta Children's Hospital. 1820 Richmond Road S.W., Calgary, Alberta T2T 5C7, Canada.

Ernest R. Schockaert, Dept. SBM, Limburgs Universitair Centrum, B-3610 Diepenbeek, Belgium.

Peter Schwank, Limnologische Flusstation des Max-Planck-Institutes für Limnologie, D-6407 Schlitz, Federal Republic of Germany.

M. S. Settembrini, Istituto di Zoologia e Anatomia Comparata, Università di Pisa, Via Volta n. 4, 56100 Pisa, Italy.

Yasuko Shirasawa, Biology Laboratory, Tokyo Medical College, 6-1-1-Shijuku, Shinjukuku, Tokyo 160, Japan.

Ronald Sluys, Instituut voor Taxonomische Zoologie, Universiteit van Amsterdam, Postbox 20125, 1000 HC Amsterdam, The Netherlands.

Julian P. S. Smith, III, Department of Zoology, Murray Hall, University of Maine, Orono, Maine 04469, USA.

Beate Sopott-Ehlers, II. Zoologisches Institut und Museum der Universität, Berliner Strasse 28, D-3400 Göttingen, Federal Republic of Germany.

Wolfgang Sterrer, Bermuda Biological Station, Ferry Reach, 1–15, Bermuda.

Sachiko Tamura, Ôsaka Prefectural Institute of Public Health, Nakamichi-1-chôme 3-69, Higashinari-ku, Ôsaka 537, Japan.

Wataru Teshirogi, Department of Biology, Hirosaki University, 3 Bunkyo-cho, Hirosaki, Aomori-ken 036, Japan.

Mary Beth Thomas, Department of Biology, University of North Carolina, Charlotte, North Carolina 28223, USA.

Seth Tyler, Department of Zoology, Murray Hall, University of Maine, Orono, Maine 04469, USA.

Gerard van der Velde, Laboratory of Aquatic Ecology, Catholic University, Toernooiveld, 6525 ED, Nijmegen, The Netherlands.

Marianne Wikgren, Institute of Biology, Åbo Akademi, Porthansgatan 3–5, SF-20500 Åbo 50, Finland.

Joan Bowman Williams, Department of Zoology, University of Canterbury, Christchurch, New Zealand.

Robert Wink, Department of Zoology, University of Aberdeen, Tillydrone Ave., Aberdeen AB9 2TN, Scotland.

Frederick J. Wrona, Department of Biology, University of Calgary, Calgary, Alberta T2N 1N4, Canada.

Comments on a phylogenetic system of the Platyhelminthes

Ulrich Ehlers
II. Zoologisches Institut und Museum der Universität, Berliner Strasse 28, D-3400 Göttingen, Federal Republic of Germany

Keywords: Platyhelminthes, phylogenetic system, frontal organ, epidermis, ciliary epidermal receptors, protonephridium

Abstract

Using the principles and methods of phylogenetic systematics, the phylogenetic relationships among the basic closed descendent communities (monophyla) of the taxon Platyhelminthes are discussed.

A phylogenetic system of the Platyhelminthes is presented and several hypothesized autapomorphies of the well-known main subordinated monophyla of free-living and parasitic platyhelminths are named. A monophylum 'Turbellaria' does not exist in reality.

Introduction

When discussing phylogenetic relationships of taxa within a group of living organisms such as the Platyhelminthes, we have to prove whether the group as well as the taxa within that group are consistent with concepts of the phylogenesis or not – that is, whether the supra-specific taxa have reality and individuality in nature.

Supra-specific groups of organisms that are consistent with phylogenesis only are 'groups of evolutionary species as closed descendent communities with one single stem species and all of its descendants' (*cit.* Ax, 1985).

What are closed descendent communities? How can we recognize them, and how do they relate to each other? Phylogenetic systematics sensu Hennig (1966) gives answers to these questions, the methods and principles of phylogenetic systematics are explained and discussed in many papers (see especially Wiley, 1981, and Ax, 1984).

Principles and methods of phylogenetic systematics

A closed descendent community is a group of organisms descendent from one stem species and including *all* of its descendants; that means the relatives of such a community have an ancestor in common only to themselves. We call such a real community of species a monophyletic taxon or a monophylum.

Does that group of species named Platyhelminthes represent a closed descendent community, a monophylum? To affirm this question we have to substantiate that the Platyhelminthes has an ancestor in common only with this group of species and that *all* the descendants of the common stem species belong to the monophylum Platyhelminthes.

How can we recognize a monophyletic taxon or monophylum? A monophylum Platyhelminthes or a monophylum Catenulida or a monophylum Euplathelminthes (Fig. 1) exists when we can substantiate that at least one new characteristic has evolved in each stem lineage of the different taxa in question.

Such an evolutionary novelty – black square 1 in Fig. 1 – is called an autapomorphy of the taxon in question – in Fig. 1, of the taxon Euplathelminthes. This autapomorphic characteristic is present in the common stem species of the taxon Euplathelminthes and can be passed on in the stem lineages of the descendants of this stem species. I have called those two taxa resulting directly from a

Hydrobiologia 132, 1–12 (1986).

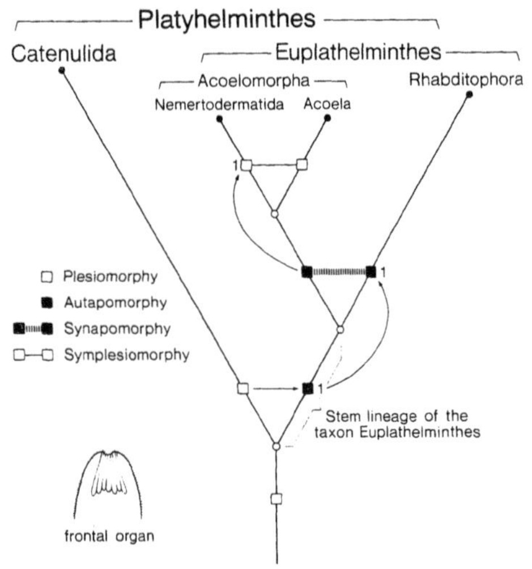

Fig. 1. Estimation of one selected characteristic (frontal organ) at different levels of hierarchy in a phylogenetic system of the Platyhelminthes.

speciation process of the stem species of the Euplathelminthes, Acoelomorpha and Rhabditophora.

Two taxa that originate at the same time form sister-groups, or, as Ax (1984, 1985) calls them, adelphotaxa. At this level, the autapomorphy 1 of the taxon Euplathelminthes forms a synapomorphy of the two taxa in question, the Acoelomorpha and the Rhabditophora. A synapomorphy always establishes a sister-group – or adelphotaxa – relationship, indicating that these two taxa are direct descendants of a stem species common only to themselves.

When the stem species of the taxon Acoelomorpha gives rise to two new taxa, namely the Nemertodermatida and Acoela, the characteristic 1 can be passed on unchanged in the stem lineages of the Nemertodermatida and Acoela. At this level, the unchanged characteristic 1 becomes a plesiomorphic one, a symplesiomorphy of the Nemertodermatida and Acoela. The unchanged characteristic 1 does not indicate a sister-group – or adelphotaxa – relationship between the Nemertodermatida and the Acoela, because the unchanged characteristic 1 can also be present in the descendants of the taxon Rhabditophora (for example in the Macrostomida and in the Trepaxonemata, see Fig. 8).

That means, if I can hypothesize, that if a given characteristic is present in an identical (unchanged) state in more than two taxa – for example in the Nemertodermatida, the Acoela, and the closest descendants of the Rhabditophora – then it is a symplesiomorphy, without evidence for establishing adelphotaxa-relationships, because I cannot favor any hypothesis that any of these taxa are adelphotaxa, either the Nemertodermatida and a descendant of the Rhabditophora or the Acoela and a descendant of the Rhabditophora or the Nemertodermatida and the Acoela. Only then, when a symplesiomorphic characteristic becomes a synapomorphic characteristic at a higher level of a system, is the characteristic relevant to indicate a hypothesized close phylogenetic relationship.

The characteristic chosen as an example in Fig. 1 is the characteristic 'frontal organ'. Such an organ situated terminally to sub-terminally in an animal and consisting of several to many mucoid gland cells, and in several taxa, of ciliary receptor cells and rhabdoid glands, occurs in the following taxa: Nemertodermatida, Acoela (*cf.* Klauser *et al.*, 1985; Smith & Tyler, 1985), Macrostomida (*cf.* Klauser *et al.*, 1985), Lecithoepitheliata, Prolecithophora, certain members of the Seriata (Fig. 2), perhaps also in certain members of the Rhabdocoela, and in the Polycladida (*cf.* Ruppert, 1978; 'apical organ' by Lacalli, 1983).

Among all those free-living platyhelminths that locomote mainly with their epidermal cilia, only the catenulids do not possess an apical glandular or glandular-sensory organ.

With that the question arises: Is the absence of a frontal organ in the Catenulida a plesiomorphic feature? That is, is this absence inherited from the stem species of the taxon Platyhelminthes, which also does not possess a frontal organ? Or does this absence represent an apomorphy, resulting from a reduction in the stem lineage of the taxon Catenulida?

We can find an answer by application of the principle of out-group comparison: 'For a given character with two or more states within a group, the state occurring in related groups is assumed to be the plesiomorphic state. If the character contains only two states, the alternative state is assumed to be apomorphic...' (*cit.* Watrous & Wheeler, 1981: 5). This statement is also conclusive to the alternative 'characteristic existent – charac-

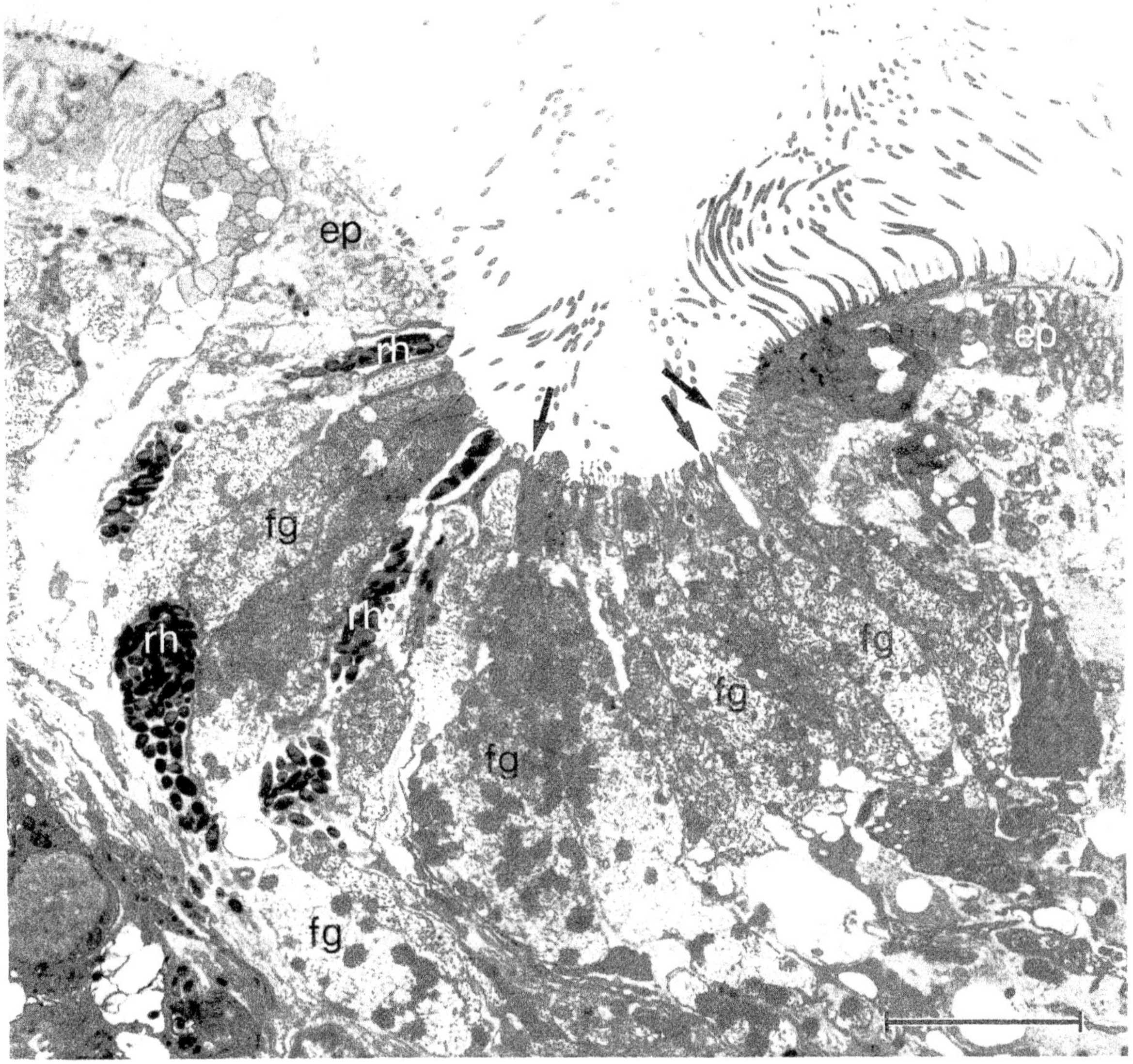

Fig. 2. Frontal organ of *Notocaryoplanella glandulosa* (Proseriata Lithophora): glandular – sensory organ with different types of glands and processes of ciliated sensory cells (arrows). Scale bar represents 5 μm. (Figs. 2+3 and 5–7: for methods see Ehlers & Ehlers, 1977). Abbreviations: cm, circular musculature; cro, caudal rootlet; cw, cell web; ep, epidermis; fg, frontal glands; iro, inner rods; lm, longitudinal musculature; n, nucleus; oro, outer rods; rh, rhabdoids; ro, rods; rro, rostral rootlet; rt, rootlet; sc, sensory cell; stc, stem cell; tc, tubule cell; we, weir.

teristic absent' – with certain reservations (see Ax, 1984: 129).

We can hypothesize an adelphotaxa-relationship between the taxon Platyhelminthes and the taxon Gnathostomulida (see Ax, 1984, 1985); the Gnathostomulida (and also other taxa of the Bilateria) do not have a frontal organ.

That means we can hypothesize that the absence of a frontal organ in the Catenulida is the plesiomorphic state and that the existence of a frontal organ within the other Platyhelminthes, the Euplathelminthes, is an apomorphic characteristic.

Because we can find frontal organs in all the Nemertodermatida, all the Acoela, all the Macrostomida, and many other taxa of the free-living platyhelminths, the most parsimonious interpretation (parsimony within the meaning of 'methodological parsimony', *cf.* Kluge, 1984) is to infer that frontal organs have not evolved several times independently but only once, as an evolutionary novelty in the stem lineage of the taxon Euplathelminthes, forming an autapomorphy of this taxon. Of course, the special structure of this characteristic 'frontal organ' can be modified in the subordinated taxa of the monophylum Euplathelminthes and can be absent secondarily, for example in adult polyclads and in the triclads.

In such a way, in which I have tried to establish the monophyly of a taxon Euplathelminthes by hypothesizing that the characteristic 'frontal organ' constitutes an autapomorphy of this taxon, I have to characterize all the other taxa named in Fig. 1 as monophyla.

Monophyly of the taxon Platyhelminthes

The monophyly of the taxon Platyhelminthes can be demonstrated by the apomorphic characteristic 'no mitosis in somatic cells.' That means that all those cells that have differentiated from blastomeres or stem cells (neoblasts) to somatic cells (e.g. epidermal cells, gastrodermal cells, muscles, nerves, glands, etc.) no longer divide (*cf.* lit. and discussion by Ehlers, 1985a; for cestodes see Gustafsson, 1977).

In post-embryonic developmental stages somatic cells are differentiated from stem cells. In Fig. 3, several cells from different taxa and in different stages of differentiation to epidermal cells are shown (the cell in Fig. 3 B seems to be more differentiated than that of Fig. 3C).

A second hypothesized autapomorphy of the taxon Platyhelminthes is the characteristic 'multiciliary epidermal cells, cilia without accessory centriole' (characteristic 1 in Fig. 4).

Only multiciliary epidermal cells occur within the Platyhelminthes, whereas the adelphotaxon of the Platyhelminthes, the Gnathostomulida, possesses monociliary epidermal cells in which each cilium is provided with an accessory centriole. By out-group comparison, the characteristic 'monociliary epidermal cells, cilium with accessory centriol' is the plesiomorphic state; this plesiomorphic characteristic is also present within the Cnidaria and several groups of the Bilateria (*cf.* Rieger, 1976) and is passed on from the stem species of the taxon Bilateria to the taxon Gnathostomulida (*cf.* Ax, 1984, 1985).

In the stem lineage of the taxon Platyhelminthes the characteristic 'each epidermal cell with more than one cilium, cilium without accessory centriole' evolved, forming an autapomorphy of this taxon.

Basic closed descendent communities of the Platyhelminthes

The Catenulida possess locomotory epidermal cilia which in several respects correspond with those of monociliary cells, in for example, the Gnathostomulida (Rieger, 1976): e.g., the basal bodies are always situated in pits of the cell (Fig. 5A), have conspicuous radial spokes (Fig. 5B), and are provided with a rostral and a caudal rootlet (Fig. 5C). Moreover, the ciliary density in the Catenulida is very low; in *Retronectes cf. sterreri* (Fig. 5), the density ranges between 0.2 and 0.6 cilia/μm^2 epidermis. In the Gnathostomulida and the Gastrotricha with monociliated epidermal cells, the density ranges between 0.15 and 0.2 cilia/μm^2 epidermis (*cf.* Rieger, 1981b; own observations). Within the other platyhelminths, the Euplathelminthes, the ciliary density usually ranges from 3 to 6 cilia/μm^2 of epidermis (*cf.* Rieger, 1981b; own observations), the arrangement of the ciliary structures within the epidermal cells is different from that in catenulid epidermal cells (*cf.* Ehlers, 1985a).

Using the principle of out-group comparison, we can hypothesize that the increase in ciliary density (characteristic 2 in Fig. 4) is an apomorphic state. The most parsimonious interpretation, that the characteristic 'increase in density to three or more cilia per μm^2 epidermis' only evolved once and in the stem lineage of the taxon Euplathelminthes and constitutes an autapomorphy of this taxon, is compatible with all the other characteristics known from the Euplathelminthes.

The monophyly of a taxon Catenulida is indicated by several hypothesized autapomorphies. One characteristic that has evolved in the stem lineage of this taxon is the feature 'aciliary spermatozoon' (characteristic 3 in Fig. 4). The Catenulida have strongly modified spermatozoa (*cf.* Rieger, 1978; Ehlers, 1985a). In an early stage of spermatogenesis

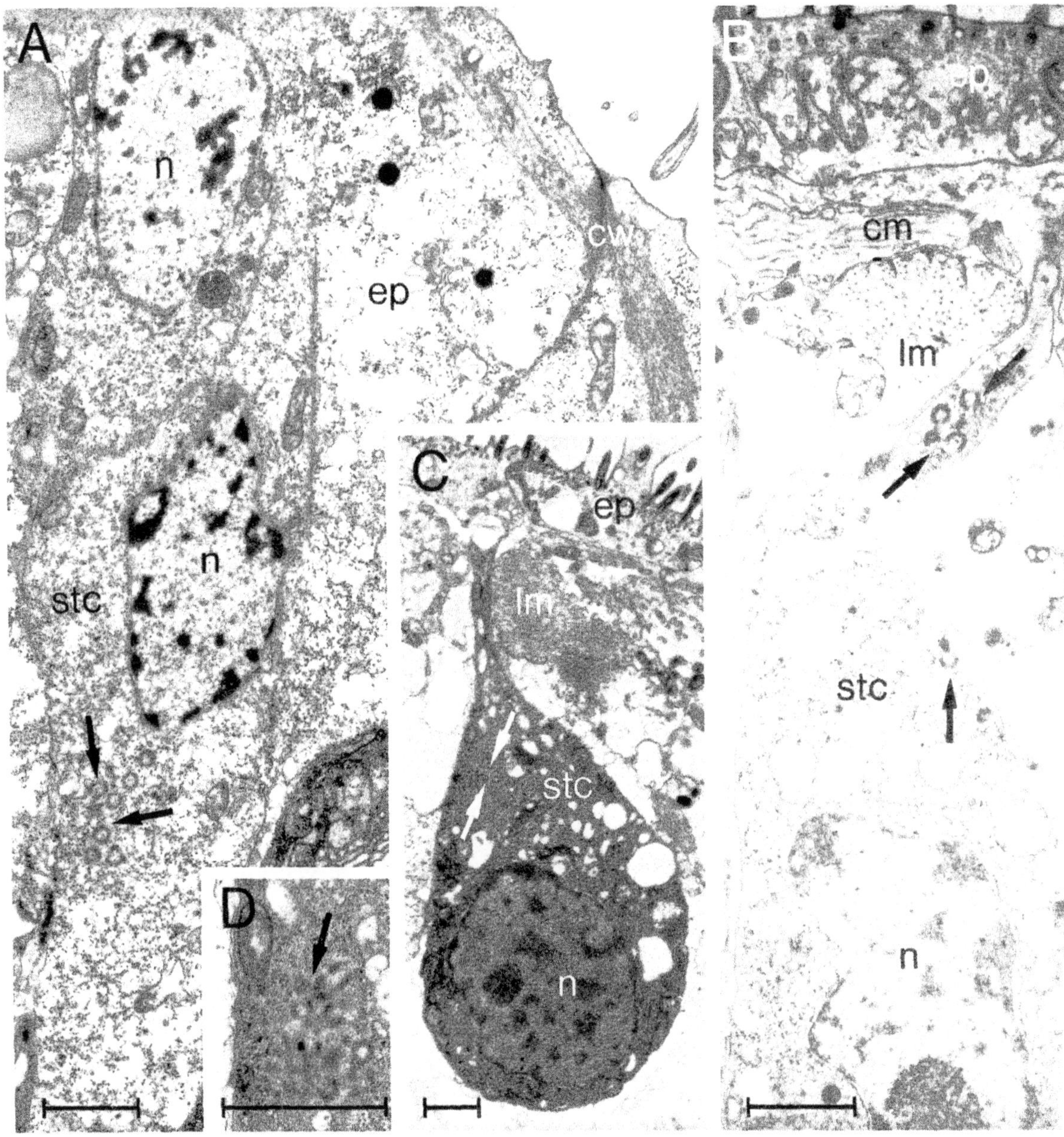

Fig. 3. Differentiating epidermal cells. A. *Retronectes cf. sterreri* (Catenulida). B. *Monocelis fusca* (Proseriata Lithophora). C. + D. *Otoplanidia endocystis* (Proseriata Lithophora), in D a section of C showing the centrioles. The cell in B seems to be more differentiated than the cell in C. The arrows point to centrioles in the differentiating cells. Scale bars represent 1 μm. For abbreviations see Fig. 2.

of *Retronectes cf. sterreri*, each spermatid possesses one rudimentary cilium which only exists transitorily (Ehlers, 1985a). This one cilium may be the relict of the cilium in the typical monociliary spermatozoon of the Metazoa and may indicate that the Catenulida are in this respect still plesiomorphous.

The monophyly of a taxon Acoelomorpha is indicated by several autapomorphies – for example by the characteristic 'modified epidermal cilia' (characteristic 4 in Fig. 4) (*cf.* Tyler, 1979; Tyler & Rieger, 1977) – and in the stem lineage of the taxon called Rhabditophora, the characteristics 'duo-gland adhesive system' and 'lamellated rhabdites'

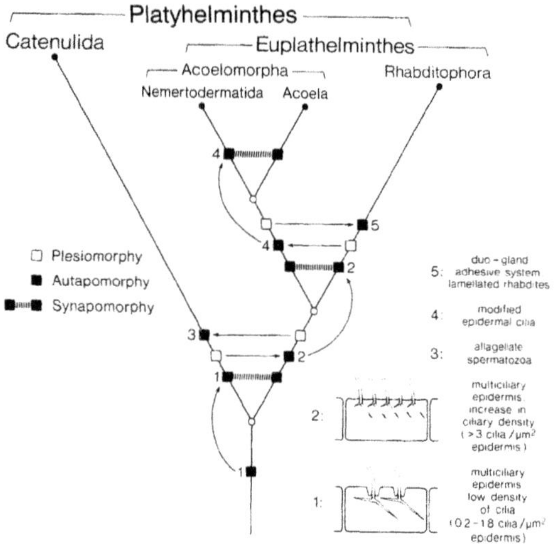

Fig. 4. Several selected characteristics upon which the phylogenetic interrelationships between the basic closed descendent communities of the Platyhelminthes are hypothesized.

(characteristics 5 in Fig. 4) evolved (*cf.* Tyler, 1976; Smith *et al.*, 1982; Ax, 1984; Ehlers, 1985a, b).

I already mentioned that I hypothesize an epidermis with a low ciliary density for the stem species of the Platyhelminthes and an increase of ciliary density in further evolution within this taxon. One can also observe such an increase in ciliary number in two other structures: epidermal sensory cells and protonephridia.

Only monociliary epidermal receptors exist in the Catenulida (Fig. 6A), the Nemertodermatida (Fig. 6C), and the Acoela (Figs. 6B, D). Whereas we can find mono- and multiciliary epidermal receptors in the taxa of the Rhabditophora (Ehlers, 1985a; however, see also Smith & Tyler, 1985).

All the taxa of the Platyhelminthes named in Fig. 8 with the exception of the Nemertodermatida and Acoela possess protonephridia. The Catenulida have an unpaired system, the proximal parts of

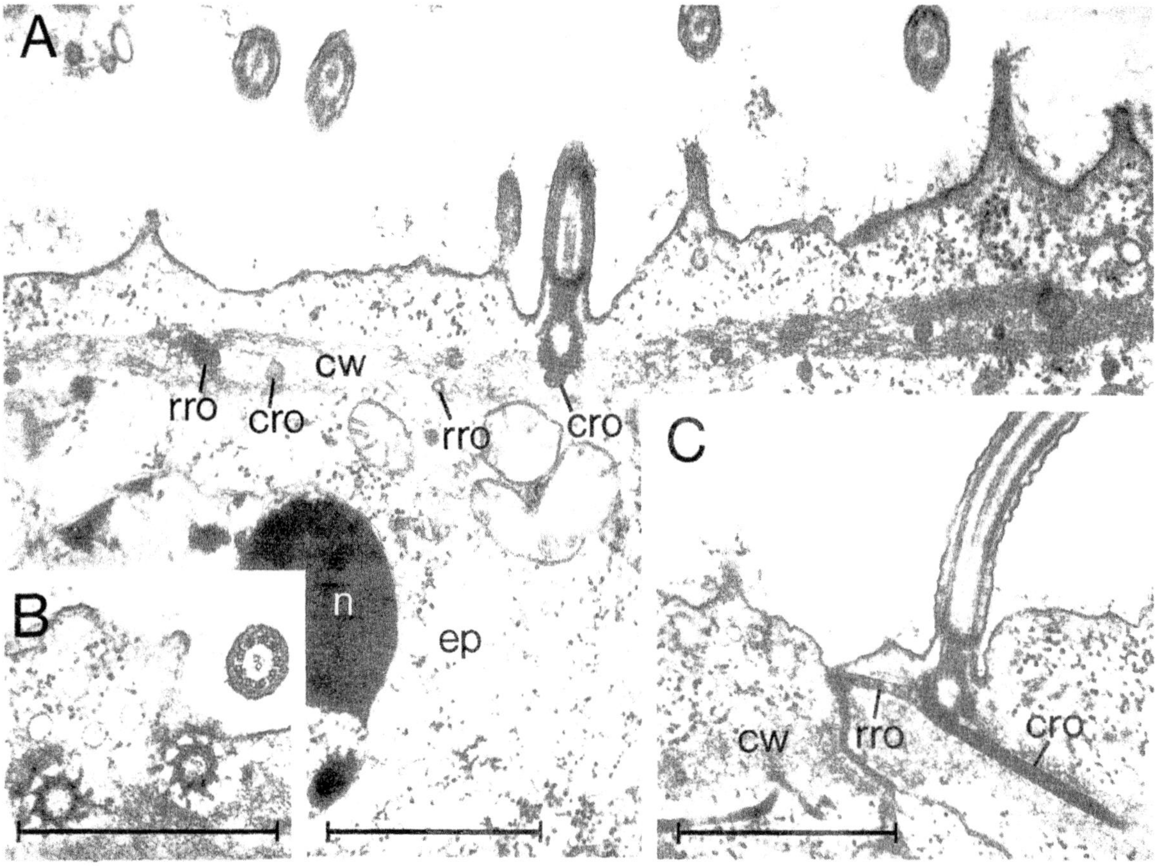

Fig. 5. Epidermis of *Retronectes cf. sterreri* (Catenulida) with low density of cilia. A. Cross-section. B. Tangential section with cross-sectioned basal bodies of cilia. C. Longitudinal section. Scale bars represent 1 μm. For abbreviation see Fig. 2.

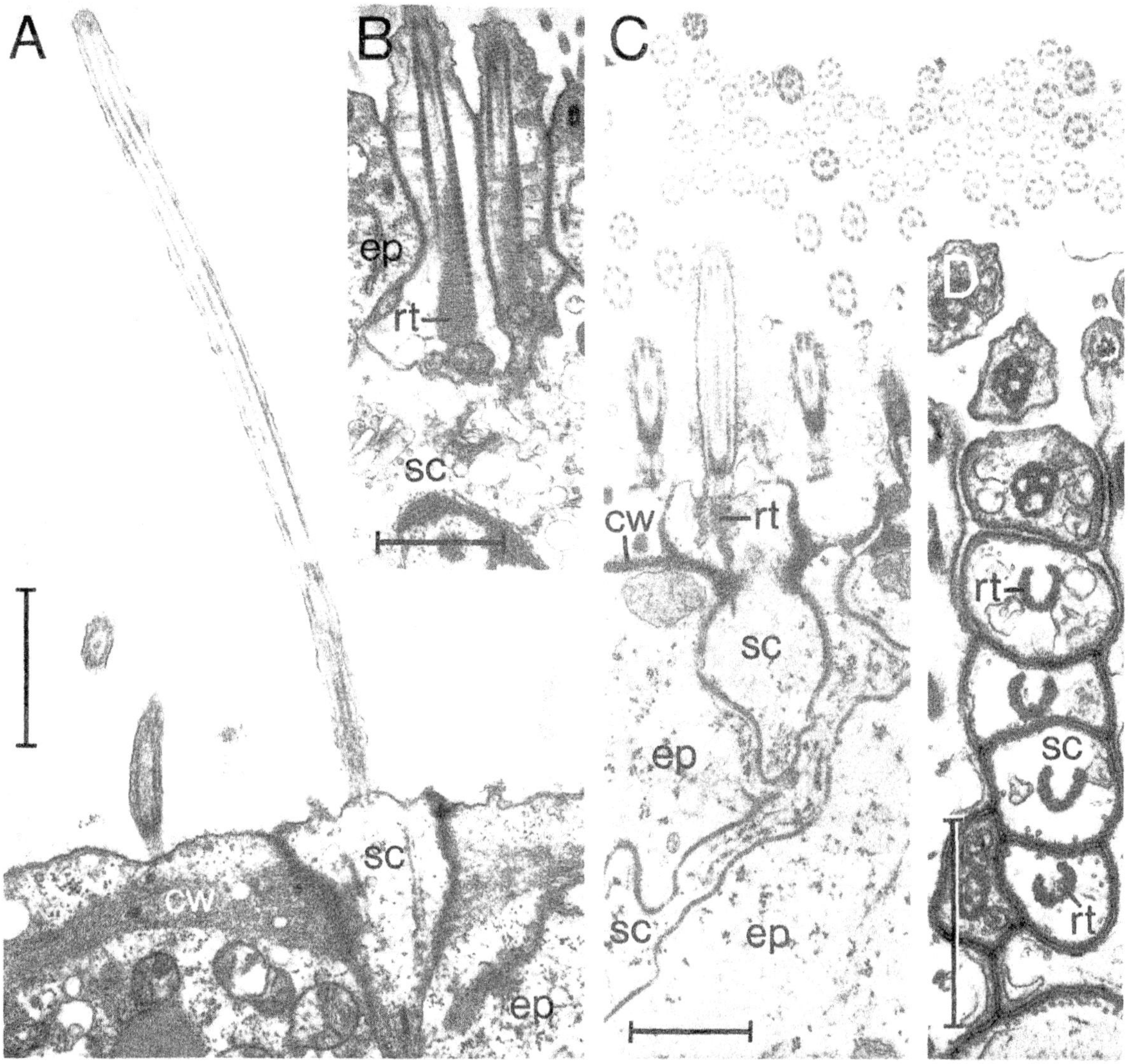

Fig. 6. Monociliary epidermal receptors. A. *Retronectes cf. sterreri* (Catenulia). B. *Mecynostomum auritum* (Acoela). C. *Nemertoderma sp.* B. (Nemertodermatida). D. *Anaparus tvaerminnensis* (Acoela). A–C. Cross-sections. D. Tangential section with cross-sectioned basal bodies and rootlets. Scale bars represent 1 μm. (Methods for Fig. 6C see Tyler & Rieger, 1977). For abbreviations see Fig. 2.

which, the cyrtocytes or flame bulbs, are formed by one cell only (Ehlers, 1985; Ax, 1984, 1985). These terminal cells bear two cilia in all the Catenulida studied until now at the EM level (*Retronectes, Catenula, Stenostomum, Xenostenostomum*); the long rootlets of these two cilia project along the narrow sides (Fig. 7A) strengthening the cyrtocyte, whereas microvilli (rods) with the weir can be found at the broad sides of the terminal cell (Ehlers, 1985a). Within the Rhabditophora we can only find cyrtocytes with more than two cilia. The lowest number hitherto observed is four cilia, found within several species of the Macrostomida (Rieger, 1981a; Ehlers, 1985a) and the Polycladida (Ruppert, 1978), whereas often more than one hundred cilia occur in a cyrtocyte of the parasitic taxa. That means we have the same situation here as with the ciliary epidermal cells and the ciliary sensory cells: multiciliary cyrtocytes, that is to say four or more cilia per terminal cell, only occur within the Rhabditophora. By the principles of out-group comparison and parsimony I hypothesize that the evolu-

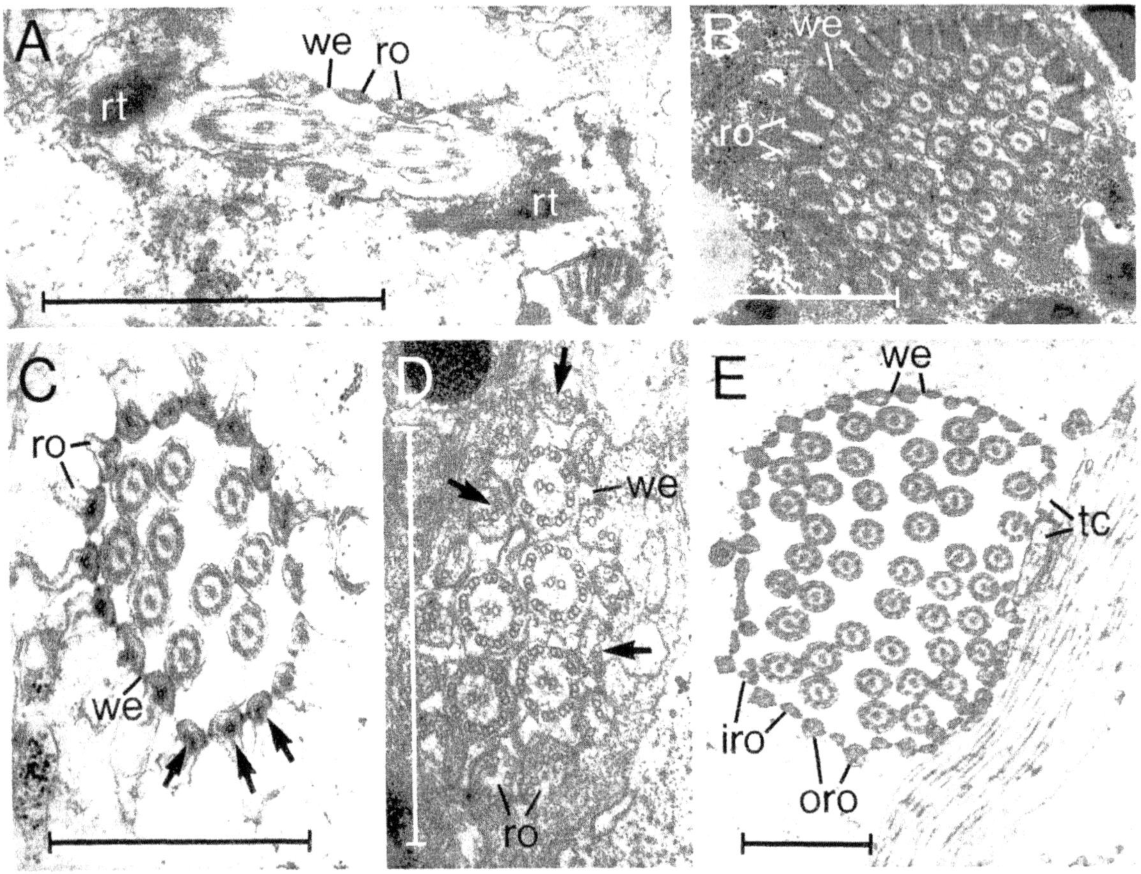

Fig. 7. Protonephridia, structure of the weir in different taxa. A. *Retronectes cf. sterreri* (Catenulida). B. *Geocentrophora sphyrocephala* (Lecithoepitheliata). C. *Mesostoma lingua* ('Typhloplanoida'). D. *Provortex psammophilus* ('Dalyellioida'). E. *Monocelis fusca* (Proseriata Lithophora). Arrows (in C + D) indicate microtubules in the rods of the terminal cell. Scale bars represent 1 μm. (D courtesy Dr. J. Brüggemann). For abbreviations see Fig. 2.

tionary novelty 'multiciliary cyrtocytes' is evolved in the stem lineage of the taxon Rhabditophora, forming an autapomorphy of this taxon.

Within the Rhabditophora the whole cyrtocyte, that is to say all the rods of the weir, can be formed only by the terminal cell (Figs. 7B, C, D), as in the Catenulida, or by the interdigitation of the terminal cell and the adjacent tubule cell (Fig. 7E), such that there are two rows of rods, an inner one projecting from the distal part of the terminal cell and an outer one projecting from the proximal part of the adjacent tubule cell. This tubule cell is in the form of a sheath wrapping around to form the tube-like protonephridial canal; the contiguous parts of the cell are joined by desmosomes (Fig. 7E) running along the whole length of the tubule cell (*cf.* Ehlers, 1985a).

Within the Rhabditophora (*cf.* Fig. 8) up to now a one-cell weir is known from species of the 'Dalyellioida' (incl. the Temnocephalidea), the 'Typhloplanoida', the Tricladida (a taxon of the Seriata), the Lecithoepitheliata (not named in Fig. 8), and perhaps also in certain marine members of the Macrostomida. A two-cell weir is found in all the Proseriata (a taxon of the Seriata) and in all the classic parasitic taxa (the Trematoda, Monogenea, and Cestoda) and probably also in the taxon *Macrostomum* (Macrostomida).

How can we substantiate whether the one-cell weir (or the two-cell weir) is the plesiomorphic or the apomorphic state? Let us attempt an out-group comparison: the Catenulida (and the Gnathostomulida, the sister-group of the Platyhelminthes) possess a one-cell weir; that means I

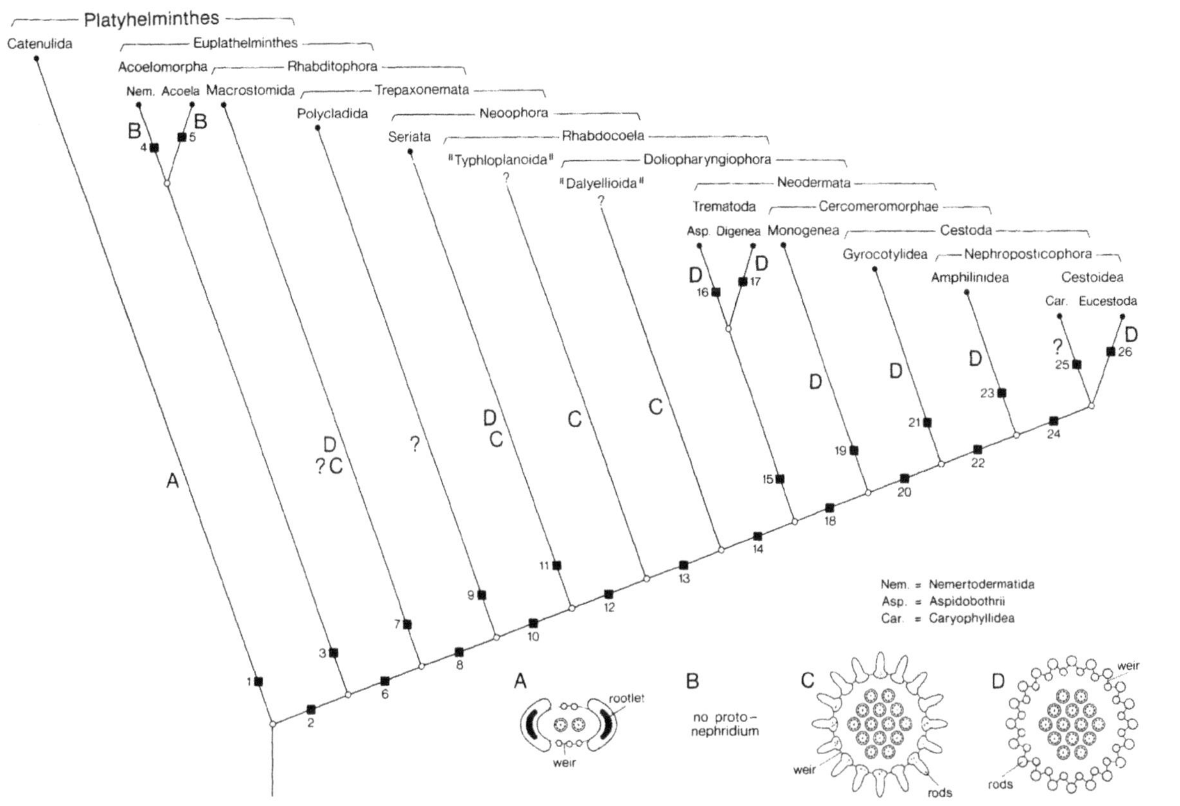

Fig. 8. Phylogenetic relationships among the main taxa of the Platyhelminthes (excluding the Lecithoepitheliata and the Prolecithophora). Black squares 1–26 are hypothesized autapomorphies, indicating the monophyly of the taxa [see text; the 'Typhloplanoida' (incl. the monophyletic Kalyptorhynchia) and the 'Dalyellioida' (incl. the Temnocephalidea) do not represent monophyla]. Capitals A–D indicate the specific structure of the protonephridial weir (or the absence of protonephridia) in the taxa in question.

favour the hypothesis that the one-cell weir found within the Rhabditophora is the plesiomorphic state [this statement differs from my earlier conclusion expressed in 1983 (see Ehlers, 1985b); at that time the fine structure of the cyrtocytes in marine Macrostomida, the 'Typhloplanoida,' and the 'Dalyellioida' was still unknown].

All the classic parasitic taxa have the apomorphic state, a two-cell weir (Fig. 8). Here I use the principle of parsimony, that is to say, I favor the most parsimonious interpretation, that such apomorphic characteristics were not evolved several times independently within the classic parasitic taxa, but only once: in a stem lineage of the taxon I have called Neodermata. The monophyly of this taxon Neodermata was established on the basis of several other autapomorphies (see Ehlers, 1985a, b; Ax 1984), the existence of a two-cell weir can be explained without conflict as another autapomorphy of this taxon. I can make this statement also, because the Proseriata and the taxon *Macrostomum* – the other taxa with a two-cell weir – do not form the sister-group of the parasitic Neodermata (see Ehlers, 1985a); that means the two-cell weir of the Proseriata was evolved independently from that of the Neodermata and probably constitutes an autapomorphy of the taxon Proseriata (and perhaps of the taxon *Macrostomum*).

In this paper I could only present very few characteristics to be found within the Platyhelminthes; it was my intention to show how to operate with characteristics when evaluating phylogenetic relationships. Only those characteristics that are proved autapomorphies or synapomorphies are relevant to establishing a monophylum or an adelphotaxa = relationship between two taxa.

When establishing a phylogenetic system we have to clarify the monophyly of the taxa in question and the adelphotaxa-relationships between these taxa in a step-by-step process; thereby we have to consider *all* the known characteristics. Of course, new findings or a better knowledge of already known features can cause modifications or alterations. A phylogenetic system has a high degree of likelihood, meaning that it only comprises closed descendent communities (and – at a lower level – evolutionary species), when it is compatible not only with the known characteristics, but also with new findings.

The 'Turbellaria' – a paraphyletic group

The phylogenetic system of the Platyhelminthes presented here (Fig. 8) corresponds in many aspects to the system of the free-living taxa published by Karling (1974); see also Rieger (1981a), and the system of the parasitic taxa published by Brooks (1982).

In the system, the name 'Turbellaria' is lacking. No autapomorphy exists to constitute a monophylum 'Turbellaria.' The diagnostic feature 'free-living' is a plesiomorphic characteristic, which can be found for example in the adelphotaxon of the Platyhelminthes, the Gnathostomulida, and which is a very ancient one. The other diagnostic feature often used is the feature 'body covered by a ciliated epidermis' (more correctly: multiciliated epidermis), this characteristic does not represent an autapomorphy of the 'Turbellaria' because a multiciliary epidermis can also be found in the larval stages of the parasitic taxa, the Neodermata. That means the 'Turbellaria' is a paraphyletic group, an assemblage of platyhelminths that in several characteristics are relatively plesiomorphic in comparison with the monophylum Neodermata and that do not have a stem species in common only to themselves.

Paraphyletic groups 'have no independent history and thus possess neither reality nor individuality' (*cit.* Hennig 1966: 146) or, as Wiley (1981b: 356) writes, 'evolution does not produce paraphyletic grades. Taxonomists produce paraphyletic grades.' We should consider that when we use the name 'Turbellaria,' we speak about a group that does not exist in reality.

Autapomorphies of the taxa named in Fig. 8

The numbers 1–26 in Fig. 8 refer to hypothesized evolutionary novelties (autapomorphies) of the taxa in question (see Ehlers, 1985a, b, and Ax, 1984, for additional autapomorphies; see Ehlers, 1985a, and Smith *et al.*, 1985, for references to characteristics):

(Black square 1) Catenulida v. Graff, 1905: lobed cerebrum; unpaired protonephridium; special organization of the cyrtocyte; special type of pharynx simplex; aberrant aciliary spermatozoa.

(2) Euplathelminthes Bresslau and Reisinger, 1928: Increase in ciliary density of the epidermis; frontal organ.

(3) Acoelomorpha Ehlers, 1984: modified epidermal cilia; reduction of protonephridia; modified intestine.

(4) Nemertodermatida Steinböck, 1931: statocyst with two statoliths.

(5) Acoela Uljanin, 1870: complex ciliary rootlet-system; special type of statocyst; biciliary spermatozoa; duet cleavage.

(6) Rhabditophora Ehlers, 1984: lamellated rhabdites; duo-gland adhesive system; multiciliary terminal cells of the protonephridia.

(7) Macrostomida (incl. the Haplopharyngida) v. Graff, 1882: special type of adhesive system; aciliary spermatozoa.

(8) Trepaxonemata Ehlers, 1984: biflagellate spermatozoa; ciliary axoneme of the spermatozoa with a special type of dense core (9+'1' pattern).

(9) Polycladida Lang, 1881: special type of pharynx plicatus; intestine with highly branched diverticula; resorption of certain blastomeres (macromeres 4A–4D and micromeres 4a–4c).

(10) Neoophora Westblad, 1948: heterocellular female gonad with germocytes and vitellocytes; ectolecithal eggs.

(11) Seriata Bresslau, 1928–33: Pharynx tubiformis; serial arrangement of follicular gonads.

(12) Rhabdocoela Ehrenberg, 1831: unspecialized pharynx bulbosus (rosulatus-type).

(?)'Typhloplanoida' (incl. the monophylum Kalyptorhynchia) v. Graff, 1905: no autapomorphy known, probably a paraphyletic group.

(13) Doliopharyngiophora Ehlers, 1984: special type of pharynx bulbosus: pharynx doliiformis; reduction of the duo-gland adhesive system.

(?) 'Dalyellioida' (incl. the Temnocephalidea and

the Udonellida) Meixner, 1924: no autapomorphy known, probably a paraphyletic group.

(14) Neodermata Ehlers, 1984: shedding of the epidermis (cells) at the end of a larval phase; post-larval stages with a syncytial neodermis; protonephridium with a two-cell weir; spermatozoa with completely incorporated ciliary axonemata.

(15) Trematoda Rudolphi, 1808: epidermal cells of the larvae separated from each other by material of the neodermis; copulatory organ in form of a cirrus; primarily parasites of molluscs.

(16) Aspidobothrii Burmeister, 1856: cotylocidium larva with a posterior-ventral disc; development of this disc to a huge ventral sucker with alveoli in the adult; neodermis with specialized microvilli.

(17) Digenea v. Beneden, 1858: tiers of epidermal cells in the miracidium larva; cercariae and adults with a ventral sucker; early developing stages (miracidium, sporocyst) without a digestive system; complex life cycle (vegetative multiplication within the mollusc hosts and changes of the hosts).

(18) Cercomeromorphae Bychowsky, 1937: larvae and (at least the younger) post-larval stages with caudal (marginal) hooks.

(19) Monogenea v. Beneden, 1858: oncomiracidium larva bears three zones of ciliated epidermal cells (one in the middle and one at each end); four (pigmented or unpigmented) rhabdomeric eyespots.

(20) Cestoda Gegenbaur, 1859: epidermis of the larvae always syncytial; all stages of development and the adults without a digestive system; copulatory organ in form of a cirrus.

(21) Gyrocotylidea Poche, 1926: apical proboscis; caudal rosette funnel.

(22) Nephroposticophora Ehlers, 1984: excretory system gets a posterior opening in later ontogeny.

(23) Amphilinidea Poche, 1922: leaf-like habitus; special type of apical organ.

(24) Cestoidea Rudolphi, 1808: only 6 caudal hooks present; neodermis with specialized microvilli (microtriches); no incorporation of mitochondria into spermatozoa mitochondria.

(25) Caryophyllidea v. Beneden (in Olsson 1893): monociliary spermatozoa, ciliary axoneme not incorporated.

(26) Eucestoda Southwell, 1930: always more than one set of gonads and genital structures per animal; ciliary or aciliary epidermis ('coracidium sheath', 'embryophore', 'inner envelope', etc.) of the oncosphaera larva with structured protein embedments.

Acknowledgments

I thank Prof. S. Tyler for the EM-material of *Nemertoderma sp.* B, Mrs. E. Hildenhagen-Brüggemann and Mrs. M. Frixe for various technical assistance, Mr. B. Baumgart for drawing the diagrams, and the Akademie der Wissenschaften und der Literatur Mainz for providing financial support.

References

Ax, P., 1984. Das phylogenetische System. Systematisierung der lebenden Natur aufgrund ihrer Phylogenese. G. Fischer, Stuttgart, New York, 349 pp.

Ax, P., 1985. The position of the Gnathostomulida and Platyhelminthes in the phylogenetic system of the Bilateria. In S. Conway Morris, D. G. George, R. Gibson & H. M. Platt (eds), The origins and relationships of lower invertebrates. University Press, Oxford: in press.

Brooks, D. R., 1982. Higher level classification of parasitic Platyhelminthes and fundamentals of cestode classification. In D. F. Mettrick & S. S. Desser (eds), Parasites – their world and ours. Elsevier Biomedical Press, Amsterdam: 189–193.

Ehlers, U., 1985a. Das phylogenetische System der Plathelminthes. G. Fischer, Stuttgart, New York, 317 pp.

Ehlers, U., 1985b. Phylogenetic relationships within the Platyhelminthes. In S. Conway Morris, D. G. George, R. Gibson & H. M. Platt (eds), The origins and relationships of lower invertebrates. University Press, Oxford: in press.

Ehlers, U. & B. Ehlers, 1977. Monociliary receptors in interstitial Proseriata and Neorhabdocoela (Turbellaria Neoophora). Zoomorphologie 86: 197–222.

Gustafsson, M., 1977. Aspects of the cytology and histogenesis in cestodes with special reference to the genus Diphyllobothrium. Academic Dissertation, Åbo Akademi, Finland.

Hennig, W., 1966 (reprinted in 1979). Phylogenetic systematics. University of Illinois Press, Urbana, Chicago, London, 263 pp.

Karling, T. G., 1974. On the anatomy and affinities of the turbellarian orders. In N. W. Riser & M. P. Morse (eds), Biology of the Turbellaria. McGraw-Hill New York: 1–16.

Klauser, M., J. P. S. Smith & S. Tyler, 1985. Ultrastructure of the frontal organ in Convoluta 'pulchra' and Macrostomum hystricinum: significance for models of the turbellarian archetype. This volume.

Kluge, A. G., 1984. The relevance of parsimony to phylogenetic inference. In T. Duncan & T. F. Stuessy (eds), The estimation of evolutionary history. Proceedings of a workshop on the theory and application of cladistic methodology. Columbia University Press, New York: 24–38.

Lacalli, T. C., 1983. The brain and central nervous system of Müller's larva. Can. J. Zool. 61: 39–51.

Rieger, R. M., 1976. Monociliated epidermal cells in Gastrotricha: significance for concepts of early metazoan evolution. Z. zool. Syst. Evolutions forsch. 14: 198–226.

Rieger, R. M., 1978. Multiple ciliary structures in developing spermatozoa of marine Catenulida (Turbellaria). Zoomorphologie 89: 229–236.

Rieger, R. M., 1981a. Morphology of the Turbellaria at the ultrastructural level. Hydrobiologia 84: 213–229.

Rieger, R. M., 1981b. Fine structure of the body wall, nervous system, and digestive tract in the Lobatocerebridae Rieger and the organization of the glio-interstitial system in Annelida. J. Morph. 167: 139–165.

Ruppert, E. E., 1978. A review of metamorphosis of turbellarian larvae. In F. S. Chia & M. Rice (eds), Settlement and metamorphosis of marine invertebrate larvae. Elsevier/North-Holland Biomedical Press, Amsterdam, New York: 65–81.

Smith, J. P. S. III & S. Tyler, 1985. Frontal organs in the Acoelomorpha: ultrastructure and phylogenetic significance. This volume.

Smith, J. P. S. III, S. Tyler & R. M. Rieger, 1985. Is the Turbellaria polyphyletic. This volume.

Smith, J. P. S. III, S. Tyler, M. B. Thomas & R. M. Rieger, 1982. The morphology of turbellarian rhabdites: phylogenetic implications. Trans. am. micros. Soc. 101: 209–228.

Tyler, S., 1976. Comparative ultrastructure of adhesive system in the Turbellaria. Zoomorphologie 84: 1–76.

Tyler, S., 1979. Distinctive features of cilia in metazoans and their significance for systematics. Tissue & Cell 11: 385–400.

Tyler, S. & R. M. Rieger, 1977. Ultrastructural evidence for the systematic position of the Nemertodermatida (Turbellaria). Acta zool. fenn. 154: 193–207.

Watrous, L. E. & Q. D. Wheeler, 1981. The out-group comparison method of character analysis. Syst. Zool. 30: 1–11.

Wiley, E. O., 1981a. Phylogenetics. The theory and practice of phylogenetic systematics. J. Wiley and Sons, New York, Chichester, Brisbane, Toronto, 439 pp.

Wiley, E. O., 1981b. Convex groups and consistent classifications. Syst. Bot. 6: 346–358.

Is the Turbellaria polyphyletic?

Julian P. S. Smith, III,[1] Seth Teyler[1] & Reinhard M. Rieger[2]
[1]*Department of Zoology, University of Maine, Orono, ME 04469, USA*
[2]*Department of Biology, University of North Carolina, Chapel Hill, NC 27514, USA*

Keywords: Turbellaria, phylogenetic systematics, Platyhelminthes, polyphyly, ultrastructure, epidermis, cilia

Abstract

Within the last two decades, syntheses of both light-microscopic and ultrastructural characters have shown that there are three well-defined monophyletic groups within the Platyhelminthes: 1) the Catenulidale, 2) the Nemertodermatida-Acoela, and 3) the Haplopharyngida-Macrostomida-Polycladida-Neoophora (+ parasitic platyhelminth classes). However, the relationships among these three groups are problematic. The possible apomorphies that would unite them are either not true homologues (i.e. frontal organ), are mutually conflicting (i.e. 9+1 axoneme in spermatozoa vs. biflagellate spermatozoa, epidermal ciliary rootlet structure, and protonephridia), or are unrooted with any outgroup and hence untestable or uncertain as apomorphies (protonephridia, mode of epidermal replacement, absence of accessory centrioles on cilia). The chief obstacle to deciphering the relationships of these groups is the lack of information on them; presently available information is insufficient to test potential synapomorphies and insufficient also to allow agreement upon a narrowly defined outgroup for the Turbellaria.

A view consistent with the present evidence (and admittedly an unsatisfactory view) is to regard the Turbellaria (and hence the Platyhelminthes) as polyphyletic, consisting of three separate and unrelatable groups.

Introduction

If one compares the various phylogenetic schemes that have been proposed for turbellarian taxa [Karling, 1940, 1974; Ehlers, 1984 (with references to earlier schemes), 1985a, 1985b], it is evident that there is considerable agreement on relationships of most of the higher taxa but that there is uncertainty and dispute concerning especially relationships of more primitive groups. As more characters have been discovered, particularly with the use of electron microscopy (see Rieger, 1981), more higher-level relationships – that is, those traditionally recognized as inter-order relationships – have been revealed. It is clear now that the neoophoran orders Proseriata, Tricladida, Rhabdocoela, and Temnocephalida share a common ancestor with the more primitive (traditionally 'archoophoran') orders Haplopharyngida, Macrostomida, and Polycladida (Tyler, 1976; Smith *et al.*, 1982; Rieger, 1981), and also with the parasitic platyhelminth classes. Relationships of the remaining orders are unclear, however. The Lecithoepitheliata and Prolecithophora are thought (at least by general agreement) to share on ancestor with other neoophoran orders (Karling, 1974; Ehlers, 1977, 1984); the Nemertodermatida and Acoela undoubtedly share a common ancestor (Karling, 1940, 1974; Tyler & Rieger, 1977), but how they are related to other turbellarians is subject to debate; and the Catenulida is particularly problematic, showing no unequivocal relationship with any other orders (Sterrer & Rieger, 1974; Rieger, 1978).

The common denominator of all recently pro-

[1, 2]The authors are not listed in alphabetical order

Hydrobiologia 132, 13–21 (1986).

posed phylogenetic schemes for the platyhelminths is recognition of three clearly monophyletic groups: one containing the Acoela and Nemertodermatida ('Acoelomorpha' of Ehlers, 1984, 1985a), one the Catenulida, and the third all other turbellarian orders together with the parasitic classes ('Rhabditophora' of Ehlers, 1984, 1985a).

The question remains, however, as to how these three groups are related (see also Hendelberg, 1977, 1983). Because we cannot find evidence for a close phylogenetic relationship among them, it is possible to argue that they are not directly related, that the Turbellaria (and thus the Platyhelminthes) is actually polyphyletic. It is already established that the Turbellaria is a paraphyletic class since the parasitic platyhelminth classes have ancestors that would be classified as rhabdocoel turbellarians (Karling, 1974). Whether the Turbellaria, and hence the Platyhelminthes, may also be polyphyletic, has not been addressed before.

Methodology of phylogenetic reconstruction

To some extent the matters of dispute between phylogenetic hypotheses arise because of differences in methodology for formulating those hypotheses. The methodology we use is an iterative, two-step process. First, homologous characters (homologues) among the taxa under consideration are identified (see Rieger & Tyler, 1979) to produce a grouping of the taxa (as would be done, essentially, for making a Venn diagram); second, we hypothesize which states of the homologous are primitive or advanced. The 'rules' by which character states can be ordered were discussed by Hennig (1950) and Remane (1956) (see also Wiley, 1981, for summary).

It is important to keep in mind that these two steps are separable but complementary. The first, identification of homologues, is an inductive process and can be performed without consideration of evolutionary theory; it springs directly from the recognition of patterns of similarity in nature. The second step, identifying given states of homologous characters as being more or less primitive, is a deductive process that posits evolution as the cause of the ordering. It is essential that once the second step has been performed, the first is rechecked to ensure that the primitive and derived states of each character linked by the postulate of evolution are actually homologous; the second step would then be repeated itself to order character states in any further homologues that may emerge.

A phylogenetic system is the most efficient means for expressing relationships. The formalized approach of phylogenetic systematics as it is now practiced, however, seems to us to fall short in one important aspect, namely that its adherents too often neglect rigorous homology analysis and commence their phylogenetic reconstruction with the ordering of character states for which homology is questionable. The important matter of distinguishing homologue from analogue similarity is often done solely by means of the 'parsimony principle' which is hardly reliable by itself as a criterion of homology. Evolutionary history, like social history, is better regarded as a series of accidents than as a series of events obeying one simple rule of parsimony, and so reliance on parsimonious distribution of characters among taxa as being the primary criterion of their homology is simplistic.

Autapomorphies for the Phylum Platyhelminthes

If the Platyhelminthes is a monophyletic group, then it must have one or more autapomorphies that defines it (synapomorphies among groups Catenulida, Acoelomorpha, and Rhabditophora – Fig. 1, solid circles). Older systems are rather vague about the definition of the phylum probably because even though it is intuitively obvious what a turbellarian is, a listing of characters that strictly distiguish the Turbellaria from other groups is difficult to construct. Recently, Ehlers (1984, 1985a, b; see also Ax,

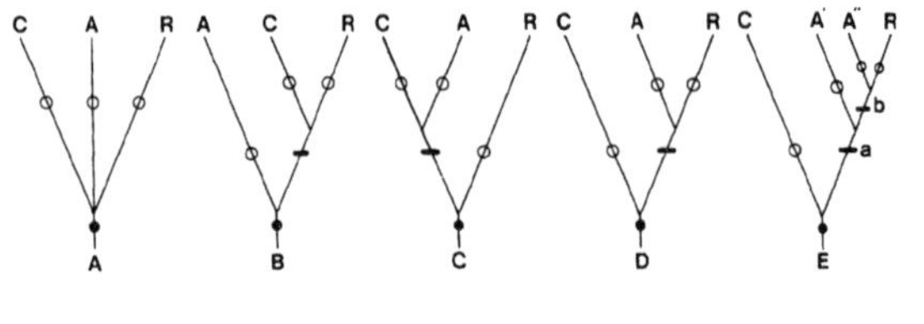

Fig. 1. Four possible cladograms relating the three most probable monophyletic groups of platyhelminth evolution, and a fifth, assuming monophyly of the Catenulida and Rhabditophora and paraphyly of the Acoelomorpha. Open circles are apomorphies for each line (see Ehlers, 1985a, b). Closed circle represents autapomorphies for the Phylum Platyhelminthes. Black bar represents synapomorphy necessary to produce groupings B–E. A = Acoelomorpha, A′ = Nemertodermatida, A″ = Acoela, C = Catenulida, R = Rhabditophora.

1984) has admirably made a concerted effort to identify autapomorphies for a phylogenetic system of the Platyhelminthes and has proposed that characters of protonephridia, of cilia in epidermal cells, and of the mode of epidermal cell replacement are autapomorphic for the phylum. As for any characters identified as autapomorphies, it must be established that these are homologous (i.e. are valid synapomorphies). Our research on ultrastructure of epidermal cells has provided and abundance of evidence that is relevant to establishing probability of homology for two of the proposed autapomorphies: 1) epidermal cells multiciliated and 2) absence of mitosis in epidermal cells and other somatic cells.

Epidermal multiciliation

The multiciliated nature of epidermal cells appears to be a homologous feature in the Rhabditophora and Catenulida especially because the ciliary structures in these cells are probably homologous. Rieger (1976) first pointed out that ciliary rootlet systems in the Catenulida and Macrostomida (in particular the genus *Myozona*) appeared to be homologous because both possess a rostral-caudal rootlet system without accessory centrioles. We have further evidence for homology of rootlets in two other macrostomids and *Haplopharynx* (Haplopharyngida). The caudally directed caudal rootlet of *Psammomacrostomum* and the vertically directed caudal rootlet of *Bradynectes* and *Haplopharynx* originate at the anterior proximal edge of the basal body, as apparently does the caudal rootlet in the monociliated epidermal cells of gnathostomulids (Figs. 2–4, compare Rieger & Mainitz, 1977: Fig. 1). At the very least, the epidermal ciliary rootlet system of *Haplopharynx* (and, we would predict, that of other members of the Rhabditophora; see also Ehlers, 1984) can be derived from the more primitive condition seen in *Psammomacrostomum* and *Myozona* (see Rieger, 1976, 1985).

Ciliary rootlets in the multiciliated epidermal cells of the Acoelomorpha are quite complex (Hendelberg & Hedlund, 1974; Tyler & Rieger, 1977), and detailed analysis of their substructure shows them to be even more complex than originally thought (Figs 5–11). The rostrally directed principal rootlet in most acoels studied is bipartite, consisting of two closely appressed portions: a posterior, cone-shaped rootlet arising proximally from the basal body and an anterior rootlet arising like a strap from the anterior face of the basal body and continuing as a trough until joining the posterior rootlet just above the knee-like bend (Figs 8–11).

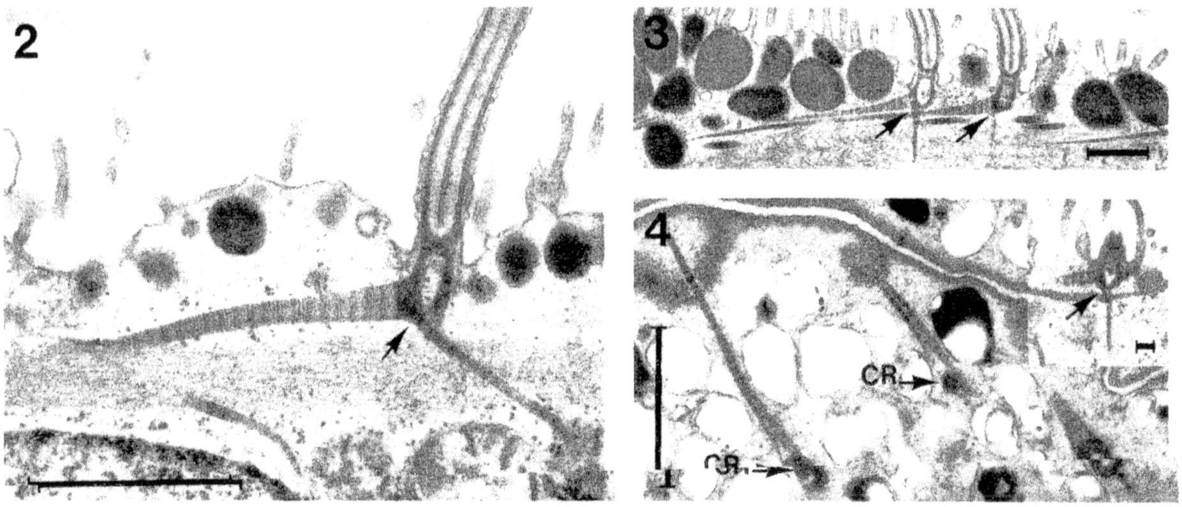

Figs. 2–4. Epidermal ciliary rootlet systems in Rhabditophora. All scales 1 μm unless noted otherwise. (2) *Psammomacrostomum* (O. Macrostomida), longitudinal section; (3) *Bradynectes* (O. Macrostomida), longitudinal section, scale = 0.5 μm; (4) *Haplopharynx* (O. Haplopharyngida), tangential section parallel to epidermal surface; inset: longitudinal section, scale = 200 nm. Note origin of caudally-directed (*Psammomacrostomum*) or vertically-directed (*Bradynectes, Haplopharynx*) caudal rootlet (CR) from the anterior proximal margin of the basal body in each case (arrows). Anterior is to the left in Figs. 2 & 3, to the upper left in Fig. 4.

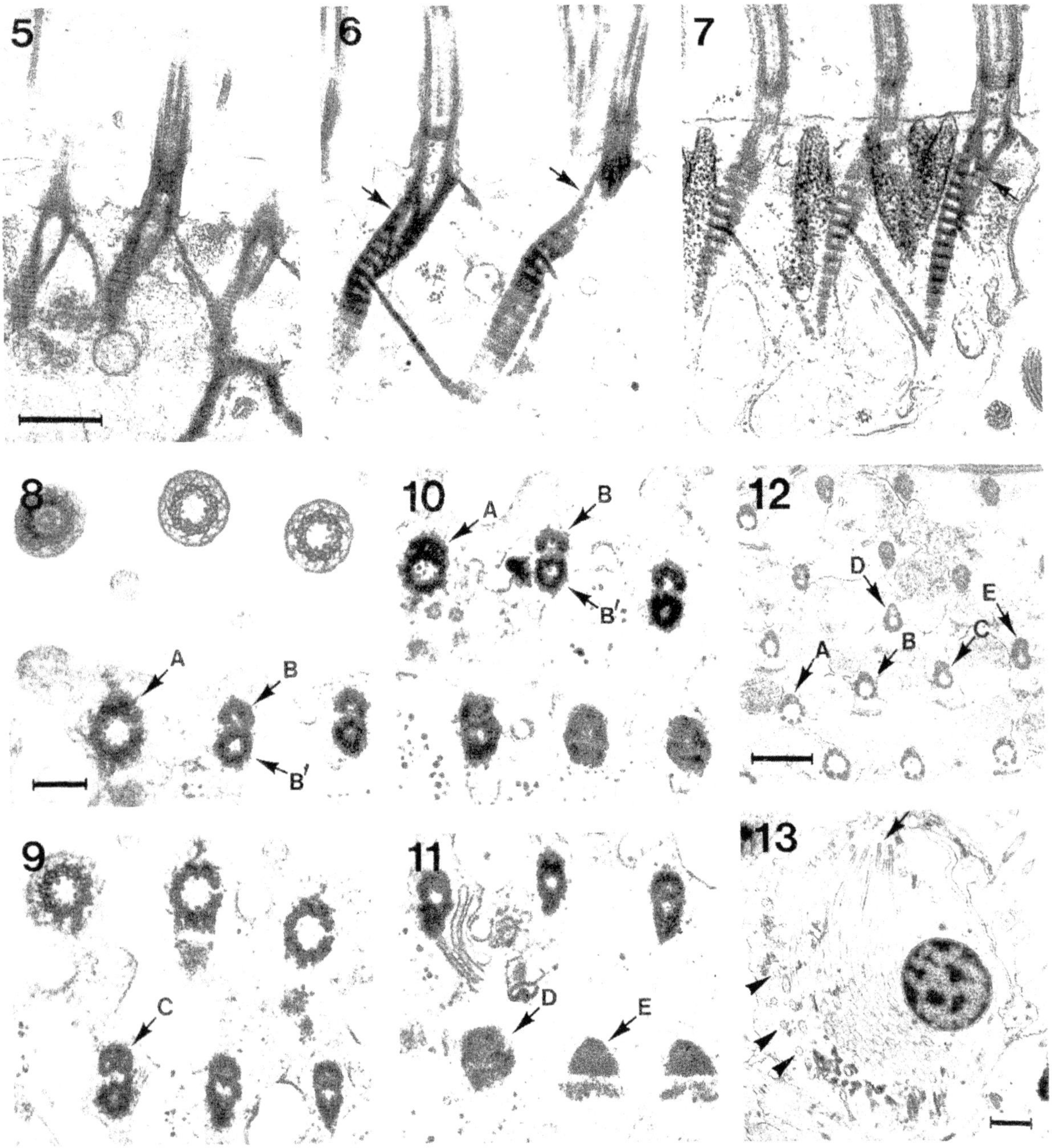

Figs. 5–7. Epidermal ciliary rootlet systems in Acoelomorpha. (5) *Flagellophora cf. apelti* (O. Nemertodermatida); (6) *Diopisthoporus* 'gymnopharyngeus' (O. Acoela); (7) *Convoluta* 'pulchra' (O. Acoela). Anterior is to the left in all figures. Scale = 500 nm.

Figs. 8–11. Serial sections, nearly parallel to epidermal surface, showing details of ciliary rootlet structure in *D.* 'gymnopharyngeus'; scale = 200 nm; see Fig. 8. Anterior is to the top of the figures, the view is from inside the animal (along cilia from base to tip). Note curved anterior portion of rootlet (Figs. 8, 10 A–B), which joins hollow rootlet arising from basal body (Figs 8, 10 B′), thickens (Fig. 9 C), and continues as principal rootlets (Fig. 11 D, E). Gap between two parts is visible in Figs. 6, 7 (arrows).

Fig. 12. Tangential section of ciliary rootlet system in *Flagellophora cf. apelti*: anterior is toward the top of the figure. Note rostral portion of rootlet that arises on the anterior face of basal body (arrows A–C), continues along it (arrow D) and fuses with it (arrow E). Scale = 400 nm.

Fig. 13. Pulsatile body, posterior epidermis of *Convoluta* 'pulchra'. Note free axonemes in cytoplasm (arrowheads) and fully formed ciliary tips (arrow). Scale = 1 μm.

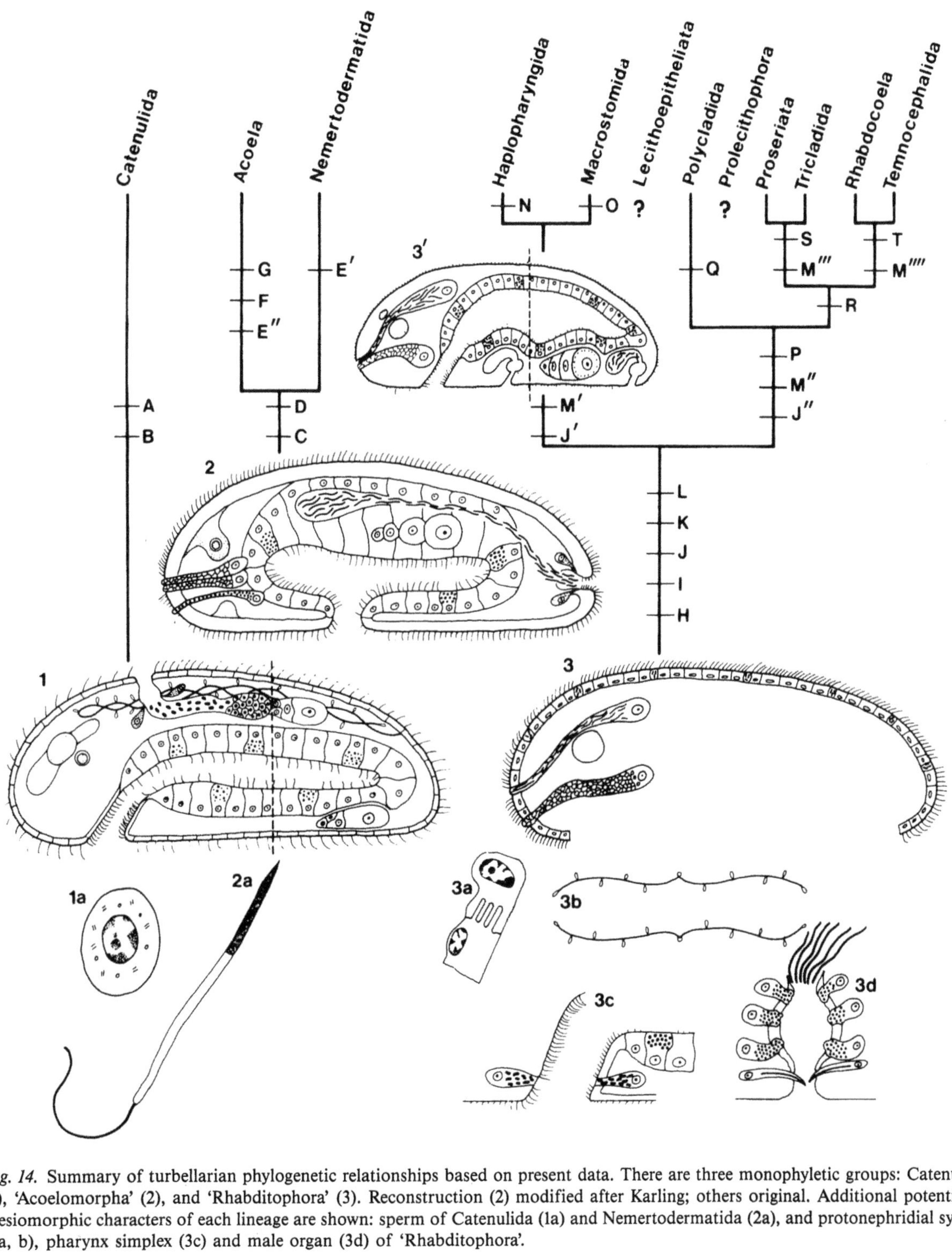

Fig. 14. Summary of turbellarian phylogenetic relationships based on present data. There are three monophyletic groups: Catenulida (1), 'Acoelomorpha' (2), and 'Rhabditophora' (3). Reconstruction (2) modified after Karling; others original. Additional potentially-plesiomorphic characters of each lineage are shown: sperm of Catenulida (1a) and Nemertodermatida (2a), and protonephridial system (3a, b), pharynx simplex (3c) and male organ (3d) of 'Rhabditophora'.

Homologues of both these parts can be identified in nemertodermatids (Fig. 12; Smith ms. in prep.) and thus the already strong probability of homology in the epidermal ciliation of the Acoela and Nemertodermatida (Tyler & Rieger, 1977) is further strengthened. It is possible even to identify some trends in rootlet evolution within the Acoelomorpha – for instance, the reduction of the laterally compressed, blade-shaped tip of the principal rootlet in Nemertodermatida to a pointed tip in some acoels follows a reading sequence (Figs 5–7) that agrees with data on the parenchyma (cf. Smith & Tyler, 1985a). Despite the abundance of characters that can be discerned in these rootlet there is still no homology that can be identified between those of the acoelomorphs and those of other turbellarians (except perhaps one comparing basal feet with posterior fiber bundles). It is still entirely possible that the multiciliated nature of these cells is homologous even if the rootlets are not. Such a possibility depends on whether the outgroup for the Turbellaria has monociliated epidermal cells, and that outgroup has not been identified unequivocally.

Epidermal replacement

A second character presumed to be apomorphic for the Platyhelminthes is the absence of mitosis in epidermal and other somatic cells (Ehlers, 1984). While epidermal cell division has been reported for catenulid turbellarians it is likely that what was actually observed was division of epidermal stem cells lying at the base of the epidermis (see also Ehlers, 1984: 22–23). Because mitosis of dedifferentiated epidermal cells is known for the Cnidaria (Martin & Thomas, 1981) but has not been (adequately) documented for any bilaterian phyla, it appears that the lack of mitosis in epidermal cells is a plesiomorphy at the level of the Platyhelminthes and apomorphic for the bilaterian stem (we believe the platyhelminth and bilaterian stems are not equivalent – see Smith & Tyler, 1985a).

The replacement of ciliated epidermal cells by cells arising and differentiating in the parenchyma is thought to occur in all platyhelminths (Lyons, 1977; Tyler, 1984; Ehlers, 1984, 1985a; Smith & Tyler, 1985a) and has been documented in rhabdocoels (Luther, 1904; Hein, 1928), in developing triclads (Skaer, 1965), and in larvae of non-turbellarian platyhelminths. Electron microscopy of presumed replacement cells in catenulids (Moraczewski, 1977, 1981, pers. commun.), haplopharyngids (Doe, 1981, Smith & Tyler unpublished observations), a microstomid (Reuter & Palmberg, 1983; Reuter personal communicattion), and triclads (Hori, 1978) shows they contain a bundle of centrioles, presumably centrioles differentiating into ciliary basal bodies. More important than the simple absence of mitosis in epidermal cells, therefore, is the process of epidermal replacement by kinetosome-containing cells that originate in the parenchyma; and we would prefer to consider this character an autapomorphy for the phylum. Certainly the prob-

For summary references and background, see Karling (1974) and Ehlers (1984, 1985a). Some proposed autapomorphies of each group (with references to comparative studies) are:
A: Position and orientation of male system (Sterrer & Rieger, 1974). Special structure of spermatozoan (Rieger, 1978).
B: Unpaired, dorsally-located protonephridial system. Fine structure of flame bulb (Kümel, 1962; Rieger, 1981).
C: Special structure of basal body-rootlet system complex and of ciliary tips (Tyler & Rieger, 1977; Tyler, 1979).
D: Fine structure of frontal organ (Smith & Tyler, 1985b).
E′: Statocyst of Nemertoderma-type (Ehlers, 1985a).
E″: Statocyst of Convoluta-type (Ferrero, 1973; Ehlers, 1985a).
F: With digestive parenchyma (Mamkaev, 1979; Smith, 1981; Smith & Tyler, 1985a).
G: Biflagellate spermatozoan with 9+2, 9+0, or 9+1 axonemes (Hendelberg, 1977).
H: With true rhabdites (Smith *et al.*, 1982).
I: Paired protonephridia with flame bulb and canal cell interdigitating to form weir (Wilson & Webster, 1974; Ehlers, 1985a).
J: Duo-gland adhesive system with viscid and releasing gland cells (Tyler, 1976).
J′: Duo-gland adhesive system with viscid and releasing glands opening within a common collar of microvilli (Tyler, 1976).
J″: Duo-gland adhesive system with viscid and releasing glands opening separately through anchor cells (Tyler, 1976).
K: Frontal organ with acidophilic rhammites (Klauser *et al.*, 1985).
L: Epitheliosomes of basic protein (Tyler, 1984).
M′: Pharynx simplex coronatus (Doe, 1981).
M″: Pharynx plicatus or bulbosus (Karling, 1940; Ax, 1961).
M″′: Pharynx tubiformis (Steinböck, 1925; Ball, 1974).
M″″: Pharynx bulbosus rosulatus or doliformis (Karling, 1940; Ax, 1961).
N: Complex glandular, muscular proboscis, ventral to brain (Karling, 1965).
O: Details of pharynx cytology (Doe, 1981).
P: 9+1 axoneme in biflagellate spermatozoan (Hendelberg, 1977).
Q: Intestine with dendritic side branches.
R: Heterocellular female gonads (Karling, 1940; Westblad, 1948).
S: For characters, see Sopott-Ehlers (1985).
T: Unpaired gonopore with atrium commune (Karling, 1940).

ability of homology of centriole-containing epidermal replacement cells appears to be rather high between the Catenulida and Rhabditophora (see also Ehlers, 1984, 1985a).

Dorey (1965) suggests that acoels have a similar mode of epidermal replacement by immigration of cells from the parenchyma. These cells, called pulsatile bodies, have only been seen, however, as fully differentiated, cilia-bearing cells (Fig. 13; see also Dorey, 1965), even sometimes in the central syncytium apparently in the process of being digested (Smith & Tyler unpublished observations on acoels and *Nemertoderma* sp. B), and therefore are more likely to be epidermal cells being withdrawn from the epidermal surface than nascent epidermal cells. This possibility needs experimental verification, but until pulsatile bodies are convincingly demonstrated to be epidermal replacement cells, it seems to us that the probability of homology between the pulsatile bodies of the Acoelomorpha and the epidermal replacement cells of the Rhabditophora and Catenulida is low. So little is known about epidermal replacement in other metazoans that it is difficult to tell, in any case, whether any common features of epidermal replacement in Acoelomorpha and other turbellarians are apomorphic or plesiomorphic.

Other autapomorphies

Because the probability of homology of both epidermal replacement mode and ciliary ultrastructure between the Acoelomorpha and other turbellarians is so low, it is necessary that other proposed automorphies of the platyhelminths be thoroughly tested. Characters relating to protonephridial structure cannot link the Acoelomorpha to other platyhelminths because all known acoelomorphs lack protonephridia entirely. The one remaining proposed autapomorphy, namely lack of accessory centrioles on ciliary basal bodies, is a negative character that cannot stand alone as an autapomorphy.

In the final analysis, there are, at present, no synapomorphies common to the three monophyletic groups of the Platyhelminthes and thus no unequivocal autapomorphies for the phylum.

Relationships among platyhelminth taxa

Assuming, as most systematists of the Turbellaria do, that all turbellarians share a common ancestor, then the three clearly identifible monophyletic groups must be related in one of the four alternative ways depicted in Fig. 1 A–D. In the absence of synapomorphies common to all three groups (Fig. 1 A, filled circle) alternative A is ruled out; similarly, alternative C can be ruled out because no synapomorphies between Acoelomorpha and Catenulida are known. However, the anterior-dorsal position of the male reproductive system in a new genus of Nemertodermatida (pers. comm. N.W. Riser and own unpub. data), a feature unknown in other turbellarians except the Catenulida, is a potentially significant link worthy of further investigation.

Present evidence indicates strongly that the group Catenulida + Rhabditophora is monophyletic (Fig. 1B, black bar), but, as noted above, synapomorphies linking this group to the Acoelomorpha are not known (Fig. 1B, filled circle). The lack of this synapomorphy makes alternative B unlikely unless the Platyhelminthes is to be diphyletic.

The remaining alternative (Fig. 1D) circumvents the lack of synapomorphies between the Acoelomorpha and the other groups by considering that lack to be a case of reduction and linking the Acoelomorpha with the Rhabditophora (Fig. 1D, black bar). This is equivalent to 'alternative 2' of Karling (1974) and to Ehlers' (1984, 1985a, b) system and seems to be the one most systematists of the Turbellaria would support. With it, protonephridia, ciliary rootlet systems, and mode of epidermal cell replacement are considered homologous in the Catenulida and Rhabditophora. Lack of a synapomorphy for the Acoelomorpha + Rhabditophora (see Smith & Tyler 1985b, Klauser *et al.*, 1985) weakens this alternative, however. Moreover, it necessitates that the biflagellate nature of spermatozoa in members of the Acoela and Rhabditophora as well as the 9+1 axonemes in some members of both these groups are not homologous.

A variant on alternative D – one that considers the characters of the spermatozoa to be homologous – is that of Fig. 1E, in which the Acoelomorpha is depicted as being paraphyletic (logical consequence of arguments raised by Hendelberg, 1977).

There is at least one compelling reason for considering biflagellate spermatozoa among turbellarians to be homologous: the Plathyhelminthes is the only metazoan group to have spermatozoa with both a biflagellate morphology and an inverted orientation (flagellar basal body leads and nucleus trials in locomotion; see Hendelberg, 1985). While these conditions can be found in spermatozoa of other metazoan groups (inverted: Marchand & Mattei, 1976; biflagellate: see Baccetti & Afzelius, 1976), they are not known to occur together in the same sperm except among platyhelminths. If these characters are homologous among platyhelminths, then the characters of protonephridia, ciliary rootlet systems, and epidermal replacement mentioned above cannot be homologous in Catenulida and Rhabditophora.

Thus, for now, we prefer to leave open the construction of a phylogenetic system for the Platyhelminthes. Although we believe the phylum is monophyletic, we cannot demonstrate that monophyly; and so rather than forcing the three monophyletic groups together, it is preferable to treat each as independent (Fig. 14). Characters that may close the gap between these groups should come from study of catenulids and nemertodermatids, in particular. We need also to understand more fully the relationships of the phylum to other invertebrate groups, for its evolutionary origin remains one of the chief puzzles of metazoan phylogeny (Smith & Tyler, 1985a).

References

Ax, P., 1961. Verwandtschaftsbeziehungen und Phylogenie der Turbellarien. Ergebn. Biol. 24: 1–68.

Ax, P., 1984. Das phylogenetische System. Systematisierung der lebenden Natur auf Grund ihrer Phylogenese. G. Fischer, Stuttgart, New York, 349 pp.

Baccetti, B. & B. A. Afzelius, 1976. The biology of the sperm cell. Monographs in Development Biology 10, 254 pp.

Ball, I. R., 1974. A contribution to the phylogeny and biogeography of freshwater triclads (Platyhelminthes: Turbellaria). In N. W. Riser & M. P. Morse (eds), Biology of the Turbellaria. McGraw-Hill, New York: 339–401.

Doe, D. A., 1981. Comparative ultrastructure of the pharynx simplex in Turbellaria. Zoomorphology 97: 133–193.

Dorey, A. E., 1965. The organization and replacement of the epidermis in acoelous turbellarians. Quart. J. microsc. Sci. 106: 147–172.

Ehlers, U., 1977. Vergleichenden Untersuchungen über Collar-Rezeptoren bei Turbellarien. Acta zool. fenn. 154: 137–148.

Ehlers, U., 1984. Das phylogenetische System der Plathelminthes. Habilitationsschrift, Universität Göttingen.

Ehlers, U., 1985a. Phylogenetic relationships within the Platyhelminthes. In S. Conway Morris, J. D. George, R. Gibson, H. M. Platt (eds), Origins and relationships of lower invertebrates. University Press, Oxford. (in press)

Ehlers, U., 1985b. Comments on a phylogenetic system for the Platyhelminthes. This volume.

Ferrero, E., 1973. A fine structure analysis of the statocyst in Turbellaria Acoela. Zool. Scr. 2: 5–16.

Hein, C., 1928. Zur Kenntnis der Regenerationsvorgänge bei den Rhabdocoelen. Mit Angaben über den feineren Bau und die Lebensäusserungen. Z. wiss. Zool. 130: 469–546.

Hendelberg, J., 1977. Comparative morphology of turbellarian spermatozoa studied by electron microscopy. Acta zool. fenn. 154: 149–162.

Hendelberg, J., 1983. Platyhelminthes – Turbellaria. In K. G. Adiyodi & R. G. Adiyodi (eds), Reproductive biology of invertebrates. Vol. II, Spermatogenesis and sperm function. J. Wiley & Sons, Ltd., Chichester: 74–104.

Hendelberg, J., 1985. The phylogenetic significance of sperm morphology. This volume.

Hendelberg, J. & K.-O. Hedlund, 1974. On the morphology of the epidermal ciliary rootlet system of the acoelous turbellarian Childia groenlandica. Zoon 2: 13–24.

Hennig, W., 1950. Grundzüge einer Theorie der phylogenetischen Systematik. Deutscher Zentralverlag, Berlin.

Hori, I., 1978. Possible role of rhabdite-forming cells in cellular succession of the planarian epidermis. J. electr. Microsc. 27: 89–102.

Karling, T. G., 1940. Zur Morphologie und Systematik der Alloeocoela Cumulata und Rhabdocoela Lecithophora (Turbellaria). Ark. Zool. 10: 331–345.

Karling, T. G., 1965. Haplopharynx rostratus Meixner (Turbellaria) mit den Nemertinen vergleichen. Z. zool. syst. Evolutionsforsch. 3: 1–18.

Karling, T. G., 1974. On the anatomy and affinities of the turbellarian orders. In N. W. Riser & M. P. Morse (eds), Biology of the Turbellaria. McGraw-Hill, New York: 1–16.

Klauser, M., J. P. S. Smith & S. Tyler, 1985. Ultrastructure of the frontal organ of Convoluta 'pulchra' and Macrostomum spp.: significance for the turbellarian archetype. This volume.

Kümmel, G., 1962. Zwei neue Formen von Cyrtocyten. Vergleich der bisher bekannten Cyrtocyten und Ergänzung des Begriffs 'Zelltyp.' Z. Zellforsch. 57: 172–201.

Luther, A., 1904. Die Eumesostominen. Z. wiss. Zool. 77: 1–273, Taf. I-X.

Lyons, K. M., 1977. Epidermal adaptions of parasitic platyhelminths. Symp. zool. Soc. London 39: 97–144.

Mamkaev, Yu. V., 1979. On histological organization of the turbellarian digestive system. [in Russian] Proc. zool. Inst. Acad. Sci. U.S.S.R. 84: 13–24.

Marchand, B. & X. Mattei, 1976. Ultrastructure du spermatozoide de Centrorhynchus milvus Ward, 1956 (Paleacanthocephala, Polymorphidae). C. r. Séanc. Soc. Biol. 170: 237–241.

Martin, V. J. & M. B. Thomas, 1981. The origin of the nervous system in Pennaria tiarella as revealed by treatment with colchicine. Biol. Bull. 160: 303–310.

Moraczewski, J., 1977. Asexual reproduction and regeneration of Catenula (Turbellaria, Archoophora). Zoomorphologie 88: 65–80.

Moraczewski, J., 1981. Fine structure of some Catenulida (Turbellaria Archoophora). Zool. pol. 28: 367–415.

Remane, A., 1956. Die Grundlagen des natürlichen Systems, der vergleichenden Anatomie und der Phylogenetik. Akademische Verlagsgeselschaft, Leipzig, 364 pp. (Reprinted 1971 by Otto Koeltz, Königstein, Taunus.)

Reuter, M. & I. Palmberg, 1983. Asexual reproduction in Microstomum lineare (Turbellaria). II. The nervous system in the division zone. Int. J. invert. Reprod. 6: 207–217.

Rieger, R. M., 1976. Monociliated epidermal cells in Gastrotricha: Significance for concepts of early metazoan evolution. Z. zool. Syst. Evolutionsforsch. 14: 198–226.

Rieger, R. M., 1978. Multiple ciliary structures in developing spermatozoa of marine Catenulida (Turbellaria). Zoomorphologie 89: 229–236.

Rieger, R. M., 1981. Morphology of the Turbellaria at the ultrastructural level. Hydrobiologia 84: 213–229.

Rieger, R. M., 1985. Asexual reproduction and the turbellarian archetype. This volume.

Rieger, R. M. & M. Mainitz, 1977. Comparative fine structural study of the body wall in Gnathostomulida and their phylogenetic position between Platyhelminthes and Aschelminthes. Z. zool. syst. Evolutionsforsch. 15: 9–35.

Rieger, R. M. & S. Tyler, 1979. The homology theorem in ultrastructural research. Am. Zool. 19: 654–666.

Skaer, R. J., 1965. The origin and continuous replacement of epidermal cells in the planarian Ploycelis tenuis (Iijima). J. Embryol. exp. Morph. 13: 129–139.

Smith, J. P. S., 1981. Fine-structural anatomy of the parenchyma in the Acoela and Nemertodermatida (Turbellaria). Ph.D. Dissertation, University of North Carolina, Chapel Hill.

Smith, J. P. S. & S. Tyler, 1985a. The acoel turbellarians: kingpins of metazoan evolution or a specialized offshoot? In S. Conway Morris, J. D. George, R. Gibson, H. M. Platt (eds), Origins and relationships of lower invertebrates. University Press, Oxford. (in press)

Smith, J. P. S. & S. Tyler, 1985b. Frontal organs in the Acoelomorpha: ultrastructure and phylogenetic significance. This volume.

Smith, J. S., S. Tyler, M. B. Thomas & R. Rieger, 1982. The nature of turbellarian rhabdites: phylogenetic implications. Trans. am. microsc. Soc. 101: 209–228.

Sopott-Ehlers, B., 1985. The phylogenetic relationships within the Seriata (Platyhelminthes). In S. Conway Morris, J. D. George, R. Gibson, H. M. Platt (eds), Origins and relationships of lower invertebrates. University Press, Oxford. (in press).

Steinböck, O. 1925. Zur systematik der Turbellaria metamerata. Zool. Anz. 64: 165–192.

Sterrer, W. & R. Rieger, 1974. Retronectidae – a new cosmopolitan marine family of Catenulida (Turbellaria). In N. W. Riser & M. P. Morse (eds), Biology of the Turbellaria. McGraw-Hill, New York: 63–92.

Tyler, S., 1976. Comparative ultrastructure of adhesive systems in the Turbellaria. Zoomorphologie 84: 1–76.

Tyler, S., 1979. Distinctive features of cilia in metazoans and their significance for systematics. Tissue Cell 11: 385–400.

Tyler, S., 1984. Turbellarian platyhelminths. In J. Bereiter-Hahn, S. Matoltsy, K. S. Richards (eds), Biology of the integument, Vol. 1, Invertebrates. Springer-Verlag, Berlin: 112–131.

Tyler, S. & R. M. Rieger, 1977. Ultrastructural evidence for the systematic position of the Nemertodermatida (Turbellaria). Acta zool. fenn. 54: 193–207.

Westblad, E., 1948. Studien über Skandinavische Turbellaria Acoela V. Ark. Zool. 41A: 1–82.

Wiley, E. O., 1981. Phylogenetics. The theory and practice of phylogenetic systematics. Wiley Interscience, New York, 439 pp.

Wilson, R. A., L. A. Webster, 1974. Protonephridia. Biol. Rev. 49: 127–160.

The major parasitic platyhelminth classes – progressive or regressive evolution?

G. Malmberg
Zoologiska Institutionen, University of Stockholm, S-106 91 Stockholm, Sweden

Keywords: parasitic, Platyhelminthes, evolution, evolutionary capacity level

Abstract

The evolution of certain ontogenetic processes appears to be of special importance to subsequent evolution of a phyletic line, and their occurrence suggests there are various 'evolutionary capacity levels' occupied by major taxa. Two groups that are at the same evolutionary capacity level probably originated in common within that level, which in turn is often shared with several other groups. Morphogenesis by invagination is characteristic of an early evolutionary capacity level common to poriferans, cnidarians, deuterostomes, and platyhelminths, *inter alia.* In the early ontogeny of the former three groups, this process produces the blastopore and archenteron, but in platyhelminths, e.g. rhabdocoels, two different processes form the digestive system: the mouth forms by invagination and the intestine by a split in the mesenchyme/parenchyma. I have selected some characters and processes to illustrate evolutionary capacity levels and to describe the possible origin of a number of invertebrate groups from various levels and their inter-relationships. My analysis on the basis of evolutionary capacity levels suggests that the deuterostomes originated early (very likely before mollusks and annelids), that tegumental absorption of nutriments is the original mode of nutriment uptake and so predates intestinal ingestion in Platyhelminthes, that ontogenetic characters show that the presumed origin of cercomeromorphaeans from rhabdocoels is improbable, and that features of ontogeny and life cycles indicate that the major parasitic platyhelminth classes arose through progressive rather than regressive evolution. Thus the non-parasitic and the parasitic platyhelminth classes very likely belong to separate evolutionary lines, all originating from the same evolutionary capacity level.

Introduction

The prevalent opinion concerning the evolution of parasitic platyhelminths is that they show regressive evolution (see e.g. Janicki, 1920; Bychowsky, 1957; Mackiewicz, 1981). Neoteny or progenesis is often claimed to be involved in that regressive evolution (e.g. Janicki, op. cit.: Mackiewicz, op. cit.). It is presumed that non-parasitic platyhelminth ancestors (turbellarians, particularly rhabdocoels) were adapted first to ectoparasitism and then to endoparasitism. In my view, however, known facts concerning the ontogeny of members of the major parasitic platyhelminth classes are not consistent with such claims (Malmberg, 1971, 1974, 1982). Rather, a progressive evolution is more likely. In this paper, I present evidence indicating there was progressive evolution in the parasitic platyhelminth classes, primarily the cercomeromorphaean classes.

In discussions of evolution, more attention is often paid to the products of evolution than to the processes behind them. In particular, similarities in the complexity of body plans and of developmental processes are sometimes mistakenly assumed to indicate close relationships. Actually such degrees of complexity simply represent levels in the evolution that given organisms have reached and that are the

Hydrobiologia 132, 23–29 (1986).

bases for further evolutionary advances to new levels; I call these 'evolutionary capacity levels' (see Figs. 1, 2, 4, 5). Because of the comparatively simple body structure in platyhelminths, the evolution from a unicellular to a multicellular capacity level is important. In this evolutionary path, tegumental absorption may be of earlier origin than intestinal ingestion. This raises the questions, how many times and for what function was an intestine evolved in the Animal Kingdom? And cannot the absence of an intestine in placozoans, cestodes, and acoels indicate an original condition shared with common ancestors that were still at an early multicellular evolutionary capacity level? Ontogenetic differences between protostomes and deuterostomes with respect to the formation of digestive systems are of special relevance to these questions. And the evolution of life cycles (and of ecosystems) is another important consideration.

Evolutionary capacity levels

In all probability the evolution of a given process preceded the appearance of features caused by such a process. For example, the evolution of a capacity to invaginate in morphogenesis preceded the presence of an archenteron formed by means of invagination (embolic gastrulation). Invagination, however, is an ontogenetic capacity used not only in forming an intestine. The same capacity causes the formation of mesoderm in deuterostomes and it is important in larval metamorphoses in the Rhynchocoela. In cestodes, the formation of the 'excretory bladder' and the apical scolex cavity take place by invagination, and in the Acoela this process forms mouth and pharynx. Furthermore, in non-parasitic platyhelminths with an intestine, at least the mouth is formed by invagination although their enteron is formed otherwise, i.e. initially as an internal mesenchymal split (Hyman, 1951). Thus, invagination is a capacity used for different ontogenetic 'purposes' in the Animal Kingdom.

The capacity for mesenchymal tissue to split internally is another capacity of ontogenetic importance not only in Platyhelminthes but also in Mollusca and Annelida. In the latter two groups the 'splitting capacity' forms the coelom. However, it is also important in Cnidaria. Here the typical formation of the archenteron is caused either by invagination or by an internal split (stereoblastula-stereogastrula). Furthermore, in *Aurelia aurita* the archenteron is formed in both ways depending on the size of the egg (see Berrill, 1949). It seems impossible to relate these processes to each other. In my view, they are two different, fully evolved processes, each indicating an evolutionary capacity level. Within the Cindaria both these levels were important to further evolution. For the evolution of the deuterostomes, however, the invagination capacity level was especially important while in the evolution of the Spiralia (*inter alia*, Platyhelminthes, Mollusca, Annelida) the 'internal split capacity' was of special importance. Thus, the Spiralia is a separate line, a product of progressive evolution (Figs. 1a and 1b).

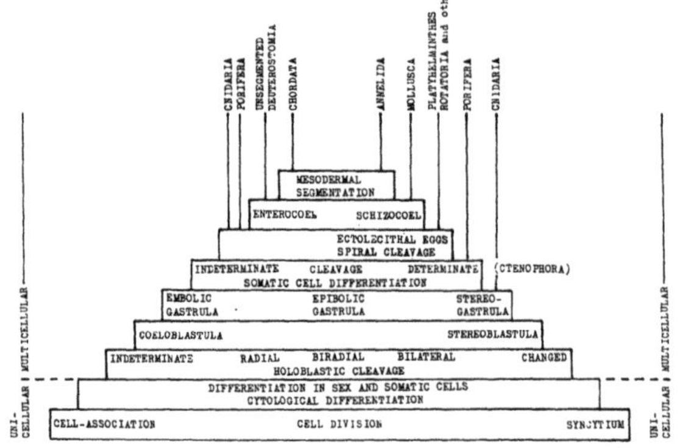

Fig. 1a. Important evolutionary capacity levels concerning Porifera, Cnidaria, Deuterostomia, and 'Spiralia' (Platyhelminthes, Mollusca, Annelida).

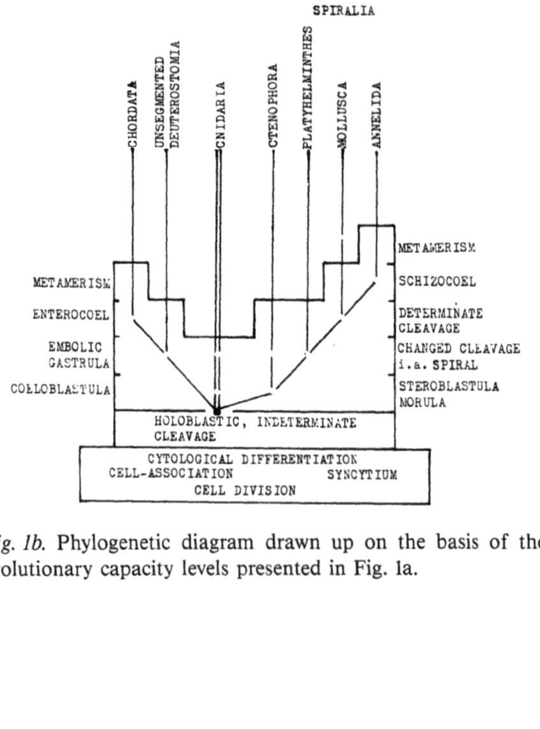

Fig. 1b. Phylogenetic diagram drawn up on the basis of the evolutionary capacity levels presented in Fig. 1a.

Platyhelminthes

The major parasitic platyhelminth classes are often claimed to have originated from a rhabdocoel-like ancestor. Evidence for this is the presence of ectolecithal eggs and development. However, this could merely indicate an origin from the same evolutionary capacity level. In fact, ectolecithal development in rhabdocoels appears to be more complex (more evolved) than in monogeneans and cestodes, while it is related to but not so complex as that in triclads. Early development in cestodes and monogeneans is in many respects, less complex than that of acoels and polyclads. After all, the absence of intestinal anlages in the Acoela and Cestoda could indicate an original platyhelminth condition. Furthermore, in terms of both form and function, the small placozoan *Trichoplax* could represent basically the first evolutionary level of hypothetical progenitors of the cercomeromorphaeans and other platyhelminth groups (Malmberg, 1974). In view of this, an intestine is presumed to have evolved more than twice in the platyhelminth line and several times among other invertebrates (Fig. 1b). If we arrange the platyhelminths on the basis of different degrees of complexity in their ontogeny, we may find indications of more progressive evolutionary lines than have been presented in evolutionary discussions hitherto. In my opinion, there are good reasons for changing the evolutionary platyhelminth 'tree' into a 'bush' (Figs. 2a and b).

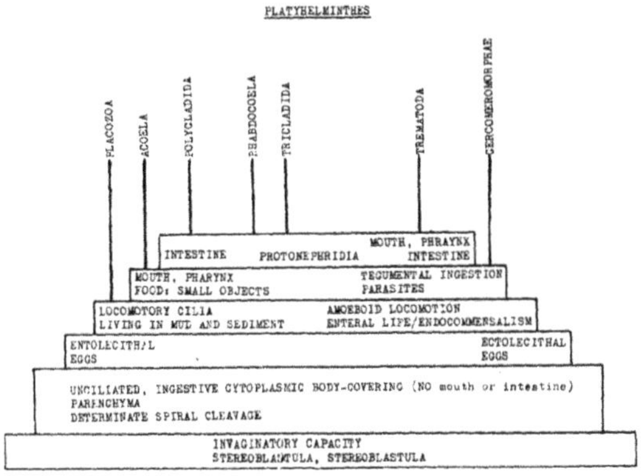

Fig. 2a. Important evolutionary capacity levels concerning non-parasitic and parasitic platyhelminths.

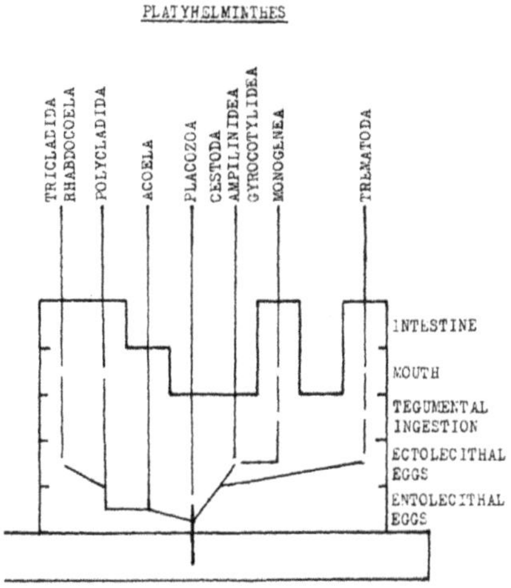

Fig. 2b. Phylogenetic diagram drawn up on the basis of the evolutionary capacity levels presented in Fig. 2a.

Cercomeromorphae

This group includes the Cestoda (monozoic and polyzoic), Amphilinidea, Gyrocotylidae and Monogenea. There appear to be several indications of progressive evolution in each group. The number of larval hooks and the shape of the protonephridial system in hatching larvae indicate different larval evolutionary levels: the cestodes exhibit the most primitive condition and the monogeneans the most advanced (Fig. 3). Monogenean adults, however, retain the bilateral organization of the protonephridial systems, while in the remaining three groups the system has developed into a reticular pattern in connection with parenteral life (in body fluids) in their hosts. The amphilinidean adults remain on this evolutionary level, morphologically

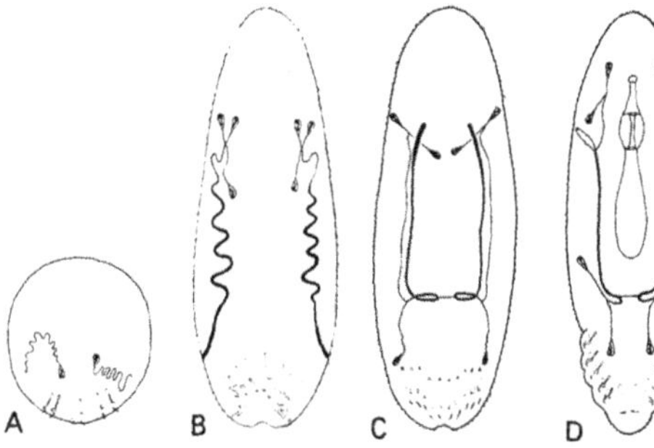

Fig. 3. The larval protonephridial systems in Cercomeromorphae (from Malmberg, 1983).

and ecologically. The adults of polyzoic cestodes and gyrocotylideans and of most monozoic cestodes, however, are adapted secondarily to an enteral life (*via* a parenteral one). The formation of strong anchoring apparatuses at the anterior end of cestodes and at the posterior end of gyrocotylideans correlate with the transition to an enteral life (Fig. 4a). Such a correlation also applies to tegumental changes in cestodes (microvilli to microtriches) and also in gyrocotylideans. My transmission-electron-microscopical investigations of gyrocotylidean postlarvae revealed that parenteral (parenchymal) postlarvae also have microvilli (Fig. 6). In enteral postlarvae, however, the microvilli were modified, but not in the parenteral postlarvae (see also Malmberg, 1979). Furthermore, in polyzoic cestodes the reticular protonephridial system develops into a *secondarily bilaterally* symmetrical system in connection with adult enteral life. In cestodes, amphilinideans, and gyrocotylideans there are very similar posterior invagination processes involving the cercomer and the changing of the body into the adult form. In monogeneans, the cercomer mostly is retained, though further evolved (see below). In my opinion the ontogeny – especially the various changes in the posterior body part and the cercomer, and changes, to different degrees, in the larval protonephridial systems – indicate a progressive evolution quite different from progressive evolutionary tends among non-parasitic platyhelminths. Cercomeromorphaean evolution involves two different parts of the life cycle: one concerns the larvae, the other the rest of the cycle. The differences concerning development of the protonephridial system during the latter part of the life cycle indicate there were three main levels of evolution in polyzoic cestodes, two main levels in monozoic cestodes, amphilinideans, and gyrocotylideans, and one in monogeneans. Thus the cercomeromorphaeans are anything but secondarily reduced animals. Also, from this point of view they are very unlikely to be of rhabdocoelan origin (Figs. 2a, b and 4a, b).

Monogenea

The monogenean haptor is probably homologous to the cercomer in cestodes and gyrocotylideans though in most cases it is retained and further developed (sometimes along with the posterior body end) in the adults. The number of larval hooks in cestodes is 3 + 3 and in amphilinideans and gyrocotylideans (3 + 3)/(2 + 2) i.e. 5 + 5 (Fig. 7). In the monogeneans, the original number is also most likely 5 + 5 (Malmberg, 1982). This number correlates with that of the most simple type of protonephridial system found in hatching monogenean larvae (Fig. 3). Regarding the number of larval hooks

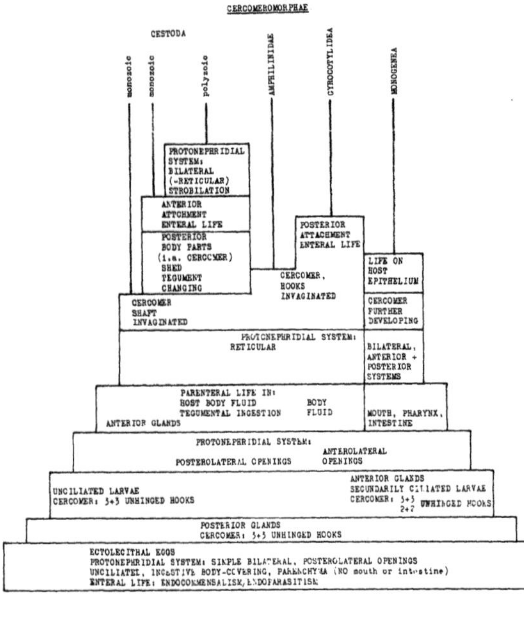

Fig. 4a. Important evolutionary capacity levels in Cercomeromorphae.

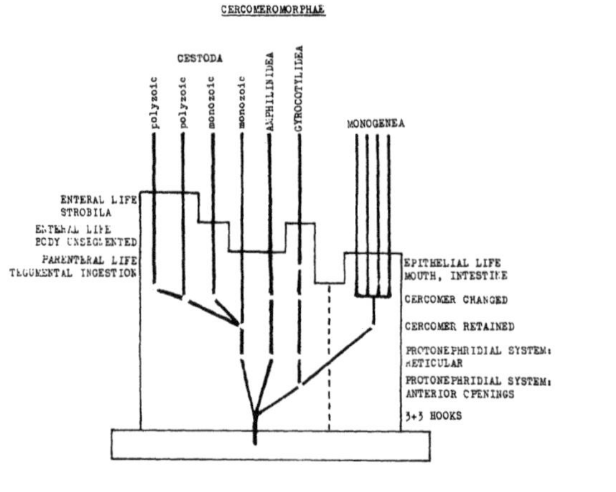

Fig. 4b. Phylogenetic diagram drawn up on the basis of the evolutionary capacity levels presented in Fig. 4a.

(marginal hooks in the adults) there are three groups of monogenean larvae: one with 5+5 hooks, another with 7+7 and a third with 8+8 hooks. In the latter two groups, the protonephridial system in the hatching larvae is more developed or evolved than in larvae with 5+5 hooks. Very likely the three groups represent three evolutionary levels. In all three groups, the ontogeny reveals the evolution of new haptoral elements in a certain order: first peripheral haptoral anchors and then (when present) central haptoral anchors. After that, in the '5+5 group', suckers and/or clamps develop.

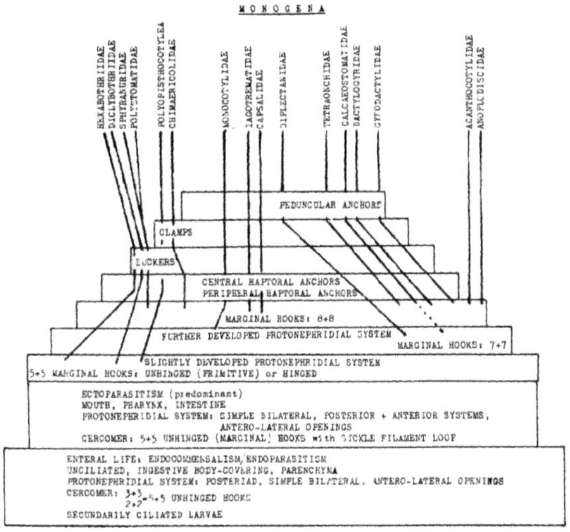

Fig. 5a. Important evolutionary capacity levels in Monogenea.

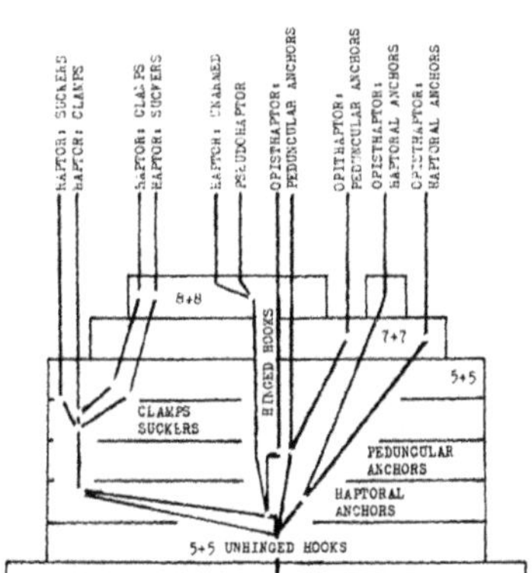

Fig. 5b. Diagram drawn up on the basis of the evolutionary capacity levels presented in Fig. 5a, illustrating possible evolution of various types of haptor in Monogenea.

Hinged marginal hooks are found only in the '5+5' and the '8+8' groups, peduncular anchors only in the '7+7' and '8+8' groups, and pseudohaptors are found in an '8+8 group' with hinged marginal hooks. However, the strict order in the monogene-

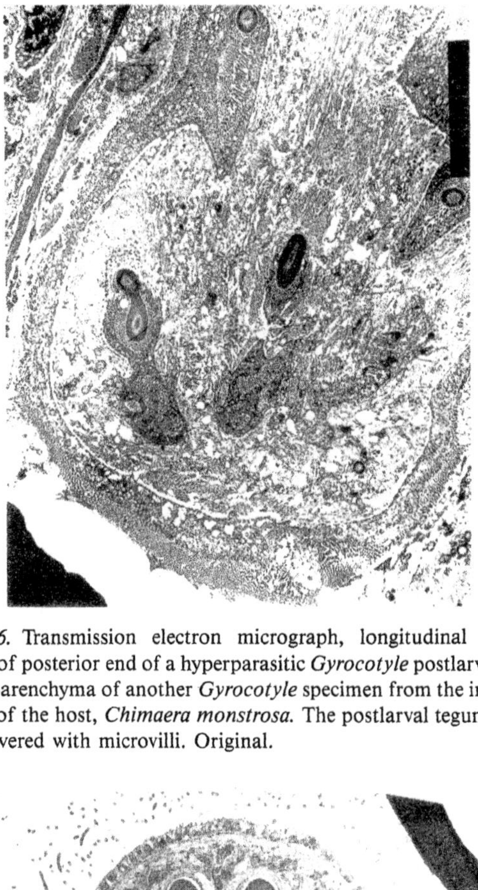

Fig. 6. Transmission electron micrograph, longitudinal section of posterior end of a hyperparasitic *Gyrocotyle* postlarva in the parenchyma of another *Gyrocotyle* specimen from the intestine of the host, *Chimaera monstrosa.* The postlarval tegument is covered with microvilli. Original.

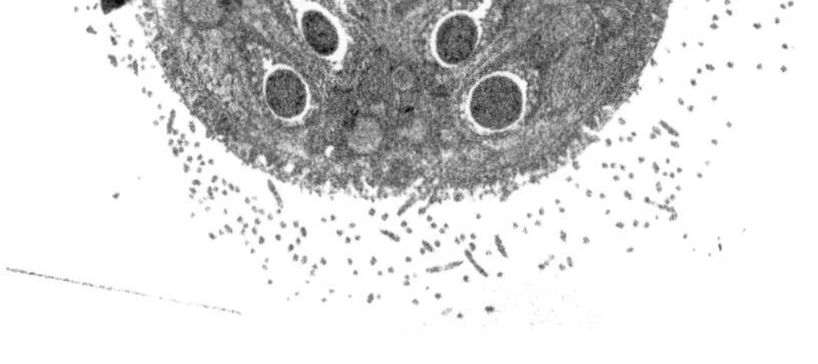

Fig. 7. Transmission electron micrograph, transverse section of posterior end of a *Gyrocotyle* lycophore/larva; host: *Chimaera monstrosa.* The transversely sectioned $\frac{3+3}{2+2}$ larval hooks (each in a pocket of a posterior invagination) are shown. Original.

an haptoral ontogeny seems to be useful for revealing different evolutionary levels and interrelationships in the Monogenea. Furthermore, comparative ontogenetical studies indicate a basically progressive evolution in monogeneans (Figs. 5a and b).

Monogenean larvae when hatching are more developed than larvae in the remaining three groups of Cercomeromorphae (Fig. 3). The ontogeny of the members of these three groups reveals no mouth, pharynx, nor intestine, but it does in monogeneans. Most monogeneans are ectoparasites; cestodes, amphilinideans, and gyrocotylideans are endoparasites. In all the latter three the larvae hatch at a distinctly earlier stage than do the most primitive monogenean larvae. This suggests that originally nutriments were absorbed tegumentally in Cercomeromorphae and that a mouth, pharynx, and intestine are secondarily evolved in the monogenean line, probably in connection with an ectoparasitic mode of life (Figs. 2b, 4b, 5b).

Lambert's (1980) interesting hypothesis concerning the evolution of Monogenea is based on findings concerning larval ciliation and numbers of sensilla. His conclusions imply a regressive evolution in monogeneans. However, the ciliation of the larvae of monogeneans, as well as those of amphilinideans and gyrocotylideans, seems to be secondarily evolved. In amphilinidean and gyrocotylidean larvae the ciliated tegument covers most of the body and overlies the tegument anlage of the adult (Malmberg, 1978, 1979; Rohde & Georgi, 1983). In monogeneans, the ciliated body covering is limited to a number of ciliated fields. According to Lambert (1980) larvae with few fields and a large number of sensilla characterize lower monogeneans while the larvae of higher monogeneans have many fields and few sensilla. If, however, a more or less complete ciliated body covering – as in amphilinidean and gyrocotylidean larvae – is a primitive condition then monogenean larvae with many ciliated fields are more primitive than larvae with few fields. Furthermore, sensilla are specialized cilia and thus the presence of few sensilla might be more primitive than sensilla in a large number. In my view, Lambert's results indicate a progressive monogenean evolution.

Life cycles, host and progenesis/neoteny

Adaptations to parasitism were probably possible at a very early stage in platyhelminth evolution, i.e. before a mouth and an intestine were evolved. Cercomeromorphaean life cycles were probably prolonged when new potential hosts had evolved. Depending on adaptation to 'new' ecological/host conditions, it could be favorable for certain parasites to retain both the former shorter and the 'new' longer cycle side by side. If so, such parasites would then have two adults with different morphology: the earlier adult possibly exhibiting a more primitive body shape and internal anatomy than the later adult. In my opinion, this applies to the monogenean *Polystomum* and the monozoic cestode *Archigetes* (see Malmberg, 1981). Thus, the sexual maturity of the gill form of *Polystomum* and of the body = cavity form of *Archigetes* reflect conditions in their original life cycles and are not caused by progenesis or neoteny. The presence of a cercomer in *Archigetes* and the absence of a strobila in monozoic cestodes can be explained similarly. The polyzoic cestodes may represent another evolutionary line where a strobila was evolved in connection with secondary adaptation to an enteral life (see above). The ontogeny of polyzoic cestodes substantiate such an opinion.

The evolutionary levels of hosts in cercomeromorphaean life cycles likewise bear on the evolution of Cercomeromorphae because the correlation 'primitive hosts, primitive parasites/more advanced hosts, more advanced parasites' can positively be applied to all four cercomeromorphaean groups.

Indications of regressive evolution

There appear to be many indications of progressive evolution not only in the cercomeromorphaeans but also among the trematodes. But even if the basic trend is progressive, results of a *secondary* regressive evolution can be found in many parasitic platyhelminth groups. A few examples within the Monogenea are the absence of haptoral anchors in *Oculotrema,* Polystomatidae parasitizing the epithelium of the eyeball of the hippopotamus; the absence of marginal hooks in Diplozoonidae (see Lambert & Denis, 1982); the unarmed posterior sucker in *Anoplodiscus* (see Ogawa & Egusa, 1981);

the absence of peduncular anchors in *Isancist.um,* Gyrodactylidae (see Malmberg, 1974); and the evolution of the protonephridial system in *Gyrodactylus* (see Malmberg, 1970).

Acknowledgments

Most of the work for this paper took place in the Department of Zoology at the University of Stockholm.

I am grateful to Y. Lilliemarck for her excellent assistance with sections for transmission electron microscopy and to A. Tapsell for his linguistic revision.

This work is supported by grants from the Swedish Natural Science Research Council, which I hereby gratefully acknowledge.

References

Berrill, N. J., 1949. Developmental analysis of scyphomedusae. Biol. Rev. 24: 393–410.

Bychowsky, B. E., 1957. [Monogenetic trematodes, their systematics and phylogeny.] (In Russian, English translation: Virginia Inst. mar. Sci. Transl. Ser. 1, AIBS, Washington D.C., 1961.) Zool. Inst. Acad. Nauk SSSR, Moscow and Leningrad.

Hyman, L. H., 1951. The Invertebrates, vol. 2. McGraw-Hill, New York, Toronto, London, 550 pp.

Janicki, C., 1920. Grundlinien einer 'Cercomer'-Theorie zur Morphologie der Trematoden und Cestoden. Festschr. für Zschokke 30: 1–22.

Lambert, A., 1980. Oncomiracidiums et phylogenèse des Monogenea (Platyhelminthes). Ann. Parasitol. Hum. Comp. 55(3): 281–325.

Lambert, A. & A. Denis, 1982. Étude de l'oncomiracidium Diplozoon nipponicum Goto, 1891. Hapteur larvaire d'un type nouveau pour le genre Diplozoon, Monogenea, Discocotylidae. Ann. Parasitol. Hum. Comp. 57(6): 533–542.

Mackiewicz, J. S., 1981. Caryophillidae (Cestoidea): Evolution and Classification. Adv. Parasitol. 19: 139–225.

Malmberg, G., 1970. The excretory systems and the marginal hooks as a basis for the systematics of Gyrodactylus (Trematoda, Monogenea). Ark. Zool. Ser. 2, 23(1): 1–235.

Malmberg, G., 1971. On the procercoid protonephridial systems of three Diphyllobothrium species (Cestoda, Pseudophyllidae) and Janicki's cercomer theory. Zool. Scr. 1: 43–56.

Malmberg, G., 1974. On the larval protonephridial system of Gyrocotyle and the evolution of Cercomeromorphae (Platyhelminthes). Zool. Scr. 3: 65–81.

Malmberg, G., 1978. Further studies on the evolution of the Cercomeromorphae. In Short Communications, Section A, p. 79. Fourth International Congress of Parasitology. 19–26 August 1978. Warszawa, Poland.

Malmberg, G., 1979a. Ontogeny and fine structure as a basis for a discussion on the relationship Cestoda – Monogenea (Platyhelminthes). Zool. Scr. 8: 315.

Malmberg, 1974b. The hyperparasitic post larval stages of Gyrocotyle. Institute of Parasitology. Åbo Akademi, Åbo, INFORMATION 15: 37–38.

Malmberg, G., 1981. Neoteny or not in the cercomeromorphaeans. Parasitology 82: 164–165.

Malmberg, G., 1982. On evolutionary processes in Monogenea, though basically from a less traditional viewpoint. In D. F. Mettrick & S. S. Desser (eds), Parasites – their world and ours. Elsevier Biomedical Press, Amsterdam, New York, Oxford: 198–202.

Malmberg, G., 1983. [The tapeworm – not harmless but always intestineless. On investigations concerning parasitic platyhelminths.] (In Swedish). Fauna och Flora 78: 183–192.

Ogawa, K. & S. Egusa, 1981. The systematic position of the genus Anoplodiscus, Monogenea, Anoplodiscidae. Syst. Parasitol. 2(4): 253–260.

Rohde, K. & M. Georgi, 1983. Structure and development of Austramphilina elongata Johnston, 1931 (Cestodaria: Amphilinidea). Int. J. Parasitol. 13(3): 273–287.

Initial morphological diversity as a criterion in deciphering turbellarian phylogeny

Yu. V. Mamkaev
Laboratory of Evolutionary Morphology, Zoological Institute Academy of Sciences, Leningrad, 199034, USSR

Keywords: Turbellaria, evolutionary morphology, phylogeny

Abstract

The most profound structural variety in morphofunctional systems and morphogenetic mechanisms, i.e. the highest morphological diversity, is observed in those groups where these systems and mechanisms are evolutionarily most primitive. Here, such variety can involve the basic body plan of a given phylum and the types of morphogenesis characteristic of it. This correlation provides a new criterion of evolutionary primitiveness, namely, the criterion of initial morphological diversity.

The highest morphological diversity among turbellarian groups is observed in the order Acoela. Acoel turbellarians are archaic in most of their features, apparently being a group near the base of the turbellarian phylogenetic tree. Among other turbellarians there are a few groups that also are archaic in some few features (above all, the Catenulida), although on the whole they are more advanced than the Acoela. The Turbellaria as a whole is notable for its morphological diversity in comparison with other classes of the Scolecida.

According to the principle of morphological radiation (or the principle of initial morphological diversity), as new systems of organisms originate, all functionally possible organizations are formed and various inter-system connections are readily established. This may lead to functionally convergent but structurally diverse solutions to biological problems. For example, the acoel turbellarians form a copulatory apparatus not only from epidermal and muscular tissues but also from pyriform glands (as in *Oligochoerus* and *Convoluta*), sagittocysts (as in *Convoluta sagittifera*, green convoluts), and glandular spines (as in *Anaperus*). In further evolution of such morphofunctional systems, the most primitive states of organization are subject less and less to fundamental restructuring and undergo only further development. On the whole, evolution of any morphofunctional system provides a pattern of morphological radiation.

An efficient method of revealing this pattern is to compare morphological differences in taxa of the same rank belonging to the same phylogenetic stock. The highest morphological diversity is often observed in groups at the base of large phylogenetic branches where a particular new pattern of body organization was formed. Here, differences can involve the basic plan of a given phylum and types of morphogenesis characteristic of it. Certain morphological features may have no or a little taxonomic significance in lower groups but are considered important taxonomic criteria in related advanced groups (e.g. pattern of segmentation in lower and higher annelids or arthropods). This approach gives a new criterion of evolutionary primitiveness, namely, the criterion of initial morphological diversity (Mamkaev 1966, 1968, 1979a, 1981, 1983).

The greatest morphological diversity among turbellarian groups is shown in the order Acoela (Mamkaev, 1967). The position of the mouth and male copulatory organ varies, being anywhere from the posterior to the anterior end of the body. In

Hydrobiologia 132, 31–33 (1986).

other turbellarian groups, the position of these organs has become more constant. Only in the Prolecithophora-Combinata is their position similarly variable, but here one is dealing with a united oral-genital complex. Rather considerable variation in position of these structures we see still in the Polycladida, Tricladida, Typhloplanoida, Dallyellioida, and Kalyptorhynchia, whereas in the Catenulida, Macrostomida, Lecithoepitheliata-Prorhynchida, Prolecithophora-Separata, and Temnocephalida the position is highly constant. The acoel turbellarians also have diverse types of epidermis (primitive, nemertine-like, or with insunk or intraepithelial nuclei). Only any one of these types develops in other groups; e.g. the Polycladida and the rhabdocoel groups have chiefly the type with intraepithelial nuclei, and parasitic platyhelminths have only the insunk type.

The nervous systems of the Acoela are also variable: both types with an external or depressed nerve plexus and types with longitudinal nerve trunks occur. The number of longitudinal nerve trunks varies from one or two pairs to five or six pairs (but more usually three pairs) (Mamkaev & Kotikova, 1972; Kotikova, this volume). In other turbellarian groups there is chiefly an orthogonal nervous system and the number of longitudinal nerve trunks is more constant (usually three pairs). Here variations affect mainly transverse nerve trunks. At the same time such an advanced nervous system does undergo some evolutionary restructuring (but only within definite limits) when the general organization undergoes considerable transformation, usually related to an increase in dimensions and alteration of body form. In such cases new diversity is exhibited, but then it is followed by some stabilization at the new level. For example, the nervous system of the Cestoda originated from the three-trunk type, exhibits significant diversity in more primitive groups, and became constant in Cyclophyllidea (Kotikova, 1979).

The order Acoela has several types of copulatory organs, pharynges, gonads, eyes (both cerebral eyes and epidermal pigment spots), spermatozoa, and extrusomes (Mamkaev, 1967). The digestive system of Acoela is variable, characterized by primitive parenchyma with formation of digestive spaces through a degenerative mode, by syncytium formation, or by a tendency for the development of an epithelial intestine (and in nemertodermatids even with differentiation of special glandular cells) (Mamkaev, 1979).

As a whole, the Acoela has a pronounced morphological diversity and thus is apparently near the base of the turbellarian phylogenetic tree, having preserved many archaic features. 'The acoel turbellarians: kingpins of metazoan evolution or a specialized offshoot?' – this is the way Smith & Tyler (1985) put the question. To my mind, it is an incorrect statement of the problem. The higher metazoans are, of course, not derived from the Acoela; rather the Acoela is a collateral evolutionary branch. Acoels have some derived features (e.g. ciliary axonemal and rootlet structure, peculiar mechanisms of extracellular digestion). At the same time they possess a number of primitive features and exhibit many stages in the formation of fundamental organ systems of metazoans. This is an extremely archaic group (although with its own evolutionary pecularities) and can be used as a model for the understanding of earlier metazoan evolution (cf. Ivanov & Mamkaev, 1973). Because of mosaic evolution among the recent lower metazoans there is no group which could be regarded as 'kingpin of metazoan evolution'. Among turbellarians there are a few groups that also are archaic in some features (above all, the Catenulida), although on the whole they are more advanced than the Acoela.

In other turbellarian groups, besides the parameters mentioned above, there are some instances of an initial morphological diversity, but it is of a narrower range. Thus, the Proseriata has diverse forms of sexual apparatus, whereas the Tricladida derived from them is more uniform (position of proseriate ovaries is variable even within related genera; triclads have ovaries only in an anterior position, etc.). The Turbellaria as a whole is notable for diversity in comparison with other classes of the Scolecida, within which the Turbellaria appears to be the most archaic group.

References

Ivanov, A. V. & Yu. V. Mamkaev, 1973. Resnichnyye chervi (Turbellaria), ikh proiskhozhdenie i evolutsiya. Filogeneticheskie ocherki. Nauka, Leningrad, 221 pp. (Turbellaria, their origin and evolution. Phylogenetic essays. In Russian.)

Kotikova, E. A., 1979. Osobennosti evolyutsii nervnogo apparata tsestod. Tr. Zool. Inst. Akad. Nauk USSR 84: 34–38. (On peculiarities of the evolution of the nerve apparatus in Cestodes. In Russian, English res.)

Kotikova, E. A., 1985. Comparative characterization of the nervous system of the Turbellaria. This volume.

Mamkaev, Yu. V., 1966. Morfologicheskoye mnogoobraziye kak pokazatyel primitivnosti gruppy. In E. A. Frolova (ed.), Nauchnaya konferentsiya, posvyashchennaya 100-letiyu so dnya rozhdeniya akademika A. N. Severtsova, Moskva, 22–24 dekabrya 1966 g., Tezisy dokladov. Nauka, Moskva: 18–20. (The morphological diversity as a primitive-group criterion. In Russian.)

Mamkaev, Yu. V., 1967. Ocherki po morfologii beskishechnykh turbellarij. Tr. Zool. Inst. Akad. Nauk USSR 44: 26–108. (Essays on the morphology of acoelous turbellarians. In Russian.)

Mamkaev, Yu. V., 1968. Sravnenie morfologicheskikh razlichij v nizshikh i vysshikh gruppakh odnogo filogeneticheskogo stvola. Zhurn. Obshchej Biol. 29: 48–56. (Comparative study of morphological differences in lower and higher groups of a uniform phylogenetic stock. In Russian, English res.)

Mamkaev, Yu. V., 1979a. Printsip morfologicheskogo mnogoobraziya kak podkhod k rasshifrovke filogeneticheskikh svyazej. In V. E. Sokolov (ed.), Sostojaniye i perspektivy razvitiya morfologii. Nauka, Moskva: 376–377. (The morphological diversity principle as a method of approach to deciphering phylogenetic relations. In Russian.)

Mamkaev, Yu. V., 1979b. O gistologicheskoj organizatsii pishchevaritel'noj sistemy turbellarij. Tr. Zool. Inst. Akad. Nauk USSR 84: 13–24. (On histological organization of the turbellarian digestive system. In Russian, English res.)

Mamkaev, Yu. V., 1981. Sravneniye taksonov po vnutrigruppovomu mnogoobraziyu kak metod evolyutsionnoj morfologii. In E. I. Vorob'eva & N. N. Iordanskij (eds.), Sovremennye problemy evolyutsionnoj morfologii zhivotnykh. Nauka, Moskva: 58–59. (The intragroup-diversity comparison as a method of evolutionary morphology. In Russian.)

Mamkaev, Yu. V., 1983. O znachenii jdej V. A. Dogelya dlya evolutsionnoj morfologii. Tr. Zool. Inst. Akad. Nauk. USSR 109: 15–36. (On significance of V. A. Dogel's ideas for evolutionary morphology. In Russian.)

Mamkaev, Yu. V. & E. A. Kotikova, 1972. O morfologicheskikh osobennostyakh nervnogo apparata beskishechnykh turbellarij. Zool. Zhurn. 51: 477–489. (On the morphological characters of nervous system in Acoela. In Russian, English res.)

Smith, J. P. S. & S. Tyler, 1985. The acoel turbellarians: kingpins of metazoan evolution or a specialized offshoot? In S. Conway-Morris, D. G. George, R. Gibson & H. M. Platt (eds.), The Origin and Relationships of the Lower Invertebrates. The Systematics Association of the United Kingdom, Oxford University Press, pp. 123–142.

Asexual reproduction and the turbellarian archetype

Reinhard M. Rieger
Department of Zoology, University of North Carolina, Chapel Hill, NC USA

Present address: Zoologisches Institut der Universität, Universitätstrasse 4, A-6020 Innsbruck, Austria

Keywords: Turbellaria, asexual reproduction, paratomy, Myomacrostomum, archetype

Abstract

The turbellarian archetype is widely believed to have been a hermaphrodite lacking asexual reproduction, and such asexual reproduction as is now seen in the Turbellaria (as paratomy and architomy) is generally assumed to have arisen secondarily several times independently. Asexual reproduction clearly prevails among the most primitive metazoans such as the placozoans, sponges, and radiates, however, and if the Platyhelminthes is indeed an early offshoot of bilaterian evolution, as some have claimed, then it is reasonable to expect asexual reproduction to be a primitive feature of the Turbellaria. Asexual reproduction by paratomy or architomy is found in all three main evolutionary lines of the Turbellaria and is most common among primitive groups such as the Catenulida and Macrostomida. The discovery of a new, apparently primitive marine genus of Macrostomida having paratomy widens the known incidence of asexual reproduction within that order. The presence of a muscle ring around the gut of several distantly related genera of the Macrostomida and similarities this ring shows with septa in the division plane of paratomizing species are evidence that paratomy was a feature of the stem species for this order – a feature only secondarily lost in most macrostomids – and suggest that asexual reproduction is a primitive feature for the Platyhelminthes as a whole.

Introduction

Presently all reconstructions of the stem group for the Turbellaria assume it comprised small, benthic organisms that moved by ciliary gliding and that had a simple life cycle without larval forms or asexual reproduction (Bresslau & Reisinger, 1928; Hyman, 1951; Ax, 1961; Ivanov & Mamkaev, 1973; Karling, 1974). Jaegersten (1972) has been the only one to argue that the turbellarian stem group had a complete life cycle with a planktotrophic larva, but his proposal has yet to be critically reviewed by specialists of the group (see Ball & Galleni, 1984). Asexual reproduction is generally believed to have arisen secondarily several times within the Turbellaria (Ax & Schulz, 1959).

On the other hand, an alternation of asexual and sexual reproduction dominates in life cycles of the primitive outgroups for the Bilateria (the Placozoa, Parazoa, and Radiata), as well as in the unicellular eukaryotes. If the turbellarians occupy a primitive position in the Bilateria, as has been reiterated recently by Ax (1985), we may expect to find evidence that such an alternation of asexual and sexual reproduction has been secondarily lost in turbellarians.

Just such evidence can be found in the order Macrostomida. Discovery of a new genus of Macrostomida with paratomy, an apparently primitive member of the order, has prompted me to reconsider the phylogenetic importance of paratomy in this group. In many respects, the Macrostomida is primitive within the rhabditophoran line of turbellarian evolution (see Ehlers, 1985; Smith *et al.*, 1985); and, therefore, a better understanding of the evolutionary origins of asexual reproduction in the

Hydrobiologia 132, 35–45 (1986).

Macrostomida is important for reconsidering the phylogenetic significance of this process in other lines of the Platyhelminthes.

Paratomy and its derivatives within the Macrostomida

In the Macrostomida, only members of the family Microstomidae have been known to reproduce asexually by paratomy. But in certain members of the Dolichomacrostomidae (*Myozonaria* and *Paramyozonaria*) and the aberrant genus *Myozona* of the Macrostomidae, what appears to be the homoiolog rudiment of a muscular division septum is manifest as a peculiar sphincter in the middle of the gut. Accordingly, I (Rieger, 1971a) have proposed that the common ancestor of these genera had paratomy similar to that of the Microstomidae and that muscles of the septum were retained as a sphincter on the gut because of the advantage this musculature offered in serving as a gizzard for breaking diatom frustules or in moving food in very elongated forms such as some species of the genera *Myozonaria* and *Paramyozonaria*. This hypothesis explains the occurrence of an intestinal muscle ring in two relatively distantly related lines of evolution in the Macrostomida. The only alternative is to consider these muscle rings to have arisen convergently, an unlikely event in that such a structure is not known in other turbellarians. (The muscular specialization in the anterior gut region of the catenulid genus *Myostenostomum* is clearly different, having a bilateral organization and a different position.)

New information on the structure of the muscle ring in the Dolichomacrostomidae further underscores the structural similarity between it and the muscle ring in *Myozona* (Fig. 1). Basement membrane of the gut is either partially or fully exposed along the muscle ring of dolichomacrostomids (Fig. 1), and the latter condition (that of *Paramyozonaria simplex,* Fig. 1b) is essentially indistinguishable from that of the muscle ring in certain species of *Myozona* (cf., e.g., Ax, 1956, and Marcus, 1948).

Furthermore, recent TEM studies have verified the existence of muscles in the division septum of paratomizing *Microstomum lineare* (Palmberg & Reuter, 1983, and Reuter & Palmberg, 1983). This septum consists of two muscular sheaths between which are nerve cells. My own reinvestigation of the

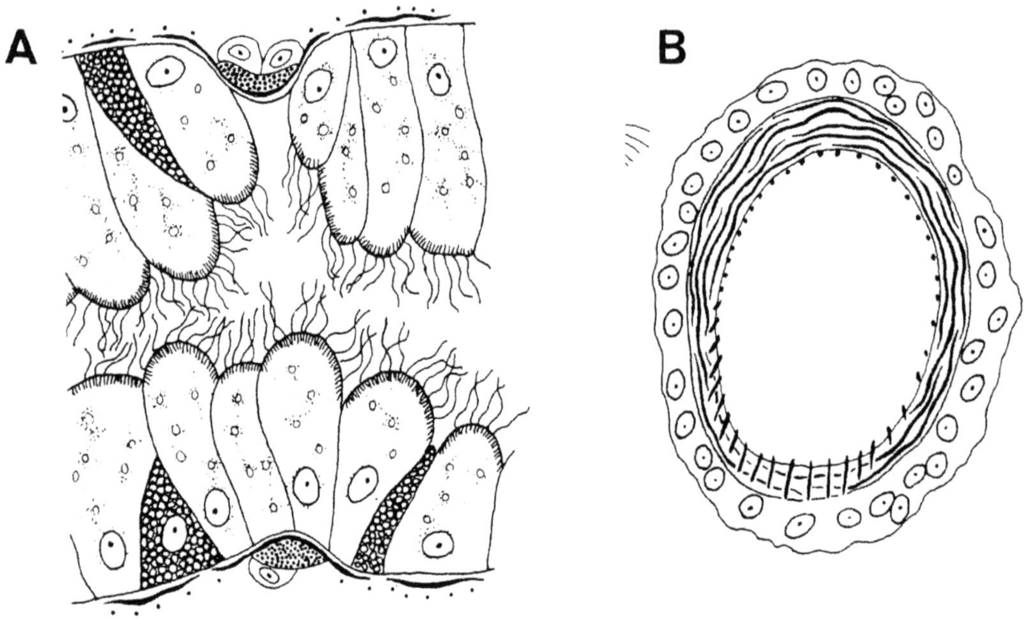

Fig. 1. Intestinal muscle ring in Dolichomacrostomidae. A) *Myozonaria bermudensis* Rieger, 1977; longitudinal section. Note that the gastrodermis is interrupted at the muscle ring only on one side. B) *Paramyozonaria simplex* Rieger & Tyler, 1977; cross section. Note that the gastrodermis is completely lacking at the ring. Nuclei to outside of circular muscles are nuclei of muscle cells. 1 μm = 1,4 mm.

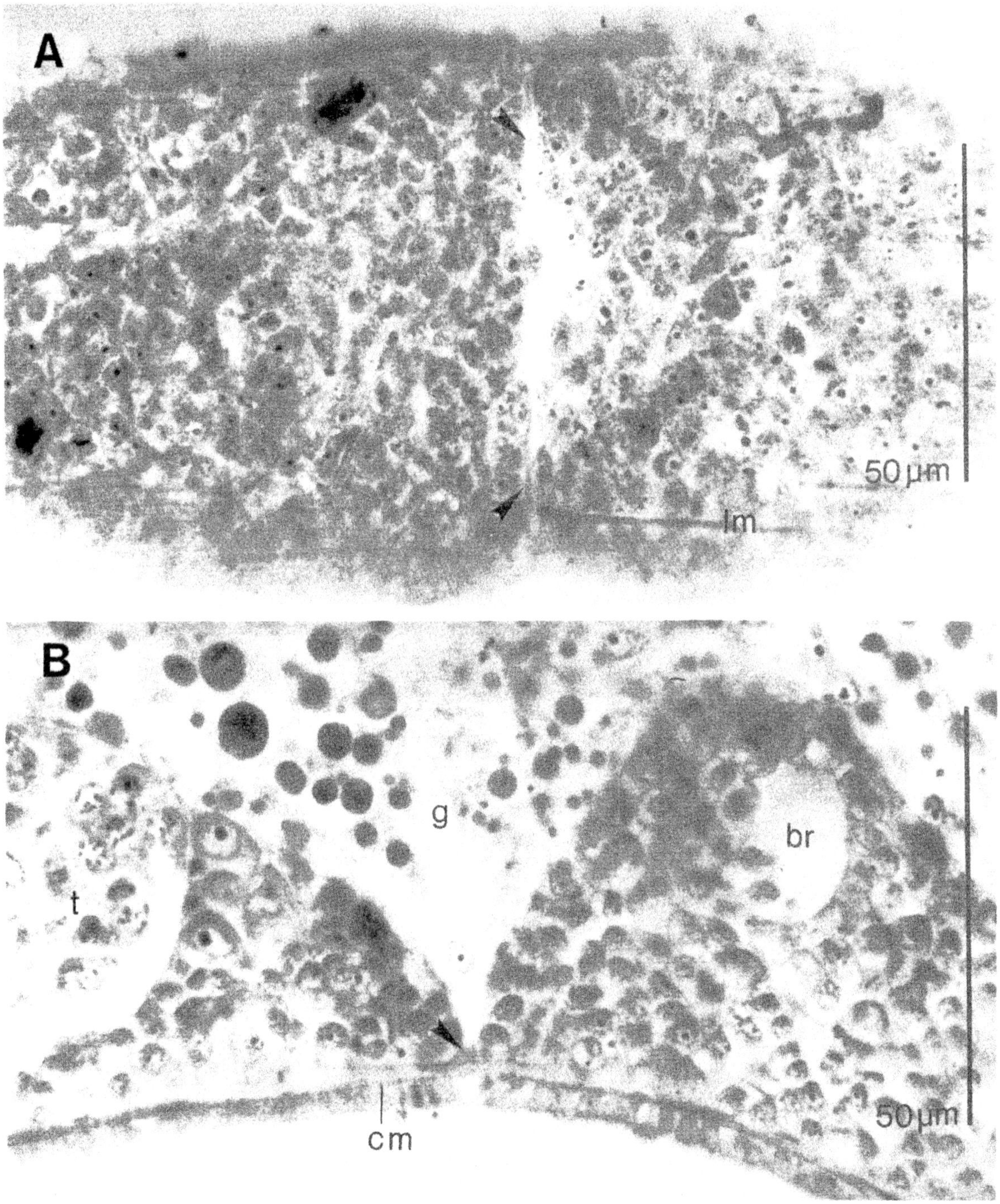

Fig. 2. Microstomum hamatum. A) Tangential section through secondary division plane (section from E. Westblad). Note the circular arrangement of muscle fibers in the plane of septum (V). B) Sagittal section through the primary division plane (ventral side). Note cross sections of muscle fibers in division plane (V). The fibers have different staining properties from the body wall musculature. br=brain, t=testis, cm=circular muscles of body wall, lm=longitudinal muscles of body wall, g=gut.

division plane of *M. hamatum* at the light microscopic level suggests muscle fibers oriented circumferentially around the gut are the dominate constituent. Palmberg & Reuter (1983) do not mention the orientation of the muscle fibers in the double septum of *M. lineare,* but their Fig. 1a suggests the fibers originate in the circular muscles of the body wall and continue in an oblique radial direction to the circular layer of the gut musculature. Clearly more TEM data is needed on structure and development of the septum in microstomids for further comparison with the intestinal muscle ring.

Most important for the interpretation of the intestinal muscle ring is the discovery of a new genus, *Myomacrostomum,* which, by virtue of the presence of a caudal sensory organ (Rieger & Tyler, 1974) and of similarities in adhesive organs (Tyler, 1977), is clearly aligned with the Dolichomacrostomidae and the genus *Bradynectes. Myomacrostomum* (see Appendix for diagnosis) is represented by two species, *M. unichaeta* and *M. bichaeta,* both of which reproduce asexually by paratomy (Fig. 3). While the possibility these two species represent two morphs in a complex life cycle cannot be excluded, the available data warrant establishment of two separate taxa. The two species are distinguished primarily by paired-vs-unpaired condition of the caudal organ, a feature that is of generic-level significance in the Dolichomacrostomidae (see Rieger & Tyler, 1974). In addition the two species differ in body size, in length of rhammites, and possibly in position of the female gonad (see Appendix for further discussion).

As for the question of the origin of intestinal muscle rings, it is significant that specimens of *M. bichaeta* have been found which have a muscle ring in the same position as other individuals having an incipient secondary division plane (Fig. 3A, B); other specimens with two zooids either have clear signs of a secondary division plane or lack such a plane entirely (Fig. 3). Preliminary studies indicate the muscle ring has the same basic structure (specifically interruption of the intestinal epithelium and exposure to the gut lumen as in *Myozona* and *Paramyozonaria simplex.* Thus the discovery of *Myomacrostomum* strongly supports the hypothesis (Rieger, 1971a) that the muscle ring in Dolichomacrostomidae is the remnant of a muscular division septum similar to that of the Microstomidae.

Phylogentic position of Myomacrostomum

The male genital organs in the new genus *Myomacrostomum* are similar to those of certain members of the Macrostomidae (*Myozona stylifera, Bradynectes sterreri*) and to those of the Microstomidae: the gonad is unpaired and is situated caudally close to the copulatory organ and the stylet points rostrally. That the vesicula granulorum and seminalis are separated suggests closer ties to members of the family Macrostomidae. However, because the Macrostomidae is rather heterogeneous as it now stands, identifying ties to it as a whole are not as meaningful as ties to individual macrostomid genera.

Myomacrostomum is clearly not, however, a member of the other remaining family in the Macrostomida, the Dolichomacrostomidae, inasmuch as there is no evidence that it has secondarily lost the complex genital apparatus characteristic of all known members of this family (see Rieger, 1971a, 1971b). At the same time, evidence from two other sets of characters – namely presence of the caudal sensory organ (Rieger & Tyler, 1974) and morphology of adhesive organs (Tyler, 1977) – shows *Myomacrostomum* belongs to a monophyletic taxon that includes the Dolichomacrostomidae and the genus *Bradynectes.* In this taxon, *Myomacrostomum* and *Bradynectes* appear as primitive sister groups to the Dolichomacrostomidae as I hypothesized earlier (Rieger, 1971c). Because *Myomacrostomum* also has characters in its male reproductive system intermediate between those of the Macrostomidae and Microstomidae (i.e. distinct v. granulorum and v. seminalis as in Macrostomidae; close spatial relationship between testis and copulatory organ as in Microstomidae), this new genus appears to be most similar to the hypothetical stem species of the Macrostomida as a whole.

Loss of paratomy in the Macrostomida and other platyhelminths

In light of the primitive position of *Myomacrostomum,* its ability to reproduce asexually may indicate that asexual reproduction is primitive in the order Macrostomida and that the absence of paratomy in most other members of the order is the

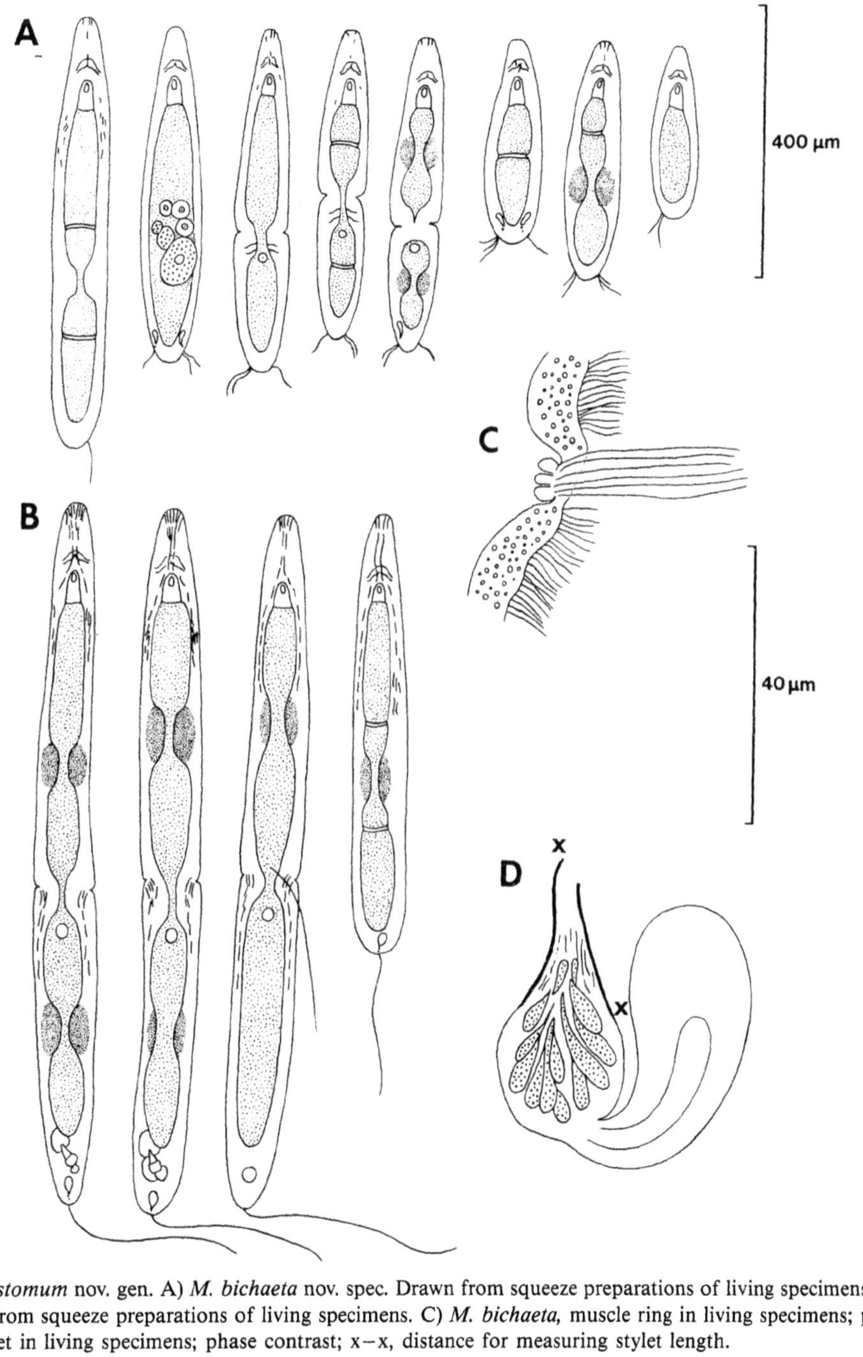

Fig. 3. Myomacrostomum nov. gen. A) *M. bichaeta* nov. spec. Drawn from squeeze preparations of living specimens. B) *M. unichaeta* nov. spec. Drawn from squeeze preparations of living specimens. C) *M. bichaeta,* muscle ring in living specimens; phase contrast. D) *M. unichaeta,* stylet in living specimens; phase contrast; x–x, distance for measuring stylet length.

result of secondary loss. The systematic distribution of an intestinal muscle ring suggests that loss of paratomy has occurred at least twice in the order, once in the Dolichomacrostomidae and once in the evolutionary line leading to *Myozona.* If asexual reproduction is primitive within the Macrostomida, it is likely that the phylum Platyhelminthes as a whole is derived from an ancestor having asexual reproduction, perhaps a primitive fission or perhaps even paratomy (see Ax & Schulz, 1959, for review of the terminology of asexual reproduction in Turbellaria). Evidence for this also comes from the fact that asexual reproduction is known in the other two major evolutionary lines of the Turbellaria (see Smith *et al.*, 1985, for definition of these lines). In the Catenulida, asexual reproduction is very common. Suppression of asexual reproduction into the juvenile stage is known in the genus *Rhynchoscolex* (Reisinger, 1924); asexual reproduction is not known in the marine family Retronectidae, but the reproductive biology as a whole of this family is enigmatic (see Doe & Rieger, 1977). In the Nemertodermatida-Acoela line of evolution, asexual reproduction is known only in the derived order Acoela, but in this group such reproduction (in the form of paratomy) occurs in a genus (*Paratomella*) which appears to be most similar to the stem species on the basis of characters of the digestive parenchyma (Smith & Tyler, 1985).

Among closer relatives of the Macrostomida, specifically in free-living groups of the Polycladida-Neoophoran line, asexual reproduction is known only in the Tricladida, and here the likelihood of secondary origin has been interpreted as high (see Ax & Schulz, 1959). It is worth noting here, however, that the frequency of occurrence of fission is highest in the Dugesiidae, next highest in the Planariidae and absent in the Dendrocoelididae (see Calow, 1985). Further study is needed, therefore, of such structures as the 'muscle septum' of certain marine proseriates (see Karling, 1966; own unpublished observations on Monocelididae from Bermuda) as the plesiomorph sister group to the Tricladida.

Asexual reproduction is, of course, very common among members of the parasitic classes of the Platyhelminthes (Hyman, 1951). Should further evidence support the hypothesis that asexual reproduction is a primitive feature of free-living acoelomates (see also, in this respect, Berg, 1985, for Nemertea), then this group could be seen as especially adapted for the evolution of parasitism.

Conclusion

While it is commonly assumed that asexual reproduction arose as a secondary specialization in the phylum Platyhelminthes, discovery of *Myomacrostomum* shows this assumption is not warranted. On the contrary, it is likely that such reproductive powers were present in the stem species of the phylum. Considering the dominance of an alternation of asexual and sexual reproduction in life cycles of the more primitive phyla Porifera, Cnidaria, and Placozoa, discovery of a bilaterian that reproduces asexually and that is as primitive a member of the Turbellaria as *Myomacrostomum* ostensibly is has far-reaching consequences for questions of the origin of the Bilateria as a whole.

Appendix

Myomacrostomum unichaeta nov. gen. nov. spec.

Material: 3 male-mature specimens and 1 immature specimen, studied alive; 1 male- and female-mature specimen, serially sectioned, holotype AMNH # 1233; 1 male-mature specimen, serially sectioned; paratype AMNH # 1234.

Locality: sandbank, 1 mile from mouth of New River Inlet at New River estuary N.C., USA, southern shore; fine sand, 0–5 cm sediment depth at LTL, summer 1972.

Etymology: 'Myomacrostomum' – habitus like *Macrostomum* and intestinal muscle ring as in *Myozona* and *Myozonaria;* 'unichaeta' – unpaired tuft of caudal sensory organ.

Description: Male-mature specimens about 1 mm long, with two zooids (see Fig. 3b); posterior zooid with male reproductive organ at U 95 and incipient second pharynx (Fig. 3b). Two specimens with a secondary division plane in process of formation; a third specimen with secondary division plane present only in anterior zooid. Male copulatory organ with simple stylet, 25 μm long (Fig. 3d, 6b). Testis small and spherical, close to the left side of the stylet. An oocyte occupying about one third of the body diameter present in both sectioned specimens immediately in front of the male organs. In one sectioned specimen, a second male reproductive system was found at the caudal end of the anterior zooid. Ciliary tuft of caudal sensory organ 250–350 μm long. In one specimen, a ciliary tuft was clearly seen also in the anterior zooid (Fig. 3b). One immature specimen found, about 600 μm in length, with developing primary division plane and a muscle ring in each of the two developing zooids; caudal sensory organ clearly unpaired

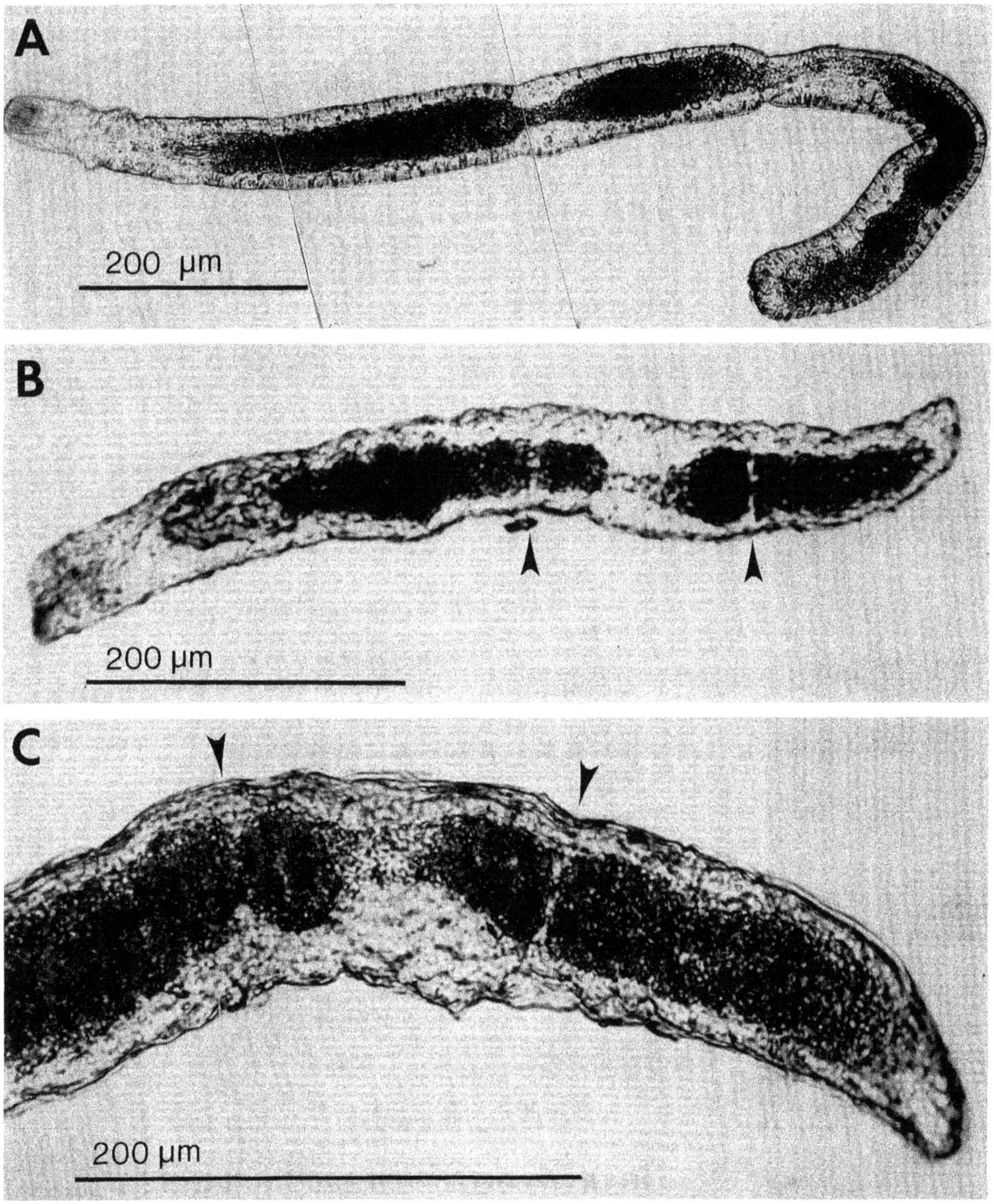

Fig. 4. Photomicrographs of squeezed animals. A) *M. unichaeta,* mature specimen; note primary division plane and the secondary division plane in each zooid. B) *M. bichaeta;* note primary division plane and two muscle rings (V). C) *M. unichaeta,* immature specimen; enlargement of caudal zooid showing muscle ring (V).

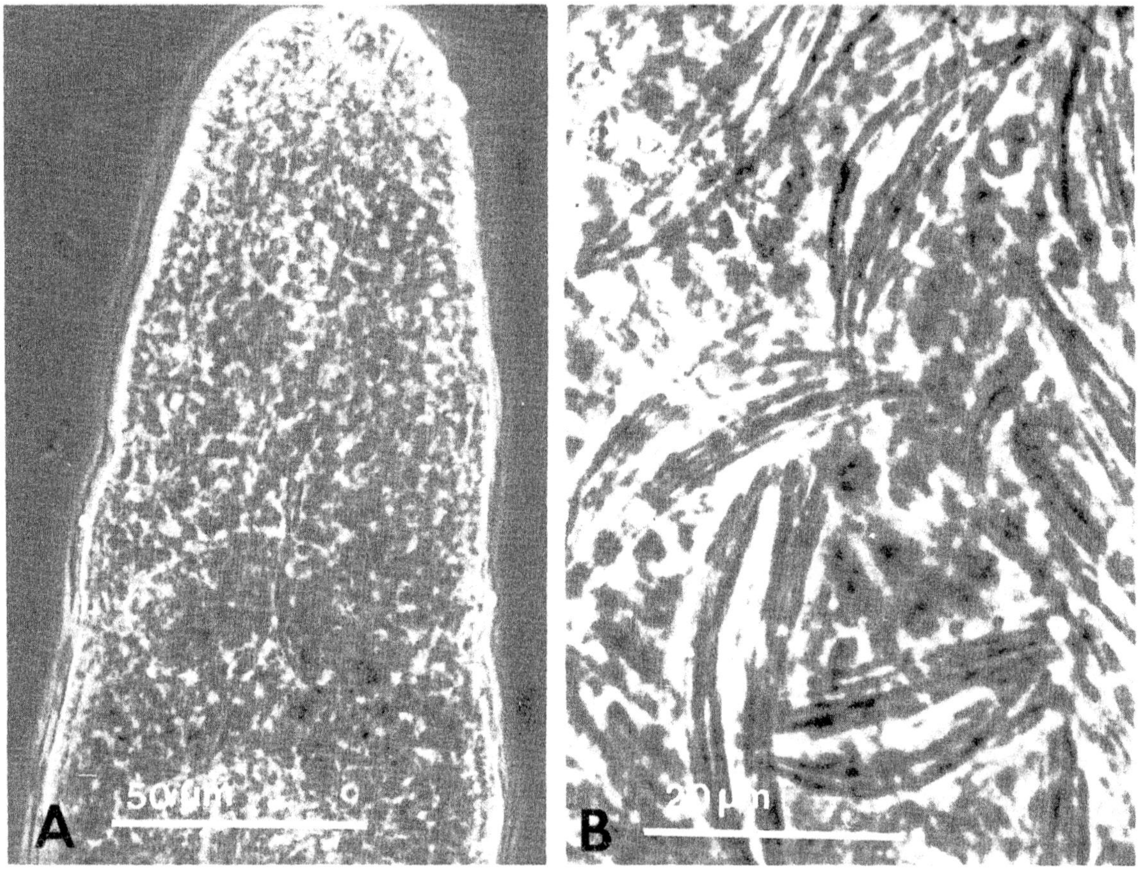

Fig. 5. A) *M. bichaeta.* Brain and rhammite strands. B) *B. unichaeta.* Rhammite glands enlarged.

in this immature specimen.

Colorless, with slightly yellowish brown intestine, and spotted appearance due to rhabdite bundles. Rhabdite glands scattered throughout the body wall and sunken below the circular muscle layer; rhabdites appear in bundles. Rhammite glands well developed, extending laterally as far posterior as the secondary division plane or muscle ring and apparently penetrating the brain. Rhammites 20–30 µm long. Gland ring of pharynx with two distinctly different gland types. Rhabdite glands clearly sunken below the circular muscle layer. Position of longitudinal nerve cords could not be determined.

Myomacrostomum bichaeta nov. spec.

Material: 7 immature specimens, 1 possibly female-mature (see Fig. 3a), studied alive; 1 immature speciment serially sectioned (holotype AMNH # 1235); 1 immature specimen serially sectioned (paratype AMNH # 1236).

Locality: same as for *M. unichaeta.*

Etymology: 'bichaeta' – paired ciliary tufts on the caudal sensory organ.

Description: The largest specimen found was 600 µm in length [paired nature of the caudal sensory organ in this specimen determined only on the basis of the asymmetrical position and length of its one ciliary tuft]; it consisted of two zooids, each with a muscle ring (Fig. 3a). Of four specimens approximately 450 µm in length, one contained structures that could possibly be maturing oocytes, one with two zooids had no signs of secondary division, one had secondary division planes developing in both of two zooids, and the third had a muscle ring

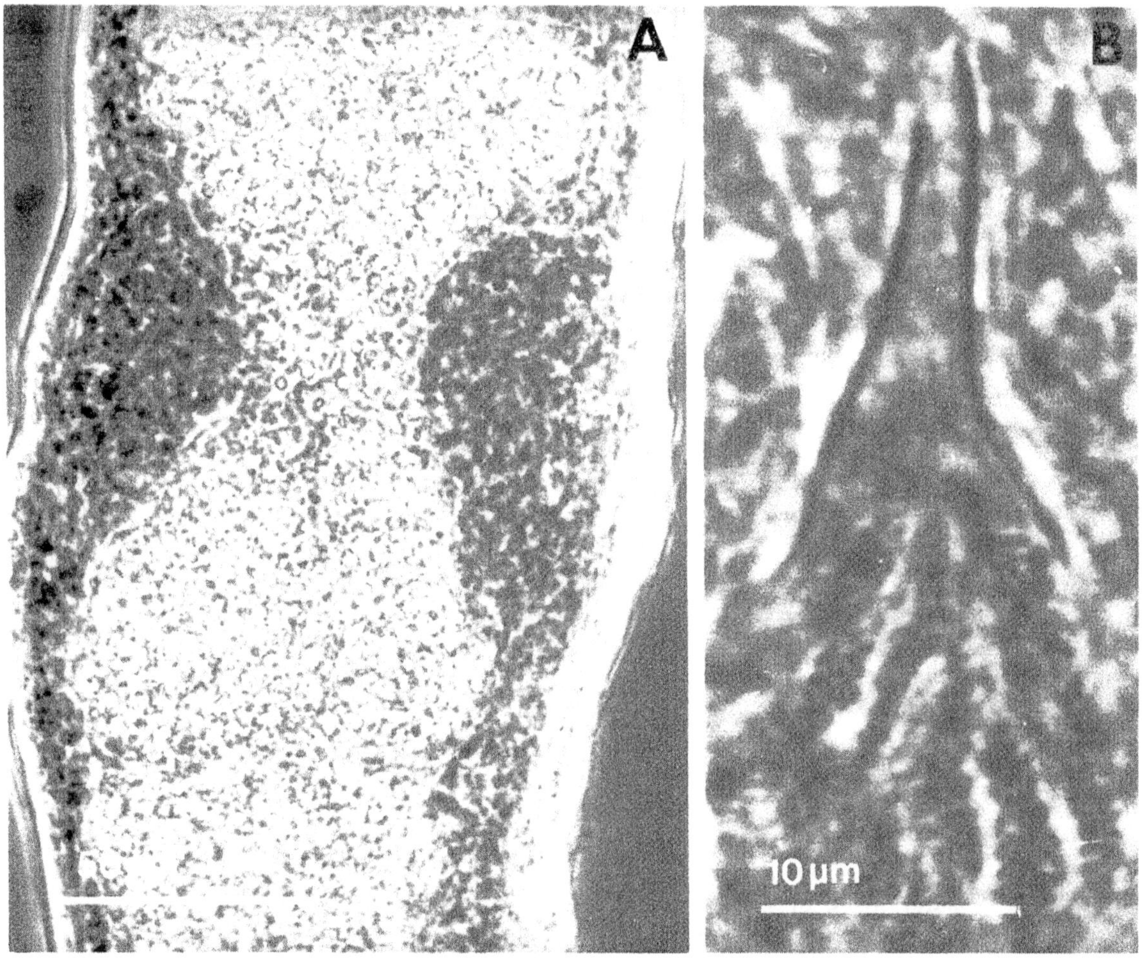

Fig. 6. A) *M. unichaeta.* Primary division plane, immature specimen. B) Stylet.

developing in a position equivalent to that of the secondary division plane in other specimens. Two other specimens (one studied alive, one studied in serial sections) approximately 350 μm in length consisted also of two zooids and had a secondary division plane in the posterior zooid and a muscle ring in the anterior zooid (Fig. 3a). Two specimens studied alive consisted only of a single zooid and were between 250 and 300 μm in length; one of these had a typical muscle ring; the other did not (Fig. 3a).

With rhabdite bundles scattered throughout the body wall, as in *M. unichaeta.* Rhammites 4–11 μm long, less than half the size of those in *M. unichaeta.* They penetrate the brain, but are less numerous than in *M. unichaeta.* Pharyngeal gland ring with two types of glands. With paired caudal sensory organ, sensory tufts 50–100 μm long. Muscle ring clearly interrupting the gastrodermis as viewed in sections, a feature not obvious in one of the live specimens (see Fig. 3c). No signs of male organs in specimens available.

Diagnosis of new genus Myomacrostomum

Macrostomida with rhammites penetrating the brain. With asexual reproduction by paratomy, caudal sensory organ, and dolichomacrostomid type of adhesive organ (with flower-like folding of anchor-cell collar). Male copulatory organ with vesicula seminalis, vesicula granulorum, and anteriad-pointing stylet. Unpaired testis just in front of the copulatory organ, slightly displaced to the left. Ovary medial, immediately in front of the testis. Female gonoducts possibly lacking. In marine sands. Type of genus: *M. unichaeta.*

Diagnoses of the new species

M. unichaeta – mature specimens 1 mm in length, with one unpaired caudal sensory organ bearing a ciliary tuft 250–300 μm long. Rhammites 20–30 μm long. With characteristically shaped copulatory stylet (see Fig. 3d, 6b) 25 μm in length.

M. bichaeta – immature specimens 250–600 μm in length, with paired caudal sensory organ bearing two ciliary tufts 50–100 μm long. Rhammites 4–11 μm long. No male organs observed. Muscle ring on gut in sections interrupting the gastrodermis.

Discussion of species differences

Presently available data indicates clearly that the two species described here are indeed distinct species, even with the lack of information on genital organs of *M. bichaeta.* Differences in other features are clearly of the sort known to be diagnostic for other species of Macrostomida [e.g., paired and unpaired nature of the caudal sensory organ appears to be a generic characteristic of Dolichomacrostomidae (see Rieger & Tyler, 1974)]. Rhammite length is significantly different in the two (by 3-fold), moreso than is expected of variation within a single species. However, because we are dealing here with a unique and new genus having asexual reproduction, we cannot entirely exclude the possibility that these two species actually belong to a single species with a complex life cycle alternating asexual and sexual phases.

Acknowledgments

The author would like to acknowledge the loan of E. Westblad's sectioned Macrostomid-material from the Naturhistoriska Riksmuseet, Stockholm. This work was supported by NSF-grants DEB 8119652 (R. M. Rieger P.I.) and BSR 8116894 (S. Tyler P.I.).

References

Ax, P., Studien über psammobionte Turbellaria Macrostomida IV. Myozona stylifera nov. spec. Zool. Anz. 157: 251–260.

Ax, P., 1961. Verwandtschaftsbeziehungen und Phylogenie der Turbellarien. Ergebn. Biol. 24: 1–68.

Ax, P., 1963. Relationships and phylogeny of the Turbellaria. In E. C. Dougherty (ed.), The lower metazoa. University of California Press, Berkeley, California: 191–224.

Ax, P., 1985. The position of the Gnathostomulida and Platyhelminthes in the phylogenetic system of the Bilateria. In S. Conway Morris, J. D. George, R. Gibson, & H. M. Platt (eds), The origins and relationships of lower invertebrates. Oxford University Press, Oxford.

Ax, P. & E. Schulz, 1959. Ungeschlechtliche Fortpflanzung durch Paratomie bie acoelen Turbellarien. Biologischen Zentralblatt 78: 613–621.

Ballarin, L. & L. Galleni, 1984. Larval development in Echinoplana celerrima. Trans. am. microsc. Soc. 103: 31–37.

Berg, B. 1985. Annulonemertes gen. nov., a new segmented hoplonemertean. In S. Conway Morris, J. D. George, R. Gibson, & H. M. Platt (eds), The origins and relationships of lower invertebrates. Oxford University Press, Oxford.

Bresslau, E. & E. Reisinger, 1928. Plathelminthes. In W. Kükenthal & T. Krumbach (eds), Handbuch der Zoologie, Vol. 2, Pt 1, pp. 34–51.

Calow, P. & D. A. Read, 1985. Ontogenetic patterns and phylogenetic trends in freshwater flatworms (Tricladida); constraint or selection? This volume.

Doe, D. A. & R. M. Rieger, 1977. A new species of the genus Retronectes (Turbellaria, Catenulida) from the coast of North Carolina, USA. Mikrofauna Meeresbodens 66: 549–556.

Ehlers, U., 1985. Phylogenetic relationships within the Platyhelminthes. In S. Conway Morris, J. D. George, R. Gibson, & H. M. Platt (eds), The origins and relationships of lower invertebrates. Oxford University Press, Oxford.

Hyman, L. H., 1951. The Invertebrates. II. Platyhelminthes and Rhynchocoela. McGraw-Hill, New York: 550 pp.

Ivanov, A. V. & Yu. V. Mamkaev, 1973. The turbellarians (Turbellaria) – Their origin and evolution. [In Russian] Akad. Nauka USSR, Leningrad, 221 pp.

Karling T. G., 1966. On the defecation apparatus in the genus Archimonocelis (Turbellaria, Monocelididae). Sarsia 24: 37–44.

Karling, T. G., 1974. On the anatomy and affinities of the turbellarian orders. In N. W. Riser & M. P. Morse (eds), Biology of the Turbellaria. McGraw-Hill, New York: 1–16.

Marcus, E., 1948. Turbellaria do Brasil. Zoologia Univ. São Paulo Brasil 13: 111–244.

Palmberg, I. & M. Reuter, 1983. Asexual reproduction in Microstomum lineare (Turbellaria). I. An autoradiographic and ultrastructural study. Int. J. Invert. Reprod. 6: 197–206.

Reisinger, E., 1924. Der Gattung Rhynchoscolex. Z. Morphol. Oekol. Tiere 1: 1–37.

Reuter, M. & I. Palmberg, 1983. Asexual reproduction in Microstomum lineare (Turbellaria). II. The nervous system in the division zone. Int. J. Invert. Reprod. 6: 207–217.

Rieger, R. M., 1971a. Die Turbellarienfamilie Dolichomacrostomidae nov. fam. (Macrostomida). I. Teil. Vorbemerkungen und Karlingiinae nov. subfam. I. Zool. Jahrb. Abt. Syst. Oekol. Geogr. Tiere 98: 236–314.

Rieger, R. M., 1971b. Die Turbellarienfamilie Dolichomacrostomidae Rieger. II. Teil. Dolichomacrostominae I. Zool. Jahrb. Abt Syst. Oekol. Geogr. Tiere 98: 569–703.

Rieger, R. M., 1971c. Bradynectes sterreri gen. nov., spec. nov., eine neue psammobionte Macrostomide (Turbellaria). Zool. Jahrb. Abt. Syst. Oekol. Geogr. Tiere 98: 205–235.

Rieger, R. M., 1981. Morphology of the Turbellaria at the ultrastructural level. Hydrobiologia 84: 213–229.

Rieger, R. M. & S. Tyler, 1974. A new glandular sensory organ in interstitial Macrostomida. I. Ultrastructure. Mikrofauna Meeresboden 42: 1–41.

Smith, J. P. S. & S. Tyler, 1985. The acoel turbellarians: kingpins of metazoan evolution or a specialized offshoot? In S. Conway Morris, J. D. George, R. Gibson, & H. M. Platt (eds), The origins and relationships of lower invertebrates. Oxford University Press, Oxford.

Smith, J. P. S., S. Tyler, & R. M. Rieger, 1985. Is the Turbellaria polyphyletic? This volume.

Tyler, S., 1977. Ultrastructure and systematics: an example from turbellarian adhesive organs. Mikrofauna Meeresboden 61: 271–286.

Ultrastructure of the frontal organ in *Convoluta* and *Macrostomum* spp.: significance for models of the turbellarian archetype

Marianne D. Klauser, Julian P. S. Smith, III & Seth Tyler
Department of Zoology, University of Maine, Orono, ME 04469, U.S.A.

Keywords: Turbellaria, frontal organ, mucus gland, sensory, Acoela, Macrostomida

Abstract

Present models of turbellarian evolution depict the organism with a frontal organ – a complex of glands whose necks emerge at the anterior tip of the body – and therefore imply that this organ is homologous throughout the Turbellaria. However, comparisons of representatives of the Acoela and Macrostomida, two putatively primitive orders of the Turbellaria, show that frontal organs in these two are not similar in ultrastructure or histochemistry. The acoel *Convoluta* 'pulchra' had a prominent cluster of frontal mucous glands whose necks emerged together in a frontal pore at the exact apical pole of the organism, and an array of smaller glands of at least five other types opened at the anterior end, separately from and ventral to this pore. The 'frontal organs' (Stirndrüsen) of two species of *Macrostomum* on the other hand, comprised an array of discretely emerging necks of at least two gland types including one with rhabdiform (rhammite) and one with globular mucous secretion granules neither of which emerge at the apical pole. In neither species did the organ appear to be sensory. Our findings indicate a low probability of homology between the frontal glands of the Acoela and Macrostomida.

Introduction

Apical or anterior glandular organs referred to as 'frontal organs' are commonly found in turbellarians and are especially prominent in the Acoela and Macrostomida, two orders believed to be primitive among the Turbellaria (von Graff, 1891; Karling, 1974). Because of the widespread systematic distribution of frontal organs, Ax (1961) and Karling (1974) postulate that such an organ was present in the turbellarian archetype, implying that all frontal organs in the Turbellaria are homologous.

A more detailed analysis of frontal organs is warranted to verify this presumed homology. Already some significant differences in details of frontal organ morphology and histochemistry are known (Karling, 1940) concerning Prolecithophora and Rhabdocoela. The orders Acoela and Macrostomida were chosen for this study because of their similarities to two archetypes: Ax's (1961) archetype stands closest to the Macrostomida, whereas Karling's (1974) is closer to the nemertodermatid-acoel line.

Materials and methods

Convoluta 'pulchra' was collected from mid-tide level (MTL) surface sediment (fine muddy sand) at Mackerel Cove, Swans Island, Maine; *Macrostomum hystricinum marinum* (form I, see Rieger, 1977) was collected from poorly sorted sand (MTL) at Hulls Cove, Mount Desert Island, Maine; *Macrostomum* sp. III (see Rieger, 1977) was collected at Bogue Bank, North Carolina on the south side, 300 m from Bogue Inlet at high tide level (HTL) from fine sand with detritus. A $MgCl_2$-anesthesia-decantation technique was used to extract the

Hydrobiologia 132, 47–52 (1986).

animals from the sediment (Hulings & Gray, 1974). *Macrostomum* sp. III was fixed in 2.5% phosphate-buffered glutaraldehyde, the other species were fixed in 'Trumps' (phosphate-buffered 4% formaldehyde and 1% glutaraldehyde); all specimens were postfixed in OsO_4, and embedded in Epon-Araldite. Serial 1-μm-thick sections were cut, mounted on glass slides and stained for light microscopy (LM) after removal of the plastic (for techniques, see Smith & Tyler, 1984). Reactivity to alcian blue and to fast green in *Macrostomum* spp. was studied in specimens fixed in 2.5% glutaraldehyde or acetic alcohol and embedded in glycol methacrylate and sectioned at 0.5–1.5 μm (Chayen *et al.*, 1969).

Thin sections for transmission electron microscopy (TEM) were stained with uranyl acetate and lead citrate.

One-micrometer-thick sections that had been mounted on glass cover slips and deresinated by immersion in absolute methanol saturated with KOH (modified after Maxwell, 1978), were critical-point dried, coated with 20 nm of gold-palladium and examined with an AMR 1000A scanning electron microscope at 5 kV.

Terminology used in this paper follows that sug-

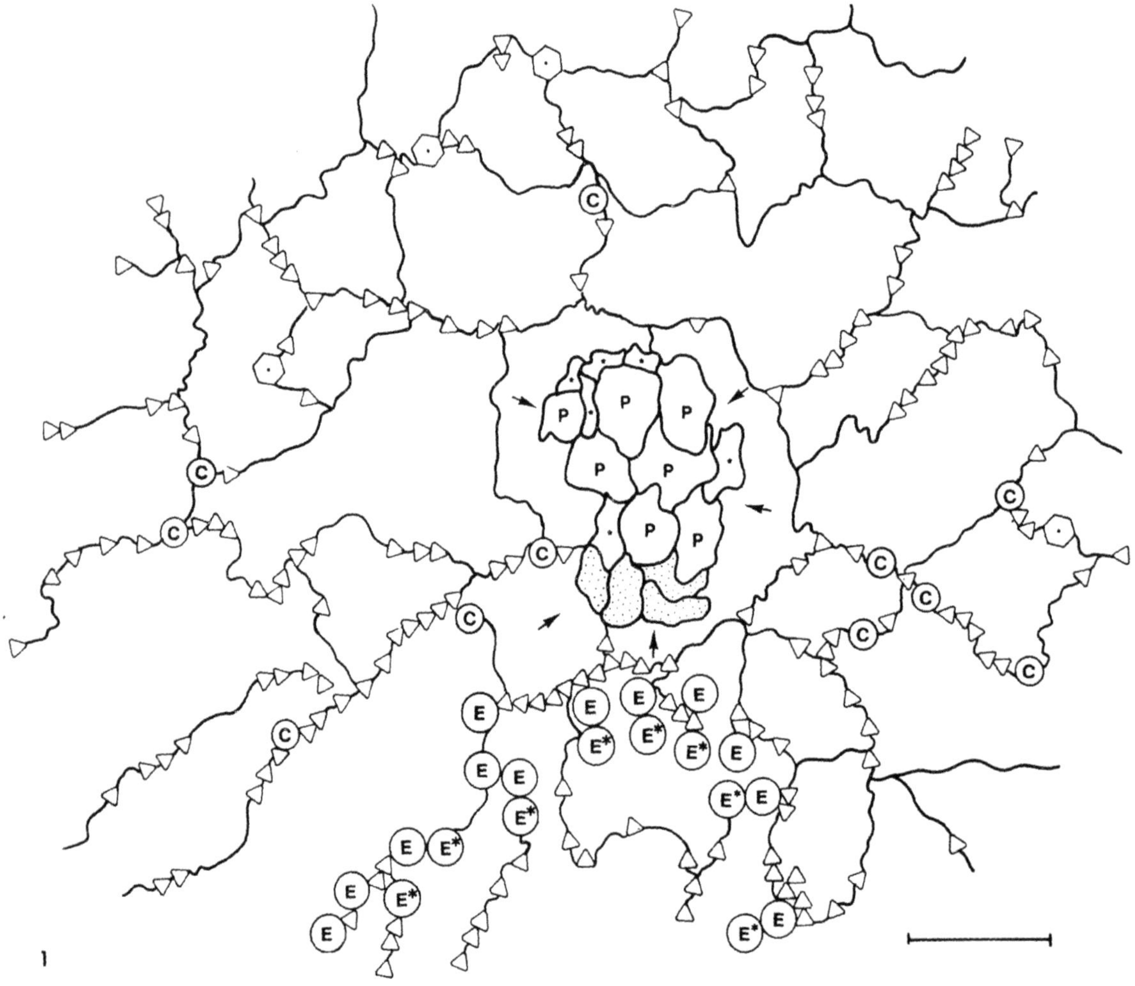

1

Fig. 1. Convoluta 'pulchra'. Schematic diagram of the anterior epidermal surface showing pores of the frontal glands and sensory receptors emerging between epidermal cells: principal mucoid glands (P), secondary mucoid glands (asterisk), alcian-blue-positive glands (stippled), colored granule glands (C), ellipsoid granule glands (E), electron-lucent granule glands (E*), and monociliated sensory cells (type I: triangles; type II: hexagons). Arrows indicate the direction of the principal (rostral) rootlets of epidermal cilia. Scale bar 5 μm.

gested by Smith & Tyler (1985). *'Frontal organ'* refers to an anterior group of gland cells whose necks project to and open at the frontal pore, a gap in the epidermis at the exact anterior tip of the body. *'Frontal glands'* or *'frontal gland complex'* are considered general terms referring to the whole suite of specialized gland cells opening at the anterior end of the body, including the frontal organ, if present.

Results

Convoluta *'pulchra'*

The prominent frontal glands of *Convoluta* 'pulchra' were three types of mucus-secreting glands whose necks opened in a frontal pore at the exact apical pole of the animal, i.e. the point towards which the rostrally-directed principal rootlets of the epidermal cells converged (Fig. 1). We designate these as mucoid glands (MG), secondary mucoid glands (SMG) and glands secreting alcian-blue-positive granules (ABG) (Figs 3, 5 and 6, respective-

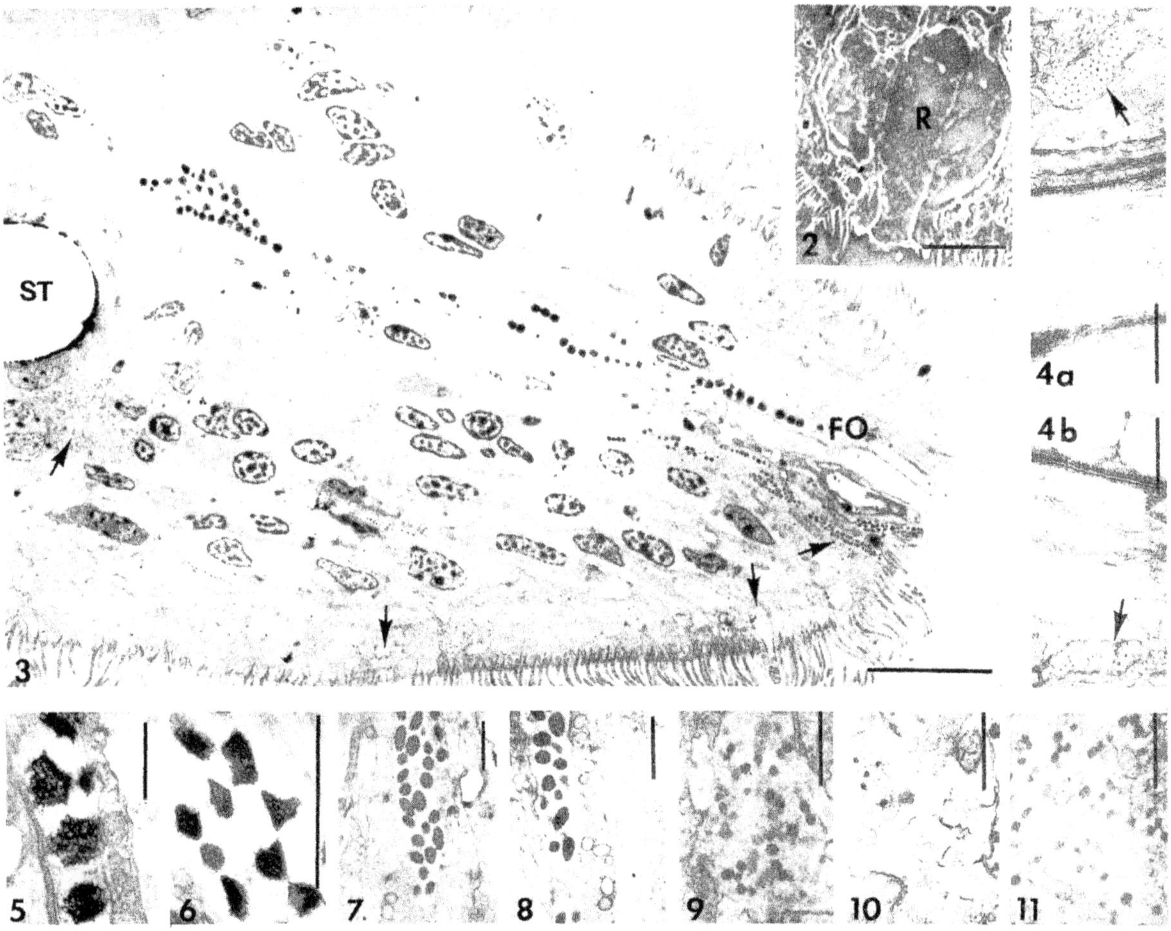

Figs 2–11. Convoluta 'pulchra'. (2) Reservoir (R) and frontal pore in a deresinated thick section; SEM; Scale bar 5 μm; (3) Sagittal section through the anterior end showing necks of the frontal organ (FO) and ventrally located necks of accessory glands (arrows); (ST) statocyst; Scale bar 10 μm; (4) Circular-muscle sphincter near frontal pore, dorsal (4a) and ventral (4b) to the gland necks (arrows); Scale bars 1 μm; (5–11) Secretion granules of the frontal glands; Fig. 5: Secondary mucoid gland secretions; Fig. 6: alcian-blue-positive gland secretions; Fig. 7: ellipsoid granules; Fig. 8: electron-lucent granules; Fig. 9: colored granules; Fig. 10: basophilic mucoid secretions; Fig. 11: target granules; Scale bars 1 μm.

ly). The frontal pore measured about 6 μm in diameter and was bounded by epidermal cells (three in our cross-sectioned specimen, see Fig. 1) with deeply sunken nuclei; other epidermal cells of the body wall typically had epimyal nuclei (*cf.* Smith, 1981). Posterior to the frontal pore, the necks of the MG expanded, forming a pear-shaped reservoir ('ampule'), which extended caudally for about 10 μm (Figs 2 and 3). Necks of the SMG and ABG lay along the dorsal and ventral sides of the reservoir, respectively. The group of gland necks at the frontal pore was surrounded by a sphincter of four circular muscle fibers, located about 4 μm below the epidermal surface (Figs 4a and 4b). Posterior to the reservoir, the gland necks diverged. The necks of the MG passed on either side of the brain, ending in large cell bodies that lay peripherally, filling the body almost entirely from the level of the testes anterior to just behind the statocyst.

The gland necks of the SMG opened dorsally and laterally in the frontal pore in a horseshoe-shaped group around the MG; a few were wedged between the necks of the MG (Fig. 1). Posterior to the reservoir, the necks of the SMG crossed those of the MG and continued rearward straight through the ventral part of the brain. Cell bodies of the SMG lay at about the same level as those of the MG, posterior to the statocyst, both dorsal and ventral.

The ABG necks opened ventral to the MG in the frontal pore (Figs 1 and 3). They extended posteriorly along the ventral side of the reservoir. There they separated; some turned ventrally, thus crossing the necks of the ellipsoid-producing glands (see below), and some passed through the ventral part of the brain. From LM data it was evident that the cell bodies of the ABG lay behind the statocyst, at about the same level as the cell bodies of the MG; however, this has not yet been confirmed by TEM.

There were five additional types of glands whose necks opened at the anterior end of the body, outside the frontal pore. Based on their secretion granule morphology, we designate these as ellipsoid (E), electron-lucent ellipsoid (E*), colored (C), basophilic mucoid (B), and target (T) glands (Figs 7–11).

The ellipsoid granules (Fig. 7) were found in glands that opened ventrally to the frontal pore in an arc. The center of curvature of this arc was towards the ventral side (Fig. 1). The gland necks extended straight posteriorly, passing ventrally to the brain and statocyst, to cell bodies that lay at the same level as the cell bodies of the MG. Even at the level of the statocyst, these gland necks were organized in a recognizable arc. Definite differences in the electron density of the E-granules (Figs 7 and 8) indicate that there might have been a second type of E-granule-secreting gland (E*) that opened in an array ventral to the first (Fig. 1).

Both B- and T-glands opened ventral to the E-glands. Their necks extended posteriorly to at least the level of the statocyst; however, we have not yet localized their cell bodies.

Glands that produce colored granules (C) (Fig. 9) opened not only in the anterior part of the animal but were found over the entire body (Smith, 1981). These were short-necked glands with cell bodies lying in or immediately beneath the epidermis. Their nuclei usually appeared pycnotic.

Even though type I and type II monociliated sensory receptors (Bedini *et al.*, 1973) were found on the anterior end of the animal, they were not associated with the frontal organ (Fig. 1). They were not different from the receptors found on the rest of the body (Smith, 1981).

Macrostomum hystricinum *and* Macrostomum *sp. III*

The so-called 'frontal organ' of *Macrostomum* spp. appeared less complex than that of *Convoluta* 'pulchra'. Its most prominent part consisted of rhammites (Fig. 14) arrayed into discrete 'Stäbchenstrassen' that were easily visible in living specimens. Unlike the glands of the frontal organ of *C.* 'pulchra', the gland necks of the 'Stäbchenstrassen' did not converge at the anterior tip of the body (Fig. 12), but instead fanned out, opening in a broad zone around the anterior margin. The rhammite gland bodies lay dorsally on either side of the pharynx and their necks extended anteriorly straight through the brain. Glands of a second type, producing mucoid globules, also opened anteriorly (Figs 12 and 13). However, these glands did not converge at a common pore but rather opened in diffuse fashion. Their necks extended through the ventral part of the brain. Posterior to the brain the necks lay dorsally. Both the rhammites and globules stained brown with iron hematoxylin and green with fast green but did not react with alcian blue.

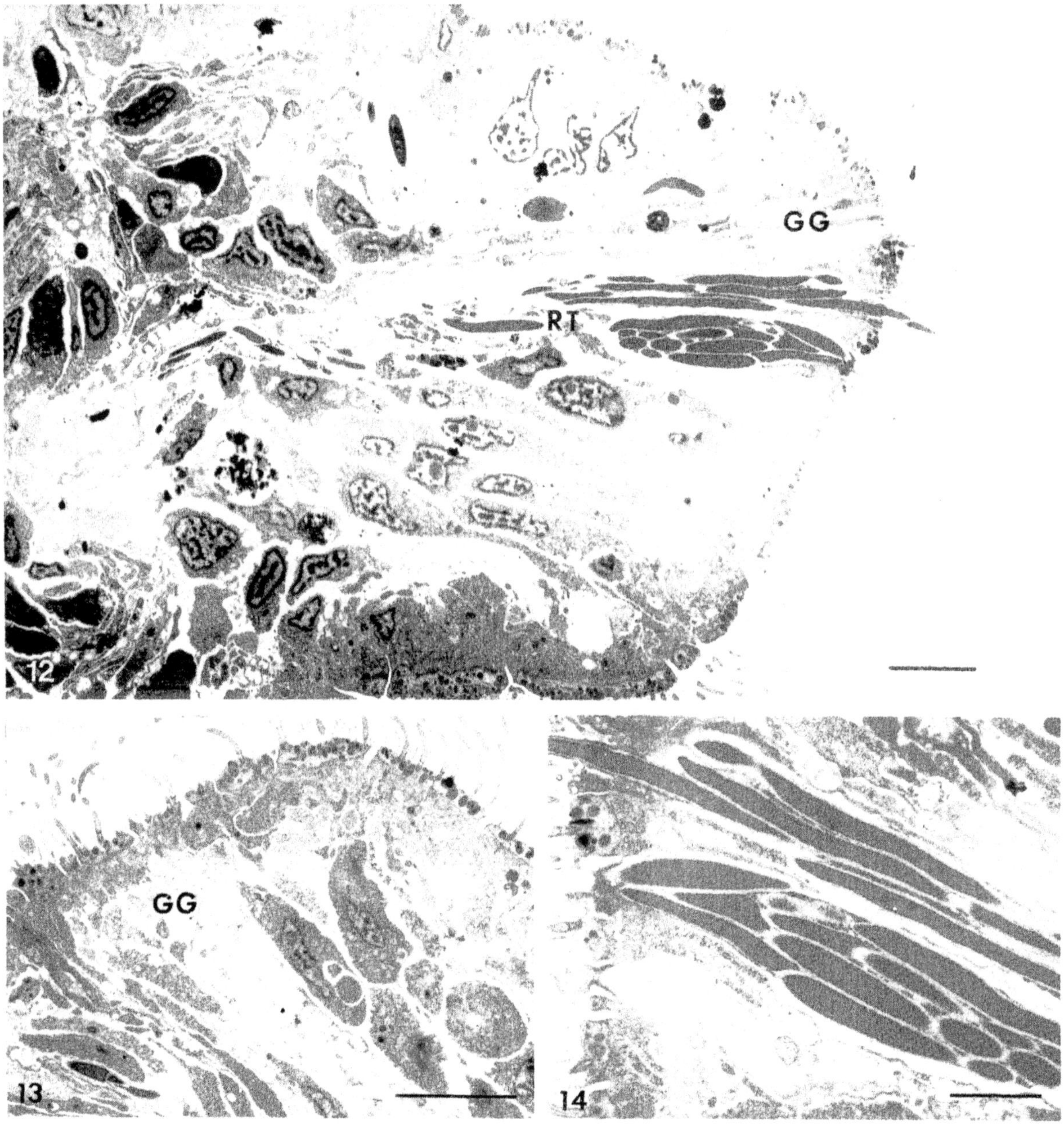

Figs 12–14. Macrostomum sp. III. (12) Sagittal section through the anterior end showing rhammite tracks (RT) and globule-secreting glands (GG). Scale bar 5 μm; (13) *Macrostomum* sp. III. Necks of globule-secreting glands. Scale bar 5 μm; (14) *Macrostomum* sp. III. Rhammite of 'Stäbchenstrasse'. Scale bar 2 μm.

Discussion

The frontal gland complex of *Convoluta* 'pulchra' comprises a true frontal organ with three types of mucoid glands that open into a frontal pore and a ventral array of five additional types of glands. This stands in sharp contrast to the frontal gland complex of the *Macrostomum* species which consists of rhammite tracks arranged into 'Stäbchenstrassen', and globule-secreting glands, which open separately across the anterior margin rather than through a common frontal pore. Further differences are apparent in histochemical properties: two gland types of the frontal organ of *C.* 'pulchra' stain positively with alcian blue, indicating the presence of acid mucopolysaccharides, while the frontal glands of *Macrostomum* spp. stain positively with fast green and iron hematoxylin as would be expected of proteinaceous material.

This lack of structural, positional, and histochemical similarities between the frontal gland complexes of *Convoluta* 'pulchra' and *Macrostomum* spp. cannot at present be bridged by intermediate forms from the Nemertodermatida-Acoela (Smith & Tyler, 1985). Similarly, present data on other characters do not reveal any synapomorphy between the Nemertodermatida-Acoela and the evolutionary line containing the macrostomids (Smith *et al.*, 1985; however, see Ehlers, 1985, for another opinion). However, because a derived position of *Macrostomum* within the Macrostomida can be inferred (from the lack of a muscle ring around the gut, Rieger, 1971), the form of the frontal glands in these two *Macrostomum* species is not necessarily pleisiomorphic for the Macrostomida as a whole. Further studies using more primitive macrostomids are warranted.

For now, however, we maintain that these organs are best regarded as non-homologous, and we suggest that the general terms '*frontal glands*' or '*frontal gland complex*' be applied to the type of arrangement found in the macrostomids and to the frontal glands in the acoels. The term '*frontal organ*', referring to the mucoid gland types opening in a common pore should be reserved for the acoels (see also Smith & Tyler, 1985).

Acknowledgments

We thank Bill Snow for printing the plates and Eddie Meisner for typing the manuscript. This study was supported by NSF grants BSR 8116894 and DEB 7706058 (S. Tyler, P.I.), and some of the costs for collection of *Convoluta* 'pulchra' was met by NSF grant GB42211 (R. M. Rieger, P.I.).

References

Ax, P., 1961. Verwandtschaftsbeziehungen und Phylogenie der Turbellarien. Erg. Biol. 24: 1–68.

Bedini, C., E. Ferrero & A. Lanfranchi, 1973. The ultrastructure of ciliary sensory cells in two Turbellaria Acoela. Tissue Cell 5: 359–372.

Chayen, J., L. Bitensky, R. G. Butcher & L. W. Poulter, 1969. A guide to practical histochemistry. J. B. Lippincott Co., Philadelphia.

Ehlers, U., 1985. Comments on a phylogenetic system of the Platyhelminthes. This volume.

Hulings, N. C. & J. S. Gray, 1971. A manual for the study of meiofauna. Smithsonian Contr. Zool. 78: 11–83. Smithsonian Inst. Press, Washington.

Karling, T. G., 1940. Zur Morphologie und Systematik der Alloeocoela Cumulata and Rhabdocoela Lecithoepitheliata (Turbellaria). Acta zool. fenn. 26: 1–260.

Karling, T. G., 1974. On the anatomy and affinities of the turbellarian orders. In N. W. Riser & M. P. Morse (eds.), Biology of the Turbellaria. McGraw-Hill, New York, pp. 1–16.

Maxwell, M. H., 1978. Two rapid and simple methods used for the removal of resins from 1.0 μm thick epoxy sections. J. Microsc. 112: 253–255.

Rieger, R. M., 1971. Die Turbellarienfamilie Dolichomacrostomidae (Rieger). II. Teil: Dolichomacrostominae. Zool. Jb. Syst. Ökol. Geogr. Tiere 98: S569–703.

Rieger, R. M., 1977. The relationship of character variability and morphological complexity in copulatory structures of Turbellaria-Macrostomida and -Haplopharyngida. Mikrof. Meeresb. 61: 197–216.

Smith, J. P. S., III, 1981. Fine structural anatomy of the parenchyma in the Acoela and Nemertodermatida (Turbellaria). Ph.D. Dissertation, Univ. North Carolina, Chapel Hill.

Smith, J. P. S., III & S. Tyler, 1985. Frontal organs in the Acoela: ultrastructure and phylogenetic significance. This volume.

Smith, J. P. S., III & S. Tyler, 1984. Serial sectioning of resin-embedded material for light microscopy: Recommended techniques for micrometazoans. Mikroskopie (Wien) 41: 259–270.

Smith, J. P. S., III, S. Tyler & R. M. Rieger, 1985. Is the Turbellaria polyphyletic? This volume.

Von Graff, F., 1891. Die Organisation der Turbellaria Acoela. Verlag von Wilhelm Engelmann, pp. 1–90.

The phylogenetic significance of sperm morphology in the Platyhelminthes

Jan Hendelberg
Dept. of Zoology, Univ. of Göteborg, Box 25059, S-40031, Göteborg, Sweden

Keywords: Turbellaria, spermatozoa, phylogeny, lower Metazoa, Trematoda, Cestoda

Abstract

The phylogenetic significance of flatworm sperm morphology is discussed against the background of general spermatology. The modified type of spermatozoon of the Nemertodermatida, a group of primitive flatworms, indicates that the Platyhelminthes evolved from forms characterized by the primitive type of metazoan sperm and by the primitive mode of fertilization, implying the release of sperm freely into sea water.

The occurrence of aberrant types of spermatozoa in most platyhelminths is obviously a consequence of early evolution of the internal mode of fertilization, which characterizes all true members of this group. It can be concluded, from the ultrastructure of these aberrant spermatozoa that 'higher' metazoans cannot have evolved from 'seriated' flatworms related to the recent Seriata (Proseriata and Tricladida). Even the seemingly primitive Acoela have such aberrant spermatozoa that evolution of 'higher' metazoans from acoels related to the recent Acoela seems highly improbable.

The ultrastructure of the spermatozoa of the parasitic groups of flatworms (Monogenea, Digenea, Cestoda) is very similar to that found in the Kalyptorhynchia, a further indication that the parasitic groups are related to the rhabdocoel turbellarians.

Introduction

According to many theories, turbellarians are considered to have played an important role in the early evolution of the Metazoa, and even in the evolution of 'higher' metazoans. Most discussions of these theories are based on comparative morphological studies of forms representing the diploid stages of the life cycle. I will discuss here the information to be derived from studies of representatives of the haploid generation, the spermatozoa. This has been done earlier, on the basis of studies by light microscopy, by e.g. Franzén (1956) and Ax (1963). Since then many new results based on studies of the ultrastructure of flatworm spermatozoa have appeared (see e.g. the reviews by Hendelberg, 1983a; Mohandas, 1983; and Davis & Roberts, 1983). At the same time our general knowledge of animal spermatology has increased.

What is the phylogenetic significance of the new information on sperm morphology? Can comparative spermatology reveal anything about the possibility of 'higher' metazoans having evolved from forms related to any of the recent groups of turbellarians, or about the relationships of parasitic groups of flatworms to turbellarians? These and some other problems are dealt with in the following.

Materials and methods

The general ideas presented below and the diagrams drawn to support them are based mainly on

Hydrobiologia 132, 53–58 (1986).

published reports. Thus generalizations about flatworm spermatozoa are based on information reviewed by Hendelberg (1983a) and on some recent papers referred to below. Conclusions regarding Kalyptorhynchia are based on unpublished studies of *Gyratrix hermaphroditus*, collected at Tvärminne, on the south coast of Finland, and *Prognathorhynchus typhlus* and *Phonorhynchus helgolandicus*, collected at Fiskebäckskil, on the west coast of Sweden. Some preliminary results were published earlier (Hendelberg, 1983b). The material was prepared and studied by means of electron microscopy according to methods described earlier (Hendelberg, 1977).

Sperm types and their phylogenetic significance in general

The nature of the primitive type of metazoan spermatozoon was deduced and described by Franzén (1956) on the basis of comparative light microscopical studies (Fig. 1). It is characterized by a short, rounded to conical head, a short middle piece containing a few mitochondria (often four mitochondrial spheres, or, as in many deuterostomians, a single mitochondrion forming a ring around the two centrioles), and a tail consisting of a flagellum with the primitive axonemal structure, the so-called '9+2' pattern (Franzén, 1977; Afzelius, 1979) (see Figs. 2–3, illustrating the ultrastructure of primitive spermatozoa of two 'lower' metazoans). This type of spermatozoon occurs in

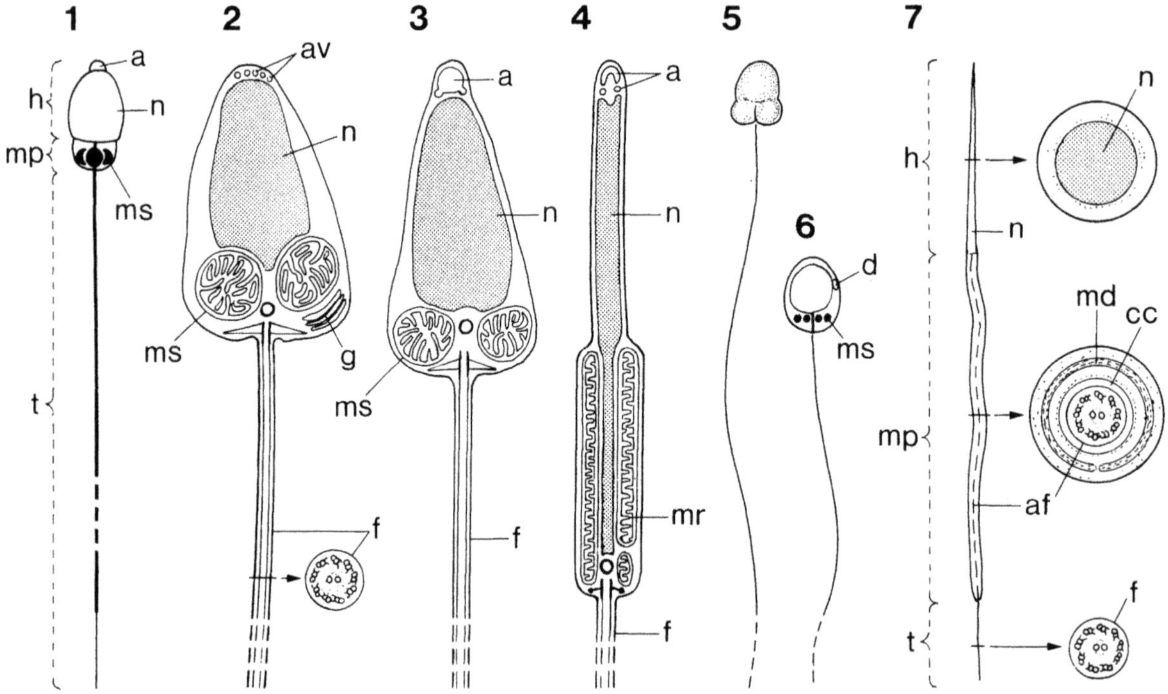

Figs. 1–7. (1) Diagram of the primitive spermatozoon in the Metazoa according to Franzén (1956). After Franzén (1956), somewhat simplified. (2–3) Diagrams of the ultrastructure of the primitive type of spermatozoon of a cnidarian (the jellyfish *Nausithoë* sp.) and a nemertean (the heteronemertean *Micrura fasciolata*), respectively; modified after Afzelius & Franzén (1971) and Franzén (1983). (4) Diagram of the modified spermatozoon of a nemertean (the bdellonemertean *Malacobdella grossa*); after Afzelius (1971) and Franzén (1983). (5–6) Spermatozoon and late spermatid, respectively, of *Xenoturbella bocki*, Xenoturbellida. After Westblad (1949) and Franzén (1956). (7) Diagram of the modified spermatozoon of a turbellarian flatworm, *Meara stichopi* (Nemertodermatida); to the left, the division of the spermatozoon into main regions (phase contrast, light microscopy), to the right the ultrastructure of cross sections of these regions. After Hendelberg (1977). a, acrosome; av, acrosomic vesicles; af, axial filament; cc, cytoplasmic canal; d, dictyosome (=Golgi complex with forming acrosome); f, flagellum; g, Golgi elements; h, head; m, mitochondrion; md, mitochondrial derivative; mp, middle piece; mr, mitochondrial rods; ms, mitochondrial sphere; n, nucleus; t, tail.

some 17 phyla, from coelenterates to annelids and molluscs among the Protostomia to chordates (lancelets and some fishes) among the Deuterostomia, always in species with what is regarded as the primitive type of sperm transfer, i.e. the release of sperm into sea-water where the spermatozoa swim freely until they reach an egg (if they do so) (Franzén, 1956; Jägersten, 1972).

One reason that this type of spermatozoon is regarded as primitive is that there are many other kinds of spermatozoa that are guided in one way or another to the eggs (=directed sperm transfer, see Afzelius, 1984). Another reason is that traces of the primitive spermatozoon can often be found in the development of other kinds of spermatozoa (Franzén, 1956, 1977). From these observations Franzén (*op. cit.*) concluded that many kinds of spermatozoa have evolved independently, in different phyla, from the primitive type of spermatozoon in connection with the evolution of new types of sperm transfer (with spermatophores, by copulation, etc.).

Spermatozoa that evolved from the primitive type and that have primitive features remaining in spermiogenesis, were called *modified spermatozoa* by Franzén (1956, 1977). An example of such primitive features in spermatids is the occurrence of mitochondrial spheres. During maturation of a modified spermatozoon these spheres are reshaped into other forms (helical, rod-like, etc.). In modified spermatozoa the middle piece containing the mitochondria (or mitochondrial derivative) is most often elongated, incorporating part of the tail flagellum as an axial filament (Fig. 7). Sometimes the mitochondrial material is elongated forward, as in the modified spermatozoa of some nemerteans, in which three to six mitochondrial rods can be found along the posterior part of the elongated nucleus (Fig. 4).

Spermatozoa that are very different from the primitive ones and in the development of which no primitive features can be traced are called *aberrant spermatozoa* by Franzén (1956). Examples of such spermatozoa are those without a flagellum. The absence of a flagellum is evidently the result of reduction-disappearance (cf., regarding turbellarians, Hendelberg, 1969; and Ehlers, 1969). According to Wirth (1984) non-flagellate spermatozoa have evolved independently approximately 25 times in different groups of the Metazoa.

From the above it is evident that the occurrence of a primitive spermatozoon does not imply that a species (or group) occupies a position near the base of the phylogenetic tree of the Metazoa, as has sometimes been assumed. What it indicates it that no change in the mode of sperm transfer has had an effect on sperm morphology. Thus, the evidently primitive spermatozoon of the Xenoturbellida (Figs. 5–6) give no clues to the phylogenetic relationships of the group (at least not until more is known about the ultrastructure of this and other primitive spermatozoa; cf. Afzelius, 1979). It only indicates that the Xenoturbellida most probably has retained the primitive mode of sperm transfer.

In the Platyhelminthes, modified spermatozoa indicating evolution from the primitive type are found in the Nemertodermatida, a group of primitive turbellarians (Tyler & Rieger, 1975, 1977; Hendelberg, 1977) (Figs. 7, 8). Thus it can be concluded that flatworm spermatozoa evolved from the primitive spermatozoon. Evidently this happened in connection with the evolution of internal fertilization which characterizes all true flatworms.

Did 'higher' metazoans evolve from the Turbellaria?

As many 'higher' metazoans have primitive spermatozoa, it is highly improbable that they evolved from forms having aberrant spermatozoa. With the exception of the Nemertodermatida (see above) all true turbellarians have aberrant spermatozoa (Fig. 8). Even the Acoela, seemingly primitive in some respects, has been found by ultrastructural studies to have spermatozoa with many apomorphies: flagellum doubled, flagella incorporated into other parts of the spermatozoon in an inverted position, many separate mitochondria (Hendelberg, 1977, 1983a with references, 1983b; Boyer & Smith, 1982). Thus, the evolution of 'higher metazoans from acoels closely related to recent Acoela is highly improbable.

According to some ideas, the 'higher' metazoans evolved from 'seriated flatworms' (Valentine, 1977). If so, this could not have occurred from forms closely related to the recent Seriata (Proseriata and Tricladida), because the aberrant type of spermatozoon characterizing these turbellarians (Fig. 8) evidently evolved before the serial repeating of inner organs in these worms.

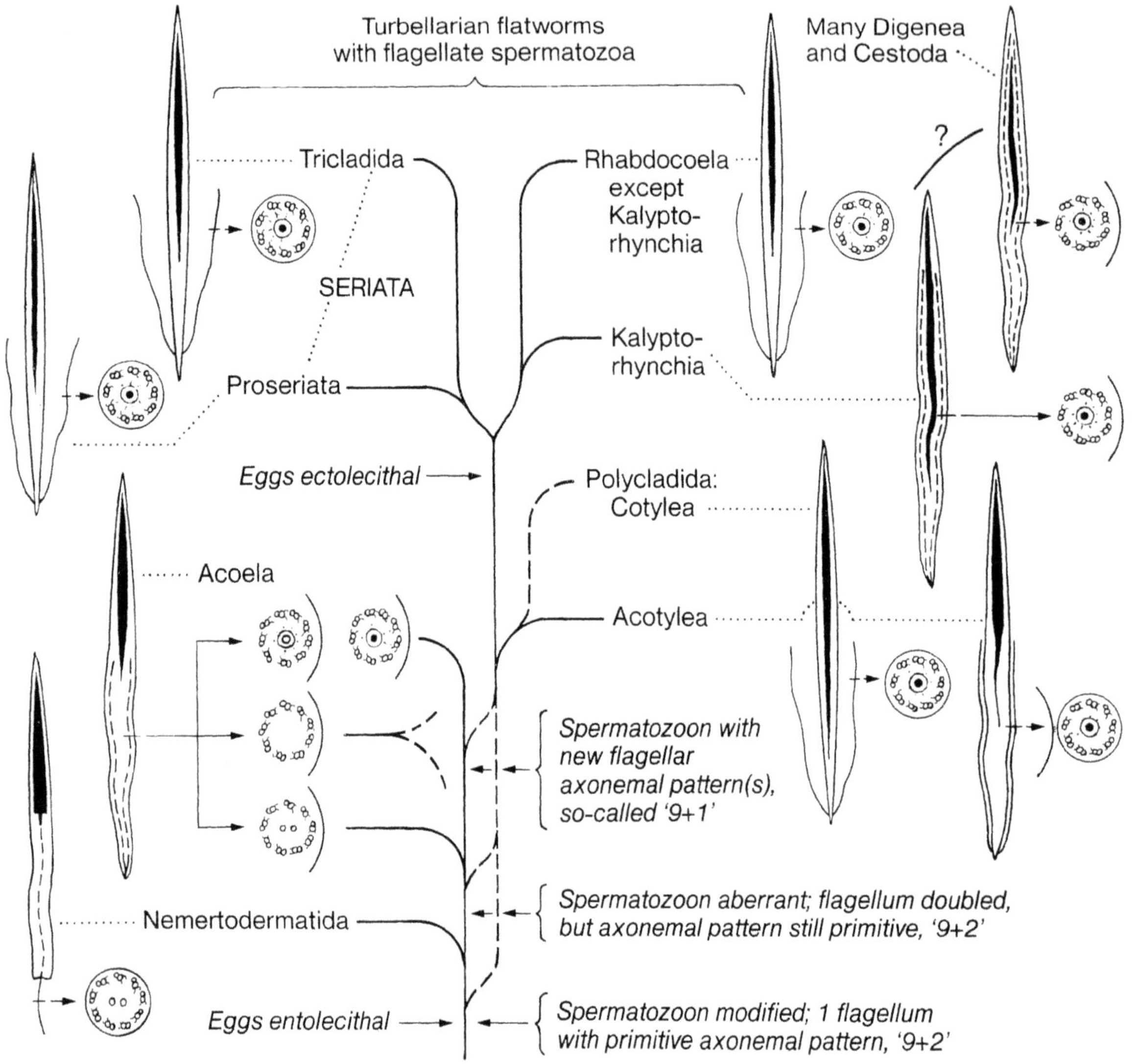

Fig. 8. Diagram demonstrating steps of evolution of flagellate spermatozoa within the Platyhelminthes. Types of spermatozoa commonly found in the sub-groups are drawn schematically with highly exaggerated thickness, demonstrating the internal position of the axonemes of the incorporated flagella in spermatozoa of some groups. Each drawing of a spermatozoon is accompanied by a cross section (×40 000) of a flagellum or internal axoneme demonstrating the axonemal pattern (9+2, 9+0, or so-called 9+1) and its position in relation to the cell membrane. The assumed relationships of the groups is indicated by the phylogenetic tree which is (mainly) after Karling (1974) and Rieger (1981) (cf. Hendelberg, 1977, 1983b). Conclusions about which sperm types are typical of the different groups are based on a great number of references most of which are listed by Hendelberg (1983a); see also Boyer & Smith (1982), Mohandas (1983) and Davis & Roberts (1983).

In fact, ultrastructural studies of spermatozoa strongly support the ideas of Franzén (1956) and Ax (1963) implying that 'higher' metazoans did not evolve from any of the recent groups of the Turbellaria.

The relationships of the parasitic groups of flatworms to the turbellarians

Ultrastructural studies of the turbellarian group Kalyptorhynchia have revealed that their spermatozoa are very similar to those described as typical for some of the parasitic groups of flatworms, i.e., monogeneans, digeneans and cestodes, Fig. 9. The

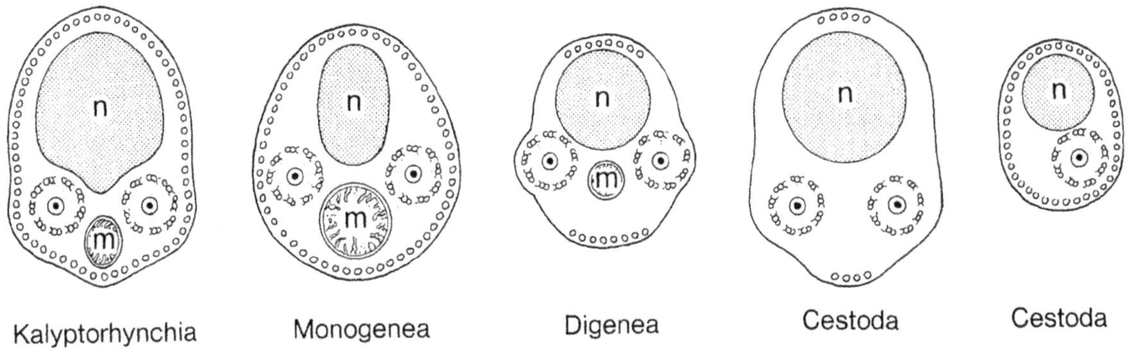

Fig. 9. Diagram of the ultrastructure of some flatworm spermatozoa (as seen in cross sections). The spermatozoa of the parasitic groups (Monogenea, Digenea, Cestoda) show great similarities to those of the turbellarian group Kalyptorhynchia (a sub-group of the Rhabdocoela). Each diagram represents a number of species, but was drawn after micrographs of the following: *Gyratrix hermaphroditus* (Kalyptorhynchia), unpublished studies by the author (cf. Hendelberg, 1983b) and personal communication from M. Reuter; *Microcotyle erythrini* (Monogenea), (after Tuzet & Ktari, 1971); *Haematoloechus medioplexus* (Digenea), (after Burton, 1972); *Lacistorhynchus tenuis*, a cestode with two sperm axonemes (after Swiderski, 1976); *Hymenolepis diminuta*, a cestode having only one sperm axoneme, (after Kelsoe *et al.*, 1977; cf. Euzet *et al.*, 1981); other references are listed by Mohandas (1983) and Davis & Roberts (1983); see also Jamieson & Daddow (1982), and Justine & Mattei (1983). Abbreviations as in Figs. 1–7. ×40000.

single axoneme of some cestode spermatozoa is evidently a derived character (Euzet *et al.,* 1981) as is the absence from the cestode spermatozoa of a mitochondrial derivative (Davis & Roberts, 1983). Morphological characteristics of the worms themselves have promoted the idea of a close relationship between the above-mentioned parasitic groups and the rhabdocoel turbellarians, and sperm morphology may be interpreted as a further indication of such a relationship.

Discussion

When using sperm morphology in discussions of phylogenetic relationships, we have to bear in mind that closely related species sometimes show great differences in sperm morphology (Franzén, 1956) and that the physiological condition of the spermatozoa as well as the preparation methods affect the appearance of the ultrastructure (Hendelberg, 1977). Are, for instance, the two types of so-called '9+1' axoneme patterns in acoel spermatozoa really different? The occurrence of convergent evolution of spermatozoa (Hendelberg, 1983b) and the fact that spermatozoan ultrastructure is known for a very low proportion of the total number of species are further problems. Caution is advised, and we may have to reconsider our views on phylogeny when new data are available.

Acknowledgments

Material was collected at the laboratories of marine biology at Tvärminne, Finland, and Kristineberg, Fiskebäckskil, Sweden. Some of the ultrastructural studies were performed at the Department of Zoology, Uppsala University. Liselotte Öhman finished the drawings. The Swedish Natural Science Research Council gave financial support.

References

Afzelius, B. A., 1971. The spermatozoon of the nemertine Malacobdella grossa. J. submicrosc. Cytol. 3: 181–192.
Afzelius, B. A., 1979. Sperm structure in relation to phylogeny in lower Metazoa. In D. W. Fawcett & J. M. Bedford (eds), The spermatozoon. Maturation, motility, surface properties and comparative aspects. Urban & Schwarzenberg, Baltimore-Munich: 243–251.
Afzelius, B. A., 1984. On the usefulness of sperm morphology in phylogeny. Proc. Ernst Haeckel-Tagung May 1984, in press.
Afzelius, B. A. & Å. Franzén, 1971. The spermatozoon of the jellyfish Nausithoë. J. ultrastruct. Res. 37: 186–199.
Ax, P., 1963. Relationships and phylogeny of the Turbellaria. In E. C. Dougherty (ed.). The lower Metazoa. Univ. California Press, Berkeley: 191–224.
Boyer, B. C. & G. W. Smith, 1982. Sperm morphology and development in two acoel turbellarians from the Philippines. Pacific Sci. 36: 365–380.

Burton, P. R., 1972. Fine structure of the reproductive system of a frog lung fluke. III. The spermatozoon and its differentiation. J. Parasitol. 58: 68–83.

Davis, R. E. & L. S. Roberts, 1983. Chapter 5. Platyhelminthes-Eucestoda. In K. G. Adiyodi & R. G. Adiyodi (eds), Reproductive Biology of Invertebrates, Vol. II, Spermatogenesis and Sperm Function. J. Wiley & Sons, Chicester, N.Y.: 131–149.

Ehlers, U., 1981. Fine structure of the giant aflagellate spermatozoon in Pseudostomum quadrioculatum (Leuckart) (Platyhelminthes, Prolecithophora). Hydrobiologia 84: 287–300.

Euzet, L., Z. Swiderski & F. Mokhtar-Maamouri, 1981. Ultrastructure comparée du spermatozoide des cestodes. Relations avec la phylogénèse. Ann. Parasitol. 56: 247–259.

Franzén, Å., 1956. On spermiogenesis, morphology of the spermatozoon, and biology of fertilization among invertebrates. Zool. Bidr. Upps. 31: 355–482.

Franzén, Å., 1977. Sperm structure with regard to fertilization biology and phylogenetics. Verh. dt zool. Ges. 1977: 123–138.

Franzén, Å., 1983. Chapter 7. Nemertina. In K. G. Adiyodi & R. G. Adiyodi (eds), Reproductive Biology of Invertebrates, Vol. II, Spermatogenesis and Sperm Function. J. Wiley & Sons, Chichester, N.Y.: 159–170.

Hendelberg, J., 1969. On the development of different types of spermatozoa from spermatids with two flagella in the Turbellaria with remarks on the ultrastructure of the flagella. Zool. Bidr. Upps. 38: 1–50.

Hendelberg, J., 1977. Comparative morphology of turbellarian spermatozoa studied by electron microscopy. Acta. zool. fenn. 154: 149–162.

Hendelberg, J., 1983a. Chapter 3. Platyhelminthes-Turbellaria. In K. G. Adiyodi & R. G. Adiyodi (eds), Reproductive Biology of Invertebrates, Vol. II, Spermatogenesis and Sperm Function. J. Wiley & Sons, Chichester, N.Y.: 75–104.

Hendelberg, J., 1983b. Trends in the evolution of flatworm spermatozoa. In J. André (ed.), The Sperm Cell. Martinus Nijhoff Publ., The Hague: 450–453.

Jägersten, G., 1972. Evolution of the metazoan life cycle. Academic Press, N.Y., 282 pp.

Jamieson, B. G. M. & L. M. Daddow, 1982. The ultrastructure of the spermatozoon of Neochasmus sp. (Cryptogonimidae, Digenea, Trematoda) and its phylogenetic significance. Int. J. Parasitol. 12: 547–559.

Justine, J.-L. & X. Mattei, 1983. Etude ultrastructurale comparée de la spermiogenèse des Monogènes. I. Megalocotyle (Monopisthocotylea Capsalidae). J. ultrastruct. Res. 82: 296–308.

Karling, T. G., 1974. On the anatomy and affinities of the Turbellarian orders. In N. W. Riser & M. P. morse (eds), Biology of the Turbellaria. McGraw-Hill, N.Y.: 1–16.

Kelsoe, G. H., J. E. Ubelaker & V. F. Allison, 1977. The fine structure of spermatogenesis in Hymenolepis diminuta (Cestoda) with a description of the mature spermatozoon. Z. Parasitenkd 54: 175–187.

Mohandas, A., 1983. Chapter 4. Platyhelminthes-Trematoda. In K. G. Adiyodi & R. G. Adiyodi (eds), Reproductive Biology of Invertebrates, Vol. II, Spermatogenesis and Sperm Function. J. Wiley & Sons, Chichester, N.Y.: 105–129.

Rieger, R. M., 1981. Morphology of the Turbellaria at the ultrastructural level. Hydrobiologia 84: 213–229.

Swiderski, Z., 1976. Fine structure of the spermatozoon of Lacistorhynchus tenuis (Cestoda, Trypanorhyncha). Sixth European Congress of Electron Microscopy, Jerusalem: 309–312.

Tuzet, O. & M.-H. Ktari, 1971; Recherches sur l'ultrastructure du spermatozoïde de quelques monogènes. Bull. Soc. zool. Fr. 96: 535–540.

Tyler, S. & R. M. Rieger, 1975. Uniflagellate spermatozoa in Nemertoderma (Turbellaria) and their phylogenetic significance. Science 188: 730–732.

Tyler, S. & R. M. Rieger, 1977. Ultrastructural evidence for the systematic position of the Nemertodermatida (Turbellaria). Acta. zool. fenn. 154: 193–207.

Valentine, J. W., 1977. Chapter 13. The evolutionary history of Metazoa. In T. Dobzansky, F. J. Ayala, G. L. Stebbins & J. W. Valentine, Evolution. Freeman, San Francisco: 397–437.

Westblad, E., 1949. Xenoturbella bocki n.g., n.sp., a peculiar, primitive turbellarian type. Ark. Zool. 1 (3): 11–29.

Wirth, U., 1984. Spermatozoa of Nematoda: diversity of form, absence of nuclear envelope, and the evoluition of non-flagellate sperms in other Metazoa. In W. Engels (ed.), Advances in Invertebrate Reproduction, 3. Elsevier, Amsterdam: 651.

Phylogenetic relationships of the Temnocephaloidea (Platyhelminthes)

Joan Bowman Williams
Department of Zoology, University of Canterbury, Christchurch, New Zealand

Keywords: Turbellaria, Temnocephaloidea, ultrastructure, epidermis, spermatozoa, parenchyma

Abstract

Temnocephala novaezealandiae (family Temnocephalidae) and *Troglocaridicola mrazeki* and *Scutariella georgica* (family Scutariellidae) were studied by electron microscopy in an attempt to reveal characters that would indicate their phylogenetic relationship to other members of the Platyhelminthes. Ultrastructural features of the epidermis in these temnocephaloideans are like those of the neoophoran turbellarians. The epidermis is syncytial, is honeycombed by a multitude of gland necks whose secretions produce an epidermal surface film, and is underlaid by a thick basement membrane. Some cells in the parenchyma are compartmentalized by intrusive cell processes from neighboring parenchymal cells in a fashion similar to parenchymal structure in the Monogenea and Digenea. The spermatozoa have a pair of free 9+1 flagella and contain aligned dense bodies. The Temnocephaloidea is evidently derived from an early rhabdocoel-turbellarian-like ancestor.

Introduction

The Temnocephaloidea, as it is defined by Baer (1953, 1961), is a systematically well-delimited taxon (see also Williams, 1981b). All members lack locomotory cilia, possess tentacles at the anterior end, and, with the exception of *Diceratocephala* (see Baer, 1953), possess an attachment organ at the posterior end. They are characteristically associated with a host.

These features have led some systematists to place the group as a separate class distinct from the Turbellaria (see Williams, 1981b, for historical review; also Brooks, 1982). Other systematists include the group in the Turbellaria, classified in, or at least allied with, the Rhabdocoela (Bresslau & Reisinger, 1933; Joffe, 1981; Ehlers, 1985).

While the Temnocephaloidea is probably a monophyletic group, with characteristics unique among flatworms, ultrastructural investigations indicate it is allied more closely to the turbellarians than to other flatworms. In this paper, I review this ultrastructural evidence and present new (and preliminary) observations on members of the temnocephaloidean family Scutariellidae.

Materials and methods

Specimens of *Temnocephala novaezealandiae* were obtained from the pincers of the New Zealand freshwater crayfish *Paranephrops neozelanicus.* They were fixed in 4% glutaraldehyde in 0.025 M phosphate buffer for 6 h at room temperature, washed in phosphate buffer (three changes), and then postfixed in 2% osmium tetroxide in 0.025 M phosphate buffer for 5 h at room temperature. They were then dehydrated in acetone and embedded in Spurr's low-viscosity epoxy resin. Specimens of the scutariellid species *Troglocaridicola mrazeki* Joffe & Djanashvili, 1981, and *Scutariella georgica* Joffe & Djanashvili, 1981, were found on

Hydrobiologia 132, 59–67 (1986).

the surface of their host, the prawn *Troglocaris anophthalmus,* collected from a cave lake in western Georgia, USSR. The animals were fixed according to the schedule detailed by Williams (1975). All sections were cut on an LKB Ultrotome with diamond knives. Sections were stained with uranyl nitrate for 30 min, then with lead citrate for 5 min, and were recorded in micrographs with a Hitachi HS-7 electron microscope.

Observations

Epidermis

All members of the Temnocephalidae that have been studied by electron microscopy have numerous epidermal glands whose cell bodies lie in the parenchyma. In *Tem. novaezealandiae,* the necks of these glands pass through the syncytial and their secretions from stable films on the body surface (a rhabdite-derived coat covers the tentacles and a mucus coat invests the general epidermal surface) or are released through the attachment disk or sucker (Williams, 1982). The surface films appear to be stabilized by epidermal microvilli (Williams, 1982) from the dense tips of which filaments of glycocalyx radiate and extend, apparently, to dense layers on the external-most surface film (Fig. 1, fe). Secretion bodies in the epidermis itself (called 'epitheliosomes' by Tyler, 1984) accumulate in an apical layer over all the body except the ventral epithelium of the sucker; this layer constitutes what has been erroneously referred to as 'cuticle' by light

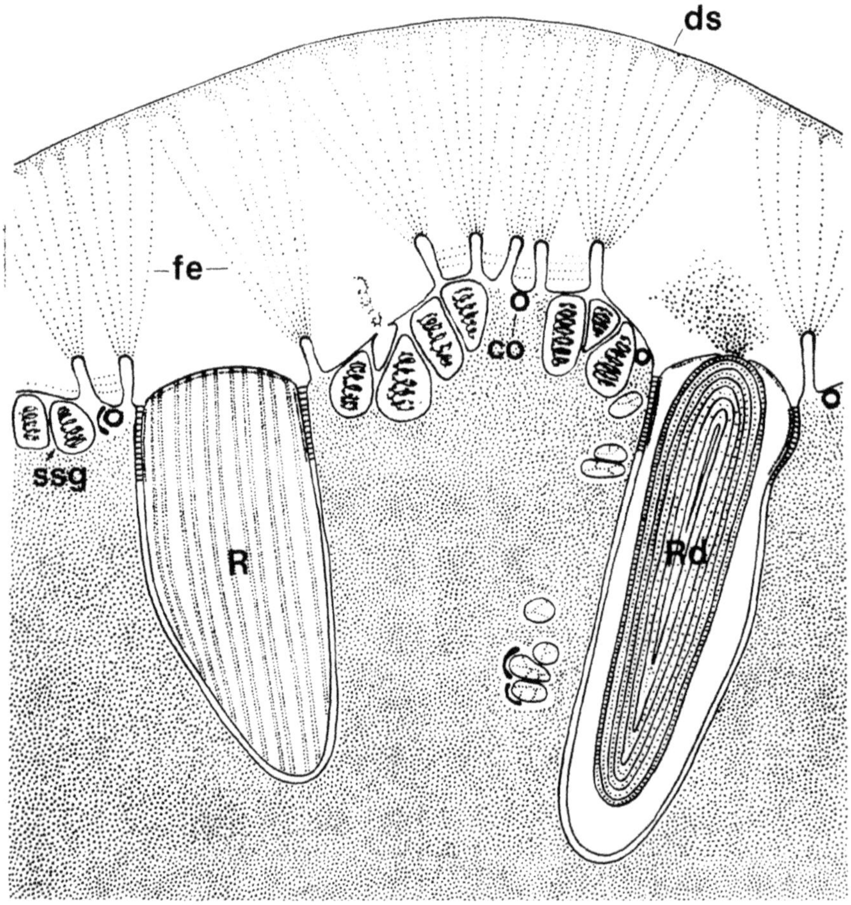

Fig. 1. Temnocephala novaezealandiae: Apical region of tentacle epidermis, drawn from micrographs. co, coated vesicle; ds, dense skin on surface of rhabdite-derived film; fe, filamentous extensions of glycocalyx stabilizing rhabdite-derived coat; Rd, discharging rhabdite; R, rhabdite gland neck; ssg, surface layer of secretion globules (epitheliosomes).

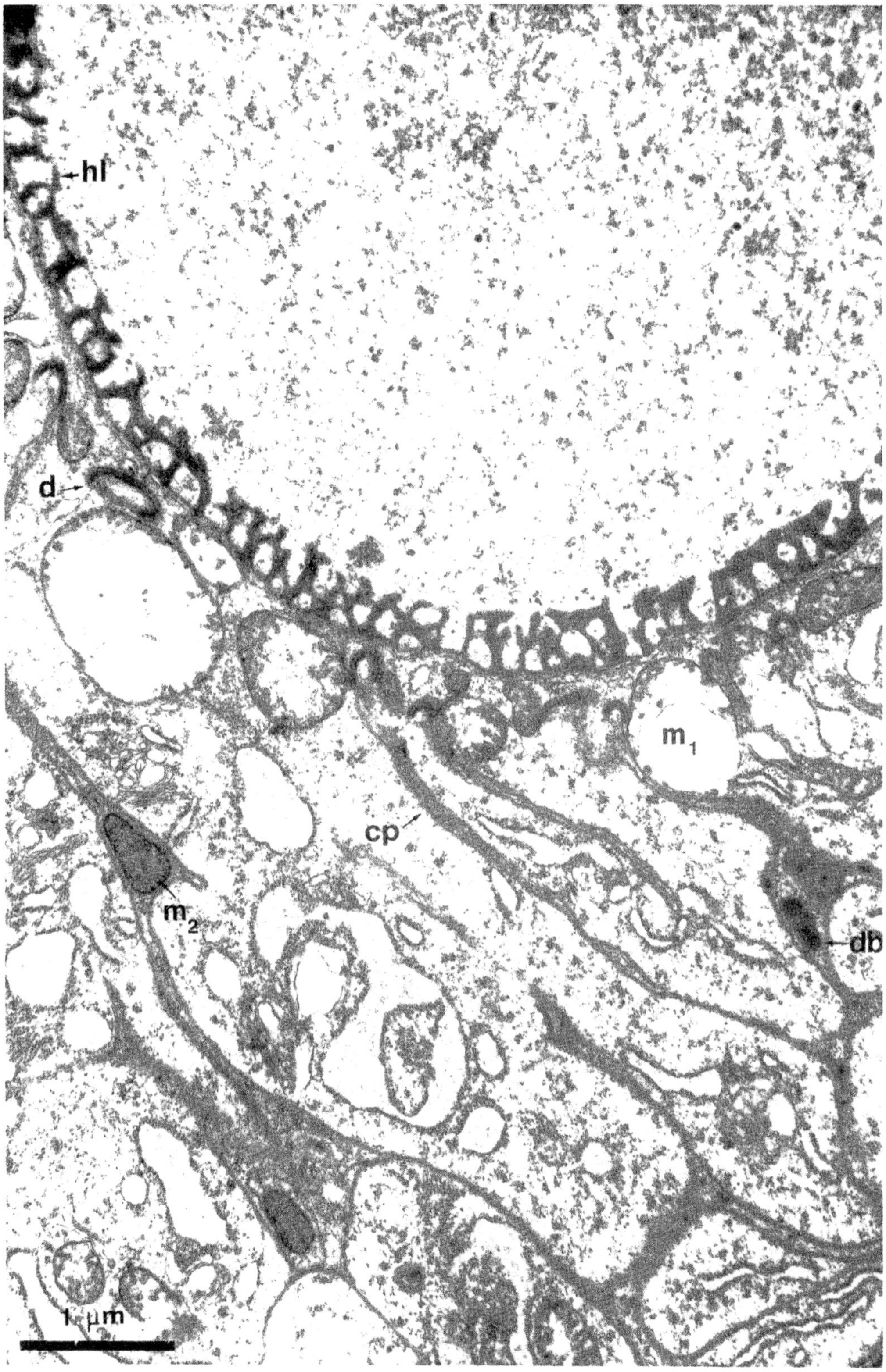

Fig. 2. Temnocephala novaezealandiae: Cell in parenchyma, with 'honeycomb' nucleus surrounded by juxtanuclear desmosomes. cp, cell process; d, desmosome; db, dense body; hl, honeycomb layer of nucleus; m_1, mitochondrion in recipient cell perikaryon; m_2, mitochondrion in cell process.

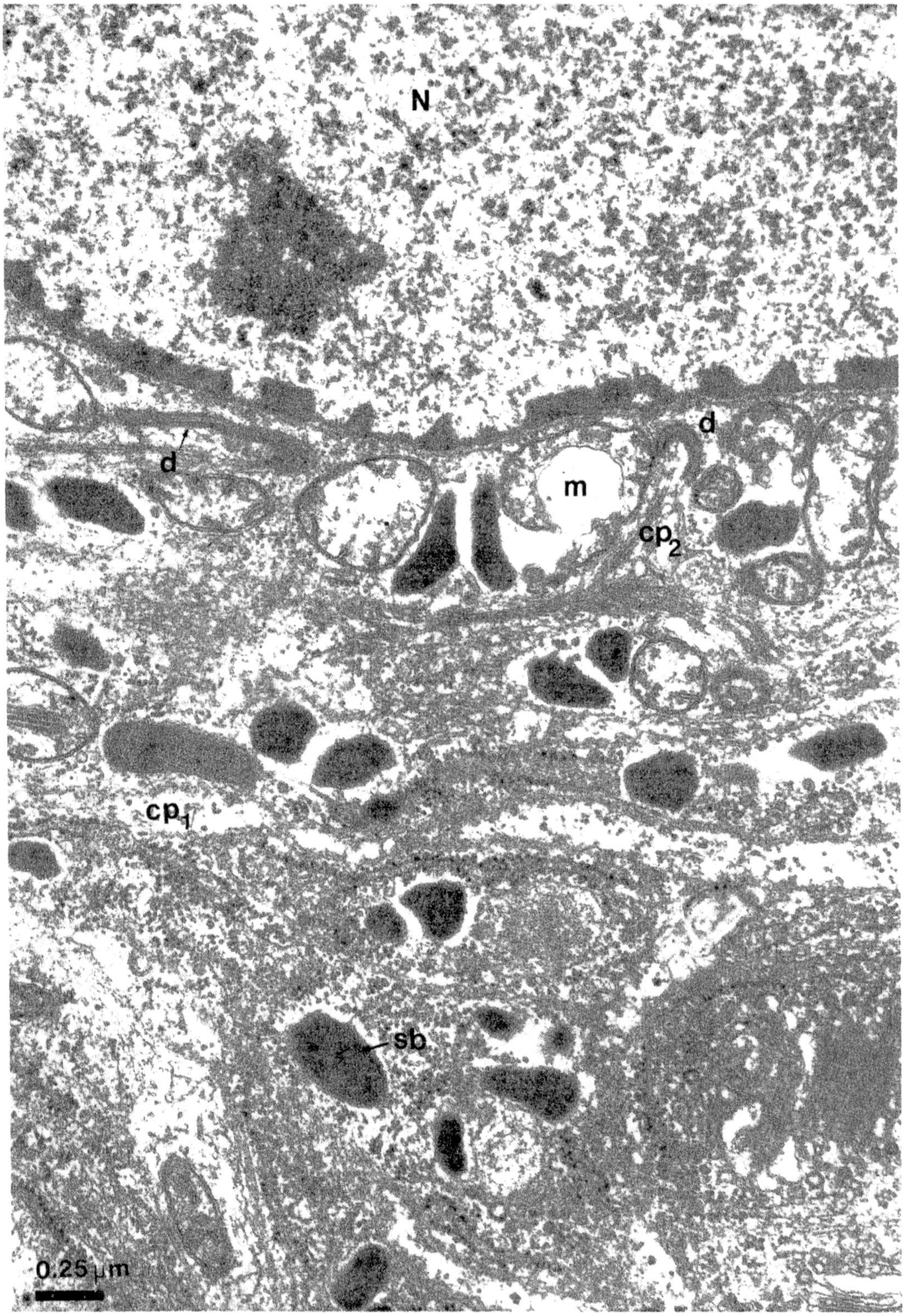

Fig. 3. Temnocephala novaezealandiae: Unidentified gland cell. cp_1, cell process; cp_2, terminal portion of cell process, containing filaments; d, desmosomes; m, juxtanuclear mitochondrion; N, nucleus; sb, secretion body.

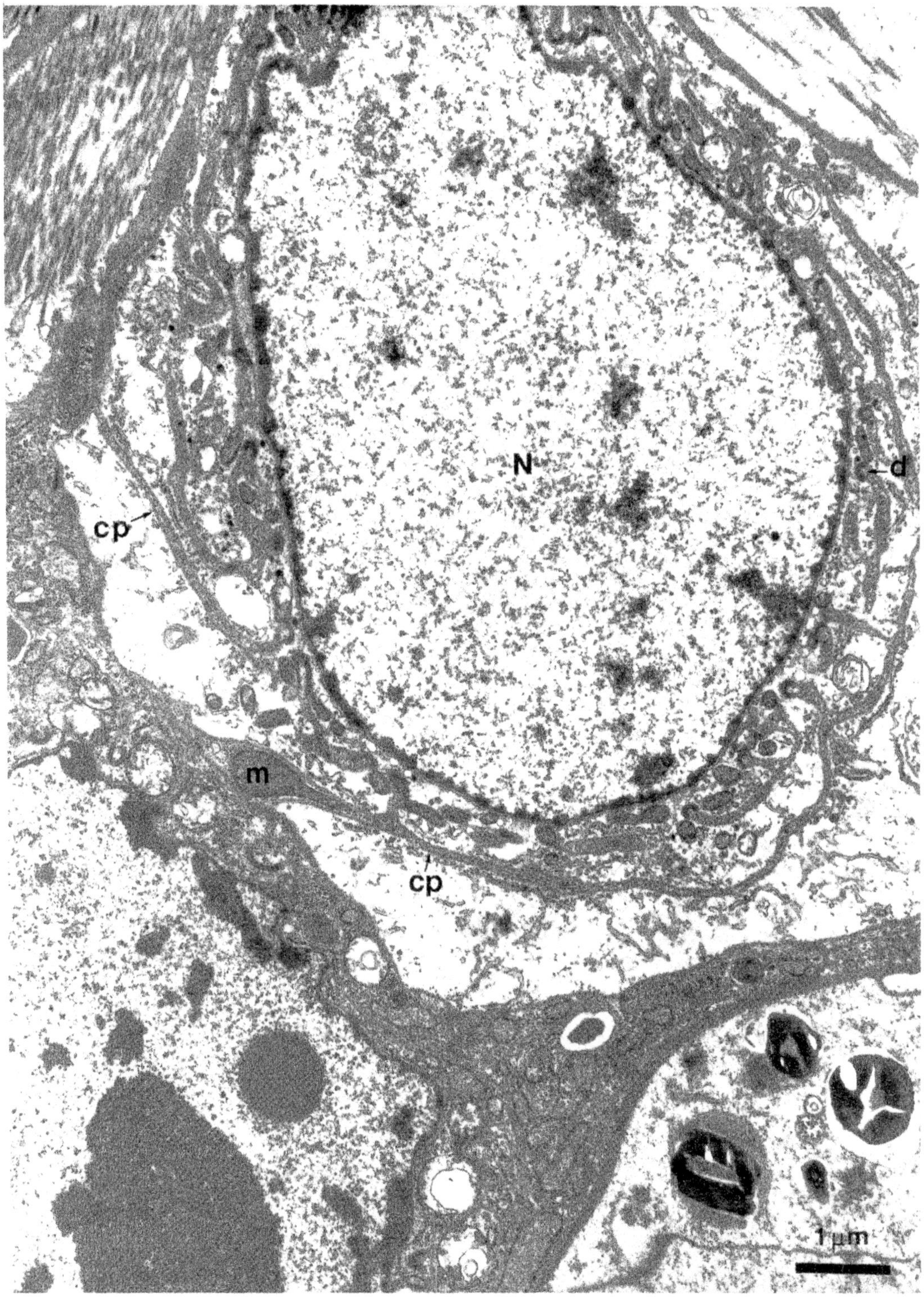

Fig. 4. Temnocephala novaezealandiae: Muscle cell perikaryon. cp, cell processes terminating at cup-shaped desmosomes; d, desmosome; m, club-shaped mitochondrion in process; N, nucleus, surrounded by juxtanuclear desmosomes and mitochondria. Nucleus at bottom left is also surrounded by desmosomes and mitochondria.

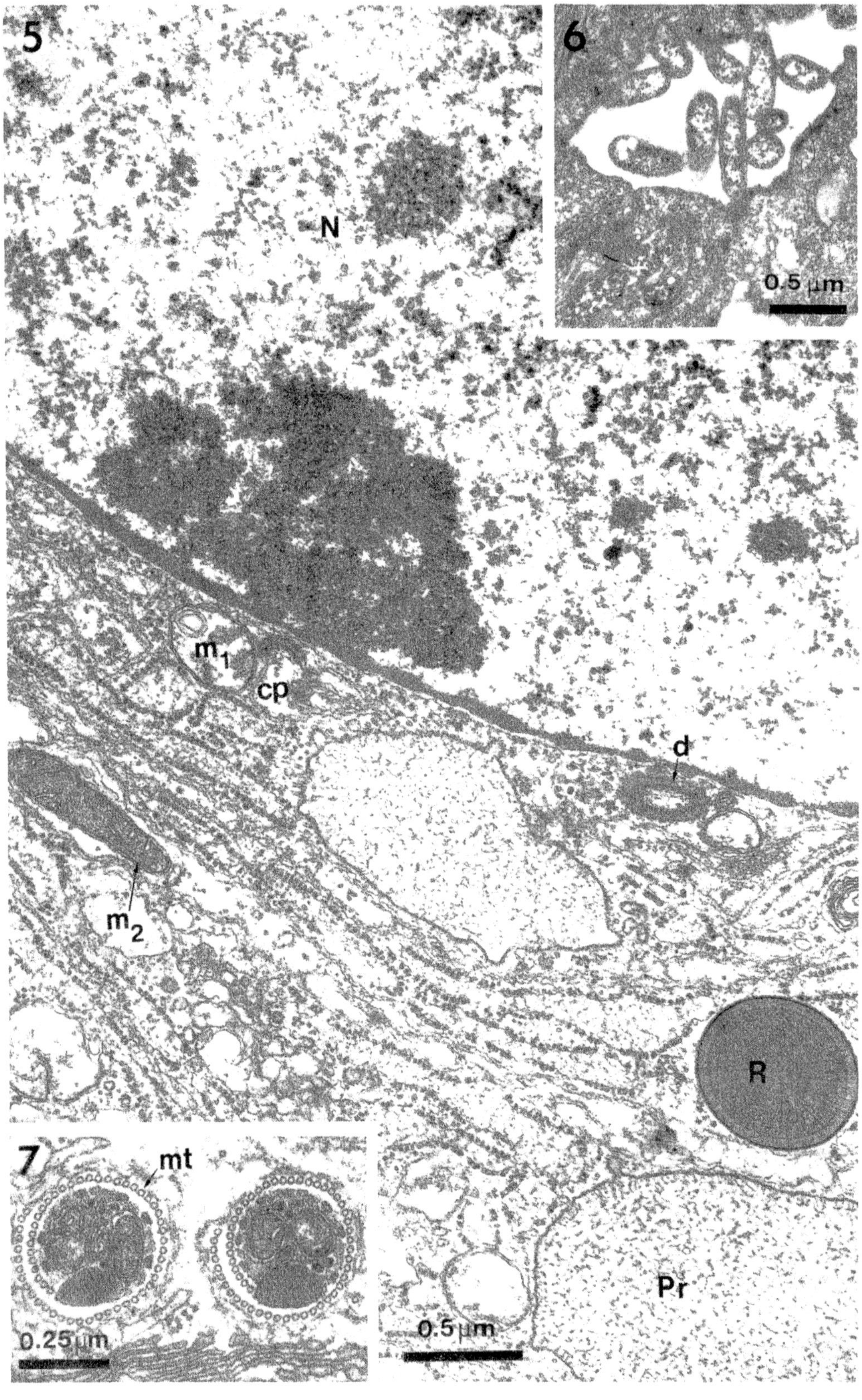
5
N
m1
cp
d
m2
R
Pr
0.5 µm
6
0.5 µm
7
mt
0.25 µm

microscopists (cf. Nichols, 1975).

The epidermis is subdivided by membranes into specialized regions (Williams, 1979, 1980a, 1982) which vary in response to changes in the environment (Williams, in press). The cell web in the epidermis is inconspicuous. The basement membrane is thick – with a prominent lamina lucida (subepidermal lucid zone) and thick lamina fibroreticularis – and is attached to the epidermis by numerous hemidesmosomes.

The suckers of specimens of *Tem. novaezealandiae* studied were copiously provided with adhesive secretion from acidophilic parenchymal glands. The presence of a second gland type with smaller secretion bodies may mean the sucker is a duo-gland type organ (cf. Tyler, 1976).

Preliminary investigation of specimens of the scutariellid *Troglocaridicola mrazeki* showed they had a similar syncytial epidermis penetrated by necks of glands and by neurons ending in uniciliated receptors. The basement membrane in these was well developed. Microvilli on the epidermis were disposed in rows, and from their tips glycocalyx filaments radiated.

Parenchyma

At least six cell types in the parenchyma – including gland cells (Fig. 2), muscle cells (Fig. 3), presumed neurosecretory cells (see Williams, 1984a), and cells similar to the neurosecretory cells but lacking secretion masses (Fig. 4) – were penetrated by intrusive cell processes from parenchymal cells that presumably have a supportive or transport function; these processes penetrated into the recipient cells as far as the perinuclear zone where they terminated at cup-shaped or canoe-shaped desmosomes.

Rhabditogen cells, also penetrated by cell processes (Fig. 5, d), bore lamellated rhabdites (as defined by Smith *et al.*, 1982). In what appeared to be early stages of development, the rhabdites were composed of a relatively lucid homogeneous material (Fig. 5, Pr) and ensheathed by microtubules.

Intestine

The intestinal lumen in specimens of both *Trog. mrazeki* and *S. georgica* contained bacteria, and bacteria in various stages of degradation were enclosed in giant lysosomes in the intestinal epithelium (Fig. 6).

Protonephridial system

The flame cell, or cyrtocyte, in *Tem. novaezealandiae* has 7–24 tightly packed cilia surrounded fence-like by a weir (Williams, 1981a; cf. *Macrostomum* in Rieger, 1981).

Spermatogenesis and sperm morphology

The spermatozoon of *Tem. novaezealandiae* is much modified into a thread-like cell with elongated nucleus and mid-piece (Williams, 1983, 1984b) and with a subplasmalemmal sheath of microtubules; and like most other platyhelminth spermatozoa it has no acrosome. The two flagella of the spermatozoon are of the 9+1 type with a helically banded central unit and are motile and free throughout development. In spermatids (Fig. 7), there is a row of oval dense bodies with paracrystalline centers, and the subplasmalemmal microtubules form a spiral array in places.

In the scutariellid *Trog. mrazeki*, the spermatozoa were thread-like and with two free flagella. The developing flagella appeared to grow at right angles to the shaft from the centrioles, between which lay a prominent lamellar body. The flagella, with a single banded central unit in the axonemes (i.e. typical 9+1 configuration), were free throughout their length in all developmental stages. Spermatids had a subplasmalemmal microtubular sheath internal to which lay rows of glycogen granules.

Fig. 5–7. (5) *Temnocephala novaezealandiae:* Rhabditogen cell. Micrograph taken by M. Ingerfeld. (6) *Troglocaridicola mrazeki:* Intestine, showing bacteria in lumen. Bacteria are enclosed in a lysosome in the intestinal epithelium, lower left. (7) *Temnocephala novaezealandiae:* Transverse sections of advanced spermatids through mid-piece, showing peripheral microtubules in spiral array. The innermost microtubules are adjacent to dense bodies. cp, cell process terminating at desmosome; d, desmosome; m_1, mitochondrion of rhabdite cell; m_2, mitochondrion in cell process; mt, microtubules; N, nucleus; Pr, protorhabdite, with striated cortex; R, rhabdite, showing lamellar structure and radial striations.

Discussion

Relationship between the Scutariellidae and the Temnocephalidae

Results of the investigations of Mrazek (1907) and Matjašič (1958) suggest that a difference in nutritional physiology distinguishes temnocephalids from many scutariellids, in particular European species such as *Scutariella didactyla* and members of the genera *Troglocaridicola* and *Subtelsonia*. These scutariellids reportedly imbibe haemolymph from their hosts, whereas temnocephalids do not feed on host tissues or body fluids. My finding bacteria in the intestine of *Trog. mrazeki* and *S. georgica*, however, indicates that at least these species feed on prey from the environment as do temnocephalids.

The ultrastructure of the epidermis, as well as major features of spermatogenesis, appear basically similar in *Trog. mrazeki* and *Tem. novaezealandiae*. Indeed, although there are significant differences between scutariellids and temnocephalids (Williams, 1981b), these similarities are consistent with the inclusion of the Scutariellidae in the Temnocephaloidea.

Phylogenetic position of the Temnocephaloidea

The epidermis of temnocephaloideans, with its thick basement membrane and syncytiality, is quite like that of the neoophoran turbellarians (see Rieger, 1981). However, it is specialized – and thus stands distinct from that of neoophorans and other turbellarians as well – in its complete lack of locomotory cilia. Also, as is characteristic of neoophorans and some other turbellarians, temnocephalids have duo-gland adhesive organs (cf. Tyler, 1976; Rieger, 1981); and true rhabdites are common to the Temnocephalidae and the turbellarian orders Proseriata, Tricladida, Rhabdocoela, Polycladida, and Macrostomida (Smith *et al.*, 1982).

Free sperm flagella with a 9+1 axonemal pattern as are found in temnocephaloideans occur also in the neoophoran orders Proseriata, Tricladida, and Rhabdocoela, and in the Polycladida (Hendelberg, 1977). Spermatozoa of *Tem. novaezealandiae* and *Trog. mrazeki* are less modified than those of certain turbellarians in that the flagella remain free during spermatogenesis. Dense bodies like those in the spermatids of *Tem. novaezealandiae* are found in spermatids of many turbellarians (cf. Henley, 1974; Hendelberg, 1977).

Affinity of the Temnocephaloideae with neoophoran turbellarians, especially the dalyellioid rhabdocoels, is also seen in light microscopic characters, namely the possession of a doliiform pharynx and the organization of the female genital complex (the common genital atrium, rather than any part of the female system alone, functions as an ootype in *Tem. novaezealandiae* – see Williams, 1981b). The habitat of dalyelliids is like that of the temnocephaloideans in that they are mostly confined to freshwater. The didymorchid dalyellioid turbellarians appear to have especially close ties with the Temnocephaloidea in that, like temnocephaloideans, they are commensal on freshwater crayfish hosts and possess a posterior attachment organ.

On the other hand, parenchymal ultrastructure in temnocephaloideans is more like that of the parasitic platyhelminth classes in the compartmentalization of cells by membranes. Compartmentalization like that found in *Tem. novaezealandiae* is reported for Monogenea (Rohde, 1974: 10; Halton *et al.*, 1974; Stranock & Halton, 1975) and Digenea (Threadgold & Irwin, 1970), but not for the Turbellaria (see Rieger, 1981).

The Temnocephaloidea is evidently a phylogenetic offshoot from early rhabdocoel-turbellarian-like forms which were characterized by a relatively unmodified epidermis without sunken nuclei or enhanced intermicrovillar glycocalyx, spermatozoa with free paired 9+1 flagella, true rhabdites, and a protonephridial system with a weir composed of a single series of processes.

Acknowledgments

I thank Mr Manfred Ingerfeld of the Department of Botany, University of Canterbury, for expert technical assistance and Mr T. P. Williams and Mrs Joan Buckley of the Department of Zoology, University of Canterbury, for assistance with photography. Scutariellid material was generously supplied by Dr B. I. Joffe of the Zoological Institute, Academy of Sciences, Leningrad, USSR.

References

Baer, J.-G., 1953. Temnocéphales. Zoological results of the Dutch New Guinea expedition 1939. No. 4. Zool. Meded., Leiden. 32: 119–139.

Baer, J.-G., 1961. Classe des Temnocéphales. In P.-P. Grassé (ed.), Traité de Zoologie, Tome 3 (Plathelminthes). Masson, Paris: 213–241.

Bresslau, E. & E. Reisinger, 1933. Temnocephalida. In W. Kuekenthal & T. Krumbach (eds), Handbuch der Zoologie. 2: 294–309.

Brooks, D. R., 1967. Higher-level classification of parasitic platyhelminths and fundamentals of cestode classification. In D. F. Mettrick & S. S. Desser (eds), Parasites, their world and ours. Elsevier Biomedical Press, New York: 189–193.

Ehlers, U., 1985. Comments on a phylogenetic system of the Platyhelminthes. This volume.

Halton, D. W., G. P. Morris & A. Hardcastle, 1974. Gland cells associated with the alimentary tract of a monogenean, Diclidophora merlangi. Int. J. Parasitol. 4: 589–599.

Hendelberg, J., 1977. Comparative morphology of turbellarian spermatozoa studied by electron microscopy. Acta. Zool. fenn. 154: 149–162.

Henley, C., 1974. Platyhelminthes (Turbellaria). In A. C. Giese & J. S. Pearse (eds), Reproduction of marine invertebrates. Academic Press, N.Y.: 267–343.

Joffe, B. I., 1981. Morphology and phylogenetic relations of the Temnocephalida (Turbellaria). Zoological Journal, Moscow 60: 661–672.

Matjašič, J., 1958. Morfologija, biologija in zoogeografija evropskih Temnocefalov in njihov sistematski položaj. Razreda Slovenske Akademije Znanosti im Umetnosti 5: 141–181.

Mrazek, A., 1907. Ein Europaïscher Vertreter der Gruppe Temnocephaloidea. Sitzungsberichte der Königlich Bohmischen Gesellschaft der Wissenschaften, Prague 11: 1–7.

Nichols, K. C., 1975. Observations on lesser-known flatworms: Temnocephala. Int. J. Parasitol. 5: 245–252.

Rieger, R. M., 1981. Morphology of the Turbellaria at the ultrastructural level. Hydrobiologia 84: 213–229.

Rohde, K., 1974. Light- and electron-microscopic studies of the pharynx and the anterior and posterior glands of Polystomoides (Monogenea: Polystomatidae). Zool. Jb Anat. 92: 1–17.

Smith, J., S. Tyler, M. B. Thomas & R. M. Rieger, 1982. Morphology of turbellarian rhabdites: phylogenetic implications. Trans. am. microsc. Soc. 101: 209–228.

Stranock, S. D. & D. W. Halton, 1975. Ultrastructural observations on Mehlis' gland in the monogeneans, Diplozoon paradoxum and Calicotyle kröyeri. Int. J. Parasitol. 5: 541–550.

Threadgold, L. T. & S. W. B. Irwin, 1970. Electron microscope studies of Fasciola hepatica. IX. Fine structure of Mehlis' gland. Z. Parasitenkde 35: 16–30.

Tyler, S., 1976. Comparative ultrastructure of adhesive systems in the Turbellaria. Zoomorphologie 84: 1–76.

Tyler, S., 1984. Turbellarian platyhelminths. In J. Bereiter-Hahn, A. G. Matoltsy & K. S. Richards (eds), Biology of the integument, Vol. 1, Invertebrates. Springer-Verlag, Berlin, Heidelberg: 112–131.

Williams, J. B., 1975. Studies on the epidermis of Temnocephala I. Ultrastructure of the epidermis of Temnocephala novaezealandiae. Aust. J. Zool. 23: 321–331.

Williams, J. B., 1979. Studies on the epidermis of Temnocephala IV. Observations on the ultrastructure of the epidermis of Temnocephala dendyi, with notes on the role of the Golgi apparatus in autolysis. Aust. J. Zool. 27: 483–499.

Williams, J. B., 1980a. Studies on the epidermis of Temnocephala V. Further observations on the ultrastructure of the epidermis of Temnocephala novaezealandiae, including notes on the glycocalyx. Aust. J. Zool. 28: 43–57.

Williams, J. B., 1980b. Morphology of a species of Temnocephala (Platyhelminthes) ectocommensal on the isopod Phreatoicopsis terricola. J. nat. Hist., Lond. 14: 183–199.

Williams, J. B., 1981a. Protonephridial system of Temnocephala novaezealandiae: structure on the flame cells and main vessels. Aust. J. Zool. 29: 131–146.

Williams, J. B., 1981b. Classification of the Temnocephaloidea (Platyhelminthes). J. nat. Hist., Lond. 15: 277–299.

Williams, J. B., 1982. Studies on the epidermis of Temnocephala VI. Epidermal topography of T. novaezealandiae and other Australasian temnocephalids, with notes on microvillar and coated vesicle function and the evolution of a cuticle. Aust. J. Zool. 30: 375–390.

Williams, J. B., 1983. The genital systems of Temnocephala I. Ultrastructural features of the differentiating spermatid of Temnocephala novaezealandiae, including notes on a possible correlation between cellular autophagy and mitochondrial function. Aust. J. Zool. 31: 317–331.

Williams, J. B., 1984a. Cells in the parenchyma of Temnocephala I. A large secretory cell, with a 'honeycomb' nucleus surrounded by juxtanuclear desmosomes: a new cell type in Temnocephala novaezealandiae (Temnocephalidae: Platyhelminthes). Aust. J. Zool. 32: 207–218.

Williams, J. B., 1984b. The genital system of Temnocephala II. Further observations on the spermatogenesis of Temnocephala novaezealandiae, with particular reference to the mitochondria. Aust. J. Zool. 32: 447–461.

Williams, J. B., in press. Studies on the epidermis of Temnocephala VII. The basal region, with comments on 'membrane flow'. Aust. J. Zool.

An ultrastructural and cytochemical study of the digestive system in *Oxyposthia praedator* (Turbellaria, Acoela)

T. G. Markosova
Laboratory of Evolutionary Morphology, Zoological Institute of Academy of Sciences, Leningrad, 199034, USSR

Keywords: Turbellaria, Acoela, ultrastructure, digestive system

Abstract

The central parenchyma in *Oxyposthia praedator* consists of multifunctional cells. These cells digest food material intracellularly, can effect extracellular digestion through release of digestive enzymes by cell lysis and clasmatosis, and synthesize reserve nutritional substances.

Introduction

Oxyposthia praedator is one of the most numerous of acoel turbellarians in the phytal zone of Far Eastern seas (Ivanov, 1952; Mamkaev, 1971). Members of this species are active predators, attacking only moving prey. Their feeding is characterized by a peculiar rush to seize the prey. They have no specialized prey-catching organs but use the whole body as a capturing apparatus (see Ivanov, 1952; Fig. 16). Prey capture is accompanied by discharge from dermal glands of secretion that attacks and destroys the prey. A drop of water containing dermal secretion kills other turbellarians in 1–2 min and destroys them in 5–10 min (Mamkaev & Markosova, 1979b).

Material and methods

Oxyposthia praedator was collected in Zaliv Petra Velikogo (Gulf of Peter the Great), fixed in 2.5% glutaraldehyde in 0.2 M Millonig buffer, and postfixed in 1% OsO_4 in the same buffer. Acid phosphatase was demonstrated by the method of Novikoff *et al.* (1971).

Results and discussion

Digestive cells of the central parenchyma arise from undifferentiated cells of the peripheral parenchyma (Mamkaev & Markosova, 1979a, b). As the cells migrate into the central parenchyma, they enlarge significantly and form numerous irregular processes that intertwine and protrude into fragments of food material.

The central parenchyma consists of multifunctional cells in which intracellular digestion and biosynthesis of nutrient reserves occur (Markosova, 1982). Electron micrographs with evidence of phagocytosis of food particles by cell processes, formation of digestive vacuoles, and accumulation of such vacuoles in the cytoplasm were observed in a number of instances. The size of the digestive vacuoles was in the range of 2–3 μm. They contained heterogeneous material in different stages of digestion, i.e. undamaged or partly lysed food fragments as well as small vesicles of various sizes containing fibrous structures. Some vesicles contained flake-like material and osmiophilic particles. Not infrequently myelin-like structures, presumably residual bodies resulting from incomplete digestion of lipids and membranous structures, could be observed. Besides digestive vacuoles, the cytoplasm contained glycogen inclusions and lipid droplets.

Hydrobiologia 132, 69–70 (1986).

Multivesicular bodies containing both smooth and coated vesicles were numerous. These multivesicular bodies were situated mainly near digestive vacuoles, and vesicles similar to those in them were often found inside the digestive vacuoles. Acid phosphatase was detected in microvesicles of Golgi complexes, in primary lysosomes, and also in many vesicles of the multivesicular bodies. All of these vesicles, by fusing with digestive vacuoles apparently take part in hydrolysis of food through formation of heterolysosomes.

Nuclei of the digestive cells were spherical (3 – 5 μm in diameter) with one compact nucleolus in each and heterochromatin clumps condensed along the periphery as well as scattered throughout the nucleus. Endoplasmic reticulum appeared as narrow rough and smooth cisternae; and there were several Golgi complexes, each consisting of five to seven closely stacked flat cisternae, a mass of vesicles, and some coated and large vacuoles. The numerous spherical and dumbbell-shaped mitochondria had an electron-dense matrix and irregular cristae.

Electron microscopy revealed what I interpret to be degeneration of the digestive cells once those cells had completed their digestive activity: increased density in the mitochondrial matrix and lysis of mitochondrial cristae, fragmentation of their endoplasmic reticulum, hypertrophy of the Golgi complex and vacuolisation of the cytoplasm. Additional evidence of this disintegration and cell lysis could be seen in a free distribution of acid phosphatase activity in the cytoplasm. The degeneration apparently ends in a rupture of the cell membrane and release of separate vesicles; after that, large parts of the cell cytoplasm become detached by clasmatosis. Where the digestive cells had disintegrated, cavities formed, and these joined into large lacunae when food was ingested. That the formation of these cavities was due to cell degeneration is demonstrated by the fact that not one but two membranes could were present between the cavities and bordering cells. One of the membranes must have belonged to a normal digestive cell, the other to a degenerated one.

Bounding the central cavities between digestive cells was a thin layer of cytoplasm where nuclei were not separated from one another by plasma membranes. This syncytium contained disintegrated cell material and apparently arose through destruction of membranes of degenerating cells. Thus, the syncytium of *O. praedator* appears to be one of the last stages in the irreversible degeneration of digestive cells.

Acid phosphatase could be demonstrated extracellularly in the cavities and also in vesicles and other structures released from degenerating cells. In light of the fact that *O. praedator* lacks specialized digestive glandular cells, one can appreciate the biological significance of this process of cell degeneration. Digestive enzymes must be released from the cells to be used in extracellular digestion. Evidently, acoel turbellarians developed this mechanism as an adaptation to feeding on comparatively large prey, prey that would be intractable to normal phagocytosis. Enzymes released from the degenerating cells must be responsible for the initial treatment of food, making it ready for intracellular digestion.

The peculiarities *O. praedator* shows for handling and digesting large prey – namely, prey capture by secretions from dermal glands and by powerful muscles and extracellular digestion in cavities which appear through autolysis of cells – are original adaptations, I believe, whose basis can be seen in the extremely primitive organization of the acoel turbellarians. I find untenable, therefore, the concept that the Acoela is secondarily simplified.

References

Ivanov, A. V., 1952. Beskishechye turbellyarii (Acoela) yuzhnogo poberezh'ya Sakhalina. In: Trudy zool. Inst. AN SSSR. Moskva, Leningrad, 12: 40 – 132.

Mamkaev, Y. V., 1971. O konvolyutakh (*Tubellaria, Acoela*) yuzhnogo Primor'ya. In: Issledovanie fauny morej. Fauna i flora zal. Pos'yeta Yaponskogo mor'ya. 8(16): 31 – 54.

Mamkaev, Y. V. & T. G. Markosova, 1979. Elektronnomikroskopicheskoe issledovanie parenkhimy predstavitelej beskishechnykh turbellyarij. In: Evolyutsionnaya morphplogiya bespozvonochnykh. Leningrad, Nauka: 7 – 12.

Mamkaev, Y. V. & T. G. Markosova, 1979. Some peculiarities of nutrition and digestion of *Oxyposthia praedator* (*Turbellaria, Acoela*). XIV. Pacific Science Congress, USSR, Khabarovak. Moskva: 90 – 91.

Markosova, T. G., 1982. Ultrastrukturnye osobennosti pishchevaritel'noj parenkhimy beskishechnoj turbellarii *Oxyposthia praedator*. Dokl. Akad. Nauk SSSR 267: 950 – 952.

Novikoff, P. M., A. B. Novikoff, N. Quintana & J.-J. Hauw, 1971. Golgi apparatus, Gerl, and lysosomes of neurons in rat dorsal root ganglia, studied by thick section and thin section cytochemistry. J. Cell Biol. 50: 859 – 886.

Frontal organs in the Acoelomorpha (Turbellaria): Ultrastructure and phylogenetic significance

Julian P. S. Smith, III & Seth Tyler
Department of Zoology, University of Maine, Orono, ME 04469 USA

Keywords: Turbellaria, Acoela, Nemertodermatida, ultrastructure, mucous gland, sensory

Abstract

Using characters discernible through electron microscopy, we redefine the organ traditionally identified as the frontal organ in acoelomorph turbellarians as being a collection of two to several large mucus-secreting glands whose necks emerge together through a frontal pore at the exact apical pole of the body, i.e. at the point where the pattern of epidermal ciliary rootlets converges. Representatives that we have studied of each of the acoel families Paratomellidae, Diopisthoporidae, Solenofilomorphidae, Convolutidae, Otocelidae, and Mecynostomidae, as well as a representative of the Nemertodermatida, have such glands. Up to five additional types of glands that open anteriorly outside of the frontal pore, some of which are indistinguishable from glands of the general body wall, could be seen in the nemertodermatid, in *Hesiolicium inops* (Paratomellidae), and in representatives of the latter four acoel families. In *Paratomella,* three different types of glands open in diffuse fashion in a frontal glandular complex reminiscent of that in the Macrostomida.

Sensory elements near the frontal pore appear to be independent of the gland necks, and so the organ cannot be considered a sensory organ.

The frontal organ, as described above, appears very likely to be homologous within the Acoelomorpha, and represents another strong (although unrooted) autapomorphy for this line of turbellarian evolution.

Introduction

The order Acoela is a large and diverse group of turbellarians for which no phylogenetic system is available. The comprehensive systematic treatment of the order by Dörjes (1968) and a subsequent emendation (Faubel, 1976) have produced 15 families defined mostly by characters of the genital organs (see Bush, 1981, for summary). These organs are quite complex and show a surprising diversity among the families; it is this very diversity that hinders phylogenetic analysis within the Acoela (*cf.* Antonius, 1968). In a search for characters that might prove more amenable to phylogenetic analysis, we have investigated the ultrastructure of a more conservative organ system characteristic of the Acoela, namely the frontal organ. This organ system has long been recognized by light microscopists (von Graff, 1891; Westblad, 1948) as a group of glands (frontal glands) whose necks reach the anterior tip of the animal through a frontal pore. By using transmission electron microscopy, we have been able to define more accurately the organ and have found that its characters prove useful in identifying major phylogenetic trends in the order and, moreover, provide yet another apomorphy for the Acoelomorpha (orders Nemertodermatida and Acoela) (see also Smith & Tyler, 1985a; Smith, *et al.,* 1985).

Materials and methods

Seven species representing six families of the Acoela and one species of the Nemertodermatida

Hydrobiologia 132, 71–78 (1986).

have been studied (Table 1). Methods for light microscopy and transmission electron microscopy are given in Smith & Tyler (1984).

Results

Glands

With the exception of *Paratomella* sp., all of the acoel species we have studied, as well as the nemertodermatid *Flagellophora* sp., had a group of anteriorly positioned mucous glands whose necks projected to a pore (frontal pore) at the apical pole of the animal's body, i.e., the point toward which the rostrally-oriented ciliary rootlets are all directed (Fig. 2). Secretion granules in these glands had a typically mucoid fibrillar ultrastructure and stained (except in *Diopisthoporus* 'gymnopharyngeus') with alcian blue. The simplest form of such a frontal organ was seen in *D.* 'gymnopharyngeus', in which there were but two such mucoid glands with their bodies situated against the body wall at the level of a nerve ring anterior to the statocyst (Figs. 1, 3). *Solenofilomorpha funilis* had a cluster of 6–8 such glands whose bodies lay farther posterior, lateral to the pharynx (and are equivalent, presumably, to the 'parenchymal strands' mentioned by Crezee, 1975); in *Hesiolicium inops* the cell bodies lay just posterior to the statocyst. *Otocelis* sp., *Paedomecynostomum* sp. (and also *Convoluta* pulchra' – see Klauser *et al.*, 1985) had more elaborate, large mucoid glands of two or more types that were distinguishable by the ultrastructure and differences in the histochemical staining of their granules. These glands virtually filled the anterior third of the body of *Paedomecynostomum* sp., one type extending back to just posterior

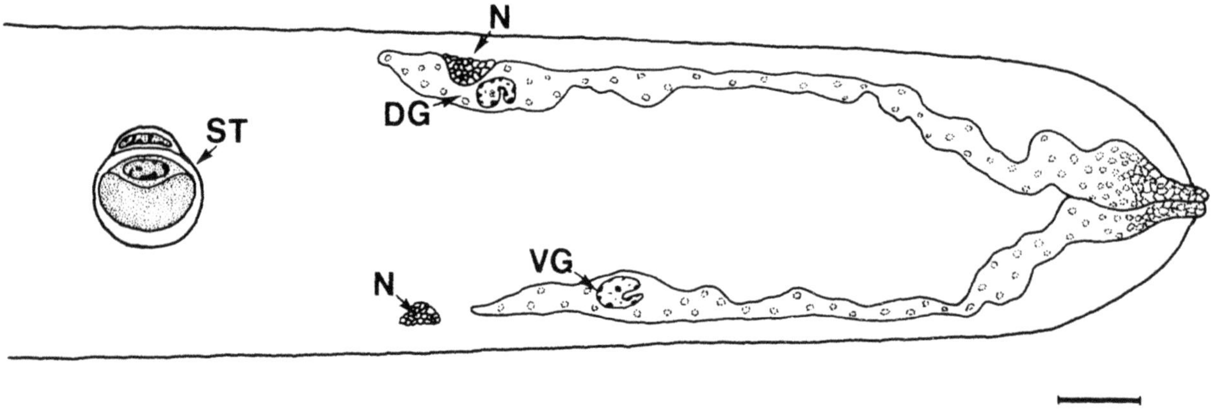

Fig. 1. Diopisthoporus 'gymnopharyngeus'. Sagittal reconstruction of the anterior end, based on sagittal serial thin sections, showing dorsally- and ventrally-located frontal gland cells (DG, VG), anterior nerve ring (N), and statocyst (ST). Scale = 10 μm (Modified after Smith & Tyler, 1985b).

Table 1. Systematic Placement of Animals Studied.

Specimen	Family	Reference or Locality
Acoela		
Diopisthoporus 'gymnopharyngeus'	Diopisthoporidae	Smith & Tyler, 1985b
Otocelis sp.	Otocelidae	Crow Neck, ME, subtidal
Paedomecynostomum sp.	Mecynostomidae	Smith, 1981
Convoluta 'pulchra'	Convolutidae	Smith & Bush, in prep.
Hesiolicium inops	Paratomellidae	Bogue Banks, NC, subtidal
Paratomella sp.	Paratomellidae	Klose, 1984
Solenofilomorpha funilis	Solenofilomorphidae	Duffy & Tyler, 1984
Nemertodermatida		
Flagellophora cf apelti		Tyler & Rieger, 1977

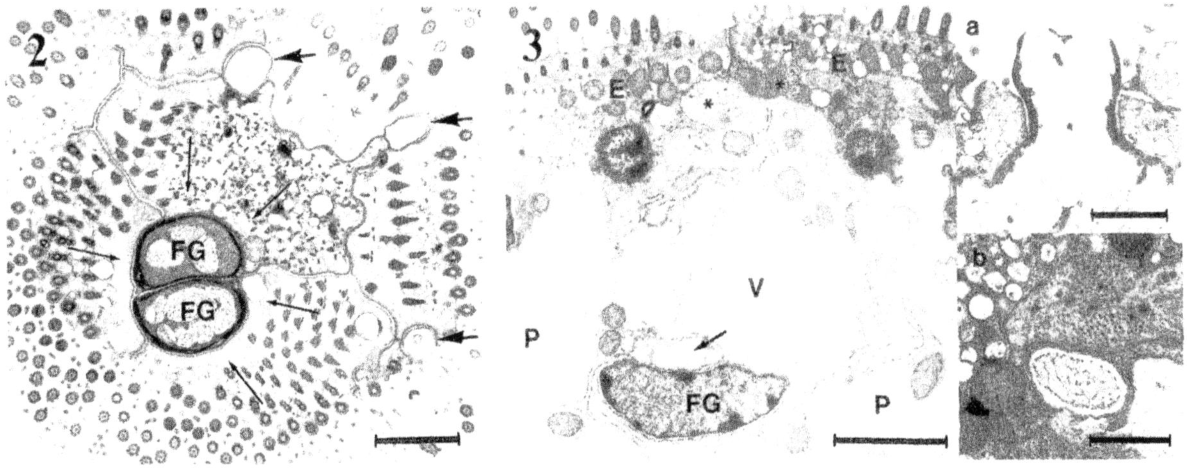

Figs. 2–3: (2) Diopisthoporus 'gymnopharyngeus.' Frontal pore in transverse section showing dorsally- and ventrally-located frontal gland necks (FG). Note that ciliary rootlets in the surrounding epidermal cells are all oriented toward the frontal pore (small arrows). Monociliated sensory cells (large arrows) open between epidermal cells outside of the frontal pore. Scale = 2 μm; (3) *D.* 'gymnopharyngeus.' Frontal gland cell (FG marks nucleus) adjacent to cells of the body wall: epidermal (E) and body-wall muscle cells (asterisks). Vacuole (V) and mature granules are visible (arrow). Parenchymal cells (P) bound frontal gland cell toward the inside of the body. Scale = 3 μm. *Insets* a, b: Body wall gland (a) and frontal gland (b) secretory granules. Scales = 1 μm.

to the statocyst and the other extending back nearly to the mouth. In *Otocelis* sp., there were approximately 38 gland necks emerging through the frontal pore; those of one type (more intensely alcian-blue positive) occupied the ventral portion of the frontal pore and so were bordered dorsally and intermixed laterally with necks of the second type (Figs. 4, 5, 6). The necks of both types passed through an anterior ring-like portion of the brain (Fig. 8).

The more elaborate frontal organs were also accompanied by a number of accessory frontal glands whose necks emerge at the anterior end of the body but outside the frontal pore. *Otocelis* sp., for instance, had an arcing array of gland necks ventral to the frontal pore with ellipsoidal electron-lucent secretion granules (Figs. 4, 7, 9), as well as scattered alcian-blue positive rhabdoid glands (Fig. 8), scattered collared mucous glands (Figs. 11, 12; probably adhesive glands – see Smith & Tyler, 1985a), and three other gland types (Figs. 10, 13, 14), all apparently comparable to the accessory frontal glands described for *Convoluta* sp. (Klauser *et al.,* 1985). *Paedomecynostomum* sp. could be seen by light microscopy to have alcian-blue-positive rhabdoid gland necks ventral to the frontal pore; *Hesiolicium inops* also had such rhabdoids (see Crezee & Tyler, 1976). In the nemertodermatid *Flagellophora,* the 'broom organ' or 'flagella-bearing organ' (see Sterrer, 1965; Faubel & Dorjes, 1978) as well as a group of ellipsoid-secretion accessory frontal glands both opened ventral to the frontal pore.

Paratomella sp., an undescribed species with asexual reproduction by paratomy, had frontal glands distinctly different from the other acoels we have examined. It had no mucoid glands comparable to those of the apical pole in other acoels but only scattered gland openings with three morphologically distinguishable gland types emerging in a belt across the anterior tip of the body (Figs. 15–18). The point of convergence of the ciliary rootlets (position of frontal pore of other acoels) was occupied by one neck of each of the three gland types; an organized frontal pore as seen in the other acoels studied was not present (Fig. 18).

Sensory cells

The anterior tip of the body in all species we have studied had many receptors (Figs. 4, 15; Klauser *et al.,* 1985, their Fig. 1; Smith & Tyler, 1985b, their Fig. 41). *Diopisthoporus* 'gymnopharyngeus,' for example, had nine types of monociliated sensory cells (Smith, unpublished) but none of these in either this or the other species was associated directly with the frontal organ and none was, at least in *Di-*

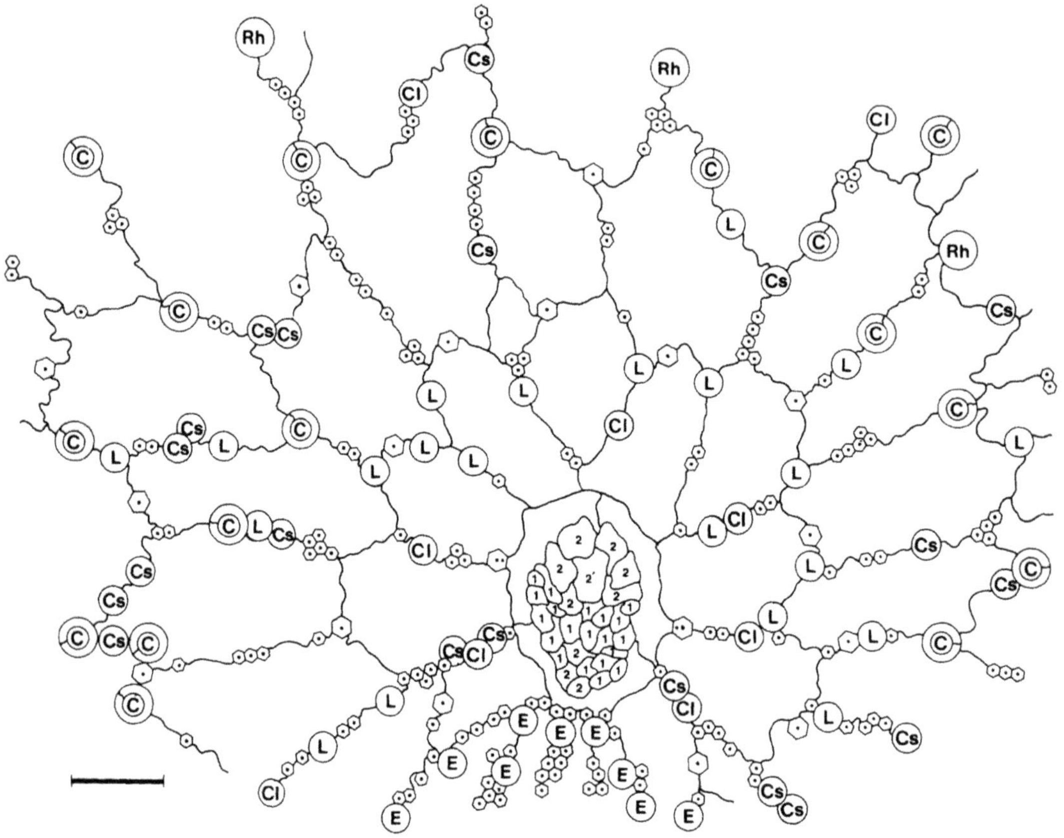

Fig. 4. Otocelis sp. Schematic map of the epidermal surface at the anterior end, reconstructed from serial transverse thin sections, showing monociliated sensory receptors (hexagons), and gland openings. (Circles: Cs = small colored granules, Cl = large colored granules, Rh = rhabdoids, C = collared gland cells, L = electron-lucent mucoid granules, E = ellipsoid granules.) Number of gland necks in frontal pore reduced for clarity; 'free nerve endings' not shown. Scale = 5 μm.

opisthoporus 'gymnopharyngeus', unique to the anterior tip, i.e., not found elsewhere on the body as well (excepting a putative photoreceptor perhaps common to platyhelminths in general – see Smith & Tyler, in 1985b; Smith, in prep.). Sensory tufts in *Paratomella* sp. were composed of clusters of receptors, each bearing one to four cilia (Figs. 19, 20) while the other species had only monociliated receptors. The only nerves associated with the frontal glands themselves were efferent (thus not sensory) nerves (see Smith & Tyler, 1985b).

Fig. 5. Otocelis sp. TEM micrograph of frontal pore in transverse section. Note two distinct types of gland necks-type 1 (= strongly alcian-blue-positive glands) and type 2 (= weakly alcian-blue-positive glands), and a third neck (2*), possibly intermediate between the two. (Compare with Fig. 6). Scale = 2 μm.

Figs. 6–8. Otocelis sp. Light microscopy of the frontal organ in transverse section. Two-μm-thick resin sections, alcian blue/iron hemotoxylin stain. Photographed with deep red filter; scale for all figures = 10 μm. (6) Frontal pore; compare gland neck distribution with Fig. 5; (7) Sagittal section of anterior end. Note cell bodies of frontal gland cells (FG), 'Ampule' (A) behind frontal pore, with darkly-staining (AB++) ventral gland necks and less-darkly-staining dorsal gland necks (comp. Figs. 5, 6). Ellipsoid gland cell bodies (E) and gland neck opening ventrally to frontal pore (arrows) are also visible. ST = statocyst, CS = central syncytium. (8) Cross-section at the level of anterior ring of neuropile (N). Note bundle of frontal gland necks passing through neuropile (arrow). Rh = rhabdoid gland neck.

Figs. 9–14. Otocelis sp. Accessory frontal glands. (9) Ellipsoids; Fig. 10: Electron-lucent mucoid granules; (11–12) Two sections through a collared gland cell opening and its ciliary collar (anterior is toward top of figs.); (13) Small colored granules; (14) Large colored granules. Scales: Figs. 9, 10, 13, 14; scale = 1 μm. Figs. 11, 12, scale = 0.4 μm.

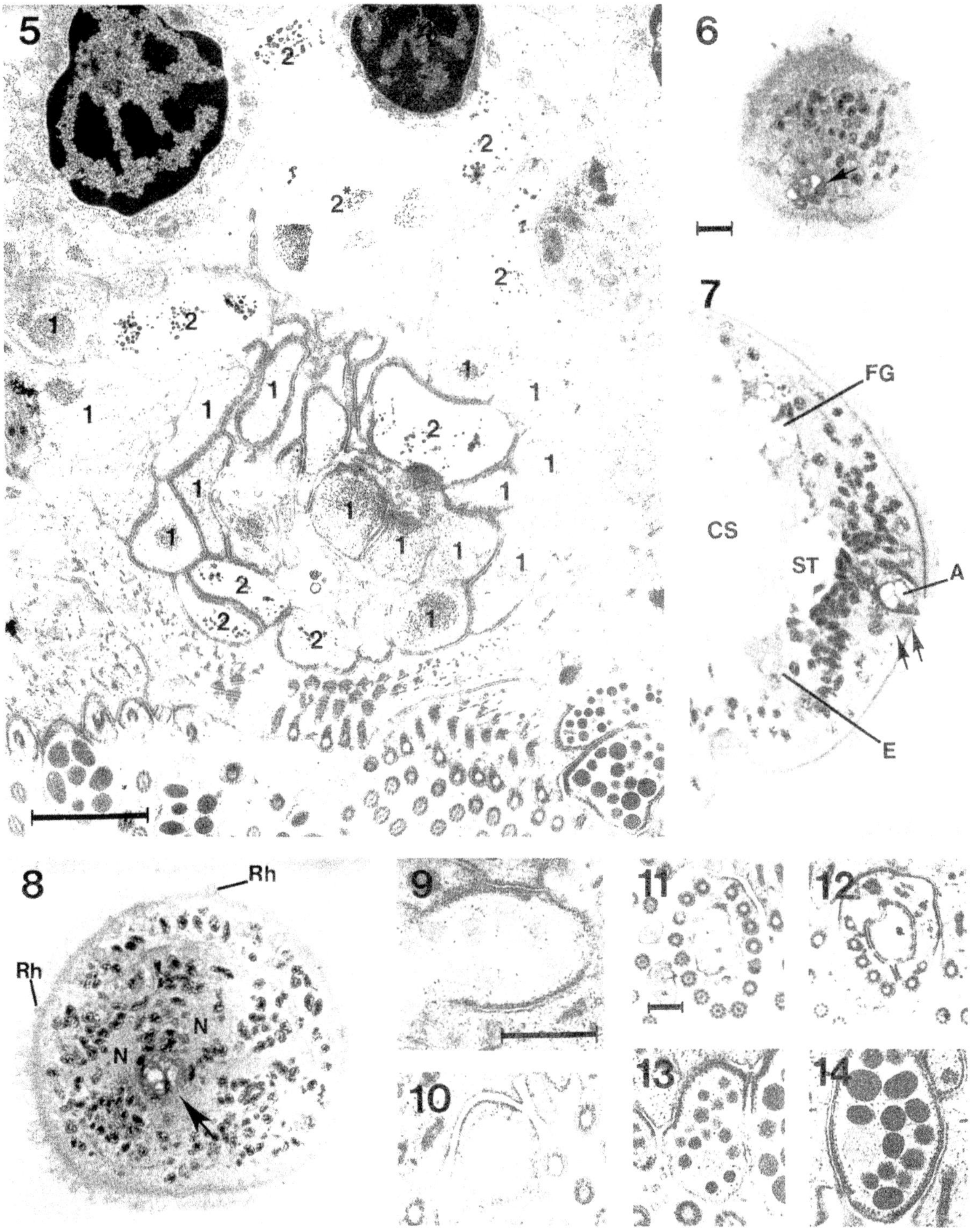
5
2
2
2
2
1
2
1
1
1
1
2
1
1
1
1
1
1
1
1
2
2
2
1
6
7
FG
CS
ST
A
E
8
Rh
Rh
N
N
9
10
11
12
13
14

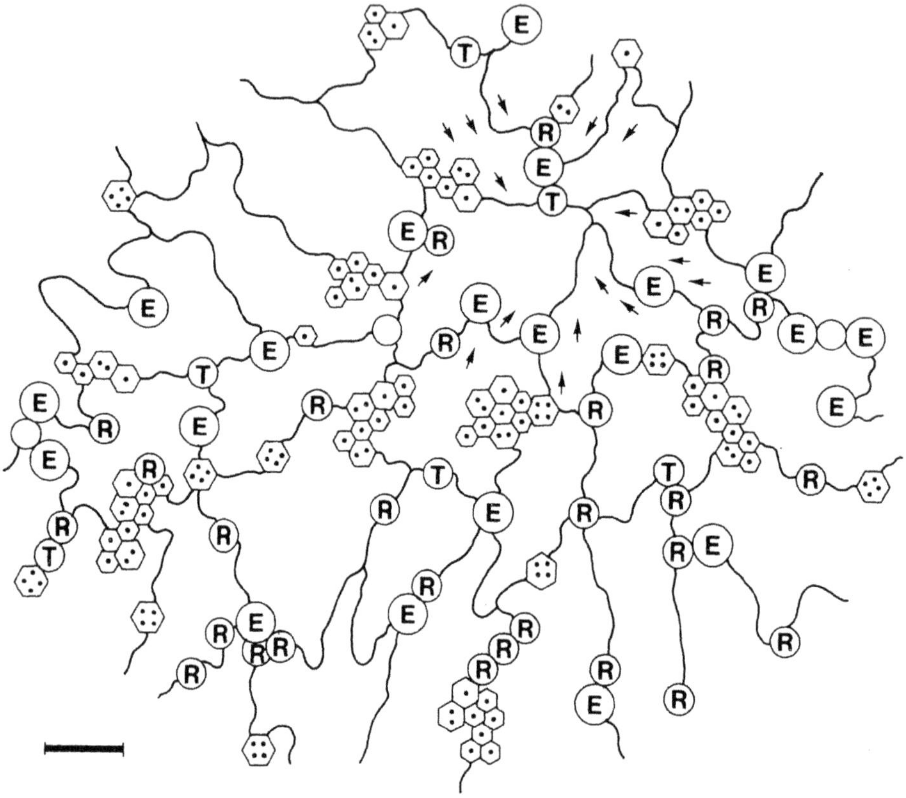

Fig. 15. Paratomella sp. Schematic map of the epidermal surface of the anterior tip, reconstructed from serial transverse thin sections, showing mono-, bi-, and tetraciliated sensory cells (hexagons), and gland cell openings (circles: T = target granules, E = ellipsoids, R = rhabdoids). Arrows indicate direction of principal ciliary rootlet in region of 'frontal pore.' Scale = 2 μm.

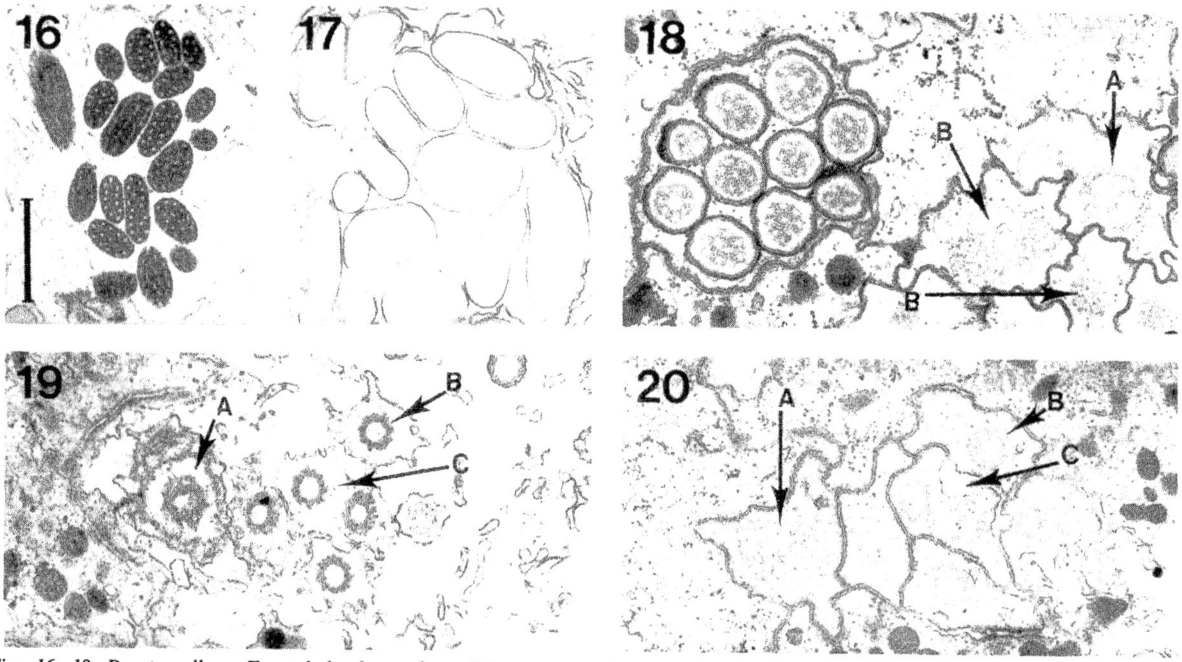

Figs. 16–18. Paratomella sp. Frontal gland secretions. (16) target granules; (17) Ellipsoids; (18) cross-sectioned rhabdoids and dendrites of two types of monociliated sensory cells (A, B). *Figs. 19, 20. Paratomella* sp. A sensory cell cluster at two levels showing two types of monociliated sensory cells (A, B) and a biciliated sensory cell (C). Ciliary rootlets of these cells are associated with mitochondria (Fig. 18, A, B). Scale = 0.5 μm for all figures.

Discussion

Acoelomorph turbellarians have a variety of frontal gland types that open at the anterior end of the body, any or all of which might be considered to constitute a frontal organ. Particularly characteristic of these lower turbellarians, however, are those mucous glands whose necks emerge at the apical pole through a pore that marks the point of convergence of the ciliary rootlet pattern of the epidermal cells, and at least some of whose necks project through a ringlike portion of the nervous system, either through the brain itself or through a more anterior nerve ring. We believe the term 'frontal organ' should refer only to these glands and that 'frontal pore' should be used only for the uniquely positioned pore through which these necks emerge; thus these terms can be considered to designate homologues, at least within the group Acoelomorpha. Other glands at the anterior tip may be called 'frontal glands' but we consider them accessory to the frontal organ proper. While some of these glands in some species are found only frontally, others are often of the same type as glands emerging at other sites on the body wall (e.g. rhabdoid and adhesive glands).

The acoelomorph frontal organ is strictly a glandular organ and evidently is not sensory. Although a sensory role would be appropriate for an organ that opens at the exact point where currents generated by the epidermal cilia first impinge on the animal, the organ itself does not have sensory elements. None of the numerous sensory receptors occurring at the anterior body tip can be considered part of the frontal organ. A strictly secretory role for the frontal organ is explicable, however, in that the frontal organ secretions emerge at the position (i.e. the frontal pore) away from which all epidermal cilia beat; thus the organ could be providing mucus for ciliary gliding.

Paratomella sp. is the only acoel we have investigated that does not possess a true frontal organ but instead has a scattered arrangement of several 'accessory' glands. Such a condition may simply reflect the fact that this species (particularly the specimens we studied) reproduces asexually by paratomy. Whether a frontal organ, as we have defined it, is compatible with this form of asexual reproduction remains an open question. The point of discontinuity in the ciliary field is probably the final point of attachment between two zooids; because the posterior zooid differentiates frontal glands before separation (see Klose, 1984), these glands must open elsewhere. The frontal glands in paratomizing individuals constitute what we call a 'frontal gland complex.' It will be necessary to examine sexually produced, pre-paratomic, individuals to ascertain whether a frontal organ is actually lacking in the species. The other known (albeit tentatively placed) genus of the Paratomellidae, *Hesiolicium,* does have a true frontal organ, and has not yet been observed to reproduce by paratomy, only sexually.

Characters of the frontal organ may serve to reveal phylogenetic relationships of the acoelomorph families if one assumes that the more simple organs, such as those of *Diopisthoporus* 'gymnopharyngeus' and *Solenofilomorpha* sp., are primitive and that the more complicated organs of the convolutids and mecynostomids are more advanced. Such an assumption actually fits well with analyses of other organ systems in these acoels, particularly the parenchyma (see Smith, 1981; Smith & Tyler, 1985a). As with the case of parenchymal structure, frontal gland arrangement places *Paratomella* in a special position apart from the other acoels. Its frontal gland complex is actually more like that of the Macrostomida in morphology than that of other acoels, but its secretions are histochemically different from those of *Macrostomum* particularly in that they are negative to fast green (Klose, 1984). Furthermore, paratomellids are the only acoels found so far to have multiciliated sensory receptors (Figs. 19, 20; *Hesiolicium inops* caudal organ: Crezee & Tyler, 1976); such receptors are known otherwise only in the rhabditophoran turbellarians (Ehlers, 1984).

The similarity of frontal organ glands with other mucous glands of the body wall supports von Graff's (1891) hypothesis that the acoel frontal organ arose in the ancestral acoel as a gathering of such glands at the anterior tip of the body (see also Westblad, 1948; Antonius, 1970). The body-wall glands in *Diopisthoporus* 'gymnopharyngeus,' for example, have secretions that are histochemically similar to those of that species' frontal organ, but the secretions in the two gland types are still morphologically distinguishable at the EM level in all species we have studied. The fact that the frontal gland necks pass through the brain or another nerve ring indicates that the frontal organ originated very early in acoelomorph evolution – perhaps appearing first as mucous glands that projected through a simple basiepithelial nerve ring – and thus show it to be a primitive feature of the Acoelomorpha.

Acknowledgments

Thanks are due to Ms. Eddie Meisner for typing the manuscript, and to Mr. William C. Snow for printing the plates. We thank Mr. Robert Klose for sharing his unpublished data on *Paratomella.* This study was supported by NSF grant BSR 81-16894.

References

Antonius, A., 1968. Faunistische Studien am Roten Meer im Winter 61/62: Teil IV: Neue Convolutidae und eine Bearbeitung des Verwandschaftskreises Convoluta (Turb.: Acoela). Zool. Jb. Syst. Biol. 92: 297–394.

Antonius, A., 1970. Sense organs in marine Acoela. Am. Zool. 10: 550.

Bush, L. F., 1981. Marine flora and fauna of the northeastern United States. Turbellaria: Acoela and Nemertodermatida. NOAA Technical Report NMFS Circular 440.

Crezee, M., 1975. Monograph of the Solenofilomorphidae (Turbellaria: Acoela). Int. Revue ges. Hydrobiol. 60: 769–845.

Crezee, M. & S. Tyler, 1976. Hesiolicium gen. n. (Turbellaria, Acoela) and observations on its ultrastructure. Zool. Scr. 5: 207–216.

Dörjes, J., 1968. Die Acoela (Turbellaria) der Deutschen Nordseekuste und ein neues System der Ordnung. Z. zool. Syst. Evolutionsforsch. 6: 56–452.

Duffy, E. & S. Tyler, 1984. Quantitative differences in mitochondrial ultrastructure of a thiobiotic and an oxybiotic turbellarian. Mar. Biol. 83: 95–102.

Ehlers, U., 1984. Das Phylogenetisches System der Plathelminthes. Habilitationsschrift, Fach Zoologie im Fachbereich Biologie, Georg-August-Universität zu Göttingen.

Faubel, A., 1976. Interstitielle Acoela (Turbellaria) aus dem litoral der nordfriesischen Inseln Sylt und Amrum (Nordsee). Mitt. Hamb. Zool. Mus. Inst. 73: 17–56.

Faubel, A. & J. Dörjes, 1978. Flagellophora apelti gen. n. sp. n.: A remarcable [sic] representative of the order Nemertodermatida (Turbellaria: Archoophora). Senckenbergiana marit. 10: 1–13.

Klauser, M., J. Smith & S. Tyler, 1985. Ultrastructure of the frontal organ of *Convoluta* 'pulchra' and *Macrostomum* spp.: significance for models of the turbellarian archetype. Hydrobiologia. This volume.

Klose, R., 1984. Organogenesis of the frontal gland complex during paratomy in *Paratomella* sp. (Turbellaria, Acoela). M.S. Thesis, University of Maine at Orono.

Smith, J. P. S., 1981. Fine-structural anatomy of the parenchyma in the Acoela and Nemertodermatida (Turbellaria). Ph.D. dissertation, University of North Carolina, Chapel Hill.

Smith, J. & S. Tyler, 1984. Serial-sectioning of resin-embedded material for light microscopy: recommended techniques for micrometazoans. Mikroskopie 41: 259–270.

Smith, J. & S. Tyler, 1985a. The acoel turbellarians: kingpins of metazoan evolution or a specialized off-shoot? In S. Conway Morris, J. D. George, R. Gibson & H. M. Platt (eds), Origins and relationships of lower invertebrates. University Press, Oxford.

Smith, J. & S. Tyler, 1985b. Fine-structure and phylogeny of the frontal organ in Turbellaria Acoela: 1. Diopisthoporus gymnopharyngeus, n. sp. Zool. Scr. 14: 91–102.

Smith, J., S. Tyler & R. M. Rieger, 1985. Is the Turbellaria Polyphyletic? Hydrobiologia. This volume.

Sterrer, W., 1965. New polylithophorous marine turbellaria. Nature 210: 436.

Von Graff, L., 1891. Die Organisation der Turbellaria Acoela. Leipzig 1891.

Westblad, E., 1948. Studien uber Skandinavische Turbellaria Acoela V. Ark. Zool. 41A: 1–82.

On the evolution of central nervous systems: Implications from polyclad turbellarian neurobiology

Harold Koopowitz
Developmental and Cell Biology, University of California, Irvine, CA 92717, U.S.A.

Keywords: Turbellaria, central nervous system, evolution, neuroanatomy, neurobiology, Polycladida

Abstract

The nervous system of the polyclad turbellarian *Notoplana acticola* consists of a series of nerve plexuses and a central ganglion, the brain. The brain contains a variety of cell types including multipolar heteropolar and bipolar neurons. These cell types are rare in other invertebrate ganglia. Individual neurons also contain a variety of different ion channels. both spiking and nonspiking neurons are found. Some neurons are multimodal interneurons. Habituation appears to be a postsynaptic phenomenon. Sensitization and long-term potentiation have not been demonstrated. Polyclads appear to represent a stage in the evolution of centralized nervous systems where much of the neuronal machinery underlying behavior occurs in the peripheral nervous system and the brain's main functions are the coordination and sequencing of peripherally placed reflexes. Even at this stage, however, the brain already contains cells that seem as advanced as those found in higher organisms.

Introduction

It is now a decade since the Libbie H. Hyman Memorial Volume on the Biology of the Turbellaria was published (Riser & Morse, editors; see Koopowitz, 1974). At that time the neurophysiology of flatworms had barely begun, but some aspects of polyclad neurobiology were presented (Koopowitz, 1974). Since then, an extensive review on progress in flatworm neurobiology has appeared (Koopowitz, 1982). With the recent advent of intracellular techniques, new facts have come to light, and it now is appropriate to consider not only how the nervous system controls behavior in turbellarians, but also how new findings have influenced our ideas about the evolution of central nervous systems. The flatworms appear to have nervous systems that are intermediate in the grade of their construction between those of the phylum Cnidaria and the other higher metazoans. It is probable that any insight into the evolution of nervous systems in the Turbellaria will have ramifications for understanding the evolution of central nervous systems for most metazoans. This is a review of the nervous system and its physiology in the acotylean polyclad turbellarians.

The nervous system

Polyclads have been considered to have the most advanced and complicated of central nervous systems among flatworms, often being compared to those of the polychaete annelids in their complexity (Bullock & Horridge, 1965). The nervous system consists of a series of plexuses that feed into pairs of longitudinal nerve cords and then into a central bilobed ganglion or brain. We consider this a 'true' brain as it carries out functions of coordinating and sequencing peripherally based reflexes (Koopowitz & Keenan, 1982) and is not merely a 'relay' for sensory information. The plexuses and

Hydrobiologia 132, 79–87 (1986).

nerve nets are a mixture of somata, neurites, and synapses. There may or may not be glial sheaths around axons or even bundles of axons. The position of cell somata is variable. Compared to the Nemertea, Annelida, and Mollusca, the Polycladida is obviously at a more primitive grade of construction. In terms of their embryology, the polyclads have close affinities with these phyla and may represent the line from which those phyla were derived. Relationships with other taxa are not so clear.

Anatomy and morphology

Most of our understandings of flatworm neuroanatomy are based on conventional light or electron microscopy. On the one hand, light microscopy tends to obscure the finer details of the nervous system, and most techniques for visualizing entire neurons do not work easily with turbellarians. Electron microscopy, on the other hand, because of its fineness for detail, is usually not helpful in relating anatomy to the entire neuron. Rieger (1981) has reviewed some aspects of the nervous system at the ultrastructural level, while Reuter and her coworkers have published a number of papers on the fine structure of *Gyratrix* (Reuter & Lindroos, 1979a, b) and *Microstomum* (Reuter *et al.*, 1980; Reuter, 1981). Our knowledge at the whole-neuron level for some orders is lacking. From light microscopy of *Notoplana* brain, we find (Keenan *et al.*, 1981) that the brain is more 'loosely' organized than previously expected. Cell bodies are not confined to a neat rind as in other invertebrates, in fact somata are scattered through the brain and even occur in the nerve trunks leading into the brain as well as in the cords of the various plexuses. Electron microscopy (unpublished) also indicates a wealth of neurites and synapses in the rind of the brain. It is now possible to investigate the morphology of individual cells. As a routine adjunct to our electrophysiology experiments we now inject dyes into individual cells and can for the first time see clearly and unambiguously at this intermediate level. This technique has led to some provocative discoveries that address several important issues:

a. Multipolar neurons

In most of the invertebrate phyla, the cells in the ganglion rind are monopolar cells. This is the major neuronal type in the animal kingdom. In polyclads, however, the cerebral ganglion contains many cell types (Hadenfeldt, 1929; Keenan *et al.*, 1981). The most significant are cells that are classically defined as multipolar heteropolar neurons (see Fig. 1). These cells differ from the neurons in coelenterate nerve nets which are multipolar isopolar cells. Isopolar cells have similar long axonal processes produced by the soma. In heteropolar cells, there are small dendritic branches as well as axons that originate at the soma. These cells are very similar to those found in the vertebrates and they may signify a relationship between the turbellarians and the deuterostomes. It is also possible that heteropolar multipolar cells evolved independently several times.

b. Cell constancy

In most invertebrate preparations there are neurons known as 'identified cells', cells that appear in each individual, and have a more or less constant morphology with similar physiological functions. The presence of these cells also implies a constancy in cell number. In vertebrate animals relatively few identified neurons have been found; examples being Mauthner cells in fish, and Rohon-Beard cells in developing spinal cord (Spitzer, 1984). Several different types of 'identified neuron' have been found in polyclads (Solon & Koopowitz, 1982; Keenan & Koopowitz, 1984; Hauser, unpublished data) but we do not know if they have a constant number of brain neurons. Some preliminary data on nuclei counts in the brain suggest that larger worms have more cells than smaller animals, but one cannot distinguish between glia and neuron in these sections and they are subject to considerable error. Using histofluorescence techniques Hauser (unpublished) has found a small circuit of dopaminergic neurons in *Notoplana acticola* brains. Small but sexually mature animals have 20 cells while large animals have 28 cells (see Fig. 2). This actually represents an addition of five new pairs of neurons and the loss of one pair. This data suggests that there are changes in the neuronal circuitry during the adult lives of the animals. Certain vertebrate lines are suspected of increasing the numbers of neurons during their adult life span – e.g. fish and reptiles that have continual growth. Once again the

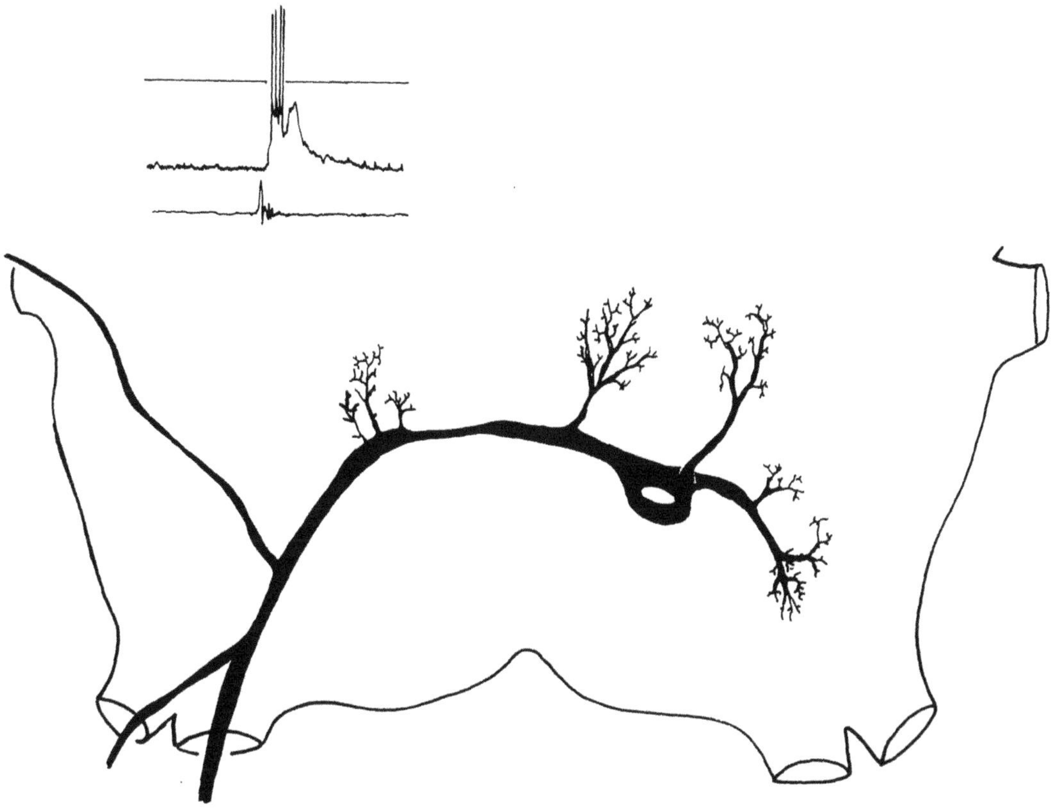

Fig. 1. Multipolar heteropolar neurons from the brain of *Notoplana acticola*. This is a bBRA cell, a multimodal interneuron. Note that this is also a decussating nerve cell. Drawn from a live cell filled with Lucifer yellow. The inset is the intracellular response to a vibration stimulus (Hauser, unpublished).

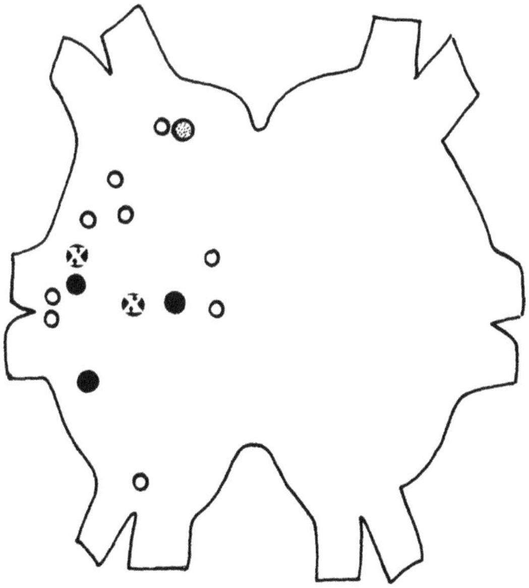

Fig. 2. Presumed dopaminergic cells in the brain of *Notoplana acticola*. Only the cells in one half of the brain are shown. Anterior is at the top of the figure. Clear circles represent position of cell bodies present in all animals. Stippled circle is a soma that is only found in small worms. Circles with 'x' appear in medium sized animals and the largest worms also have the three pairs indicated by the dark, filled circles. Drawn from wholemounts (Hauser, unpublished).

polyclads have similarities with both vertebrates and protostomes.

c. Decussating neurons

In all bilaterally symmetrical animals there appear to be decussating neurons in the central nervous system (Bullock & Horridge, 1965). It is no surprise therefore to find these kinds of cells in polyclad flatworms as well (see Fig. 1). We think that the functional significance of this type of cell may have to do with ensuring that the two sides of the body do not indulge in mutually antagonistic kinds of behavior.

d. Dye-coupled neurons

Another unusual feature of *Notoplana* brain is the widespread occurrence of dye-coupled neurons. Dye coupling occurs when one cell is penetrated by an intracellular electrode and then filled with dye which is carried into an adjacent cell or series of cells (see Fig. 3). Where this happens there are cytoplasmic connections between the cells, and they are electrically coupled so that they act in unison.

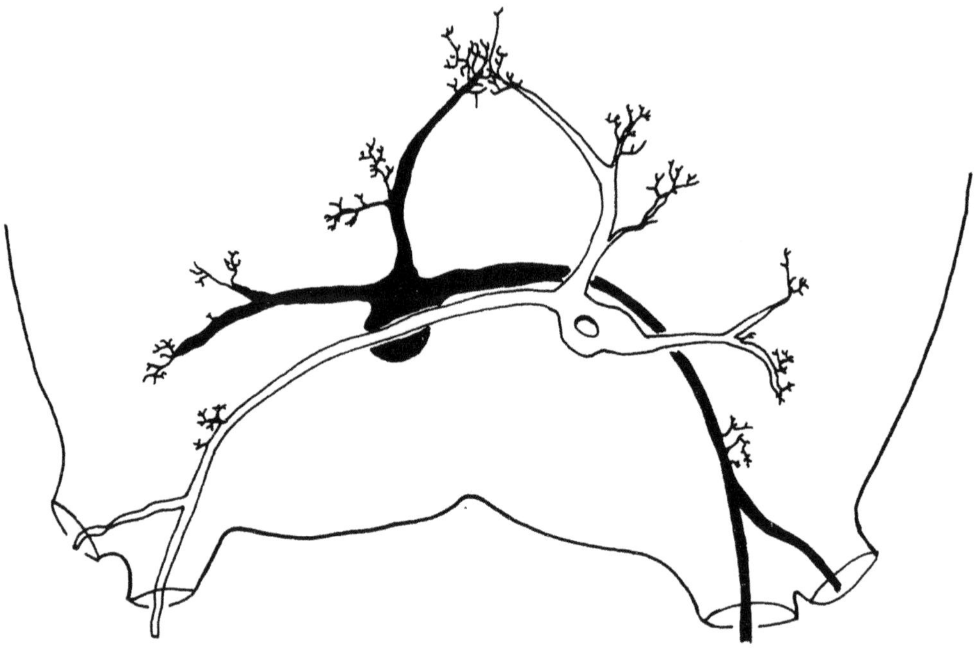

Fig. 3. Dye-coupled neurons. wBRA cells from the brain of *Notoplana acticola*. Note that these are also decussating, heteropolar multipolar neurons. Drawn from live cells filled with Lucifer Yellow dye.

In polyclads, symmetrical pairs of cells have been discovered that lie across the mid-line of the brain (Solon & Koopowitz, 1982; Keenan & Koopowitz, 1984). These are the 'BRA' cells, mutimodal interneurons, that are now being studied intensively in my laboratory. In other invertebrate preparations, dye coupling is found but is relatively rare and usually unpredictable. In *Notoplana* and *Alloeoplana*, dye coupling is invariant in certain cell types such as the BRA cells. In hydrozoans, where synchronous contractions of the swimming muscles are required, there are dye-coupled networks of neurons that make up the nerve rings (Spencer & Satterlie, 1980). It is possible that in turbellarians, activities that require synchronous activity on the two sides of the worm will also need dye-coupled neurons.

d. Bipolar neurons

These are nerve cells that are common components of coelenterate nerve nets. Bipolar cells have a spindle shaped soma from which extend long neurites. This configuration is rare in the protostome invertebrate phyla, but is found in the deuterostomes. Several bipolar cells occur in the rind of the brain in *Notoplana*. One type that has been investigated (Keenan, unpublished) are the so-called 'Burst' cells (see Fig. 4). Upon being de-

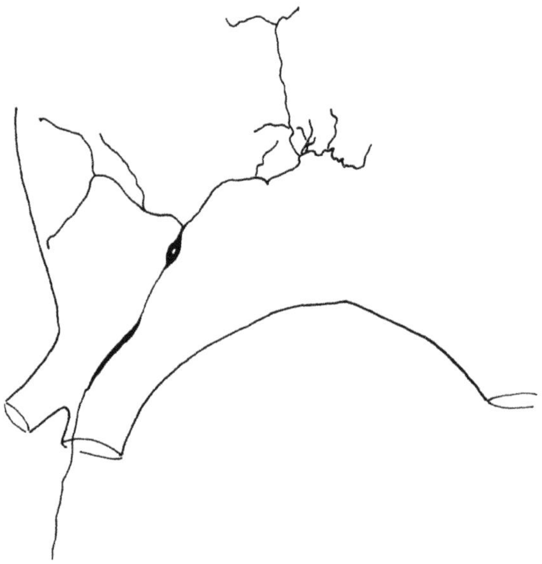

Fig. 4. Bipolar cerebral cell from *Notoplana acticola*. This is a 'Burst' cell that acts as a spontaneous pacemaker neuron. Drawn from Lucifer-Yellow fill (Keenan, unpublished).

polarized, these cells fire in bursts and act like pacemaker cells. Bipolar cells in the rind of ganglia are uncommon in other invertebrates.

e. Central cells

There are a series of cell types in the brain that have unusual morphology and are confined entirely to the brain. They do not seem to have processes that enter or leave the ganglion. Most have received little attention but among these are the 'BUG' and 'H' cells (Fig. 5). Hauser (unpublished) has found that BUG cells are non-spiking interneurons that respond to waterborne vibration. Unlike other vibration-sensitive interneurons (BRA and SC cells), the BUG cells have a graded response that can reach considerable size. Nonspiking interneurons are found in a variety of other brains (Pearson, 1976), and we suspect that they are sophisticated analog devices for processing information.

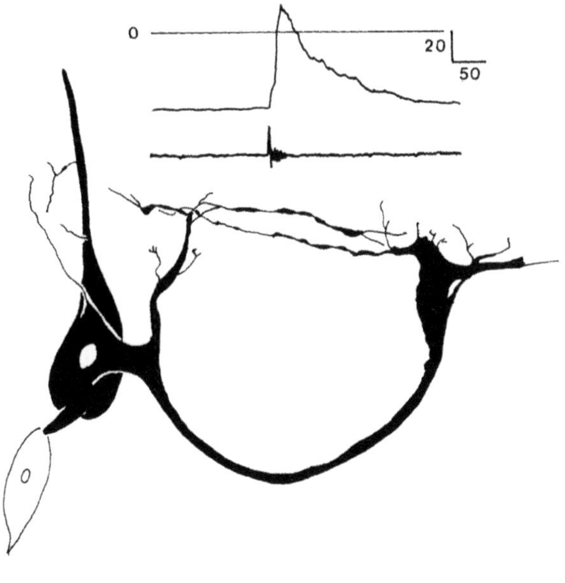

Fig. 5. Non-spiking BUG cell from the brain of *Notoplana acticola.* Note the strange extensions coming off the cell and the smaller dye-coupled neuron. Neurite emerging from the soma does so as a flattened sheet rather than a cylindrical branch. Inset, is the response to a vibration stimulus. This is a long lasting potential that reaches nearly 20 mV positive (Hauser, unpublished).

Physiology and behavior

a. Ion channels

A wide variety of ion channels occur in neurons of flatworms. We find at least two sodium channels, one of which is sensitive to TTX (tetrodotoxin) (Keenan & Koopowitz, 1981) and they may occur in different cells or appear in combination in the same cell. There is at least one calcium-activated potassium channel and one voltage-dependent potassium channel. We know that at least one calcium channel occurs, and there may be others (Keenan & Koopowitz, 1984).

b. Integration

The power of the nervous system depends on its ability to integrate information. In the simplest situation the system merely relays information, e.g. reflexes where sensory systems drive motor systems directly. In a more sophisticated system it may be necessary to compare sensory inputs from several different modalities before reacting. The BRA cells in the brain of *Notoplana* respond to many modalities, including waterborne vibrations, tactile stimuli at the front and sides of the anterior part of the worm, the onset and offset of light, and even salt solutions applied to the tail. These cells also change their responses to repeated stimulation by habituating. This decreased responsiveness to repetition is the simplest plastic change known and lasts for several minutes, in the BRA cells perhaps as long as 0.25 h (Hauser, unpublished). Events underlying habituation are now being unravelled (Keenan & Koopowitz, 1982, 1983; Hauser, unpublished). In *Notoplana* habituation is accompanied by a change in the post-synaptic membrane. There is a decrease in cell resistivity by several megohms and the time constant decreases 34.1% (S.E. 6.8%). This is a different process than the one proposed by Kandel and his co-workers for *Aplysia* (see Kandel & Schwartz, 1982; Carew, 1984, for reviews) where habituation is considered to be due to presynaptic decreases in transmitter release. In crayfish, habituation also appears to be a presynaptic event (Zucker, 1972). The evidence, in vertebrates is not clear but many of the synaptic events thought to be presynaptic in *Aplysia* may be postsynaptic in vertebrates (Lynch & Baudry, 1984). This is currently an area of active research and the next few years should indicate whether the flatworms fall with deuterostomes or protostomes as regards habituation.

c. Arousal

While most organisms display simple reflexes, there is usually a central gating mechanism that can override the reactions. For example, if food is presented to a hungry *Notoplana* that has been sitting in a quiet dish for several hours the food will be ignored. However, if the dish is tapped several times before the food is introduced the animal will feed readily. It has to be aroused first. Even at the turbellarian level, the animal is not simply a mass of several reflexes; there are higher orders of integration.

d. Redundancy

As far as we can ascertain, the brain is bilaterally symmetrical. How well can the animals function with only half a brain? *Notoplana* does not appear capable of regenerating excised portions of the brain. It is fairly easy to extirpate half of the brain (Faisst *et al.*, 1980) and observe behavioral responses. Following hemicerebration, severed nerves grow back into the remaining portions of the ganglion and establish connections. Much of the behavior of half-brained animals is indistinguishable from that of normal worms. This includes behaviors that are involved with left-right orientation as well as locomotion where alternate waves of contraction pass down along the two sides of the body. Some behaviors such as avoidance-turning are recovered. Animals with only one side of the brain can correctly avoid noxious stimuli to either side of the animal (see Fig. 6).

e. Neuronal controls of behavior

Flatworms without brains can carry out a considerable array of muscle activities that are initiated by peripheral sensory receptors (Gruber & Ewer, 1962; Koopowitz, 1970, 1982). This suggests that much of the neuronal circuitry directly underlying behavior is peripheral to the brain. One can demonstrate that habituation and dishabituation to waterborne vibration stimuli occurs not only in the

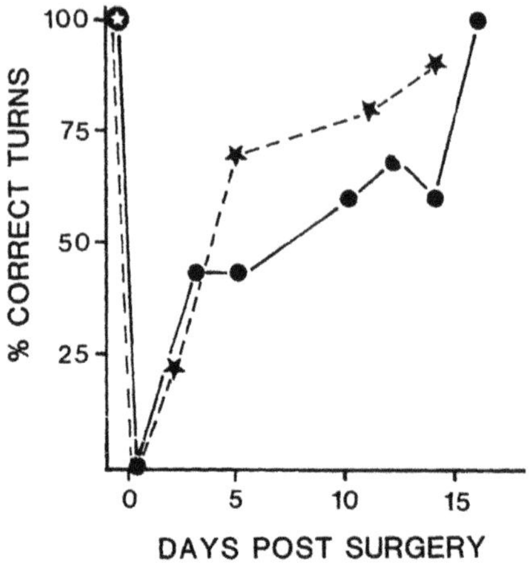

Fig. 6. Recovery of avoidance behavior by *Notoplana acticola*. Graph shows the recovery performance for two individuals following hemidecerebration. (Data redrawn from Faisst *et al.*, 1980).

brain but also in the longitudinal nerve cords (Bernardo *et al.*, 1977). Animals without brains are slower in carrying out most activities (Loeb, 1907). Decerebrates cannot coordinate their activities in terms of temporal sequencing, in addition they are unable to inhibit peripheral reflexes such as feeding.

f. Transplants

We have noted above that the brain acts as a central control that overrides or augments peripheral reflex pathways occurring in the plexus. When this kind of system is built, it requires that specific central cells recognize and make connections with specific peripheral cells and vice versa. Are these recognition factors genetically determined for a species or individual, or does this merely come about during development? It is possible to transplant brains between individual worms; and when they heal, behavioral controls are re-established (unpublished data). This suggests that there are species-wide specific recognition sites for individual neurons.

g. Learning

There are several important features of nervous-system physiology that we have *not* yet been able to demonstrate. We have been unable to demonstrate

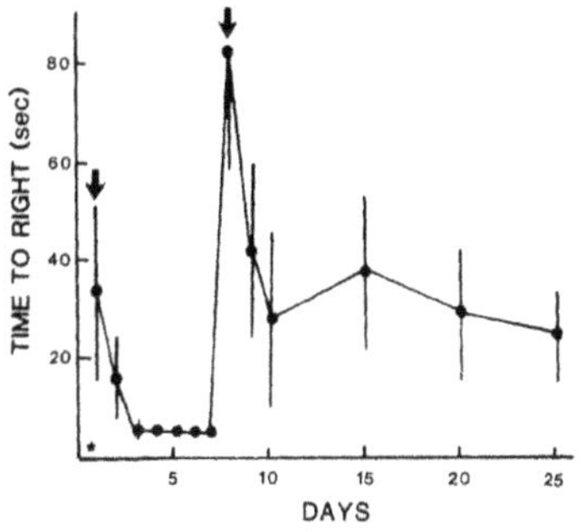

Fig. 7. Recovery of righting behavior. Small star in the lower left hand corner of the graph is the mean righting time in seconds for normal *Notoplana acticola.* Following brain transplantation (first arrow), righting behavior recovers and approaches the normal time. At second arrow, the brain was removed. Note the improvement in performance with time. Bars are standard errors (Gallemore, unpublished).

sensitization (facilitation) at the cellular level and have not found post-tetanic or long-term potentiation either. At the behavioral level, we have been totally unable to condition a polyclad in a convincing manner. It is possible that associative learning does not occur in polyclads and this would be a major qualitative difference between them and the higher phyla. *Notoplana* can show improvement in behavior: decerebrate flatworms 'learn' to right more efficiently with time (Gallemore, unpublished) but whether or not this improvement (see Fig. 7) is merely a recovery from surgery or a long-term change in neuronal efficacy still remains to be determined.

Discussion

In the last few years it has become evident that many of the earlier views held about neurons and nervous system components, in general, were simplistic. The multiplicity of cell types, both from physiological and morphological points of view, within a single organism was not realised. It behooves us now to reconstruct phylogenetic trends very cautiously. One should also bear in mind that comparisons are based on present-day living representatives that have had in excess of 600 million years to perfect themselves and consequently may bear little resemblance to the ancestral conditions. In addition, many of the simplest turbellarians may be the result of reductionism, a loss of cells, producing smaller animals that 'fit' more readily into microhabitats such as the spaces between sand grains or detritus. To equate this secondary simplification with primitiveness may be mistaken.

Our findings point to several important features about the early evolution of central nervous systems:

a) Most of the basic cellular-level machinery that we find in higher metazoans already exists in the polyclads. There is a multiplicity of membrane channels, transmitter substances, cell types, etc. These must all have been early inventions in the evolution of nervous systems.

b) Sophisticated cell types already exist too and include non-spiking interneurons and multimodal interneurons. Either there are only a few ways by which neurons can process information – and by necessity these have been forced to evolve independently many times – or else the original patterns evolved in the turbellarians (or their progenitors).

c) Perhaps the most surprising features we have found are the multipolar heteropolar cells that occur in the brain. Among invertebrates, these appear to be unique and perhaps make flatworm brains better models for vertebrate nervous systems than the other currently popular invertebrate systems. In an evolutionary sense they tie the turbellarians closer to the deuterostome line than previously thought. It might be pointed out that monopolar 'invertebrate-type' neurons are found in vertebrates: the fish tectum is rich in them (Coss & Globus, 1978; Vanegas *et al.*, 1974). Presumably the early progenitors of both protostomes and deuterostomes had a mixture of cell types like those found in modern polyclads.

There are several bits of 'missing' information that would be helpful in reconstructing a phylogenetic sequence for nervous system evolution within the Turbellaria. What is the distribution of multipolar heteropolar neurons among the various orders? Light micoscopy is misleading. In the light microscope with conventional stains *Notoplana* looks as if it has a rind of monopolar cells. It would be interesting to find out how widespread the occurrence of multipolar heteropolar cells is in the other turbellarian orders. Perhaps morphologists and anatomists should be encouraged to try to fill individual nerve cells with various dyes in an effort to gain a better understanding of the distribution of cell types within the Turbellaria.

It is curious that the polyclads lack statocysts and difficult to believe that they lost these important sense organs through evolution. If the brain evolved around the statocyst originally, it should still be there. If they never had a statocyst, it could point to the archaic position of their ancestors within the class.

Polyclads appear to have a naked nerve net that runs between the bases of the ectodermal cells. This is reminiscent of the Cnidaria. Such a structure is difficult to determine and visualize, and one wonders how widespread this feature is in the class. If the ectodermal net is confined to the polyclads, then it ties them between Cnidaria and higher metazoans. What does this say about the phylogeny of the other orders?

Acknowledgments

This work is based to a large extent on the efforts of many students in my laboratory. In particular I would like to express my gratitude to Dr. C. Larry Keenan who did the pioneering work with intracellular techniques and Melinda Hauser who continues in the same fashion. Much of this work was carried out during the tenure of a grant from PHS, NS 13713-01 through -07.

References

Bernardo, K., G. C. Stone & H. Koopowitz, 1977. Primitive nervous systems: habituation in decerebrate flatworms. J. Neurobiol. 8: 141–150.

Bullock, T. H. & G. A. Horridge, 1965. Structure and function in the nervous systems of invertebrates. 1. Freeman, San Fransisco, 798 pp.

Carew, T. J., 1984. An introduction to cellular approaches used in the analysis of habituation and sensitization in Aplysia. In H. V. S. Peeke & L. Petrinovich (eds), Habituation, Sensitization and Behavior. Academic Press, N.Y.: 205–249.

Coss, R. G. & A. Globus, 1978. Spine stems on tectal interneurons in jewel fish are shortened by social stimulation. Science 200: 787–790.

Faisst, J., L. Keenan & H. Koopowitz, 1980. Neuronal repair and avoidance behavior in the flatworm, Notoplana acticola. J. Neurobiol. 11: 483–496.

Gruber, S. A. & D. W. Ewer, 1962. Observations on the myoneural physiology of the polyclad Planocera gilchristi. J. exp. Biol. 39: 459–477.

Hadenfeldt, D., 1929. Das Nervensystem von Stylochoplana maculata und Notoplana atomata. Z. wiss. Zool. 133: 586–638.

Kandel, E. & J. Schwartz, 1982. Molecular biology of learning: modulation of transmitter release. Science 218: 433–443.

Keenan, L. & H. Koopowitz, 1981. Tetrodotoxin-sensitive potentials from the brain of the polyclad flatworm, Notoplana acticola. J. exp. Zool. 215: 209–213.

Keenan, C. L. & H. Koopowitz, 1982. Postsynaptic habituation in identified flatworm neurones. Soc. Neurosci. Abst. 8: 161.

Keenan, C. L. & H. Koopowitz, 1983. Postsynaptic-cell membrane events accompanying habituation. Soc. Neurosci. Abst. 9: 733.

Keenan, C. L. and H. Koopowitz, 1984. Ionic bases of action potentials in identified flatworm neurones. J. comp. Physiol. A. 155: 197–208.

Keenan, L., Coss, R. G. & H. Koopowitz, 1981. Cytoarchitecture of primitive brains: Golgi studies on flatworms. J. comp. Neurol. 195: 697–716.

Koopowitz, H., 1970. Feeding behavior and the role of the brain in the polyclad flatworm, Planocera gilchristi. Anim. Behav. 18: 31–35.

Koopowitz, H., 1974. Some aspects of the physiology and organization of the nerve plexus in polyclad flatworms. In N. W. Riser & M. P. Morse (eds), Biology of the Turbellaria. McGraw-Hill, N.Y.: 198–212.

Koopowitz, H., 1982. Free-living Platyhelminthes. In G. A. B. Shelton (ed.), Electrical Conduction and Behaviour in 'Simple' Invertebrates. Oxford University Press, Oxford: 359–392.

Koopowitz, H. & L. Keenan, 1982. The primitive brains of the Platyhelminthes. Trends Neurosci. 5: 77–79.

Loeb, J., 1907. Comparative physiology of the brain and comparative psychology. G. P. Putman's Sons, N.Y.: 72–83.

Lynch, G. & M. Baudry, 1984. The biochemistry of memory: a new and specific hypothesis. Science 224: 1057–1063.

Pearson, K. G., 1976. Nerve cells without action potentials. In J. C. Fentress (ed.), Simpler networks and behavior. Sinauer Assoc. Inc., Sunderland: 99–110.

Reuter, M., 1981. The nervous system of Microstomum lineare (Turbellaria, Macrostomida). II. The ultrastructure of synapses and neurosecretory release sites. Cell Tissue Res. 218: 375–387.

Reuter, M. & P. Lindroos, 1979a. The ultrastructure of the nervous system of Gyratrix hermaphroditus (Turbellaria, Rhabdocoela). I. The brain. Acta zool. 60: 139–152.

Reuter, M. & P. Lindroos, 1979b. The ultrastructure of the nervous system of Gyratrix hermaphroditus (Turbellaria, Rhabdocoela). II. The peripheral nervous system and the synapses. Acta zool. 60: 153–161.

Reuter, M., M. Wikgren & I. Palmberg, 1980. The nervous system of Microstomum lineare (Turbellaria, Macrostomida). Cell Tissue Res. 211: 31–40.

Rieger, R. M., 1981. Morphology of the Turbellaria at the ultrastructural level. In E. R. Schockaert & I. R. Ball (eds), The Biology of the Turbellaria. Dr. W. Junk, The Hague: 213–229.

Solon, M. H. & H. Koopowitz, 1982. Multimodal interneurons in the polyclad flatworm, Alloeoplana californica. J. comp. Physiol. 145: 171–178.

Spencer, A. N. & R. A. Satterlie, 1980. Electrical and dye-coupling in an identified group of neurons in a coelenterate. J. Neurobiol. 11: 13–19.
Spitzer, N. C., 1984. What do Rohon-Beard cells do? Trends Neurosci. 7: 224–225.
Vanegas, H., M. Laufer & J. Amat, 1974. The optic tectum of a perciform Teleost I. General configuration and cytoarchitecture. J. comp. Neurol. 154: 43–60.
Zucker, R., 1972. Crayfish escape behavior and central synapses. II. Physiological mechanisms underlying behavioral habituation. J. Neurophys. 35: 621–637.

Comparative characterization of the nervous system of the Turbellaria

E. A. Kotikova
Laboratory of Evolutionary Morphology, Zoological Institute, Academy of Sciences, Leningrad, 199034, USSR

Keywords: Turbellaria, nervous system, comparative morphology

Abstract

Initial stages of the centralization of the nervous apparatus in the Turbellaria can be traced through a comparison of the structure of the nervous system in various representatives of the class. The most primitive state, found in the Acoela, is predominantly plexiform with a varying number of longitudinal trunks. Three, and in some cases four, longitudinal trunks are found in the Proseriata and Temnocephalida. Commissures appear in the Macrostomida and all higher orders and form an orthogon. Brain shape varies from ring-shaped in the Acoela to bilobed in the Neorhabdocoela. While the nervous system of the Polycladida is peculiar, having numerous lateral trunks and separation of dorsal and ventral parts of the nervous system, the development of the nervous system in Müller's larvae of polyclads shows it is of an orthogonal type comparable to other platyhelminths. Transition to parasitism is accompanied by some progressive transformations in the structure of the nervous system.

It is possible to retrace the initial stages of the centralization of the nervous system in the Turbellaria, a group of multicellular animals on a relatively low organizational level. The main features of the process of centralization in the nervous system of Turbellaria are the formation of longitudinal trunks, then commissures, and finally the oligomerisation (partial reduction) of the commissures. The presently published data on morphology of the nervous system were obtained through a variety of procedures (Hyman, 1951; Bullock & Horridge, 1965; Reisinger, 1972, 1976), resulting sometimes in only fragmentary information. I have developed a method of histochemical staining for cholinesterase, permitting the nervous system of these animals to be studied in wholemounts (Gerebtzoff, 1959).

The most primitive state of the nervous system – a plexus with only weakly differentiated longitudinal trunks – is to be observed in Acoela. In this order there is also a varying number (from one to five pairs) of longitudinal trunks (Fig. 1) (Mamkaev & Kotikova, 1972). The trunks are almost always paired, an unpaired number of trunks having been observed only in one representative of Nemertodermatida (Faubel & Dorjes, 1978). In one acoel from the White Sea (*Baltalimania* sp.), I detected an unpaired dorsal trunk, as well as four pairs of longitudinal trunks.

Four pairs of trunks are characteristic of the Proseriata (Kotikova, 1976). In the Temnocephalidae, forms with three pairs of trunks are most common, although it is possible to detect a fourth pair along the body margin of some species (Joffe, 1981a).

In some acoels, some strengthened transverse strings can be distinguished, and these represent the first step in the development of circular commissures. Among lower turbellarians, it is in the Macrostomida that commissures appear first. They are present in all higher orders, forming an orthogon (cf. Reisinger, 1972). The number of commissures may vary from more than thirty in some

Hydrobiologia 132, 89–92 (1986).

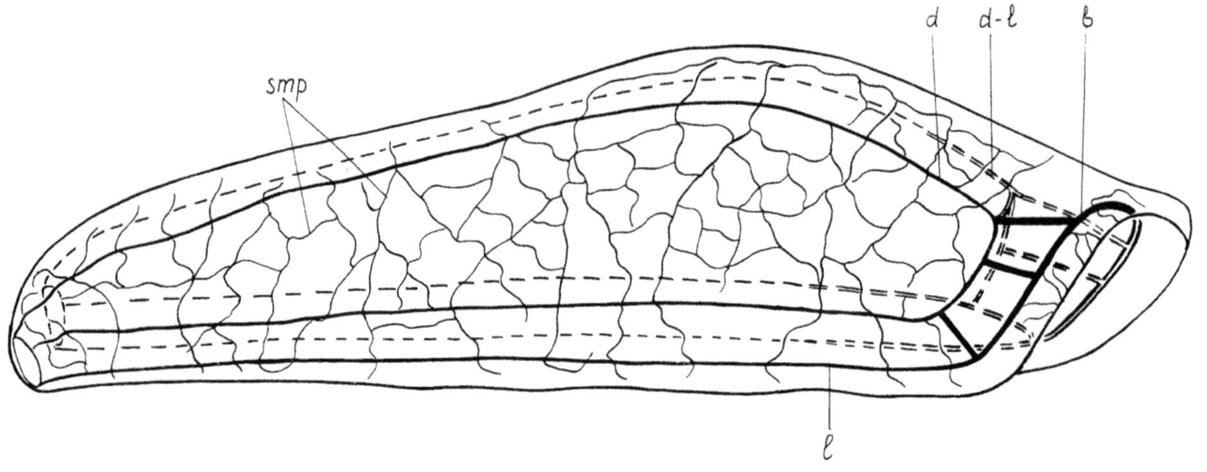

Fig. 1. Nervous system of *Convoluta bifoviolata* (Acoela) schematic side view. b: brain; d, l, d-l: dorsal, lateral, and dorso-lateral nerve trunks; smp: submuscular plexus.

proseriates to very few in some microstomids, for example, and even only one in many Dalyellioida (Fig. 2).

The brain of turbellarians is usually of a bilobed shape, but it may be more globular in cases where it is surrounded by a capsule. In Acoela, the brain is more often ring-shaped. As a rule, trunks arising from a bilobed brain are ventral (and in some cases ventro-lateral) and sometimes lateral. Dorsal trunks (and in most cases lateral trunks as well) are connected with the brain by nerve strands.

The nervous system of the Polycladida has a peculiar organization. The ontogeny of the nervous system in Müller's larva is of particular interest for the evolution of the nervous system. The rudiments of the definitive nervous system – for example, the brain, ventral longitudinal connectives, and the ventral and dorsal longitudinal trunks – appear at early stages of development and are gradually transformed into corresponding parts of the adult worm's nervous system. In the course of growth up to eight pairs of lateral trunks form at the expense of the commissure elements, and the ventral and dorsal areas of the nervous system become separated (Kotikova, 1983; see also Lacalli, 1982, 1983). The subepithelial plexus lies directly over the submuscular one and is in contact with it; additionally, an infra-epithelial plexus has been described (Chien & Koopowitz, 1977). There is also an internal plexus, located in the parenchyma, linking the submuscular plexus of the dorsal and ventral sides of the body. I consider the nervous system of polyclads to be an altered orthogon. The alterations are connected both with the dorsoventral flattening of the body and with the development of the caky shape of the body.

These data speak against the claim that the polyclad nervous system is non-homologous with that of other platyhelminths (see Minichev & Pugovkin, 1979). The primitive polyclads, like the larvae examined so far, could have had a nervous system of an orthogonal type, with four or more distinct longitudinal trunks, although possibly without any strongly marked circular commissures. Systems like these are found, for example, in the Acoela (Mamkaev & Kotikova, 1972). Thus, the polyclad nervous system, with its origin easily traceable to the primitive orthogon, gives no ground for distinguishing the Polycladida from other flatworms.

The confusion in terminology makes it difficult to relate the four nerve plexuses of polyclads to nerve plexuses in representatives of other orders. But both my data and that of the literature (e.g. Hyman, 1951; Reisinger, 1972; Baguña & Ballister, 1978) leave no doubt that such plexuses are generally widespread in the Turbellaria. In all cases, the plexuses are linked with each other, forming a unified nervous net.

In various groups of the Turbellaria, some morphologically similar types of nervous systems have arisen independently, demonstrating striking parallelisms. For example, representatives of the

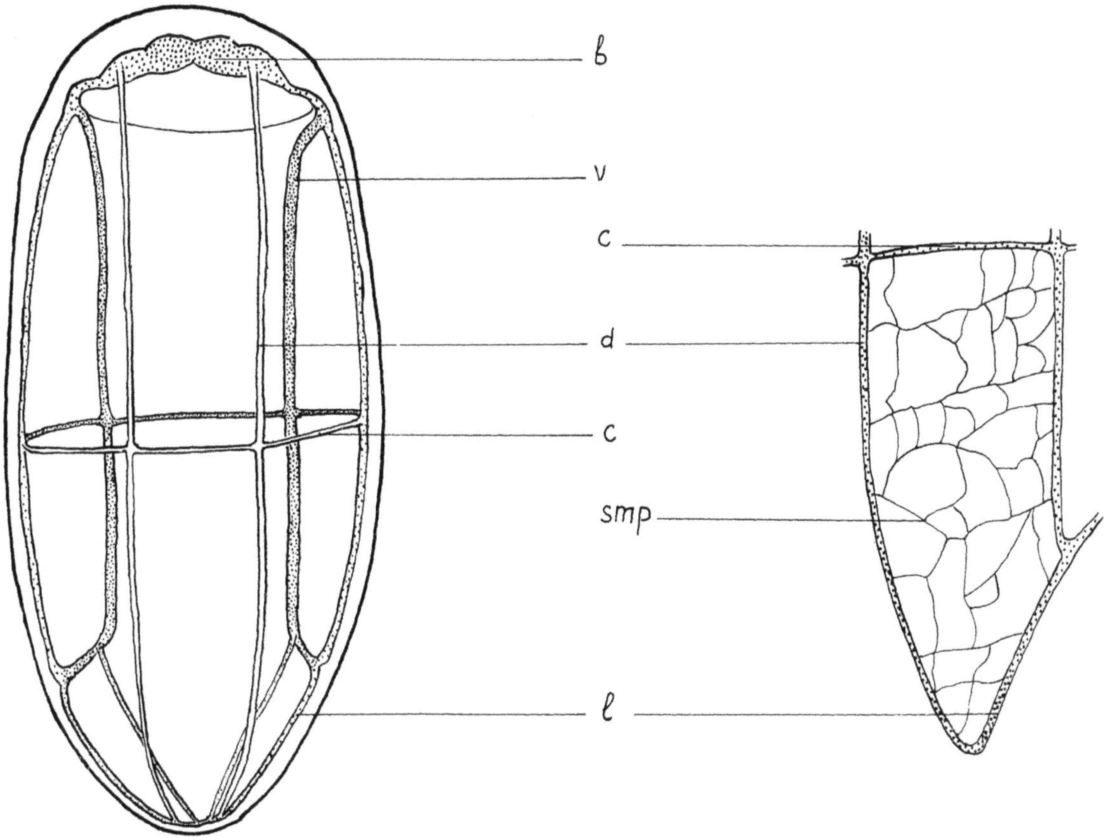

Fig. 2. Nervous system of *Halammovortex macropharynx* (Dalyellioida) dorsal view at left; submuscular plexus is shown at right for posterior dorso-lateral 'cell' (drawn by Kotikova and Joffe). b: brain; d, l, v: dorsal, lateral, and ventral nerve trunks; c: ring commissure; smp: submuscular plexus.

families Scutariellidae, Graffillidae, Anoplodiidae, and Udonellidae all have a bilobed brain, three pairs of longitudinal nervous trunks, and one circular commissure (Joffe & Kotikova, 1983; Kotikova, 1980; Pugovkin *et al.*, 1977). Since these families are phylogenetically quite distantly related, the form of their nervous systems apparently arose independently in each (Ivanov, 1952; Joffe, 1981b). The anatomical similarity does not result from reduction and is not associated with body size, which is essentially different in these families. All of these forms illustrate one route of the progressive development of the nervous system related to the transition to parasitism.

It has been repeatedly noted that in the Platyhelminthes the transition from a free mode of living to a parasitic one is accompanied by some progressive transformations in the structure of the nervous system (Reisinger, 1972; Kotikova, 1979; Kotikova, 1980; Joffe, 1981a). These modifications may proceed both in oligomerization of the nervous system and by way of developing new nervous structures as has happened, for example, in cestoids. On the whole, however, the nervous system of the parasitic platyhelminths retains the organizational level typical of higher turbellarians.

References

Baguna, J. & R. Ballester, 1978. The nervous system in planarians: peripheral and gastrodermal plexuses, pharynx innervation, and the relationship between central nervous system structure and the acoelomate organization. J. Morph. 155: 237–252.

Bullock, T. H. & G. A. Horridge, 1965. Structure and function in the nervous systems of invertebrates 1. Freeman and Co., San Francisco, London, 798 pp.

Chien, P. K. & H. Koopowitz, 1977. Ultrastructure of nerve plexus in flatworms. 3. The infra-epithelial nervous system. Cell and Tissue Res. 176: 335–347.

Faubel, A. & J. Dörjes, 1978. Flagellophora apelti gen. n. sp. n.: A remarcable representative of the order Nemertodermatida (Turbellaria: Archoophora). Senkenb. Marit. 10: 1–13.

Gerebtzoff, M. A., 1959. Cholinesterases. London, 195 pp.

Hyman, L. H., 1951. The Invertebrates. II. Platyhelminthes and Rhynchocoela. McGraw-Hill, New York, 550 pp.

Ivanov, A. V., 1952. Stroenie Udonella caligorum Johnston, 1835 i polozhenie Udonellidae v sisteme ploskikh chervej. Parasitologicheskij sbornik Zool. Inst. Akad. Nauk SSSR 14: 112–163. [The structure of Udonella caligorum Johnston, 1835, and the position of Udonellidae in the systematics of flatworms. In Russian].

Joffe, B. I., 1981a. Stroenie temnotsefalid i morfologicheskie izmeneniya pri perekhode k parasitismu u ploskikh chervej. Parazitolodiya 15: 209–218. [Morphology of Temnocephalids and morphological alterations during the transition to parasitism in the Platyhelminthes. In Russian].

Joffe, B. I., 1981b. Stroenie i filogeneticheskie svyasi temnosfalid (Turbellaria). Zool. Zh. 60: 661–672. [Morphology and phylogenetic relations of the Temnocephalida (Turbellaria). In Russian].

Joffe, B. I. & E. A. Kotikova, 1983. Nervnaya sistema Scutariellidae (Turbellaria, Temnocephalida). Parazitologiya 17: 101–106. [Nervous system of the Scutariellidae (Turbellaria, Temnocephalida). In Russian].

Kotikova, E. A., 1976. Nekotorye osobennosti nervnogo apparata Proseriata (Turbellaria). Zool. Zh. 55: 508–514. [Some features of the nervous system in Proseriata (Turbellatia). In Russian].

Kotikova, E. A., 1979. Osobennosti evolutsii nervnogo apparata tsestod. V Kn.: Evolutsionnaya morfologiya bespozvonochnykh. Tr. Zool. Inst. Akad. Nauk SSSR, Leningrad, 84: 34–38. [On the peculiarities of the evolution of the nervous system in Cestodes. In Russian].

Kotikova, E. A., 1980. Nervnyj apparat parasiticheskikh turbellarij semejstva Graffillidae iz morskikh bryukhonogikh mollyuskov. V sb.: Voprosy parazitologii vodnykh bespozvonochnykh zhivotnykh, Vilnyus: 53–54. [Nervous system of the parasitic turbellarians of the family Graffillidae from marine gastropods. In Russian].

Kotikova, E. A., 1981. Morfologicheskie osobennosti razvitiua nervnogo apparata myullerovskoj lichinki polikladid. Biol. morya 6: 30–36. [Morphological peculiarities of the development of the nervous system in Müllers larva (Turbellaria, Polycladida). In Russian].

Kotikova, E. A., 1983. Nervnaya sistema turellarii Acerotisa sp. (Polycladida, Cotylea). Zool. Zh. 62: 1148–1154. [The nervous system of Acerotisa sp. (Polycladida, Cotylea, Turbellaria). In Russian].

Lacalli, T. C., 1982. The nervous system and ciliary band of Müller's larva. Proc. r. Soc. Ser. B, London, 217: 37–58.

Lacalli, T. C., 1983. The brain and central nervous system of Müller's larva. Can. J. Zool. 61: 39–51.

Mamkaev, Yu. V. & E. A. Kotikova, 1972. O morfologicheskikh osobennostyakh nervnogo apparata beskishechnykh turbellarij. Zool. Zh. 51: 477–489. [On morphological characters of the nervous system in Acoela. In Russian].

Minichev, Yu. S. & A. P. Pugovkin, 1979. Nervous system of the polyclad flatworm Notoplana atomata (O. F. Müller). Cah. Biol. mar. 20: 181–188.

Pugovkin, A. P., Yu. S. Minichev & T. A. Timofeeva, 1977. Nervnaya sistema Udonella caligorum Johnston (Turbellaria, Udonellida). Parazitologiya 11: 3–8. [The nervous system of Udonella caligorum Johnston (Turbellaria, Udonellida). In Russian].

Reisinger, E., 1972. Die Evolution des Orthogons der Spiralier und das Arachicölomatenproblem. Z. zool. Syst. Evolutionsforsch. 10: 1–43.

Reisinger, E., 1976. Zur Evolution des stomatogastrischen Nervensystems bie den Plathelminthen. Z. zool. Syst. Evolutionsforsch. 14: 241–253.

Neuropeptides in free-living and parasitic flatworms (Platyhelminthes). An immunocytochemical study

Marianne Wikgren, Maria Reuter & Margaretha Gustafsson
Institute of Biology, Åbo Akademi, Porthansg. 3–5, 20500 Åbo, Finland

Keywords: Turbellaria, Cestoda, Neuropeptides, Serotonin, Immunocytochemistry

Abstract

The nervous systems of the turbellarians *Microstomum lineare* and *Polycelis nigra* and of the cestodes *Diphyllobothrium dendriticum* and *Schistocephalus solidus* were studied by means of the peroxidase-antiperoxidase (PAP) immunocytochemical method, with the use of antisera to the neuropeptides FMRF-amide, vasotocin, leu-enkephalin, met-enkephalin, neurotensin, somatostatin, and VIP, and to the bioamine serotonin. Anti-FMRF-amide positive perikarya and fibers occurred in all species, while the occurrence of the vertebrate brain-gut peptides and serotonin varied between the species. Anti-somatostatin and anti-VIP gave a negative result. Anti-FMRF-amide and anti-vasotocin positive immunoreactivity was found in the brain and gut of *M. lineare,* and in the CNS and the peripheral nerve net of the cestodes. We suggest that the brain-gut peptides of free-living flatworms act on the subtegumental region in the cestodes, which lack a gut but absorb their nutrients directly through the tegument.

Introduction

Within the last ten years, a large number of biologically active peptides in vertebrates have been described. A host of publications have established the dual localization of many of these peptides in the central nervous system (CNS), and in endocrine cells and enteric nerves of the gastrointestinal tract (the 'brain-gut peptides') (for review, see e.g. Krieger, 1983; Polak & Bloom, 1982). During recent years, a number of reports show that homologues to vertebrate peptide families are distributed throughout the invertebrate phyla and that many of them have biological activities. The occurrence of these 'naturalized' as well as 'native' peptides in invertebrates, and the curious irrelation between their physiological roles have been reviewed by Greenberg & Price (1983). The wide distribution of peptide neuroregulators indicates that primitive nervous systems are essentially peptidergic.

Several homologues to vertebrate peptides have been demonstrated by IC (immunocytochemistry) in the primitive nervous system of *Hydra* (for references, see Greenberg & Price, 1983). In the flatworms, which lack endocrine glands and a proper circulatory system, neurosecretion plays an important integrative role. A few reports from planarians show the occurrence of cells that react positively with antisera to somatostatin, the neurophysins, ACTH, met-enkephalin, and neurotensin (Bautz *et al.,* 1980; Schilt *et al.,* 1981; Carraway *et al.,* 1982; Venturini *et al.,* 1983).

In the present IC (peroxidase-antiperoxidase, PAP) study, the occurrence of a number of neuropeptides and of the bioamine serotonin (5-hydroxytryptamine, 5-HT) in the CNS and in the peripheral nervous system (PNS) of the free-living turbellarians *Microstomum lineare* and *Polycelis nigra* and the endoparasitic cestodes *Diphyllobothrium dendriticum* and *Schistocephalus solidus* are compared. Our aim is to get more information about the role of neuropeptides – especially brain-gut

Hydrobiologia 132, 93–99 (1986).

peptides – in flatworms. A comparison between turbellarians and cestodes is motivated by the fact that turbellarians have a gut, while cestodes are totally devoid of an alimentary tract and absorb all their nutrients through the tegument. Also, from a phylogenetic point of view this comparison may give interesting results. Previously, the cytology and distribution of peptidergic neurons in *M. lineare* (Reuter *et al.,* 1980; Reuter, 1981; Reuter & Palmberg, 1983), and in *D. dendriticum* (Gustafsson & Wikgren, 1981a, b, c) have been studied.

The results show that positive reactions with antisera to the vertebrate peptides vasotocin, leu-enkephalin, met-enkephalin and neurotensin, and to the bioamine serotonin could be demonstrated in flatworms, but that the occurrence of these substances varied between the species studied. An antiserum to a 'native' invertebrate peptide, FMRF-amide, originally isolated from the clam *Macrocallista nimbosa* (Price & Greenberg, 1977) gave a positive reaction in all species.

Materials and methods

Specimens of *Microstomum lineare* (Müller 1774) (Turbellaria, Macrostomida) were caught in shallow brackish water at Stortervo, Pargas (SW Finland). Specimens of *Polycelis nigra* (Turbellaria, Tricladida) were collected from a stock culture maintained in containers with tap water at the Institute of Biology, Åbo Akademi. Plerocercoid larvae of *Diphyllobothrium dendriticum* (Cestoda, Pseudophyllidea) were excised from the stomach wall of whitefish (*Coregonus lavaretus*) from Lake Pyhäjärvi (SW Finland). Plerocercoids of *Schistocephalus solidus* (Cestoda, Pseudophyllidea) were obtained from the body cavity of the stickleback (*Gasterosteus aculeatus*) from the Baltic. For details of fixation and embedding, see Reuter *et al.* (1984).

Immunocytochemical procedure. The sections were stained according to the unlabelled antibody enzyme peroxidase-antiperoxidase (PAP) technique (Sternberger *et al.,* 1970). The following antisera were used (dilutions in parentheses): anti-FRMF-amide (1:500, 1:1 000); anti-vasotocin (1:500); anti-leu-enkephalin (1:500, 1:1 000); anti-met-enkephalin (1:1 000, 1:2 000, 1:3 000, 1:4 000), anti-neurotensin (1:500, 1:2 000); anti-somatostatin (1:500, 1:1 000, 1:2 000); anti-VIP (1:10 000, 1:20 000) and anti-serotonin (1:10 000, 1:20 000). The antiserum against the molluscan cardioactive peptide FMRF-amide was raised in rabbits by L.P.C. Schot at Biologisch Laboratorium, Vrije Universiteit, Amsterdam. Anti-leu-enkephalin was kindly donated by Dr. M. Pelto-Huikka at the University of Tampere, Finland, and anti-neurotensin by Dr. J. M. Polak at the Royal Postgraduate Medical School, London. All other antisera were purchased from Immunonuclear. The detailed procedure for the immunocytochemical methods and the controls are described in Reuter *et al.* (1984).

Results

FMRF-amide

Turbellarians

A detailed description of the distribution of anti-FMRF-amide immunoreactivity in *Microstomum lineare* is given by Reuter *et al.* (1984). Anti-FMRF-amide-positive perikarya and fibers were observed both in the CNS and the PNS of the worm. Positive perikarya were revealed in the brain and in the immediate surroundings of the ciliated pits. Numerous nerve fibers occur in the neuropil and in the longitudinal nerve cords. In the stomatogastric part of the PNS, positive perikarya are regularly distributed among the basal cells of the intestinal epithelium, and positive fibers were observed in the periphery of the alimentary tract. Nerve fibers of the pharyngeal nerve ring show a strongly positive reaction (Fig. 1). In addition, positive perikarya were observed symmetrically in the mesenchyme on both sides of the pharynx.

In the planarian *Polycelis nigra,* immunoreactive fibers were observed along the muscle layer of the pharynx (Fig. 2), in the main nerve cords, and in the intestinal wall. No anti-FMRF-amide-positive perikarya were observed, however, in the intestinal epithelium.

Cestodes

A positive reaction with anti-FMRF-amide was found in the CNS and the PNS of *Diphyllobothrium dendriticum.* The scolex is abundantly innervated by anti-FMRF-amide-positive fibers. They surround the lateral cerebral ganglia, and extend to

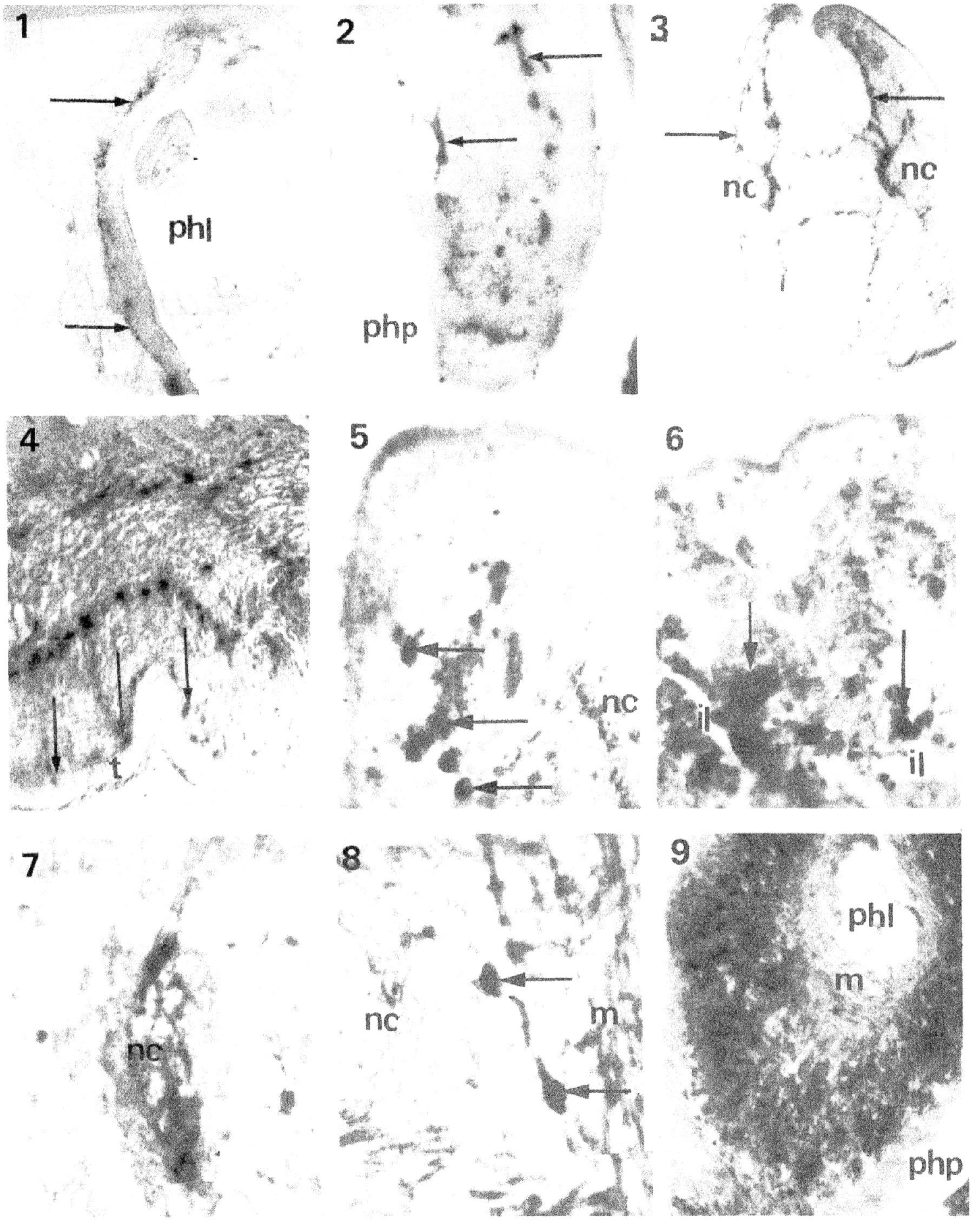

Figs. 1–9. (1) *M. lineare.* Anti-FMRF-amide-positive fibers (arrows) in pharyngeal nerve ring. ph l – pharyngeal lumen ×450. (2) *P. nigra.* Anti-FMRF-amide-positive fibers (arrows) in pharynx. ph p – pharyngeal pouch ×930. (3) *D. dendriticum,* scolex. Anti-FMRF-amide-positive fibers surrounding main nerve cords (nc) and forming a peripheral nerve net in bothridia (arrows) ×310. (4) *D. dendriticum,* bothridium. Anti-FMRF-amide-positive fibers. Note positive terminals (arrows) beneath tegument (t) ×1 200. (5) *M. lineare.* Anti-vasotocin-positive perikarya (arrows) in brain. nc – nerve cord ×930. (6) *M. lineare.* Anti-vasotocin-positive perikarya (arrows) in gut. il – intestinal lumen ×1 100. (7) *D. dendriticum.* Anti-leu-enkephalin-positive fibers in main nerve cord (nc) ×650. (8) *S. solidus.* Anti-met-enkephalin-positive perikarya (arrows) near main nerve cord (nc). m – muscles ×930. (9) *P. nigra.* Anti-serotonin-positive fibers in pharynx. ph p – pharyngeal pouch, ph l – pharyngeal lumen, m – muscles ×930.

form a nerve net around the bothridial muscles (Fig. 3). No positive perikarya were observed in or around the lateral ganglia, but – beginning from the basal part of the scolex – they occur along the main nerve cords throughout the entire length of the worm. Positive fibers were abundant in the nerve cords. Immunoreactive perikarya were observed in association with the peripheral nerve cords, and the longitudinal muscle layer is interspersed with positive fibers in the peripheral nerve net. The main nerve cords and the peripheral nerve net are interconnected by way of distinct anti-FMRF-amide-positive fibers. Positive fibers form terminals immediately beneath the basal lamina of the tegument. The terminals occur most frequently at the inner side of the bothridia (Fig. 4).

In *Schistocephalus solidus,* anti-FMRF-amide-positive fibers were observed within the CNS and the PNS, but the distribution of anti-FMRF-amide immunoreactivity has not been studied in detail.

Vasotocin

Immunocytochemical staining with anti-vasotocin has been performed on *M. lineare* and *D. dendriticum.*

Turbellarians

In *M. lineare,* a positive reaction with anti-vasotocin was obtained in perikarya in the brain (Fig. 5) and in the alimentary tract (Fig. 6). The positive cells in the gut seem to be facing the intestinal lumen. The distribution of anti-vasotocin-positive cells resembles that of the anti-FMRF-amide-positive ones. However, no positive reaction was detected in the fibres of the pharyngeal nerve ring nor in the anti-FMRF-positive perikarya on either side of the pharynx. The number of anti-vasotocin-positive nerve fibers was distinctly smaller than the number of fibers revealed by anti-FMRF-amide.

Cestodes

In *D. dendriticum,* the anti-vasotocin-positive fibers show a distribution similar to that of anti-FMRF-amide positive fibers. The number of anti-vasotocin-positive fibers is, however, smaller. Only a few positive perikarya were observed.

Leu- and met-enkephalins

Turbellarians

In *M. lineare* no immunoreactivity was obtained either in the brain or in the alimentary tract with antibodies to leu-enkephalin and met-enkephalin.

In *P. nigra,* anti-met-enkephalin immunoreactivity was detected in the longitudinal nerve cords.

Cestodes

In *D. dendriticum,* a positive reaction with anti-leu-enkephalin was obtained in perikarya along the main nerve cords and in the peripheral nerve net. Positive fibers surround the lateral ganglia, and were observed within the ganglionic commissure, the main nerve cords (Fig. 7), and the peripheral nerve cords. Some positive fibers form terminals beneath the basal lamina of the tegument. Anti-met-enkephalin-positive fibers were observed in the main nerve cords and in the scolex.

Positive reactions with anti-leu-enkephalin and anti-met-enkephalin were found in perikarya near the main nerve cords in *S. solidus* (Fig. 8).

Neurotensin

Turbellarians

No immunoreactivity with anti-neurotensin was obtained in *M. lineare* or in *P. nigra.*

Cestodes

In *D. dendriticum,* positive nerve fibers were found in the main and peripheral nerve cords. No immunoreactive perikarya were observed.

Somatostatin, VIP

With anti-somatostatin and anti-VIP (vasoactive intestinal polypeptide) no immunoreactivity was observed in either the turbellarians or the cestodes studied.

Serotonin

Turbellarians

An antibody to the bioamine serotonin gave a negative result in *M. lineare*[1]. (See note, page 98.)

In *P. nigra* anti-serotonin positive immunoreactivity was found in abundance within the nervous system. Positive perikarya and fibers occur more frequently in the gastrointestinal nervous system than in the CNS, and the pharynx, especially, is richly innervated by anti-serotonin-positive fibers (Fig. 9).

Cestodes

A large number of positive perikarya and fibers were found in the CNS and PNS of *D. dendriticum*. The perikarya were located in the ganglionic commissure, and along the main nerve cords. The distribution of anti-serotonin-positive perikarya and fibers in *S. solidus* resembles that in *D. dendriticum.*

Discussion

The present study, together with earlier IC investigations made on planarians (Bautz *et al.*, 1980; Schilt *et al.*, 1981; Venturini *et al.*, 1983), reveals the presence in flatworms of peptides showing sequential similarities with several vertebrate peptides. While great caution is needed in the interpretation of results obtained by IC methods alone, it is possible to compare the occurrence of immunoreactive substances in the different species used in this study.

Anti-FRMF-amide was the only antiserum that gave a positive reaction in all species studied, while anti-vasotocin was tested only on *D. dendriticum* and *M. lineare* and gave a positive reaction in both species. These substances have not been reported from other flatworms, but Grimmelikhuijzen (1983) found anti-FMRF-amide immunoreactivity in all classes of coelenterates. Our results support the view that FMRF-amide and related peptides are phylogenetically old neuroregulative substances that play an important role in primitive nervous systems.

We found anti-enkephalin positive immunoreactivity in the cestodes and in *P. nigra* but not in *M. lineare.* Met-enkephalin-positive immunoreactivity has previously been demonstrated in neurons and neuropil of the planarian *Dugesia gonocephala* (Venturini *et al.*, 1983). Neurotensin-positive fibers were present only in the cestodes. Yet neurotensin-like immunoreactivity has been reported from a planarian (Carraway *et al.*, 1982) and *Hydra* (Grimmelikhuijzen *et al.*, 1981). Our failure to detect immunoreactivity with anti-somatostatin in either cestodes or the turbellarians is surprising, since Bautz *et al.* (1980) describe the distribution of somatostatin-positive cells in *Dendrocoelum lacteum* and *Dugesia lugubris.* It is notable that antisera to these vertebrate peptides gave no positive reaction in *M. lineare,* a fact that may reflect the low phylogenetic level occupied by this microturbellarian.

Serotonin has been shown to be the predominant bioamine present in cestodes (for references, see Gustafsson & Wikgren, 1981a), and the occurrence of serotonin in *P. nigra* has also been reported (Lur'e, 1975). In *M. lineare*, only the occurrence of catecholamines was detected by the Falck-Hillarp method (Reuter *et al.*, 1980).[1]

Little is known about the function of vertebrate peptides in invertebrates. However, in molluscs and planarians a relationship between enkephalins and the dopaminergic system has been demonstrated (Stefano *et al.*, 1981; Juel, 1982; Venturini *et al.*, 1983). Interestingly, the cestodes gave a positive reaction with the anti-enkephalins but lack dopamine, while *M. lineare* contains catecholamines but gave a negative result for the enkephalins.

The status of vasotocin, neurotensin, somatostatin, and the enkephalins, as well as of the bioamine serotonin, as brain-gut peptides in vertebrates and various invertebrates is established (see e.g. Polak & Bloom, 1982; Greenberg & Price, 1983; Krieger, 1983). Furthermore, FMRF-amide or related substances are present in the CNS and alimentary tract of molluscs (Schot & Boer, 1982) and vertebrates (Dockray *et al.*, 1981). In planarians, cells showing immunoreactivity with antisera to vertebrate neuropeptides have been described only in association with the CNS (Bautz *et al.*, 1980; Schilt *et al.*, 1981; Venturini *et al.*, 1983). The present study reveals the dual location of immunoreactivity to FMRF-amide and vasotocin in the CNS and the gut of *M. lineare.* According to Grimmelikhuijzen (1983), the distribution of the anti-FMRF-amide positive neurons in the coelenterates indicates a role of FMRF-amide in food capture and ingestion, and in molluscs FMRF-amide and related substances have been shown to act as muscular neurotransmitters and neuromodulators (Greenberg & Price, 1980; Cottrell *et. al.*, 1983; Schot *et al.*, 1983). Probably FMRF-amide has similar functions in *M. lineare.*

In the cestodes, adapted to a parasitic mode of life and totally devoid of an alimentary tract, the tegument acts as an important absorptive surface. While a positive reaction with anti-neurotensin was restricted to the nerve cords, immunoreactivity with anti-FRMF-amide, anti-vasotocin, anti-leu-enkephalin, anti-met-enkephalin, and anti-serotonin was found in both the CNS and the peripheral nerve net. Only anti-serotonin-positive perikarya, however, were observed within the lateral ganglia. The musculature and the subtegumental region were richly innervated, especially by anti-FMRF-amide- and anti-vasotocin-positive fibers. The absence of peptide-positive perikarya in the lateral ganglia emphasizes the mainly peripheral (paracrine?) nature of this type of neurosecretion. It is thus tempting to draw a parellel between the action of brain-gut peptides in the alimentary tract of the free-living flatworms on one hand, and in the subtegumental region of the cestodes on the other hand.

The demonstration of a peptide by the IC method has to be related to the release of the active substances and to a possible physiological function. Reuter (1981) describes release sites of neurosecretory material in the nervous system of *M. lineare,* including the pharyngeal nerve ring and the division zone. Release sites have also been observed in the gut (Reuter, unpublished). Release from peptidergic axon terminals in the subtegumental region of *D. dendriticum* occurs in connection with the change of host (Gustafsson & Wikgren, 1981c). In order to get a better picture of the neuroregulative substances involved, we are in the process of applying immunocytochemistry at the ultrastructural level.

Note added in proof

Recently, whole mount IC using anti-serotonin (dilution 1:100) gave a positive reaction in the NS of *M. Lineare.*

References

Bautz, A., J. Schilt, J. P. Richoux & M. P. Dubois, 1980. Détection immunocytologique, dénombrement et localisation des cellules à somatostatine (SRIF) chez deux espèces de Planaires, Dugesia lugubris et Dendrocoelum lacteum (Turbellariès, Triclades). C.r. Acad. Sci., Paris. Ser. D. 291: 833–836.

Carraway, R., S. E. Ruane & H.-R. Kim, 1982. Distribution and immunochemical character of neurotensin-like material in representative vertebrates and invertebrates: apparent conservation of the COOH-terminal region during evolution. Peptides 1: 115–123.

Cottrell, G. A., L. P. Schot & G. J. Dockray, 1983. Identification and probable role of a single neurone containing the neuropeptide Helix FMRF-amide. Nature 304: 638–640.

Dockray, G. J., C. Vaillant, R. G. Williams, R. J. Gayton & N. N. Osborne, 1981. Vertebrate brain-gut peptides related to FMRF-amide and Met-enkephalin Arg^6Phe^7. Peptides 2: 25–30.

Greenberg, M. J. & D. A. Price, 1980. Cardioregulatory peptides in molluscs. In F. E. Bloom (ed.), Peptides: Integrators of Cell and Tissue Function. Raven Press, New York: 107–126.

Greenberg, M. J. & D. A. Price, 1983. Invertebrate neuropeptides: native and naturalized. Ann. Rev. Physiol. 45: 271–288.

Grimmelikhuijzen, C. J. P., 1983. FMRF-amide immunoreactivity is generally occurring in the nervous system of coelenterates. Histochemistry 78: 361–381.

Grimmelikhuijzen, C. J. P., R. E. Carraway, Å. Rökaeus & F. Sundler, 1981. Neurotensin-like immunoreactivity in the nervous system of hydra. Histochemistry 72: 199–209.

Gustafsson, M. K. S. & M. C. Wikgren, 1981a. Peptidergic and aminergic neurons in adult Diphyllobothrium dendriticum Nitzsch, 1824 (Cestoda, Pseudophyllidea). Z. Parasitenkd 64: 121–134.

Gustafsson, M. K. S. & M. C. Wikgren, 1981b. Activation of the peptidergic neurosecretory system in Diphyllobothrium dendriticum (Cestoda, Pseudophyllidea). Parasitology 83: 243–247.

Gustafsson, M. K. S. & M. C. Wikgren, 1981c. Release of neurosecretory material by protrusions of bounding membranes extending through the axolemma, in Diphyllobothrium dendriticum (Cestoda). Cell Tissue Res. 220: 473–479.

Juel, C., 1982. Enkephalin increases the efficacy of dopaminergic transmission in Helix pomatia. Neurophysiology 21: 1301–1303.

Krieger, D. T., 1983. Brain peptides: what, where, and why? Science 222: 975–985.

Lur'e, B. L., 1975. Monoamine-containing neurons of the planarian Polycelis nigra (Turb.). Moscow Univ. Biol. Sci. Bull. 30: 69–76.

Polak, J. M. & S. R. Bloom, 1982. Localization of regulatory peptides in the gut. Br. med. Bull. 38: 303–307.

Price, D. A. & M. J. Greenberg, 1977. Structure of a molluscan cardioexcitatory neuropeptide. Science 197: 670–671.

Reuter, M., 1981. The nervous system of Microstomum lineare (Turbellaria, Macrostomida). II. The ultrastructure of synapses and neurosecretory release sites. Cell Tissue Res. 218: 375–387.

Reuter, M., T. Karhi & L. P. C. Schot, 1984. Immunocytochemical demonstration of peptidergic neurons in the central and peripheral nervous system of the flatworm Microstomum lineare with antiserum to FMRF-amide. Cell Tissue Res. 238: 431–436.

Reuter, M. & I. Palmberg, 1983. Asexual reproduction in Microstomum lineare (Turbellaria). II. The nervous system in the division zone. Int. J. Invert. Reprod. 6: 207–217.

Reuter, M., M. Wikgren & I. Palmberg, 1980. The nervous system of Microstomum lineare (Turbellaria, Macrostomida). I. A fluorescence and electron microscopic study. Cell Tissue Res. 211: 31–40.

Schilt, J., J. P. Richoux & M. P. Dubois, 1981. Demonstration of peptides immunologically related to vertebrate neurohormones in Dugesia lugubris (Turbellaria, Tricladida). Gen. Comp. Endocrinol. 43: 331–335.

Schot, L. P. C. & H. H. Boer, 1982. Immunocytochemical demonstration of peptidergic cells in the pond snail Lymnaea stagnalis with an antiserum to the molluscan cardioactive tetrapeptide FMRF-amide. Cell Tissue Res. 225: 347–354.

Schot, L. P. C., H. H. Boer & J. Vijdenes, 1983. Localization of neurons innervating the heart of Lymnaea stagnalis studied immunocytochemically with anti-FMRF-amide and anti-vasotocin. In J. Lever & H. H. Boer (eds), Molluscan Neuroendocrinology. Mon. Royal Neth. Academy of Arts and Sciences, North Holland Publ. Comp., Amsterdam, Oxford, New York: 203–208.

Stefano, G. B., E. J. Catapane & R. M. Kream, 1981, Characterization of the dopamine stimulated adenylate cyclase in the pedal ganglia of Mytilus edulis: Interactions with etorphine, β-endorphin, DALA, and methionine enkephalin. Cell Mol. Neurobiol. 1: 57–68.

Sternberger, L. A., P. H. Hardy Jr., J. J. Cuculis & H. G. Meyer, 1970. The unlabelled antibody enzyme method of immunochemistry. Preparations and properties of soluble antigen-antibody complex (Horseradish peroxidase-antihorseradish peroxidase) and its use in identification of spirochetes. J. Histochem. Cytochem. 18: 315–333.

Venturini, G., A. Carolei, G. Palladini, V. Margotta & M. G. Lauro, 1983. Radioimmunological and immunocytochemical demonstration of met-enkephalin in planaria. Comp. Biochem. Physiol. C. 74: 23–25.

Aspects of photoreceptor structure and phototactic behavior in Platyhelminthes, with particular reference to the symbiotic turbellarian *Paravortex*

Alan W. Pike & Robert Wink
Department of Zoology, University of Aberdeen, Aberdeen, Scotland, U.K. AB9 2TN

Keywords: Turbellaria, Platyhelminthes, Paravortex, symbiosis, photoreceptor, phototaxis

Abstract

Photoreceptor structure and function in the Platyhelminthes has traditionally been treated separately in the Turbellaria on one hand and the conventional parasitic classes on the other. In this paper, an attempt is made to bring together data from the literature and to highlight deficiencies and areas where a more integrated approach would be beneficial. This is done with particular reference to the endosymbiont genus *Paravortex* which belongs to the Turbellaria but which functionally has more similarities to the parasitic platyhelminths especially with regard to the host-finding requirement of the larval stages.

Introduction

The literature on platyhelminth biology generally separates along taxonomic lines into that on the Turbellaria (both free-living and symbiotic) and that on the conventionally recognized parasitic classes of platyhelminths. Such a dichotomy has unfortunate consequences in that the benefits of a broader view of the phylum are lost. In particular, the literature on symbiotic turbellarians rarely overlaps with that of the parasitic platyhelminths although exceptions do occur – for instance, the work of Jennings (1974), Calow & Jennings (1974) and Jennings & Calow (1975) which has made a valuable contribution to the thinking on the evolution of parasitism.

The symbiotic *Paravortex* spp., which inhabit various species of bivalves, are metabolically dependent upon their hosts, especially with regard to nutritional and respiratory physiology (Jennings & Phillips, 1978; Jennings & LeFlore, 1979). The life cycle of these organisms has more in common with that of some parasitic platyhelminths than with that of other turbellarians, particularly as far as the biology of the larva is concerned. The host-locating requirement and capability of larval *Paravortex* spp. also have many similarities to those of the miracidium (Digenea) and oncomiracidium (Monogenea).

It is evident that a more inclusive view is necessary in considering aspects of the biology of parasitic and free-living platyhelminths. Accordingly, in this review we discuss the structure and function of photoreceptors in both larval and adult stages of the major platyhelminth groups and, by means of comparison, attempt to expose the artificial gap that stands between the turbellarians and the parasitic classes.

Photoreceptor occurrence and structure in platyhelminths

The presence of cerebral photoreceptors is a characteristic, though not universal, feature of the adult platyhelminths (Hyman, 1951). They are absent from some turbellarians, and monogeneans, most digeneans and all cestodes. They occur more commonly on the larval stages of all groups except the Cestoda. A single pair of laterally oriented,

Hydrobiologia 132, 101–104 (1986).

antero-dorsally situated rhabdomeric photoreceptors is typical of most species, although Monogenea often possess two pairs, and numerous variations are seen in the Turbellaria (Hyman, loc. cit.). The situation in larval stages is frequently more complex especially when one considers non-pigmented ciliary photoreceptors.

The cerebral photoreceptors of adult turbellarians typically consist of a multicellular pigment cup with numerous microvillous rhabdomeres. In larval stages these photoreceptors comprise a single pigment cup cell and a small number of rhabdomeres (Lanfranchi *et al.*, 1981; Eakin & Brandenburger, 1981). There have been no similar studies of adult monogeneans or digeneans but the larvae possess a unicellular pigment cup containing only one or two rhabdomeres (Fournier & Combes, 1978). While cilia have been found in developing cerebral photoreceptors of some turbellarians (Lanfranchi *et al.*, *loc. cit.*; Eakin & Brandenburger *loc. cit.*) they have not been reported for either larval digeneans or monogeneans. It would be interesting to know whether cilia are entirely absent from developmental stages of the parasitic platyhelminths.

Whereas cerebral photoreceptors are usually present in both adult and larval turbellarians, their occurrence in parasitic platyhelminths is less consistent. Monogenean larvae have been divided in the past (Llewellyn, 1963) into those that possess two pairs of photoreceptors containing spherical lenses and those possessing a single pair of photoreceptors usually without lenses but sometimes with oil droplets that may serve a similar function. Newer studies by Kearn (1978) and Fournier (1975) suggest that this distinction may be less reliable than hitherto believed. However, the description by Kearn (*loc. cit.*) of a second pair of photoreceptors in *Diplozoon paradoxum* larvae adds weight to Lyons' (1972) suggestion that two pairs of photoreceptors are the primitive condition in the Monogenea.

Photoreceptors may be retained in adult monogeneans or undergo partial or complete breakdown with subsequent resorption (Kearn, 1971; Fournier *loc. cit.*). Some monogeneans are reported to be without photoreceptors in larval and/or adult stages although ascertaining such a thing is difficult in view of the increasing number of reports of non-pigmented photoreceptors.

The situation is slightly less complex in the Digenea. With the exception of the ectoparasitic genus *Transoversotrema* which has a single pair of pigment cup photoreceptors (Whitfield & Wells, 1973), all adult digeneans are endoparasitic and lack photoreceptors. The two free swimming larval stages, the miracidium and the cercaria, may possess a single pair of pigment cup photoreceptors (Brooker, 1972; Iseroff & Cable, 1968). Additional structures resembling non-pigmented photoreceptors have also been reported in some cercariae (Erasmus, 1972). In those species such as the heterophyid cercariae, in which photoreceptors are present, these often persist in the metacercarial stage and only disappear as the metacercaria develops to the adult stage in the definitive host.

The most obvious variations in the structure of platyhelminth photoreceptors is seen in the Monogenea, in which, in addition to the spherical lenses, there is (in *Polystoma integerrimum* and several other polystomids) an unusual pigment-cup cell containing reflective plates which reflect light to a focal point on the rhabdomeres (Fournier & Combes, 1978).

Ciliary photoreceptors have now been reported from several turbellarians (Ehlers & Ehlers, 1977a, b; Sopott-Ehlers, 1982) as well as from larval monogeneans (Lyons, 1972), larval digeneans (Brooker, 1972; Short & Gagné, 1975), and one cestode larva (Xylander, 1984). There appears to be variation in the degree of modification of the ciliary axoneme.

Epidermal photoreceptors have not been reported from any parasitic platyhelminths.

Function of photoreceptors

Despite the absence of direct physiological evidence, it is generally accepted, by analogy with other invertebrates for which experimental evidence is available, that the photoreceptors of platyhelminths are responsible for the way in which these organisms respond to light. For most adult turbellarians there is a photonegative response, whereas many larval stages are photopositive (Hyman, 1951). Such behavior would be consistent with habitat preference in adults and dispersal in the larvae. There is more variation in the larval stages of parasites since the requirement for locating a host may demand differing phototactic responses.

Little is known about the behavior of monogenean larvae (Llewellyn, 1963) except that the oncomiracidium of *Entobdella soleae* has complex responses to light (Kearn, 1980). More studies have been made on the miracidia and cercariae of digeneans (Smyth & Halton, 1983) but generally without control over the quality of light being used.

It is very likely that phototactic responses of larvae play a role in dispersal, especially for monogeneans (Llewellyn, 1972; Kearn, 1980). A further complexity is the existence of diurnal periodicity of egg hatching in which light is known to be involved (Kearn, 1973).

The variety of photoreceptor structures present, particularly in larval platyhelminths, is not presently satisfactorily explained by existing knowledge of their behavior. In general, morphological studies have not been matched with behavioral studies that might explain the presence of several photoreceptor types in one individual.

Photoreceptors and behavior in *Paravortex* spp.

The four known species of this endosymbiotic genus all possess a pair of antero-dorsally situated pigment cup photoreceptors. In *P. karlingi*, the photoreceptor pigment cup is usually irregularly shaped. Diameters of pigment-cup cells in fixed worms are approximately 0.010 to 0.015 mm ($\bar{x}$ of 22, 0.013 mm). In juvenile worms, the pigment cup cell is more regularly shaped and measures 0.003 to 0.008 mm long and 0.003 to 0.007 mm broad ($\bar{x}$ of 12, 0.006 by 0.004 mm). There is, therefore, an approximately twofold increase in the size of the pigment cup from juvenile to adult. The size of pigment granules also increases from juvenile to adults.

The existence of clear phototactic responses in the adults of *P. cardii* and *P. karlingi* has already been reported (Pike & Burt, 1981), indicating no loss of photoreceptive capability despite the endosymbiotic existence of these organisms in the alimentary canal of the bivalve *Cerastoderma edule*. It is noteworthy that despite similarity in habitat and way of life, these species respond differently to light.

Discussion

In free-living turbellarians, photoreceptors are functionally as important in the adults as in the larvae, although their function may differ in each stage. This is not the case for the Digenea and probably not for most monogeneans. It is not surprising, therefore, to see the disappearance of photoreceptors from adult digeneans and their frequent reduction and even disappearance in monogeneans, even though the latter may be attached to the host's skin where light can still be perceived (Fournier & Comber, 1978). It is difficult to recognize a selective advantage to the parasites in losing photoreceptors unless there were metabolic penalties for maintaining them. Sometimes the photoreceptors break down rapidly; for example, in *Euzetrema* (Monogenea) the rhabdomeres begin to degenerate within 48 h of the parasite establishing itself in the host (Fournier, 1975).

A similar reduction of photoreceptors could be expected for the endosymbiotic *Paravortex*, but this is not the case: Pike & Burt (1981) reported clear phototactic responses in both *P. cardii* and *P. karlingi*. The persistence and considerable growth of the photoreceptors between larvae and adult of this genus re-affirm the continuing importance of the photoreceptors. Despite this the adult worms are not known to leave the host, and it seems unlikely that light reaches them through the host tissues.

While dark adaptation in turbellarian photoreceptors is widely reported (Röhlich & Török, 1962; Röhlich & Tar, 1968; Carpenter *et al.*, 1974; Bedini *et al.*, 1977; and Lanfranchi & Bedini, 1982) it has been ignored in the parasitological literature despite the fact that it may be of interest to the process of degeneration of photoreceptors. One must presume that most, if not all, studies on parasite larvae have been on light-adapted worms, although this is never stated.

We hope to encourage other researchers to seek common solutions to what have been hitherto regarded as separate problems of the respective platyhelminth groups. In particular, a more integrated approach is needed to rectify the patchiness in our knowledge of the development of platyhelminth photoreceptors.

Acknowledgments

The financial support of the Carnegie Trust for Scotland, the University of Aberdeen, and the Royal Society is gratefully acknowledged by A.W.P. Robert Wink acknowledges the support of the University of Aberdeen and NERC. We gratefully acknowledge the excellent services of Irene Johnston who typed the manuscript at very short notice.

References

Brooker, B. E., 1972. The sense organs of trematode miracidia. In E. U. Canning & C. A. Wright (eds), Behavioural aspects of parasite transmission. Zool. J. linn. Soc. 51 suppl.: 171–180.

Bedini, C., E. Ferrero & A. Lanfranchi, 1977. Fine structural changes induced by circadian light-dark cycles in photoreceptors of Dalyellidae (Turbellaria: Rhabdocoela). J. ultrastruct. Res. 58: 66–77.

Calow, P. & J. B. Jennings, 1974. Calorific values in the phylum Platyhelminthes: the relationship between potential energy, mode of life and the evolution of entoparasitism. Biol. Bull. 147: 81–94.

Carpenter, K. S., M. Morita & J. B. Best, 1974. Ultrastructure of the photoreceptor of the planarian Dugesia dorotocephala. II. Changes induced by darkness and light. Cytobiologie 8: 320–338.

Eakin, R. M. & J. L. Brandenburger, 1981. Fine structure of the eyes of Pseudoceros canadensis (Turbellaria, Polycladida). Zoomorphologie 98: 1–16.

Ehlers, B. & U. Ehlers, 1977a. Die Feinstruktur eines ciliaren Lamellarkörpers bei Parotoplanina geminoducta Ax (Turbellarian, Proseriata). Zoomorphologie 87: 65–72.

Ehlers, B. & U. Ehlers, 1977b. Ultrastruktur perecerebraler Cilienaggregate bei Dicoelandropora atriopapillata Ax und Notocaryoplanella glandulosa Ax (Turbellaria, Proseriata). Zoomorphologie 88: 163–174.

Erasmus, D. A., 1972. The biology of trematodes. Arnold, London, 312 pp.

Fournier, A., 1975. Les taches pigmentaires de l'oncomiracidium d'Euzetrema knoepffleri (Monogenea): ultrastructure et évolution au cours du cycle biologique. Z. Parasitkde 46: 203–209.

Fournier, A. & C. Combes, 1978. Structure of photoreceptors of Polystoma integerrimum (Platyhelminthes, Monogenea). Zoomorphologie 91: 147–155.

Hyman, L. H., 1951. The Invertebrates: Platyhelminthes and Rhynchocoela. The acoelomate bilateria. Vol. II. McGraw-Hill, N.Y., 550 pp.

Isseroff, H. & R. M. Cable, 1968. Fine structure of photoreceptors in larval trematodes. Z. Zellforsch. 86: 511–534.

Jennings, J. B., 1974. Symbiosis in the Turbellaria and their implications in studies on the evolution of entoparasitism. In W. B. Vernberg (ed), Symbiosis in the Sea. University of South Carolina Press: 127–160.

Jennings, J. B. & P. Calow, 1975. The relationship between high fecundity and the evolution of entoparasitism. Oecologia 21: 109–115.

Jennings, J. B. & W. B. LeFlore, 1979. Occurrence and possible adaptive significance of some histochemically demonstrable dehydrogenases in two entosymbiotic rhabdocoels (Platyhelminthes: Turbellaria). Comp. Biochem. Physiol. 62B: 301–304.

Jennings, J. B. & J. I. Phillips, 1978. Feeding and digestion in three entosymbiotic graffillid rhabdocoels from bivalve and gastropod molluscs. Biol. Bull. 155: 542–562.

Kearn, G. G., 1971. The attachment site, invasion route and larval development of Trochopus pini, a monogenean from the gills of Trigla hirundo. Parasitology 63: 513–525.

Kearn, G. C., 1973. An endogenous circadian hatching rhythm in the monogenean skin parasite, Entobdella soleae, and its relationship to the activity rhythm of the host (Solea solea). Parasitology 66: 101–122.

Kearn, G. G., 1978. Eyes with, and without, pigment shields in the oncomiracidium of the monogenean parasite Diplozoon paradoxum. Z. Parasitkde 57: 35–47.

Kearn, G. G., 1980. Light and gravity responses of the oncomiracidium of Entobdella soleae and their role in host location. Parasitology 81: 71–89.

Lanfranchi, A. & C. Bedini, 1982. The ultrastructure of the sense organs of some Turbellaria Rhabdocoella: 1. The eyes of Polycystis naeglii (Eukalyptorhynchia, Polycystididae). Zoomorphologie 101: 95–102.

Lanfranchi, A., C. Bedini & E. Ferrero, 1981. The ultrastructure of the eyes in larval and adult polyclads (Turbellaria). Hydrobiologia 84: 267–275.

Llewellyn, J., 1963. Larvae and larval development of monogeneans. Advances in Parasitology 1: 287–326.

Llewellyn, J., 1972. Behaviour of monogeneans. In E. U. Canning & C. A. Wright (eds), Behavioural aspects of parasite transmission. Zool. J. Linn. Soc. 51 suppl.: 19–30.

Lyons, K. M., 1972. Sense organs of monogeneans. In E. U. Canning & C. A. Wright (eds), Behavioural aspects of parasite transmission. Zool. J. linn. Soc. 51 suppl.: 181–199.

Pike, A. W. & M. D. B. Burt, 1981. Paravortex karlingi sp. nov from Cerastoderma edule L., in Britain. Hydrobiologia 84: 23–30.

Röhlich, P. & E. Tar, 1968. The effect of prolonged light deprivation on the fine structure of planarian photoreceptors. Z. Zellforsch. mikrosk. Anat. 90: 507–518.

Röhlich, P. & J. Török, 1962. The effect of light and darkness on the fine structure of the retinal clubs of Dendrocoelum lacteum. Q. J. microsc. Sci. 104: 543–548.

Short, R. B. & H. T. Gagné, 1975. Fine structure of a possible photoreceptor in cercariae of Schistosoma mansoni. J. Parasitol. 61: 69–74.

Smyth, J. D. & D. H. Halton, 1983. Physiology of Trematodes. 2nd Ed. Cambridge University Press.

Sopott-Ehlers, B., 1982. Ultrastructure of potential photoreceptor cells of a different organization in a proseriate (Platyhelminthes). Zoomorphologie 101: 165–176.

Xylander, W. E. R., 1984. A presumptive ciliary photoreceptor in larval Gyrocotyle urna Grube and Wagener (Cestoda). Zoomorphology 104: 21–25.

Whitfield, P. J. & J. Wells, 1973. Observations on the ectoparasitic digenean Transversotrema patialensis. Parasitology 67: xxvii–xxviii.

Embryology of the Turbellaria and its phylogenetic significance

Mary Beth Thomas
Department of Biology, The University of North Carolina at Charlotte, Charlotte, N.C. 28223, USA

Keywords: Turbellaria, phylogeny, embryology, cleavage, gastrulation, eggshell granules

Abstract

Developmental characters – including oocyte and yolk cell structure, patterns of cleavage, and modes of gastrulation – are presented and examined in relation to the phylogeny of the Turbellaria. Eggshell granules, which have been demonstrated to occur in the oocytes of entolecithal eggs and the yolk cells of ectolecithal eggs, are compared among species, and their potential value as a taxonomic character is discussed. The quartet 4d spiral cleavage of the entolecithal egg of polyclads is described as reminiscent of the primitive pattern of early development for the Turbellaria. This is compared to duet spiral cleavage of acoels, and possible phylogenetic schemes involving the two types of spiral cleavage are reviewed. The link between the precise spiral cleavage, which characterizes development of most archoophorans, and blastomere separation (*Blastomeren-Anarchie*), which occurs in several neoophoran orders, is established by the occurrence of quartet 4d spiral cleavage in one neoophoran order, and of both quartet spiral cleavage and *Blastomeren-Anarchie* in different species of a second neoophoran order. The epibolic gastrulation of polyclads is described as primitive for the Turbellaria because of its similarity to that of other members of the Spiralia. Although no identical process occurs in neoophoran development, the earlier event of formation of the hull membrane in some neoophorans, and the later event of formation of the definitive epidermis in all neoophorans studied are presented as processes of possible homology to the epibolic gastrulation of polyclads. The lack of correspondence between polyclads and neoophorans in the relationship of the definitive body axes to the egg axis is discussed, and an hypothesis is advanced to account for the differences. The phylogenetic relationships indicated by known developmental phenomena differ only slightly from the scheme presented by Karling in 1974.

Introduction

The notion of using developmental features to elucidate phylogenetic relationships among the Turbellaria is compelling because of the large number of developmental characters and the wide variation in these characters. In an attempt to illustrate their potential value to phylogenetic studies, I have selected for emphasis morphological and morphogenetic characters concerned with 1) oocyte and yolk cell structure, 2) patterns of cleavage, and 3) modes of gastrulation, including the associated phenomena of the origin of the hull membrane and the establishment of body axes.

Oocyte and yolk = cell structure

Ectolecithy involves the separation of the germ line into oocytes and yolk cells, the latter of which are packaged in the eggshell for the nutrition of the embryo. It occurs in the neoophoran turbellarians as well as the trematodes and cestodes. Because ectolecithy has arisen within the Turbellaria, it is

Hydrobiologia 132, 105–115 (1986).

generally accepted that the primitive condition was the entolecithal egg in which yolk stores were integral to the oocyte, as exhibited by the archoophoran turbellarians (including the Polycladida). Although the polyclads appear to be more highly differentiated than the other archoophoran orders, their pattern of quartet spiral cleavage is considered to represent a retention of a primitive trait (Ax, 1961, 1963). Since polyclads exhibit two primitive traits, entolecithy and quartet spiral cleavage, the structure of their oocytes is of particular interest with regard to the primitive condition. Fortunately, the oocytes of polyclads have been studied both ultrastructurally and cytochemically in both cotyleans (Boyer, 1972) and acotyleans (Domenici *et al.*, 1975; Ishida *et al.*, 1981, 1985). Examination of the oocytes reveals a feature which may be particularly useful for phylogenetic studies, the eggshell granule (Fig. 1).

In all cases in which these granules have been subjected to cytochemical tests for polyphenols (Domenici *et al.*, 1975; Ishida *et al.*, 1981), the tests have been positive. Their chemical composition strongly supports the view first advanced for the polyclads by v. Hofsten (1912) and reiterated by Kato (1940) that the granules function in the elaboration of the sclerotin-like capsule which surrounds the polyclad egg.

Synthesis of eggshell granules, which involves the rough endoplasmic reticulum and the Golgi apparatus, begins at the time of breakdown of the germinal vesicle (see Kato, 1940) and precedes the initiation of yolk synthesis. As the egg matures and yolk granules accumulate centrally, the shell granules assume a peripheral location, at which time they average 1.25 – 2.5 μm in diameter. Despite the fact that authors differ somewhat in their descriptions of the ultrastructure of the granules, comparisons of micrographs strongly suggest that they are quite similar among the species studied, especially given differences in preparative techniques and planes of section and given the fact that, at least in *Gnesioceros floridana* (P. Espinosa & M. B. Thomas, in preparation), the granules go through a series of morphological changes as the oocyte matures.

Since all turbellarians possess egg coverings that, with a few exceptions such as the summer egg of *Mesostoma ehrenbergii* (see Domenici & Gremigni, 1977), are compositionally quite similar (see Smyth & Clegg, 1959), it appears likely that shell granules would be found in entolecithal oocytes of other archoophorans, and in either the oocytes or yolk cells of the ectolecithal eggs of the neoophorans. Among the archoophoran orders, eggshell granules have been observed by light microscopy in various macrostomids (v. Hofsten, 1912; Seilern-Aspang, 1957; Reiger, 1981), but their ultrastructure and cytochemistry are not known.

Studies to date of neoophorans indicate that granules from the yolk cells, rather than from the oocyte, give rise to the shell, as first demonstrated by v. Hofsten in response to a similar finding for trematodes by Henneguy and Goldschmidt (see v. Hofsten, 1912).

Shell granules that can be compared closely with those of polyclads are those of the rhabdocoels *Microdalyellia fairchildi* (see Bunke, 1979, 1981, 1982) and *Mesostoma ehrenbergii* (see Domenici & Gremigni, 1977) and the triclads *Dugesia lugubris* (see Gremigni & Domenici, 1974) and *Dendrocoelopsis ezensis* (see Yanagita and Yamamoto, 1981). The granules in each of these species have been shown to contain polyphenols and to be involved in shell formation. Perhaps of significance for phylogenetic studies, the shell granules of the yolk cells of *Dugesia*, the only triclad in which the yolk cells have been examined ultrastructurally, are remarkably similar morphologically to those described for the oocyte of the polyclad (Fig. 1A & B). By contrast, the eggshell granules of the two rhabdocoels, one a dalyellioid and the other a typhloplanoid, are morphologically distinct from the granules in the polyclads and triclads, but do resemble each other (Fig. 1C & D). The occurrence of two ultrastructurally distinct forms of a complex organelle which appear, from the limited number of cases studied, to sort out along taxonomic lines is of potential value to taxonomic and phylogenetic studies. Of interest in this regard is the newly discovered species *Acanthiella chaetonotoides* (Rieger & Sterrer, 1975a & b). It was described as a primitive neoophoran and, on the basis of other characters, tentatively assigned to the Prolecithophora. The observations that the egg itself contains shell granules (Rieger, 1981) and that the granules resemble those of polyclads, support the conclusion that the species is primitive. Comparative studies of the oocytes and yolk cells of other species of Prolecithophora and other primitive neoophorans

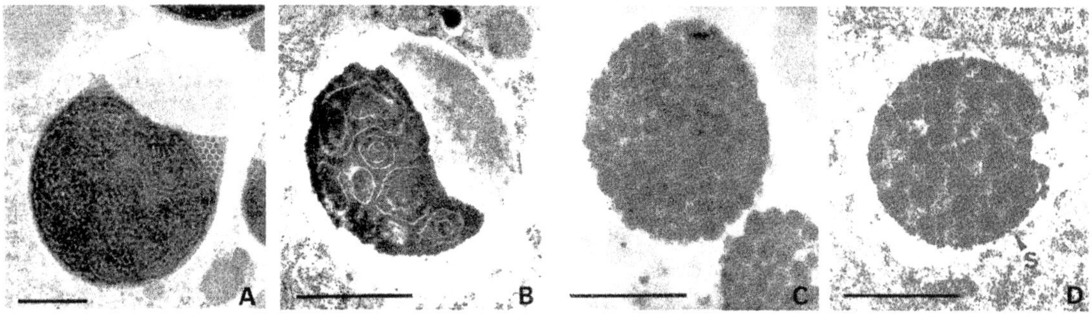

Fig. 1. Eggshell granules from A) oocyte of the polyclad *Gnesioceros floridana*, B) yolk cell of the triclad *Dugesia lugubris*, C) yolk cell of the typhloplanoid rhabdocoel *Mesostoma ehrenbergii*, D) yolk cell of the dalyellioid rhabdocoel *Microdalyellia fairchildi*. Bar=1 μm. (A. original; B. from Gremigni & Domenici, 1974; C. from Domenici & Gremigni, 1977; D. from Bunke, 1981; B–D reproduced with the permission of Springer-Verlag).

may shed additional light on their taxonomic position.

In addition, analysis of other features of oocyte and yolk cell structure in the Turbellaria may prove illuminating as suggested by Bunke (1981). Of particular note in this respect are the cortical granules and yolk granules that have been observed in the oocytes of certain triclads (Gremigni & Domenici, 1975; Gremigni, 1979; Gremigni & Nigro, 1983). These organelles, which occur in addition to shell and yolk granules of the yolk cells in these forms, have already provided important information concerning phylogeny within the order (Gremigni, 1979) and may prove relevant in clarifying relationships outside the order.

Cleavage

Because the presence of external yolk has affected developmental processes, it is logical to look for remnants of the more primitive developmental schemes in forms with entolecithal eggs. Among the archoophorans examined, the predominant cleavage pattern is quartet spiral cleavage which has been described for species in the Macrostomida, Catenulida, and Polycladida (for a review, see Ax, 1961). Although the polyclads are not themselves thought to be a primitive group, their mode of cleavage is considered to be plesiomorphic (Ax, 1961, 1963). Their development has been well studied, both morphologically (Lang, 1884; Surface, 1907; Kato, 1940) and, recently, experimentally

Table 1. Comparison of spiral cleavage in the Polychaeta, Polycladida, and Lecithoepitheliata[1].

Taxon	Type of cleavage	Relative size of quadrants	Fate of 2d	Fate of 4d	Size of 4A–4D	Fate of 4A–4D
Polychaeta (*Nereis limbata* and others)	Quartet spiral	Unequal	Primary somatoblast	Mesoderm	Large	Endoderm
Polycladida (several species)	Quartet spiral	Equal	Not distinguished as primary somatoblast	Mesoderm Endoderm	Small	Disintegrate and are absorbed
Lecithoepitheliata (*Xenoprorhynchus steinboecki*)	Quartet spiral	Unequal	Not distinghuished as primary somatoblast	Mesoderm	Large	Endoderm

[1] Summarized from Wilson, 1892; Okada, 1968; Kato, 1940, 1968; Reisinger *et al.*, 1974a & b.

(Boyer, 1981, 1985). Although the early development of polyclads is different in a number of ways from that of polychaetes (Table 1), which have been proposed as a stem group in the phylogeny of the Spiralia (see Siewing, 1980), the similarities, especially when contrasted with development in other groups of organisms, leave little doubt of a relationship among all forms with spirally cleaving eggs (Hyman, 1951; Ax, 1961, 1963; Reisinger *et al.*, 1974b; Ivanova-Kasas, 1981).

Cleavage in the polyclads is homoquadratic, or only slightly heteroquadratic, whereas the D quadrant is generally larger than the other three in the polychaetes and most other forms with spiral cleavage (Table 1). In polychaetes – but not polyclads – micromere 2d, the first somatoblast, is larger than the other cells of the second quartet and gives rise to the ectoderm of the body wall; in polyclads the origin of the body wall is not so restricted. In polychaetes, micromere 4d is the second somatoblast, dividing bilaterally to give rise to all the mesendoderm of the developing organism. Micromere 4d has the same fate in polyclads. In most spirally cleaving forms, the micromeres of the fourth quartet (4a–4d) are smaller than the macromeres (4A–4D); in the polyclads, however, the reverse is true. In addition, the macromeres 4A–4D together with the three micromeres 4a–4c, which form endoderm in other Spiralia, disintegrate and are assimilated in polyclads, and both endoderm and mesoderm originate from 4d. Experimental work on polyclads (Boyer, 1981, 1985) indicates that, as in polychaetes, early development is determinate. Cleavage in polychaetes and polyclads culminates in the formation of a stereoblastula.

Important to considerations of phylogeny is the existence of an alternate form of spiral cleavage within the Turbellaria, the duet spiral cleavage of acoels (for a review, see Henley, 1974). Experimental studies involving destruction of specific blastomeres of the acoel *Neochildia fusca* suggest, quite surprisingly, that development is regulative rather than mosaic (Boyer, 1971).

Questions of the evolution of spiral cleavage are intriguing (see Schmidt, 1966; v. Salvini-Plawen, 1978; Siewing, 1980; Ivanova-Kasas, 1959, 1981). Siewing's (1980) archicoelomate theory of the origin of the Bilateria argues that the cleavage exemplified by polychaetes would most closely resemble that of the primitive spirally cleaving egg, with that of polyclads – and, of course, acoels – being derived. Ivanova-Kasas (1981) takes strong exception to the view that the highly regimented, precise process in polychaetes could have evolved directly and suggests that a more likely origin would have been by way of a primitive regulative duet spiral cleavage which would subsequently have given rise to the duet spiral cleavage of acoels as well as the mosaic quartet spiral cleavage of the Spiralia. The implication of such a proposal causes serious problems since it requires either independent evolution of quartet 4d spiral cleavage in the Turbellaria and other Spiralia or a phylogenetic separation of the Acoela from all other turbellarians. Knowledge of development in the Nemertodermatida, which show close alignment with the Acoela, could be most instructive.

A third view, presented by Ax (1961, 1963), suggests that both the turbellarians and other spiralians arose from a common ancestor with primitive quartet spiral cleavage and that duet spiral cleavage is apomorphic. Such a proposal requires that not only the cleavage pattern but also the indeterminate development of acoels is secondarily derived. It is possible, however, that these two changes did not arise independently, but were both results of a single change, perhaps in the orientation of the spindle at first cleavage. Whereas with quartet spiral cleavage the first two cleavages (to produce the four macromeres) are clearly spiral, or oblique to the animal-vegetal axis of the egg (see Costello, 1955), the first cleavage in acoels, which results in formation of the two macromeres, is often described as meridional. In this case, cytoplasmic determinants that might be organized symmetrically along the animal-vegetal axis of the egg would not become segregated at first cleavage in the acoels. Experimental work to test such an hypothesis could be useful. Of related interest are studies which have reported the occurrence of quartet cleavage in acoels (see Henley, 1974) and duet cleavage in macrostomids (Seilern-Aspang, 1957) and polyclads (see Henley, 1974). Although Apelt (1969) has demonstrated convincingly that the cases of quartet cleavage he observed in acoels were strictly abnormal, the situation for the other two orders should be reinvestigated.

Within certain neoophoran orders, cleavage is highly modified as compared to the spiral quartet or duet form characteristic of most archoophorans [for reviews, see Hyman (1951) and Kato (1968)]. This is especially true with the triclads, the proseriate *Bothrioplana semperi* (see Reisinger *et al.*, 1974b), and the prolecithophorans *Plagiostomum girarde* (see Bresslau, 1904) and *Hydrolimax grisea* (see Newton, 1970), in which the phenomenon of *Blastomeren-Anarchie* occurs. As cells cleave, often equally, the resulting blastomeres lose contact with each other and, with continued division, become successively isolated by yolk cells. Subsequently, the isolated cells aggregate, usually in the center of the yolk mass, to form a solid blastema.

The development of a sufficient number of neoophoran species has been studied to demonstrate that the phenomenon of blastomere separation is not an obligate condition of ectolecithy; rather, as pointed out by Ax (1961, 1963), remnants of spiral cleavage persist in the Lecithoepitheliata, Rhabdocoela, and Proseriata. Furthermore, insights into trends within the neoophoran orders are beginning to emerge (Reisinger *et al.*, 1974a & b).

Perhaps the most striking example of the retention of spiral cleavage has come from the studies by Reisinger *et al.* (1974a) of the lecithoepitheliate *Xenoprorhynchus steinboecki* (Table 1). Of the four macromeres, D is largest, an observation which supports the view that homoquadratic cleavage in polyclads is derived. Cleavage proceeds identically to that of other Spiralia. Most striking is the occurrence of the 4d mesoblast which divides bilaterally and gives rise to a mesodermal streak. These cells subsequently disperse as mesenchymal cells with no trace of coelom formation. Until this divergence, however, the similarity of development to that of the polychaetes supports the idea that quartet 4d spiral cleavage is primitive for the Turbellaria.

In the proseriates *Monocelis fusca* (see Giesa, 1966), *Minona trigonopora* (see Reisinger *et al.*, 1974a), and *Otomesostoma auditivum* (see Reisinger *et al.*, 1974b) early development clearly occurs by equal to slightly unequal quartet spiral cleavage. With the formation of the second quartet, a coeloblastula appears, perhaps as a consequence of the paucity of yolk within the embryo. There is no indication in the Proseriata of the presence of 4d; rather, at least in *Minona*, cells cleaved into the blastocoel from several cells seem to account for all mesoderm.

In the dalyellioid rhabdocoel *Paravortex gemellipara* (see Ball, 1916) and the typhloplanoid rhabdocoels *Mesostoma ehrenbergii* and *Bothromesostoma personatum* (see Bresslau, 1904), blastomeres do not become separated from one another and cleavage is unequal. Nevertheless, the pattern is not totally fixed and certainly lacks the observable precision and symmetry of conventional spiral cleavage.

Blastomeres of the prolecithophoran *Plagiostomum giradi* become separated by yolk during cleavage, but unequal cleavage leads to a regular pattern in which micromeres occur peripherally and macromeres centrally (Bresslau, 1904). The freshwater prolecithophoran species *Hydrolimax grisea* cleaves in a seemingly more random pattern which, in that respect, resembles closely the process of *Blastomeren-Anarchie* in triclads (Newton, 1970). The same is true of the proseriate *Bothrioplana semperi* (see Reisinger *et al.*, 1974b).

A phenomenon that occurs in certain other proseriates may foreshadow the occurrence of *Blastomeren-Anarchie* in *Bothrioplana* and members of the Prolecithophora and Tricladida. In *Otomesostoma* (see Reisinger, 1974b) and *Monocelis* (see Giesa, 1966), small 'abortive blastomeres' with compact chromatin and sparse cytoplasm are produced from the micromeres. These blastomeres, which appear to be totally inactive, either adhere to the periphery of the embryonic mass or, more commonly, go out among the yolk cells. Approximately half these abortive blastomeres are subsequently reincorporated and participate in the formation of the body epithelium, whereas the remaining abortive blastomeres die. Abortive blastomeres of the proseriates bear a striking resemblance to all blastomeres in forms which undergo *Blastomeren-Anarchie*. In both situations, undifferentiated blastomeres separate from other embryonic cells and move into the yolk before aggregating with other blastomeres to undergo differentiation.

It would appear that the phenomenon of blastomere separation arose within the Proseriata. From a phylogenetic perspective, the occurrence of *Blastomeren-Anarchie* in both the triclads and proseriates is not unexpected since numerous characters link the two orders (see Ax, 1961, 1963; Karling, 1974). The significance of its occurrence in the Prolecithophora is not known, but is consistent with the hypothesized relationships of this order (Karling, 1974).

Gastrulation

Gastrulation encompasses the processes that lead to the concentric positioning of the germ layers. Often associated with gastrulation, though perhaps not causally related in some organisms, are formation of the archenteron and establishment of the body axes. All three aspects will be examined here. In addition, since mechanisms for yolk utilization in the Turbellaria may be established during gastrulation, that topic will be included.

Because gastrulation in polyclads resembles that of other spiralians, it will be described in some detail, primarily from accounts by Kato (1940, 1968). The positioning of the ectoderm in polyclads is accomplished by epiboly, in which just after the bilateral division of 4d, micromeres from the first three quartets spread from the animal pole over the surface of the remaining blastomeres (the large micromeres 4a – 4c and $4d^1$–$4d^2$, and the four small macromeres) toward the vegetal pole. In doing so, they cover the surface of the embryo. Subsequent proliferation of $4d^1$ and $4d^2$ produces endoderm and mesoderm in their definitive concentric locations.

The leading edge of the advancing ectodermal micromeres forms the blastopore as the cells converge at the vegetal pole. After closure of the blastopore, the ectoderm, at or near the site of closure, either invaginates or ingresses and cavitates to form the mouth and pharynx. The cavity subsequently becomes continuous with a cavity in the endoderm to produce the gastrovascular cavity (Fig. 2A).

The definitive body axes become established in polyclads by a gradual shift of the animal-vegetal axis. The animal pole assumes an anterior position and the brain arises there from an invagination of the ectoderm. The vegetal pole, as indicated by the position of the mouth, is ventro-caudally located (Fig. 2B).

As far as the other archoophoran orders are concerned, gastrulation has been described in detail only for the acoels and macrostomids. In the Acoela, the process is similar to that described for polyclads, occurring by invagination and epiboly (Apelt, 1969). The description of gastrulation in *Macrostomum appendiculatum* differs, however, from that of the other two orders (Seilern-Aspang, 1957). In *M. appendiculatum*, a process of inverted

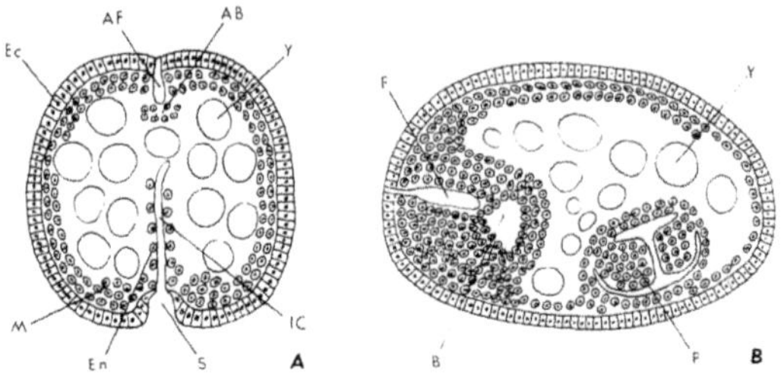

Fig. 2. Generalized scheme for late development in the Polycladida. A) Embryo shortly after epiboly. Anlage of the frontal organ lies at the animal pole, stomodeum at the vegetal pole. B) Embryo just before hatching. An axial shift due to unequal growth positions the former animal pole, marked by the position of the frontal organ, at the anterior end of the embryo while the vegetal pole (indicated by the pharynx) occupies a ventral location. AB = anlage of brain, AF = anlage of frontal organ, B = brain, Ec = ectoderm, En = endoderm, F = frontal organ, IC = intestinal cavity, M = mesoderm, P = pharynx, S = stomodeum, Y = yolk. (Modified from Kato, 1940).

epiboly is described in which, at the 16-cell stage, 12 micromeres sink into four large, yolk-filled, vegetal blastomeres. The four large blastomeres then flatten to surround the remaining blastomeres completely and form a yolk mantle or hull membrane (a term normally used in neoophoran development to describe a cellular layer which occurs beneath the eggshell or capsule and encloses the yolk cells within the embryonic sphere). Following this event, development resembles strongly the processes that occur in most neoophorans studied to date (Fig. 3). First, a blastema, which includes the anlagen of the brain-pharynx and of the genital apparatus, forms along the future ventral surface of the embryo. The brain anlage is at the anterior end which is coincident with the animal pole, while the pharynx and genital apparatus, which will lie on the ventral surface, are meridionally located. Subsequently, epidermal cells from the ventral surface of the blastema spread laterally and dorsally along the inner surface of the hull membrane to enclose some large, yolk-laden blastomeres that lie in the dorsal half of the embryo.

Among the neoophorans that have been described developmentally, the species that compares most closely with the polyclads and other spiralians is the lecithoepitheliate *Xenoprorhynchus steinboecki* (see Reisinger *et al.*, 1974a & b). In this form, which so closely resembles other spiralians in its cleavage, Reisinger and coworkers also describe an epibolic

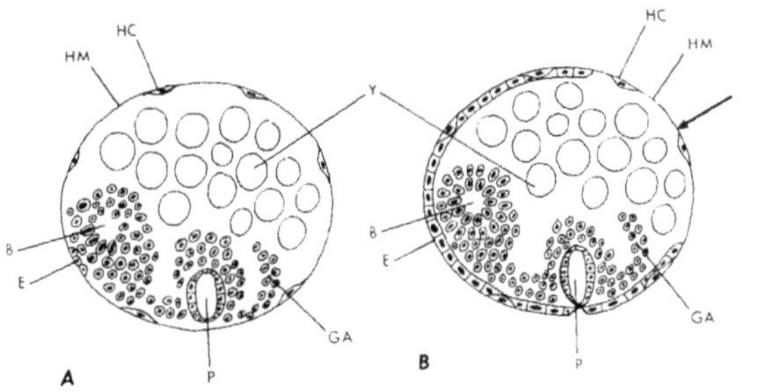

Fig. 3. Generalized scheme for late development in *Macrostomum appendiculatum* and most neoophorans. A) Embryo before spreading of epidermis. B) Embryo in which the epidermis has begun spreading from a site near the developing brain. Arrow indicates point of future convergence of epidermis, homologous, perhaps, to a blastopore (BP^2 in Fig. 5). In both A and B, the hull cells may be derived from micromeres, macromeres, or yolk cells, depending upon species. The yolk may be contained in blastomeres (*M. appendiculatum*) or yolk cells (Neoophora). B=brain, E=epidermis, GA=genital apparatus, HC=hull cells, HM=hull membrane, P=pharynx, Y=yolk. (Modified from Bresslau, 1904).

gastrulation (Fig. 4). In this process products of primarily the second and third micromere quartets spread not only over the remaining blastomeres, but on into the adjacent mass of yolk cells. When the blastopore cells (derived from the 2nd quartet) converge to close the blastopore, this external epithelium, termed the primary hull membrane, isolates the embryo and a portion of the yolk from

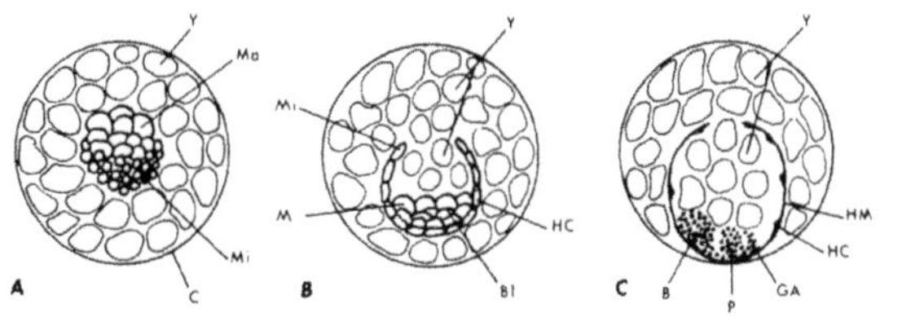

Fig. 4. Diagram of major features of hull-membrane formation in certain lecithoepitheliate and proseriate egg capsules. A) Micromeres mark the animal pole. B) They spread over the remaining blastomeres, including the macromeres, and extend into the mass of yolk cells. C) They subsequently flatten and expand to form the hull membrane which surrounds both blastema and yolk cells. The opening in the hull membrane is possibly homologous to a blastopore (BP^2 in Fig. 5), but is located dorsally. Note that animal pole becomes coincident with ventral surface of blastema. B=brain, Bl=blastema, C=capsule, GA=genital apparatus, HC=hull cell, HM=hull membrane, Ma=macromeres, Mi=micromeres, Y=yolk cells. (Modified from Giesa, 1966, and Reisinger *et al.*, 1974a & b).

the remaining yolk. The primary hull membrane persists until the enclosed yolk is assimilated, after which the cells become incorporated into the surface layer of the embryo.

As the primary hull membrane disappears, a second one forms, originating also from the blastomeres, and incorporates the remaining yolk. During this time, the embryonic mass becomes organized into the anlagen of the brain and pharynx. The convex surface of the blastema differentiates into the ventral surface of the developing worm; the yolk mass occupies the dorsal hemisphere. Subsequently, the epidermis spreads laterally and dorsally beneath the hull membrane from the edges of the blastema to enclose the yolk and form the definitive body wall.

An epibolic movement of micromeres to form a hull membrane has been described as part of gastrulation of the proseriates *Minona trigonopora* (see Reisinger *et al.*, 1974a), *Otomesostoma auditivum* (see Reisinger *et al.*, 1974b) and *Monocelis fusca* (see Giesa, 1966), and is quite similar among the three species. In *Minona* the cells that undergo epiboly are derived, as in *Xenoprorhynchus*, from the cells of the second and third micromere quartets. In *Otomesostoma* and *Monocelis*, in which abortive blastomeres as well as hull cells are formed from micromeres, the number of prospective hull cells is fewer than in *Minona.*

In *Minona*, the cellular movements are similar to those shown diagrammatically in Fig. 4. The hull cells encompass the embryo and extend into the mass of yolk cells. The hull membrane expands as all the yolk cells enter through the opening in the hull membrane, the blastopore. Certain descendants of the macromeres, which are considered endodermal cells, become vitellophagocytes and engulf the yolk cells that enter the embryonic sphere. The process differs in *Monocelis* and *Otomesostoma* chiefly in that the blastopore closes before much of the yolk is incorporated. The yolk outside the hull membrane enters at the vegetal pole in both species, but passes *between* cells of the closed blastopore in *Otomesostoma*, and *through* blastopore cells by phagocytosis in *Monocelis.* In all three species, subsequent development involves the formation of the blastema with the convex surface equivalent to the ventral surface of the worm. The formation of the anlagen of the brain, pharynx, and genital apparatus, and the spreading of the epidermis over the dorsal yolk occur as described for *Macrostomum* and *Xenoprorhynchus* (Fig. 3).

Development in the Rhabdocoela studied to date (Bresslau, 1904; Ball, 1916) differs from that described for the Lecithoepitheliata and Proseriata in that formation of a hull membrane from blastomeres does not occur; instead, either the hull membrane is absent, as in *Paravortex gemellipara* (see Ball, *op. cit.*), *Bothromesostoma personatum* and the winter egg of *Mesostoma ehrenbergii* (see Bresslau, *op. cit.*), or it is derived from yolk cells as in *M. productum, M. lingua* and the summer egg of *M. ehrenbergii* (see Bresslau, *op. cit.*). Where there is no hull membrane, the yolk cells, which lie dorsal to the developing embryo, remain unincorporated until late in development. They become enclosed by the dorsal spreading of the definitive epidermis from the blastema only after the formation and partial differentiation of the blastema (Fig. 3). In *Paravortex* certain derivatives of the macromeres termed primary endoderm cells become phagocytic and engulf a large part of the yolk (Ball, *op. cit.*). These cells eventually die and are engulfed by secondary endodermal cells which subsequently participate in the formation of the digestive system.

Early development has been described for the marine prolecithophoran *Plagiostomum girardi* (see Bresslau, *op. cit.*) and the freshwater form *Hydrolimax grisea* (see Newton, 1970). Although both species undergo blastomere separation during cleavage, and developmental processes in *Hydrolimax* appear to be unique, development in *Plagiostomum* is remarkably similar to that described for the Lecithoepitheliata, Proseriata, and Rhabdocoela in terms of orientation of the body axes, formation of the anlagen, and spreading of the epidermis. The hull membrane originates from yolk cells as it does in certain rhabdocoels.

Development in triclads appears to differ from that in other turbellarians to such an extent that comparisons are meaningless. It can be recognized, however, that the organization of the blastema and the formation of the definitive epidermis are similar to those processes in the other neoophorans. The events prior to that time, such as the formation of the provisional ectoderm, the embryonic pharynx, and the embryonic intestine, are developmental innovations with no obvious homologies with other turbellarians.

Comparisons of gastrulation and associated phenomena among the orders can be instructive. It becomes obvious, for example, that the structure termed the hull membrane is not necessarily homologous within the class since it may arise from vegetal macromeres (Macrostomida), animal micromeres (Lecithoepitheliata and several Proseriata), or yolk cells (some Rhabdocoela and Prolecithophora). Additional studies may demonstrate that variations in its origin from group to group may, in fact, provide a useful taxonomic character.

If the formation of the hull membrane is disregarded, a pattern of late development can be seen to be basic to both the archoophoran *Macrostomum appendiculatum* and to all neoophorans described above (Fig. 3): 1) the blastomeres form a blastema at the future ventral surface of the embryo; 2) the blastema becomes organized into the primordia of the brain, pharynx, and genital apparatus; and 3) epithelial cells at the surface of the blastema just peripheral to the brain anlage spread over the ventral and lateral surfaces of the blastema and then continue dorsally over the yolk cells (blastomeres in *Macrostomum* or external yolk in the Neoophora) to form the definitive body wall.

Such a process may appear to bear little resemblance to the development of polyclads. Ball (1916), however, pointed out the similarity between the spreading of cells to form the definitive epidermis in neoophorans, and the epiboly of micromeres to form the epidermis in polyclads. In both cases, cells from the anterior (neoophoran) or future anterior (polyclad) end of the embryo move over the remaining blastomeres to form the outer epithelium. The primary difference appears to be only one of timing, such that the process occurs at the end of cleavage in the polyclads, but not until after formation of the blastema in the neoophorans.

If the spreading of the epidermis is homologous to the epibolic gastrulation of polyclads, then the significance of the spreading of micromeres to form the hull membrane in certain members of the Lecithoepitheliata and Proseriata must be addressed. This phenomenon has been interpreted by both Giesa (1966) and Reisinger *et al.* (1974a & b) as a process homologous to gastrulation in polyclads. In these forms, the hull cells, like the ectodermal cells of polyclads, are derived from the first three micromere quartets, and spread over the remaining blastomeres to form an outer epithelium (covering yolk cells as well as blastomeres, however). This process occurs prior to the stage of epiboly in polyclads, beginning as early as the 20-cell stage in *Xenoprorhynchus* (see Reisinger *et al.*, 1974a).

The possibility must be considered that in *Xenoprorhynchus* and the three proseriates the formation of the hull membrane and of the epidermis are but two temporally separated phases of ectodermal epiboly. Each process would create a blastopore at the site of convergence of the spreading cells (see Fig. 5C). The first blastopore (BP^1) would occur during formation of the hull membrane; the second (BP^2) would appear at the site of closure of the epidermis. According to this interpretation, both processes would be homologous to gastrulation in the polyclads, but only the second would constitute gastrulation in the conventional sense of positioning germ layers of the embryo proper.

Consideration of either or both processes as homologous to gastrulation in polyclads poses some difficulties if the relationship of the animal-vegetal axis to the definitive body axes of the embryo is compared to that in polyclads. The problem lies in the fact that the blastopore (formed by

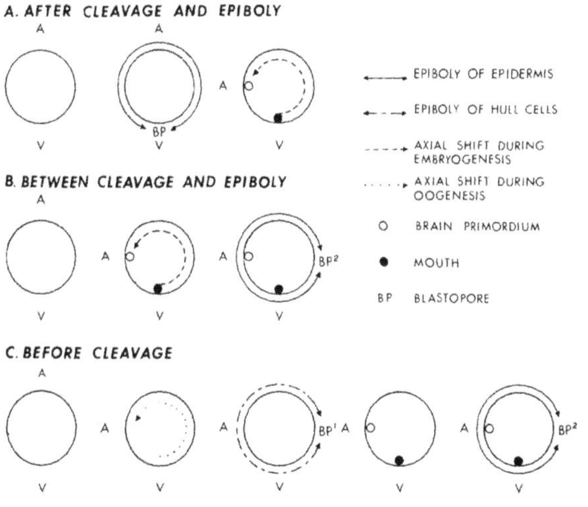

Fig. 5. Possible effects of timing of an axial shift on the polarity of an embryo. A) Sequence in polyclads in which epiboly precedes shift of the animal pole to anterior and results in a ventral position of the blastopore. B) Delay in epiboly until after the axial shift would be sufficient to account for dorsal site of convergence of epidermis (BP^2) in *M. appendiculatum* and Neoophora. C) Homology of opening in hull membrane (BP^1) with blastopore of polyclads requires that axial shift occur prior to cleavage.

epidermal spreading) in *Macrostomum* and all neoophorans, and *both* blastopores in *Xenoprorhynchus* and the three proseriates occur on the dorsal surface of the embryo, whereas in polyclads the blastopore is formed on the future ventral surface. Furthermore, in the Neoophora the animal pole becomes coincident with the ventral surface rather than the anterior end of the embryo.

As shown in Fig. 5A & B, the dorsal location of the blastopore formed in *Macrostomum* and the Neoophora during spreading of the epidermis (BP^2) could be explained simply by the fact that epidermal epiboly is delayed. Epibolic gastrulation in polyclads occurs before the establishment of the organ anlagen and the definitive body axes (Fig. 5A), and that process in *Macrostomum* and neoophorans occurs after those events (Fig. 5B). Such a temporal shift in epiboly could produce the spatial shift of the blastopore to its observed position on the dorsal surface of the embryo in *Macrostomum* and the Neoophora (Fig. 5A & B).

The axial shift during development proposed above to account for the dorsal location of the blastopore during formation of the epidermis occurs too late to explain the dorsal location of the blastopore during formation of the hull membrane in *Xenoprorhynchus* and the three proseriates. If, however, the shift in the animal-vegetal axis, which occurs during embryogenesis in the polyclads, were to occur during oogenesis or after fertilization but before cleavage, then the relationships seen in *Macrostomum* and the Neoophora would result (Fig. 5C). The proposal here is that the essential developmental difference between these forms and the polyclads and other spiralians resulted from a shift in ooplasmic determinants before cleavage such that the animal pole, which marks the anterior end of the embryo, became located *ca.* 90° from the mouth opening (which had primitively been associated fortuitously with the blastopore).

Conclusions

Before comprehensive conclusions can be drawn, it will be necessary to obtain descriptions of embryogenesis in orders that have not been examined developmentally (Nemertodermatida, Catenulida, Haplopharyngida) or for which information is restricted to one or two species (Lecithoepitheliata,

Prolecithophora). Nevertheless, from the limited number of studies available, certain common patterns emerge.

One general pattern is the widespread occurrence of spiralian early development, including quartet spiral cleavage, within the class. This suggests that these ontological processes were primitive for the Turbellaria. Additionally, the demonstration of an almost unmodified spiralian development in the neoophoran order Lecithoepitheliata (Reisinger *et al.*, 1974a & b) helps establish the developmental link between the Archoophora and the Neoophora. This relationship is extended to neoophorans with highly modified development (characterized by blastomere separation) by the occurrence of both modified spiral cleavage and *Blastomeren-Anarchie* in the same order, the Proseriata. The shared occurrence of *Blastomeren-Anarchie* itself suggests a possible relationship of the Prolecithophora as well as the Tricladida to the Proseriata. The unique cleavage of the Rhabdocoela (resembling neither spiral cleavage nor *Blastomeren-Anarchie*), together with the unusual origin of the hull membrane (Bresslau, 1904; Ball, 1916), suggests that this order has diverged markedly from the line encompassing the Proseriata and Tricladida. This idea may be supported by the occurrence of a type of eggshell granule not described outside the order, but more information must be obtained to evaluate the validity of the trait as a character for phylogenetic studies.

The unity of the neoophorans is demonstrated by the common mechanisms of blastema formation and epidermal epiboly. The occurrence of similar processes in the archoophoran *Macrostomum appendiculatum* (see Seilern-Aspang, 1957) is of possible significance. It would seem, in fact, that the most prominent late-developmental differences occur not between Archoophora and Neoophora, but between certain macrostomids and the other archoophorans. Since it is during this late phase of development that a dorsal yolk mass is incorporated in both *Macrostomum* (yolk-filled blastomeres) and the Neoophora (yolk cells proper), it is possible that development in *Macrostomum* represents a preadaptation to ectolecithy.

The developmental patterns observed are generally compatible with the phylogenetic diagram of the orders of the Turbellaria proposed by Karling (1974) and based primarily on morphological features of adult organisms. Major differences from Karling's scheme are limited here to a more central role for the macrostomids in the origin of the Neoophora, and a greater degree of divergence between the Rhabdocoela on one hand and the Proseriata and Tricladida on the other.

Acknowledgments

I am grateful to N. C. Edwards, S. E. Peters, and J. P. S. Smith III for their suggestions in the preparation of the manuscript and to S. F. Zane for assistance with photography. The work was supported in part by the Center for International Studies, UNCC, and in part by the UNCC Foundation.

References

Apelt, G., 1969. Fortpflanzungsbiologie, Entwicklungszyklen und vergleichende Frühentwicklung acoeler Turbellarien. Mar. Biol. 4: 267–325.

Ax, P., 1961. Verwandtschaftsbeziehungen und Phylogenie der Turbellarien. Ergebn. Biol. 24: 1–68.

Ax, P., 1963. Relationships and phylogeny of the Turbellaria. In E. C. Dougherty (ed.), The Lower Metazoa. U. Calif. Press, Berkeley; Los Angeles: 191–224.

Ball, S. C., 1916. The development of Paravortex gemellipara (Graffilla gemellipara Linton). J. Morph. 27: 453–558.

Boyer, B. C., 1971. Regulative development in a spiralian embryo as shown by cell deletion experiments on the acoel Childia. J. exp. Zool. 176: 97–105.

Boyer, B. C., 1972. Ultrastructural studies of differentiation in the oocyte of the polyclad turbellarian, Prostheceraeus floridanus. J. Morph. 136: 273–295.

Boyer, B. C., 1981. Experimental evidence for the origins of mosaic development in the polyclad Turbellaria. Biol. Bull. 161: 318.

Boyer, B. C., 1985. Experimental evidence for the origin of determinative development in the polyclad Turbellaria. This volume.

Bresslau, E., 1904. Beiträge zur Entwicklungsgeschichte der Turbellarien. I. Die Entwicklung der Rhabdocölen und Alloiocölen. Z. wiss. Zool. 76: 213–332.

Bunke, D., 1979. Ein Dopa-oxidierendes Ferment in den Schalendrüsen von Microdalyellia fairchildi (Turbellaria, Neorhabdocoela). Zoomorphologie 193: 153–169.

Bunke, D., 1981. Ultrastruktur-Untersuchungen an Vitellocyten von Microdalyellia fairchildi (Turbellaria, Neorhabdocoela). Zoomorphology 99: 71–86.

Bunke, D., 1982. Ultrastruktur-Untersuchungen zur Eischalenbildung bei Microdalyellia fairchildi (Turbellaria). Zoomorphology 101: 61–70.

Costello, D. P., 1955. Cleavage, blastulation and gastrulation. In B. H. Willier, P. A. Weiss, and V. Hamburger (eds), Analysis of Development. W. B. Saunders, Philadelphia, Pa.: 213–229.

Domenici, L., L. Galleni & V. Gremigni, 1975. Electron microscopical and cytochemical study of egg-shell globules in Notoplana alcinoi (Turbellaria, Polycladida). J. submicr. Cytol. 7: 239–247.

Domenici, L. & V. Gremigni, 1977. Fine structure and functional role of the coverings of the eggs in Mesostoma ehrenbergii (Focke) (Turbellaria, Neorhabdocoela). Zoomorphologie 88: 247–257.

Giesa, S., 1966. Die Embryonalentwicklung von Monocelis fusca Oersted (Turbellaria, Proseriata). Z. Morph. Ökol. Tiere 57: 137–230.

Gremigni, V., 1979. An ultrastructural approach to planarian taxonomy. Syst. Zool. 28: 345–355.

Gremigni, V. & L. Domenici, 1974. Electron microscopical and cytochemical study of vitelline cells in the fresh water triclad Dugesia lugubris s.l. Cell Tiss. Res. 150: 261–270.

Gremigni, V. & L. Domenici, 1975. Genesis, composition, and fate of cortical granules in the eggs of Polycelis nigra (Turbellaria, Tricladida). J. Utrastruct. Res. 50: 277–283.

Gremigni, V. & M. Nigro, 1983. An ultrastructural study of oogenesis in a marine triclad. Tiss. & Cell 15: 405–416.

Henley, C., 1974. Platyhelminthes (Turbellaria). In A. C. Giese & J. S. Pearse (eds), Reproduction of Marine Invertebrates, 1. Academic Press, N.Y.: 267–343.

Hyman, L. H., 1951. The invertebrates, 2. McGraw-Hill, N.Y., 550 pp.

Ishida, S. T., T. Gotoh & W. Teshirogi, 1981. Oogenesis and eggshell formation in polyclads. Rep. Fukaura Mar. Biol. Lab., Hirosaki Univ. 9: 32–48 (in Japanese).

Ishida, S. T. & W. Teshirogi, 1985. Eggshell formation in polyclads (Turbellaria). This volume.

Ivanov-Kasas, O. M., 1959. Entstehung und Evolution der Spiralfurchung. Sowjetwiss. Naturwiss. Beiträge 12: 1267–1279.

Ivanova-Kasas, O. M., 1981. Phylogenetic significance of spiral cleavage. Biol. Morya (Vladivost) 5: 3–14 (in Russian).

Karling, T. G., 1974. On the anatomy and affinities of the turbellarian orders. In N. W. Riser & M. P. Morse (eds), Biology of the Turbellaria. McGraw-Hill, N.Y.: 1–16.

Kato, K., 1940. On the development of some Japanese Polyclads. Jap. J. Zool. 8: 537–573.

Kato, K., 1968. Platyhelminthes. In M. Kumé & K. Dan (eds), J. C. Dan (trans.), Invertebrate embryology. NOLIT, Belgrade: 125–143.

Lang, A., 1884. Die Polycladen des Golfes von Neapel. Flora Fauna Golfes von Neapel 11: 1–688.

Newton, W. D., 1970. Gastrulation in the turbellarian Hydrolimax grisea (Platyhelminthes; Plagiostomidae): formation of the epidermal cavity, inversion and epiboly. Biol. Bull. 139: 539–548.

Okada, K., 1968. Annelida. In M. Kumé & K. Dan (eds), J. C. Dan (trans.) Invertebrate embryology. NOLIT, Belgrade: 192–241.

Papi, F., 1953. Beiträge zur Kenntnis der Macrostomiden (Turbellarien). Acta zool. fenn. 78: 2–32.

Reisinger, E., I. Cichocki, R. Erlach & T. Szyskowitz, 1974a. Ontogenetische Studien an Turbellarien: ein Beitrag zur Evolution der Dotterverarbeitung im ektolecitalen Ei. I. Z. zool. Syst. Evolut.-forsch. 12: 161–195.

Reisinger, E., I. Cichocki, R. Erlach & T. Szyskowitz, 1974b. Ontogenetische Studien an Turbellarien: ein Beitrag zur Evolution der Dotterverarbeitung im ektolecithalen Ei. II. Z. zool. Syst. Evolut.-forsch. 12: 241–278.

Rieger, R. M., 1981. Morphology of the Turbellaria at the ultrastructural level. Hydrobiologia 84: 213–229.

Rieger, R. M. & W. Sterrer, 1975a. New spicular skeletons in Turbellaria and the occurrence of spicules in marine meiofauna, I. Z. zool. Syst. Evolut.-forsch. 13: 207–248.

Rieger, R. M. & W. Sterrer, 1975b. New spicular skeletons in Turbellaria and the occurrence of spicules in marine meiofauna, II. Z. zool. Syst. Evolut.-forsch. 13: 249–278.

Schmidt, G. A., 1966. Evolutionäre Ontogenie der Tiere. Akademie-Verlag, Berlin, 364 pp.

Seilern-Aspang, F., 1957. Die Entwicklung von Macrostomum appendiculatum (Fabricius). Zool. Jb. Anat. 76: 311–330.

Siewing, R., 1980. Das Archicoelomatenkonzept. Zool. Jb. Anat. 103: 439–482.

Smyth, J. D. & J. A. Clegg, 1959. Egg-shell formation in trematodes and cestodes. Exp. Parasit. 8: 286–323.

Surface, F. M., 1907. The early development of a polyclad, Planocera inquilina. Proc. Acad. nat. Sci. Philad. 59: 514–559.

v. Hofsten, N., 1912. Eischale und Dotterzellen bei Turbellarien und Trematoden. Zool. Anz. 39: 111–136.

v. Salvini-Plawen, L., 1978. On the origin and evolution of the lower Metazoa. Z. zool. Syst. Evolut.-forsch. 16: 40–87.

Wilson, E. B., 1892. The cell-lineage of Nereis. J. Morph. 6: 361–480.

Yanagita, Y. & T. S. Yamamoto, 1981. Formation and chemical property of the cocoon in a freshwater planarian Dendrocoelopsis ezensis Ichikawa et Okugawa (Turbellaria, Tricladida). Zool. Anz. 207: 130–142.

Experimental evidence for the origins of determinative development in the polyclad turbellarians

Barbara C. Boyer
Department of Biological Sciences, Union College, Schenectady, NY 12308, U.S.A.

Keywords: Turbellaria, Polycladida, development, spiral cleavage, mosaicism

Abstract

Cell-deletion experiments were carried out on the embryo of the polyclad turbellarian *Hoploplana inquilina* to examine further the nature of development in primitive spiralians. The polyclads are of particular interest because they provide a link between the regulative development of acoels and the determinative development of annelids and molluscs. Single blastomeres were deleted at the two- and four-cell stages by puncture through the eggshell membrane with tungsten needles. All deletions resulted in abnormal larvae with consistent characteristics representing half or three-quarter Müller's larvae. The number of larval eyes was a particularly useful character in revealing mosaicism. This study establishes the polyclad embryo as determinative, but with important cell interactions also occurring during early development, and provides evidence that mosaicism became associated with spiral cleavage in the quartet form during the evolution of the Turbellaria.

Introduction

The association between spiral cleavage and mosaic development (early embryonic determination) has been firmly established in the embryos of annelids and molluscs. Spiral cleavage is also characteristic of the entolecithal turbellarians, specifically the acoel and polyclad orders, which probably represent the most primitive spiralians. However early development in acoels, which cleave in duet pattern rather than the more common quartets, is regulative, indicating that spiral cleavage and mosaicism are not always linked (Boyer, 1971). The polyclad embryo is strikingly similar to annelid and mollusc embryos both in its quartet cleavage pattern and cell lineage (Surface, 1907) and therefore may provide a link between the acoel indeterminate development and mosaicism, which is more typically spiralian. However, in the absence of experimental studies on polyclad embryos, the nature of development in this group has remained unknown. The purpose of this study is to provide further information on the relationship between spiral cleavage and mosaicism by examining the developmental capacity of partial embryos of polyclads.

Materials and methods

The experiments were done at the Marine Biological Laboratory, Woods Hole, on the polyclad *Hoploplana inquilina*. Specimens were collected from the mantle cavity of the whelk *Busycon* and maintained in finger bowls on the sea table where they deposited eggs. Single blastomeres were deleted randomly through the eggshell membrane with sharp tungsten needles at the two- and four-cell stages. Intact eggs from the same mass served as controls and both experimental and control larvae, raised in pasteurized sea water, were observed for developmental defects on day six.

Hydrobiologia 132, 117–119 (1986).

Results and discussion

1. Half embryos

Fifty-one complete deletions of one cell at the two-cell stage resulted in partial larvae which were abnormal in body shape, lobe development and symmetry, eye formation, and swimming behavior. Typically the larvae exhibited a general morphology suggestive of the normal Müllers larva, but lobe development was always rudimentary and frequently the ventrolateral lobes were asymmetric. The great majority were either eyeless or one-eyed and many swam in circles or rotated around the antero-posterior axis as they moved. However, behavior was not as consistently affected as morphology.

Since 33% of the half larvae had no eyes and 55% had one eye, the developmental potentials of the blastomeres appear to differ even at the two-cell stage. (Four percent of these larvae had two eyes). Incomplete deletions, in which a portion of the destroyed cell did not cytolyze, increased the proportion of one-eyed (64%) and two-eyed (19%) larvae, with only 14% having no eyes. These results suggest an interaction in normal development between the blastomeres at the two-cell stage to form two eyes.

2. Three-quarter embryos

Ninety-one complete deletions of one cell at the four-cell stage gave rise to larvae in which development was considerably more like normal than in the case of half embryos. They looked quite similar to the Müller's larva controls, though lobe development still was not complete and deficiencies in the anterior, posterior, right or left quadrants were common. All larvae had either one or two eyes and swimming behavior was very similar to that of the controls.

Approximately half of the larvae from complete deletions were bilaterally symmetrical but frequently exhibited incomplete development of the oral hood or mid-dorsal lobe. Asymmetric animals typically had a markedly truncated right or left ventrolateral lobe.

One eye was found in 73% of the experimental larvae and 27% had two eyes. If the morphogenetic determinants for eyes are localized in only two blastomeres at the four-cell stage as would characterize a strictly mosaic system, random deletions should result in a 1:1 ratio of one-eyed to two-eyed larvae. The 3:1 ratio observed suggests that deletion of any of three blastomeres results in loss of an eye. This is consistent with the hypothesis that two of the blastomeres at the four-cell stage are in the direct eye lineage but the presence of a third is necessary for the development of two eyes. Such cell interactions are found in other spiralian embryos where the D quadrant may play a particularly significant role in development, even serving as an inducer (Clement, 1962).

In 24 of 27 larvae (89%) with one eye and asymmetric ventrolateral lobes, the eye was located in the normal position on the side in which lobe development was most complete. This correlation between absence of eye and lobe deficiencies on the same side further supports the concept of specific cytoplasmic determinants in each of the blastomeres.

This study demonstrates that the embryo of *Hoploplana* is mosaic, as half and three-quarter embryos hatch with consistent characteristics that represent half or three-quarter Müller's larvae. However, as is common in developmental systems with early determination, cell interactions also seem to play an important role, probably as early as the two-cell stage, indicating that the concept of strict mosaicism is an oversimplification.

Establishment of the polyclad embryo as mosaic suggests that early embryonic determination and quartet spiral cleavage are always associated and probably represent a primitive condition that is highly conservative. However, early determination is not characteristic of all spiral cleavage patterns as demonstrated by the regulative acoels. Thus the association of spiral cleavage and mosaic development perhaps became established in the quartet pattern during the evolutionary history of the Turbellaria and has become a permanent feature of the turbellarian-annelid-mollusc lineages.

Acknowledgments

This work was supported by an MBL Steps Toward Independence Fellowship, a Research Corporation Grant and the Union College Faculty Research Fund.

Note added in proof:

A more complete presentation and discussion of these results is in press (Int. J. Invert. Reprod. Devel.)

References

Boyer, B. C., 1971. Regulative development in a spiralian embryo as shown by cell deletion experiments on the acoel, Childia. J. exp. Zool. 176: 97–106.

Clement, A. C., 1962. Development of Ilyanassa following removal of the D macromere at successive cleavage stages. J. exp. Zool. 149: 193–216.

Surface, F. M., 1907. The early development of a polyclad Planocera inquilina. Proc. Acad. nat. Sci. Philad. 59: 514–559.

Electron microscopic study of larval eye development in Turbellaria Polycladida

Alberto Lanfranchi & Celina Bedini
Istituto di Biologia Generale, Univ. di Pisa, Via Volta 6, I-56100 Pisa, Italia

Keywords: Turbellaria, Polycladida, development, epidermal eye, cerebral eye, ultrastructure

Abstract

The appearance and development of the embryonic and larval eyes of the polyclad turbellarian *Stylochus mediterraneus* were studied. In the embryo, the left epidermal eye appears first. Subsequently, the right epidermal eye appears, and within hours it sinks into the parenchyma and turns into a cerebral eye. Newly hatched Götte's larvae possess both the left epidermal and the right cerebral eye. Three days after hatching, an incomplete eye appears adjacent to the left epidermal eye. The left cerebral eye then originates from this incomplete eye as it sinks into the parenchyma. This third eye is believed to originate through a process of induction.

Introduction

The Müller's larvae of *Thysanozoon brocchii* and the Götte's larvae of *Stylochus mediterraneus* possess both epidermal and cerebral eyes (Lanfranchi *et al.,* 1981). The epidermal eye consists of a cup-shaped pigmented cell with a cavity filled by lamellae of ciliary origin and a small cell that covers the cup-like opening. The cerebral eye is made up of three cells: one pigmented cell with ciliary projections and two rhabdomeric-type photoreceptor cells.

Lang (1884) and Kato (1940) proposed that cerebral eyes in polyclad larvae arise from epidermal eyes that sink into the mesoderm. However, Ruppert (1978: 80) believed 'The detailed structure of the eyes in ... [the Müller's and Götte's] larvae.. indicates that both epidermal and cerebral eyes are definitive organs with different structural bases. The cerebral eyes are not derived from epidermal eyes...'. This opinion is also shared by Eakin & Brandenburger (1981).

In studying the ultrastructure of eyes in larvae and young of three polyclad species, we found evidence that epidermal eyes can change into parenchymal eyes (Lanfranchi *et al.,* 1981), and we support that view here with a study of the sequence of appearance of the two types of eye in the embryo and larvae of *S. mediterraneus.*

Materials and methods

Adult specimens of *Stylochus mediterraneus* Galleni, 1976, collected in the Ligurian Sea near Leghorn were maintained in the laboratory in artificial sea water at 18 °C and 33–35‰ salinity. They deposited egg plates on the glass walls of the aquarium. The eggs were collected and isolated at the time of deposition and then fixed 5, 6, 6.5, and 7 days after deposition. The larvae, which generally hatched on day 8, were fixed 1, 3, 5, 6, and 9 days after hatching. The larvae were kept only until day 15 because the artificial environment was not suitable for the completion of metamorphosis.

Fixation was carried out in 1% osmium tetroxide in 0.1 M phosphate buffer and followed by dehydration and embedding in an Epon-Araldite mix-

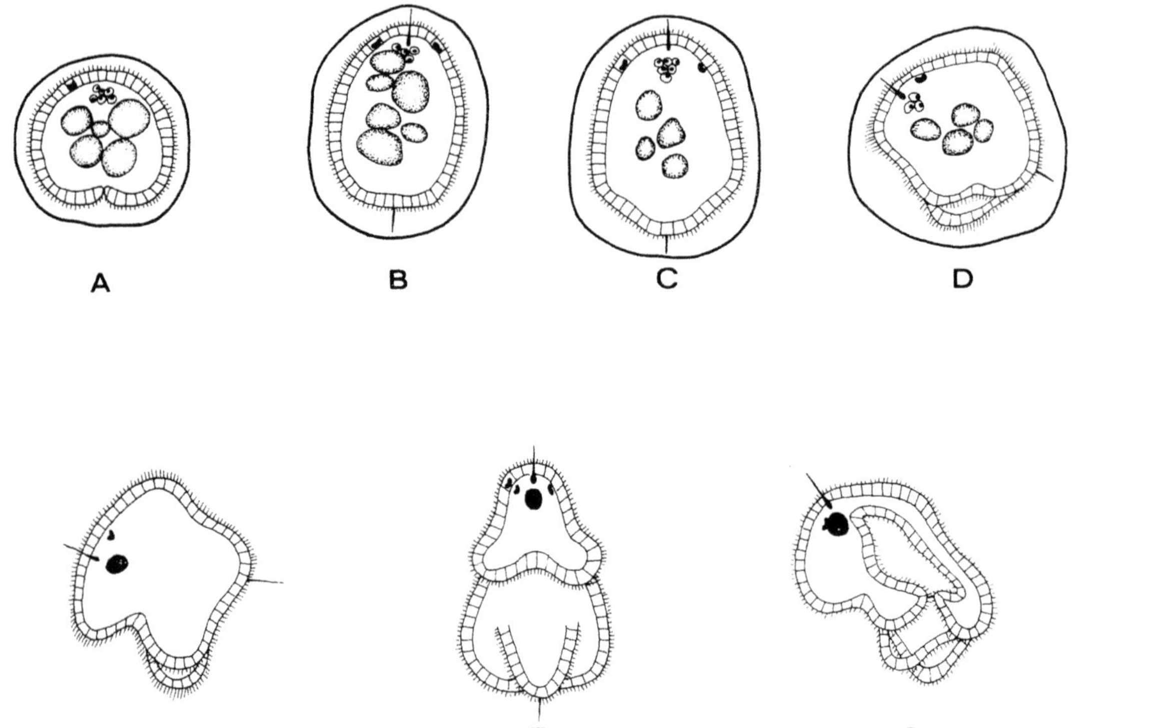

Fig. 1A–G. Scheme of the embryonic and larval development of eyes in *Stylochus mediterraneus.* A. 5 d after deposition. Embryo showing the first epidermal eye on the left side. Longitudinal section. B. 6 d after deposition. Embryo showing two epidermal-type eyes. Frontal section. C. 6.5 d after deposition. Embryo with left epidermal and right cerebral eye. Frontal section. D. 7 d after deposition. Two-eye stage embryo, showing only the right cerebral eye (lateral view). E. 8 d after deposition. Newly hatched Götte's larva, showing the right cerebral eye (lateral view). F. 6 d after hatching. Götte's larva with 3 eyes: one epidermal and two cerebral eyes (dorsal view). G. 9 d after hatching. Götte's larva with cerebral eye near the cerebral ganglia (lateral view).

ture using routine methods. Thin sections stained with both uranyl acetate and lead citrate were examined through a Siemens Elmiskop 1A electron microscope.

Results

The first embryonic eye appeared five days after egg deposition. It was located on the left side of the embryo and lay between the epithelial cells and the basal lamina. Thus, it was an epidermal-type eye (Figs. 1 and 2).

On day 6, the second eye appeared on the right side of the embryo. This too was of the epidermal type (Fig. 3B). Its appearance here was very short,

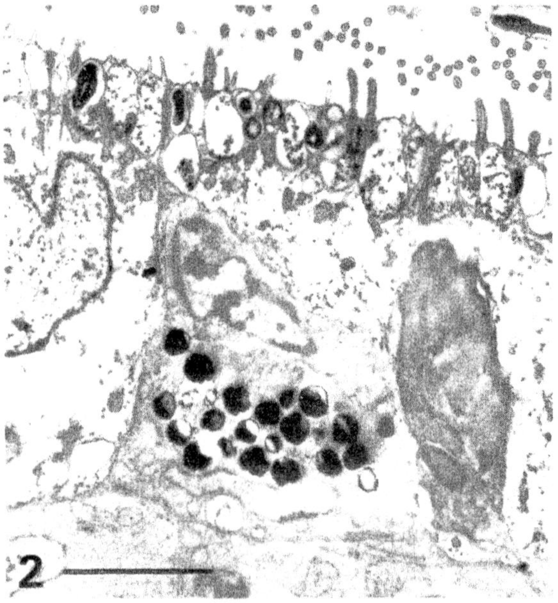

Fig. 2. Egg 5 d after deposition. Embryo with the initial left epidermal eye (scale = 5 μm).

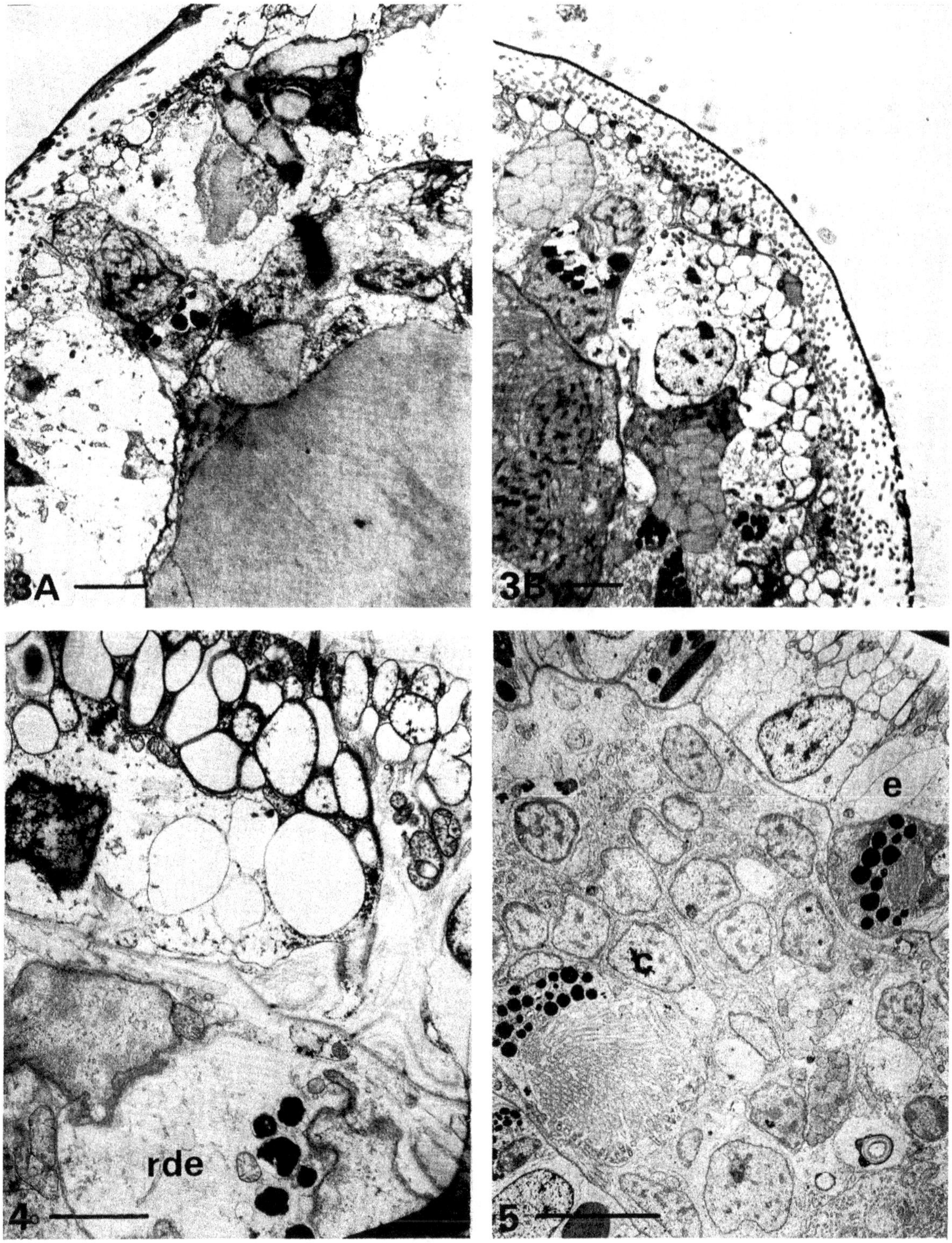

Figs. 3 – 5.(3A-B) Egg 6 d after deposition. Embryo with left (A) and right (B) epidermal eye (scale=5 μm). (4) Egg 6.5 d after deposition. The right eye (rde), which has descended into the parenchyma, is in the process of becoming a cerebral eye (scale=2 μm). (5) Egg 7 d after deposition, and newly hatched Götte's larva (on deposition day 8). The left epidermal (e) and the right cerebral eye (c) are shown (scale=5 μM).

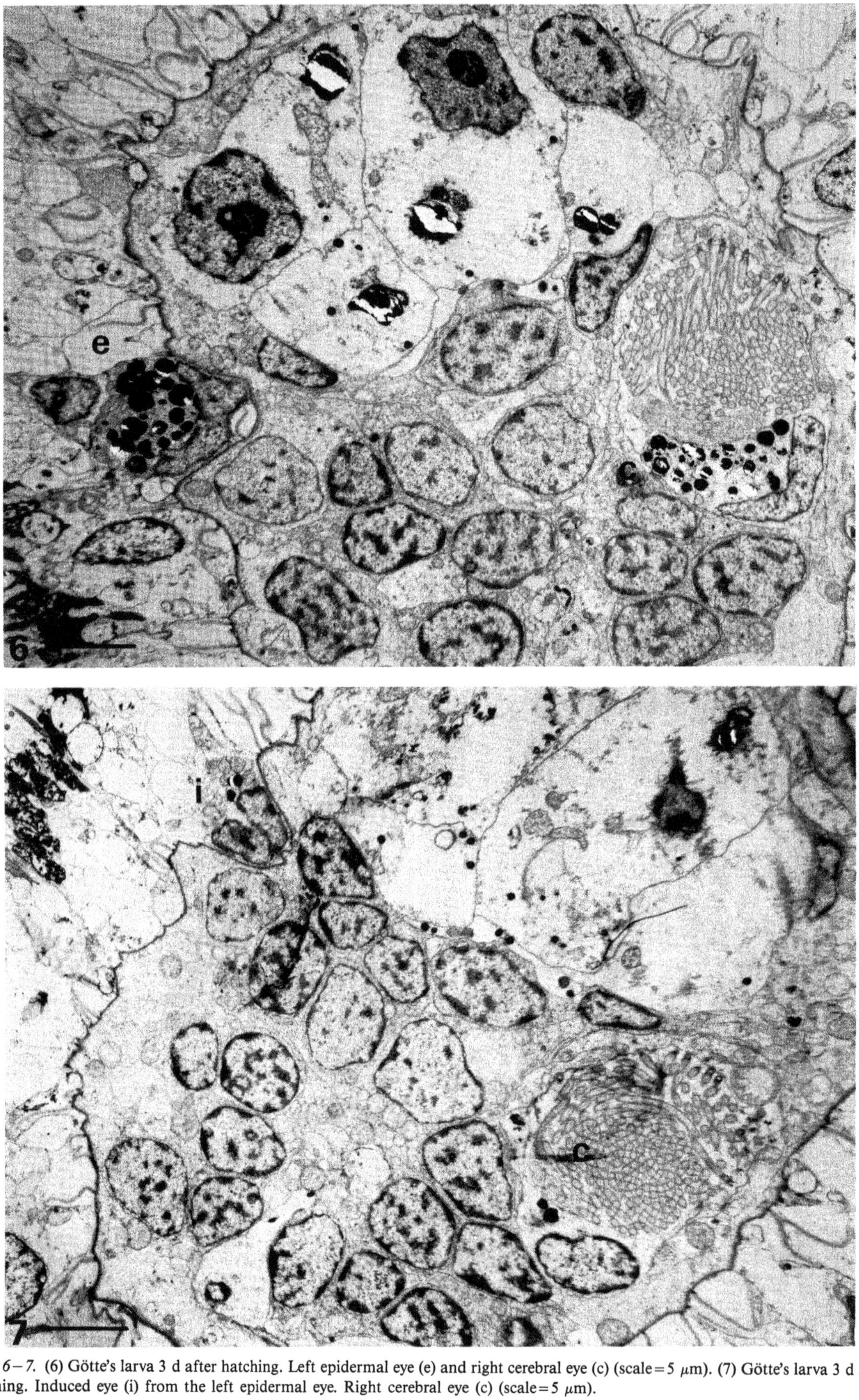

Figs. 6–7. (6) Götte's larva 3 d after hatching. Left epidermal eye (e) and right cerebral eye (c) (scale = 5 μm). (7) Götte's larva 3 d after hatching. Induced eye (i) from the left epidermal eye. Right cerebral eye (c) (scale = 5 μm).

as the right epidermal eye soon (within hours) sank into the parenchyma and turned into a cerebral eye (Fig. 4). Initially, this eye (rde) appeared to be made up of a pigmented cell without photosensitive lamellae, and was characterized by cytoplasm with very few organelles. The ciliary projections and the two rhabdomeric-type photoreceptor cells were not present. The space left in the overlying epidermis by the descent of the eye was initially filled with an epidermal cell with an extensive cell web anchoring on the basal lamina by hemidesmosomes.

Seven days after deposition, the right cerebral eye was fully formed (Fig. 5), while the left epidermal remained in the epidermis. Upon hatching (8 days after deposition) the Götte's larva possessed a left epidermal and a right cerebral eye.

Three days after hatching, the Götte's larva showed an incomplete eye adjacent to the left epidermal eye (Figs. 6 and 7). This eye displaced the basal lamina inwards as if it were sinking into the parenchyma (Fig. 7). At this stage, it contained no photoreceptive organelles, only pigment granules.

Five days after hatching, the Götte's larva possessed a newly formed cerebral eye under the basal lamina, which appeared interrupted at the point between this eye and the left epidermal eye (Fig. 8, arrows). The epidermal and cerebral eyes were very near each other. At this stage, the cerebral eye was already cup shaped; in its still-small cavity, some ciliary projections and two rhabdomeric processes of reduced dimension could be seen.

Six days after hatching, a continuous basal lamina separated the epidermal eye and the now completely formed cerebral eye (Fig. 9).

Over the subsequent few days (until 9 days after hatching), this third eye moved away from the epidermal one and approached the cerebral ganglia (Fig. 1G).

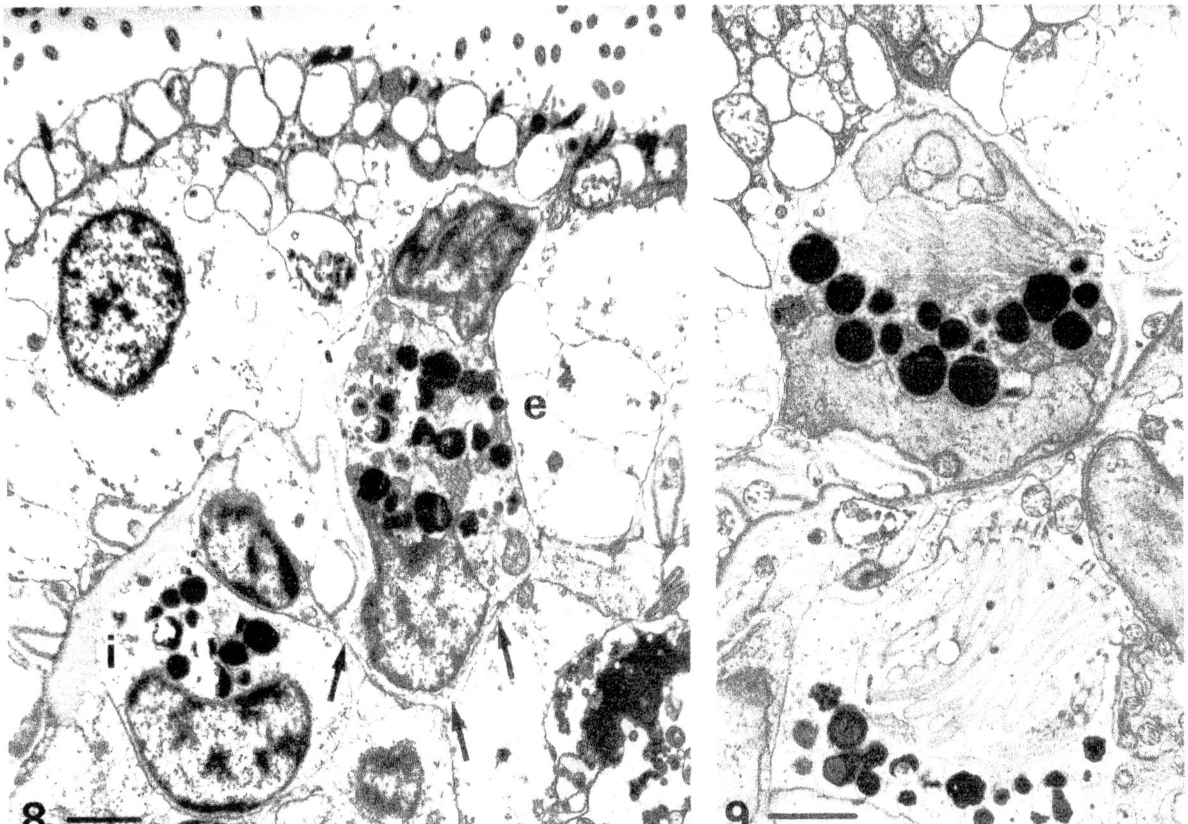

Figs. 8–9. (8) Götte's larva 5 d after hatching. Induced eye (i), located under the basal lamina, and left epidermal eye (e). Note the interrupted basal lamina (arrows) (scale=2 μm). (9) Götte's larva 6 d after hatching. The left epidermal eye and the left cerebral eye are separated by the continuous basal lamina (scale=2 μm).

Discussion

According to Kato (1940), the left epidermal eye appears first in the embryo of *Planocera multitentaculata, Stylochus uniporus,* and *S. aomori.* His observations are similar to our findings in *S. mediterraneus.* Kato (1940: 562) reports that the appearance of the right epidermal eye in *P. multitentaculata* follows that of the left epidermal eye, while in *S. uniporus* and *S. aomori* the left eye seems 'to sink deeply into the mesoderm from the ectoderm' and 'on the following day after the formation of the left eye-spot, another eye-spot is formed at the right side of the embryo' (1950: 563). According to Lang (1884), the first single eye in *Thysanozoon* and in *Yungia* divides into two, giving rise to the left and the right eye. Kato (1940: 560), however, states that, in this material the first two eyes 'are formed without doubt independently and separately.' Our observations demonstrate that the second eye appears in the ectoderm on the right side of the body independent of the first eye. Within a few hours, this right epidermal eye sinks into the parenchyma turning into a cerebral eye, as Kato thought was the case in *P. multitentaculata.* According to Kato, the three-eye stage in this species results from the division of the left epidermal eye. In contrast, we find that the left epidermal eye does not undergo division, but a presumptive eye appears *de novo* in the ectoderm near it. This new eye, upon sinking into the parenchyma, differentiates into a cerebral-type eye.

In conclusion, we can affirm that the left epidermal eye is a definitive organ, while the right epidermal eye is a transitory structure which later differentiates into a cerebral eye. The proximity of the third eye to the neighboring, the pre-existent left epidermal eye suggests it is induced by that epidermal eye.

Acknowledgments

We are grateful to Dr L. Ballarin for raising and maintaining the animals.

This work was supported by Consiglio Nazionale delle Ricerche, Gruppo Nazionale Scienze Neurologiche.

References

Eakin, R. M. & J. L. Brandenburger, 1981. Fine structure of the eyes of *Pseudoceros canadensis* (Turbellaria, Polycladida). Zoomorphology 98: 1–16.

Kato, K., 1940. On the development of some japanese polyclads. Jap. J. Zool. 8: 537–573.

Lanfranchi, A., C. Bedini & E. Ferrero, 1981. The ultrastructure of the eyes in larval and adult polyclads (Turbellaria). Hydrobiologia 84: 267–275.

Lang, A., 1884. Die Polycladen des Golfes von Neapel. In Fauna und Flora des Golfes von Neapel. Zoologischen Station zu Neapel. 11: 1–688.

Ruppert, E. E., 1978. A review of metamorphosis of turbellarian larvae. In F.-S. Chia & M. E. Rice, Settlement and metamorphosis of marine invertebrate larvae, Elsevier North-Holland Biomedical Press, N.Y.: 65–81.

Eggshell formation in polyclads (Turbellaria)

Sachiko Ishida & Wataru Teshirogi
Department of Biology, Faculty of Science, Hirosaki University, Hirosaki 036, Japan

Keywords: Turbellaria, polyclads, eggshell, shell formation, sclerotin

Abstract

Eggshell formation in polyclads was studied by means of transmission electron microscopy and histochemistry. Shell-forming granules (SFG) in the egg, as well as secretions of shell glands (SGS), play roles in eggshell formation. As the oocytes pass through the portion of the female tract where the shell glands open, they are surronded by a two-layered envelope of SGS. This envelope prevents the dispersion of SFGs discharged after oviposition, and its inner layer participates in eggshell formation with the SFGs. In *Pseudostylochus* sp., most SFGs consist of five parts. Similarities in staining between the parts of the SFGs and the parts of the eggshell indicate that discrete parts of the shell are derived from specific SFG components. Hardening of the eggshell and egg-plate matrix takes place through primary tanning of a sclerotin-like protein.

Introduction

Flatworm eggs are enveloped by a hard eggshell. The structure of this shell and its formation have been studied most intensively in species of the Trematoda and Cestoidea (Smyth & Clegg, 1959; Wittrock, 1982; and others). As far as the Turbellaria is concerned, several investigators reported that the eggshells of triclads and rhabdocoels are formed with globules extruded from yolk cells (Yanagita & Yamamoto, 1981; Bunke, 1982; and others). However, in polyclads, which produce endolecithal eggs lacking yolk cells, the mechanism of eggshell formation is not clear yet. The eggshell is thought to originate either from secretions of the shell gland (Lang, 1884), or from shell-forming substance produced in egg cells (Hofsten, 1912; and others). We have used transmission electron microscopy and histochemical analysis to evaluate these two hypotheses and to clarify the nature of eggshell structure and formation in the Polycladida.

Materials and methods

Pseudostylochus sp. and *Planocera multitentaculata* Kato, 1944, were collected on the coast of Fukaura and Mutsu Bay, Aomori Prefecture, Japan.

Transmission electron microscopy. Samples were fixed in 3% glutaraldehyde in seawater for two hours and, after washing in seawater, postfixed in 1% osmic acid in seawater for two hours. They were then dehydrated with ethanol and QY-1 and embedded in epoxy resin. Ultrathin sections were prepared, stained with 3% uranyl acetate and lead citrate (Sato, 1968), and observed with transmission electron microscopes (Hitachi HU-12A and HS-7D).

Histochemistry. Worms and egg masses were fixed with Bouin's, Zenker's, or 10% formalin solution. The fixed specimens were sectioned and prepared with the following histological procedures (all from Okamoto *et al.*, 1965, Takeuchi *et al.*, 1967, and Sano, 1976): periodic acid-Schiff reac-

Hydrobiologia 132, 127–135 (1986).

tion (PAS reaction, McManus, 1948) for polysaccarides; ninhydrin-Schiff reaction (Yasuma-Ichikawa, 1953) and Bonhag's mercury bromphenol blue method for protein; naphthol yellow S reaction (Deitch, 1955) for basic protein; Millon's tyrosine reaction (Bensley & Gersh, 1933) for tyrosine; Fitzpatrick's method (1954) for tyrosinase; Bruemmer, Carver & Thomas's modified method (1957) for indole radical; Masson's (1914) modified method for melanin and ferric ferricyanide method for melanin (Lillie, 1957); and azan staining and Delafield's haematoxylin-eosin staining for routine observations.

Some resin-embedded samples were sectioned at 1 μm in thickness, and the sections were stained by the technique of Kurotaki (1972), or with each staining method listed above, and examined under a light microscope.

Results

1. The structure of the eggshell and egg-plate matrix

Polyclad eggs are enclosed within the eggshell and embedded together in a thin plate of gelatinoid material. The egg-plate of *Pseudostylochus* sp. was irregular in shape and had one egg in each eggshell; that of *P. multitentaculata* was in the form of a tight spiral and had several eggs in each eggshell. The gelatinoid matrix was homogenous in the irregular egg-plate (Fig. 1A); in the spiral egg-plate it was covered by a thin-wrinkled layer and contained scattered groups of shell-gland granules (Fig. 1B). Spermatozoa were often observed in the space between the eggshell and the egg.

The eggshells were ornamented with a network of circular ridges and some irregular sculpturing (Fig. 1C). At the ridges (BACS), the shell was 1–1.3 μm thick and the enclosed areas (CSA) were about 0.6 μm in thickness and 40 μm in diameter in the shells of *Pseudostylochus* sp. at the two cell stage, about 10 hours after oviposition. Three layers were apparent in the shell, the innermost being the thinnest, about 80 nm. Electron density and histochemical reactivity (Fig. 2; Table 1) of the CSA and BACS differed. The shells ruptured along the BACS at hatching.

The egg-plate gelatinoid matrix (EPGM) of

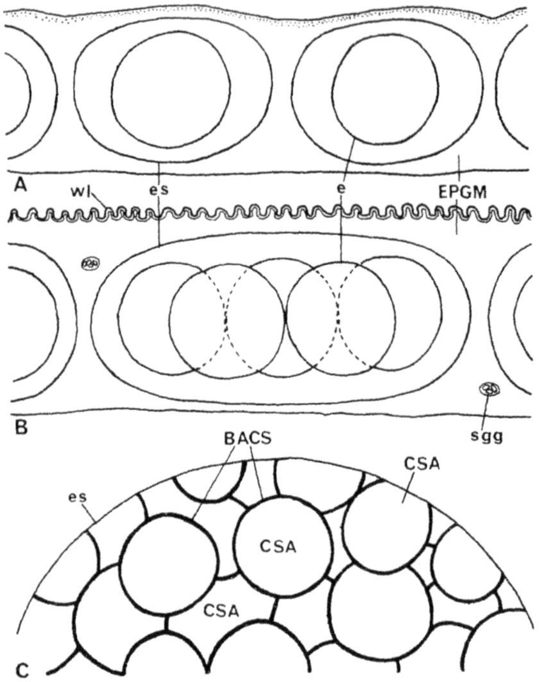

Fig. 1. Scheme of the cross section of egg-plate (A, B) and scheme of eggshell surface (C). A: *Pseudostylochus* sp. B: *Planocera multitentaculata*. BACS – boundary area between circular sculptures; CSA – circular sculpture area; e – egg; es – eggshell; EPGM – egg-plate gelatinoid matrix; sgg – shell-gland granule; wl – wrinkled layer.

Pseudostylochus sp. contained long, striated fiber-bundles (bamboo-like fiber-bundles), small fiber = bundles, and a network of fine fibrils (Fig. 2). These fibers arose from the shell envelope layer of the gelatinoid matrix covering the eggshells. The bamboo-like bundles were made up of units about 0.3 μm in length in which about six striae were found. The intervals between the striae were not regular, ranging from 24 nm to 66 nm. The small fiber bundles were crowded more or less in an area 1 μm thick adjacent to the outer shell layer.

2. Structure of shell-forming granules (SFGs)

The SFGs located in a row in the cortical layer varied in size and shape. In *Pseudostylochus* sp., larger granules were about 6 μm in diameter and smaller ones 1.8 μm. Most SFGs consisted of five parts: (1) an innermost part of high electron density occupying about 2/3 of the bulk of the granule, (2) a fibrous part of higher electron density, (3) a layer

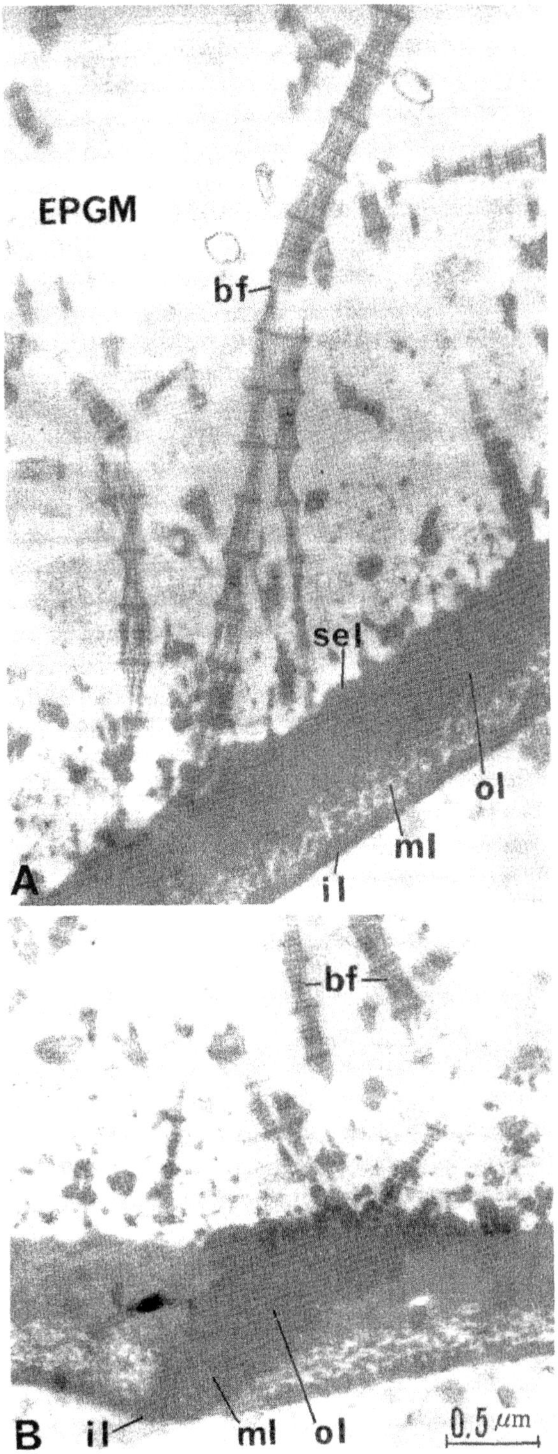

Fig. 2. Pseudostylochus sp. Electron micrograph of cross section of CSA (A) and BACS (B) in eggshell about 10 hours after oviposition. CSA is distinguished in three layers, i.e. inner, middle, and outer layers, and bamboo-like fiber bundles protrude from shell envelope layer; bf – bamboo-like fiber bundles; il – inner layer; ml – middle layer; ol – outer layer; sel – shell envelope layer; BACS, EPGM – as in Fig. 1.

about 40 nm in thickness with the highest electron density, (4) a part of medium electron density occupying 1/3 – 1/4 of the granule, and (5) a part of minute scattered granules (Fig. 3B). Apparently SFGs originated in vesicles derived from Golgi complex and to which material produced in rER (rough endoplasmic reticulum) was added (Fig. 3A). SFGs also appeared to be fused with mitochondria and autophagosomes during their maturation process. Some SFGs lacking parts (1) and (3) were apparently formed without input from the Golgi complex.

3. Shell = gland granules and shell = gland cells

In *P. multitentaculata*, three kinds of shell-gland granules constituting the shell-gland substance (SGS) could be distinguished by differences in reactivity with azan stain, etc.; in *Pseudostylochus* sp., only two kinds were observed. By electron microscopy, the granules of *Pseudostylochus* sp. appeared ovoid and membrane-bounded and had a striated substructure with bands 0.05 – 0.1 μm in width separated by 0.1 – 0.3 μm (Fig. 4). Some granules were only inconspicuously striated, and some had scattered or irregular protein crystalline substructures. Gland cells producing these shell-gland granules had an irregular ameboid shape and well-developed rER. The secretory substance appears to be synthesized on rER and condensed in the Golgi complex.

4. Eggshell formation

Eggs coming down the oviduct from the seminal receptacle were surrounded either singly (*Pseudostylochus* sp.) or in groups (*P. multitentaculata*) by shell-gland substance (SGS) at the shell-gland opening. The envelope formed had an outer granular layer and a thin homogeneous inner layer. The latter was silver-stained by Masson's modified method for melanin. The shell-gland granules fused with each other to form a colloid at the shell-gland opening and the silver-reactive substance became localized at the inside to form this inner layer of the egg-envelope. The eggs are then oviposited from the female genital pore. Soon after oviposition, the envelope of SGS swelled through absorption of water and separated from the egg. Then the SFGs were discharged by the egg, and the eggshell was formed. In immature eggs, SFGs were not dis-

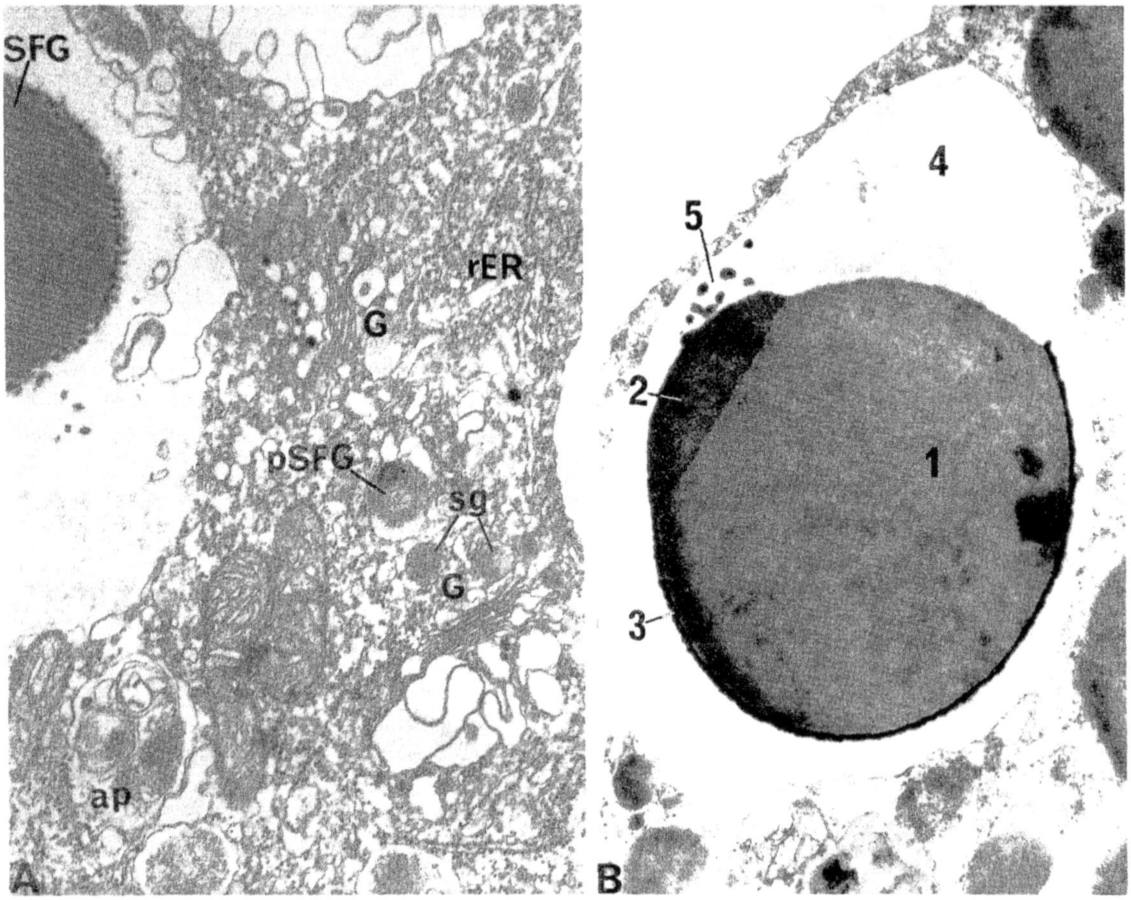

Fig. 3. Pseudostylochus sp. Electron micrograph of shell-forming granule (SFG). A: early stage of SFG formation in ovarian oocyte. Material produced in rER is added to the secretory granules derived from Golgi complex, and primordial SFGs are formed. B: a typical SFG seen in cortex of seminal receptacle oocyte. It consists of five parts (1–5). ap – autophagosome; G – Golgi complex; pSFG – primordial SFG; rER – rough surfaced endoplasmic reticulum; sg – secretory granule.

charged and therefore no eggshell was formed.

The eggshell and the egg-plate just after oviposition were white, soft, and weak; but over time they became yellowish, hard, and strong. Around the egg just after oviposition, there was a thin transparent layer, which seemed to be the inner layer of the egg-envelope secreted by the shell gland. The SFGs discharged from the egg by exocytosis soon after oviposition adhered to or entered the layer. Discharged granules stained deeply with Janus green; the layer itself stained only slightly just after oviposition but became deeply stainable after SFGs had fused with it. Then the circular sculptures peculiar to the eggshell appeared.

By transmission electron microscopy it was apparent that the thickness of eggshells increased after oviposition until a maximum of about 0.6 μm was reached. Even in eggshells where the BACS had not yet formed, three layers of CSA were apparent. The contents of SFGs discharged from eggs were added as the components penetrated from the inside into the eggshell. Soon after formation, the thickness of the BACS was almost the same as that of the CSA, but one hour after oviposition the BACS was thicker.

The outer layer of CSA in the eggshells was thinner, by about 0.04 μm, in eggs one and ten hours after oviposition than in eggs just oviposited, and the middle layer was thicker in these same eggs by about 0.02 μm. The inner layer of CSA was not

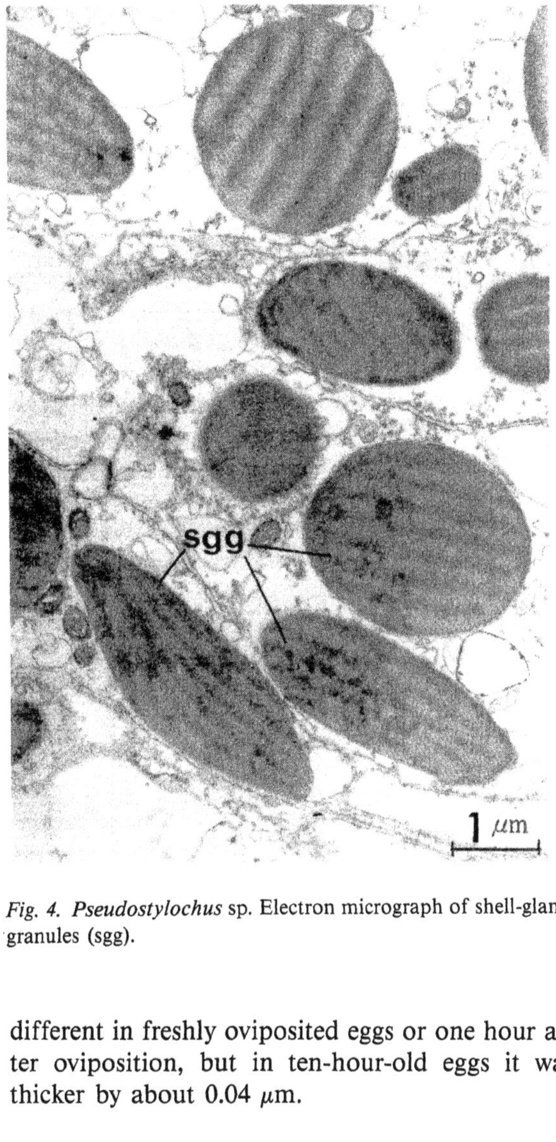

Fig. 4. Pseudostylochus sp. Electron micrograph of shell-gland granules (sgg).

different in freshly oviposited eggs or one hour after oviposition, but in ten-hour-old eggs it was thicker by about 0.04 μm.

5. Stainability and histochemistry

Table 1 summarizes the results of the histochemical tests on the SGS, on the granules within the eggs, and on the eggshell of *Pseudostylochus* sp. Because the inner layer of both the CSA and the BACS stain the same, reactions for this layer of the BACS are omitted from the Table. These results were observed in 1-μm resin sections and confirmed with staining of paraffin sections; details of substructure were drawn from comparison with electron micrographs. Parts 1 and 3 of SFGs stained with malachite green as did the outer layer of CSAs in the eggshell; part 2 stained intensely with azocarmin and in this respect resembled the middle and outer layers of BACS; and part 4 stained with aniline blue as did the inner and middle layers of CSA, while part 5 did not stain to any noticeable extent with any of the dyestuffs we used. Shell-gland granules stained in several tests in ways that resembled staining of both the outer layer of CSA and EPGM.

Such a difference between the two species was not seen in stainability of the SFGs, shell-gland granules, yolk granules, and eggshell. However, while in *Pseudostylochus* sp. the EPGM stained purple with azan, in *P. multitentaculata* it stained bright blue, indicating it is somewhat different in these species. The wrinkled structure of the egg-plate surface of *P. multitentaculata* stained red with azan, and was positive to Masson's modified method for melanin and to the ferric ferricyanide method for melanin.

Discussion

The origin of SFGs

Boyer (1972) and Domenici *et al.* (1975) reported that SFGs in *Prostheceraeus floridanus* and *Notoplana alcinoi* are derived from the Golgi complex. Our observations on *Pseudostylochus* sp. indicate that components derived from rER, mitochondria, and autophagosomes are also added to the granules.

Eggshell formation

While the two seemingly divergent views in the literature on the origin of the eggshell in polyclads hold that it arises either as secretion of the shell gland (Lang, 1884) or as secretion of the egg cells (Hofsten, 1912; Kato, 1940), our findings make it clear that both the egg and the shell glands play a role (Fig. 5).

SFGs are formed in the oocytes prior to formation of yolk granules and move to the egg cortex. The oocytes ovulated from the ovary begin the first meiotic division on the way to the seminal receptacle. Meiosis is suspended at metaphase in the seminal receptacle (Fig. 5A). On reaching the shell-gland opening via the oviduct, the oocytes are surrounded by SGS, either singly or in groups depend-

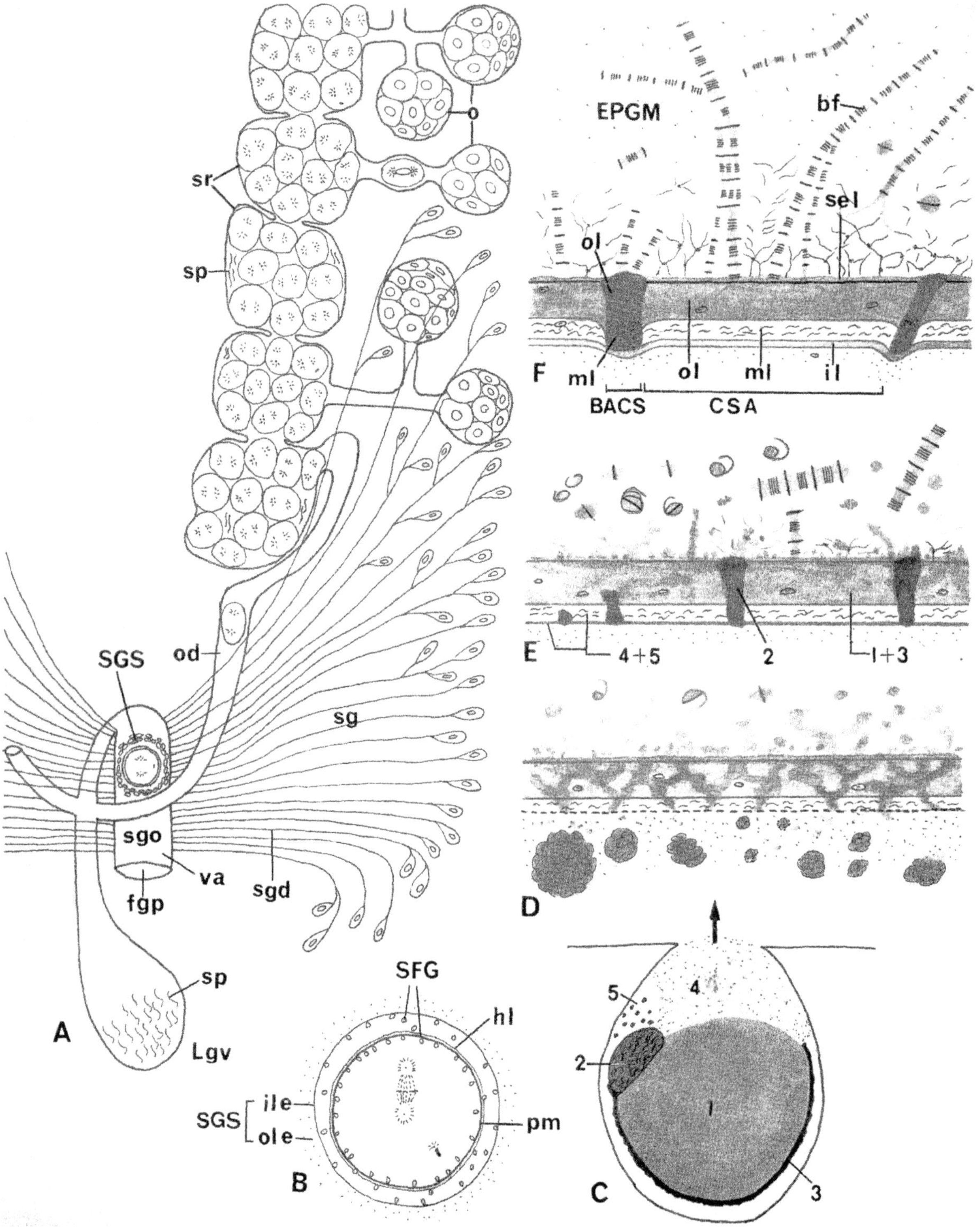

Fig. 5. Scheme of eggshell formation in *Pseudostylochus* sp.; (A) structure of female genital organs and process from ovulation through oviposition; (B) egg just after oviposition. (C) secretion of SFG; (D) early stage of eggshell formation in which BACS is not formed yet; (E) formation of BACS; (F) eggshell formed completely. 1–5 – parts of SFGs; fgp – female genital pore; hl – hyaline layer; ile – inner layer of egg-envelope; Lgv – Lang's glandular vesicle; o – ovary; od – ovovitelline duct; ole – outer layer of egg-envelope; pm – plasma membrane; sg – shell gland; sgd – shell-gland duct; sgo – shell-gland opening; SGS – shell-gland substance; sp – sperm; sr – seminal receptacle; va – vagina; BACS, bf, CSA, EPGM, il, ml, ol, sel, SFG, as in Figs. 1–4.

ing on the species. The SGS comprises the two layers (inner and outer) of the egg envelope (Fig. 5B). The inner layer is thought to have a function similar to that of the vitelline coat of other animals such as echinoderms and amphibians: it prevents the dispersion of SFGs discharged after oviposition and participates in eggshell formation as a substrate for the eggshell. The outer layer is thought to form the EPGM.

The derivation of eggshell components from specific parts of SFGs is suggested by similarities in electron density and the histochemical reactions. After oviposition, the egg envelope of SGS swells by absorption of water and separates from the egg to some extent (Fig. 5B). Parts 1 and 3 of SFGs adhere to and fuse along its inner layer, forming the outer layer of CSA of the eggshell (Fig. 5C-E). Then parts 4 and 5 form the middle and inner layers underlying this. Part 2 of SFGs arrives at the eggshell somewhat later and enters it, forming the BACS (Fig. 5E). And then, the outline of the circular figure peculiar to the eggshell becomes clear (Fig. 5F). The fact that the inner layer of egg-envelope, derived from shell-gland granules, plays some part in forming the outer layer of CSA is indicated by the common positive reaction to the melanin-detection method.

Some differences were found in the thickness of eggshells among eggs soon after oviposition, but after a thickness of about 0.6 μm was reached, there was no noticeable increase with age. Presumably once nearly all the SFGs have discharged (within about 40 minutes after oviposition), the eggshell does not become any thicker. We believe changes in the thickness of component layers of the CSA during the first 10 hours after oviposition reflect an increase in density of proteins of the outer layer of CSA and swelling of the inner and the middle layers by absorption of water in proteins constituting these layers. Protrusion of the BACS on both sides of the CSA (Fig. 2B) is also believed to have been caused by swelling through water absorption.

Three schemes of the cross section of the eggshell by light microscopy shown previously by us (Ishida *et al.*, 1981) are contradicted by the present electron microscopical findings.

Hardening of the eggshell and EPGM

Since Pryor (1940) found a tanned protein combined with quinone in cockroach oothecae and called this protein sclerotin, similar proteins have been found in the cuticles of insects and egg cases of various invertebrates. Various substances from which quinone is derived are free O-diphenol, β-glucoside, protocatechuic acid, 3:4-dihydroxybenzoic acid, tyrosine, p-tyrosine, etc. (cf. Smith & Clegg, 1959; Locke & Krishnan, 1971). In the Platyhelminthes, the detection of diphenol and protein in the composite eggs of Trematoda, Cestoidea, and Turbellaria suggests the existence of sclerotin, and the presence of sclerotin precursors in yolk cells has been reported (Nurse, 1950; Burton, 1963; Nollen, 1971; Bunke, 1972, 1981; Gremigni & Domenici, 1974; and others). In polyclads, which oviposit endolecithal eggs, Vialli (1934), Domenici *et al.* (1975) and others observed that previtellogenic oocytes contain granules with di- or polyphenols and protein which are distributed in the cortex of mature oocytes. These granules have been considered to be shell globules forming shell sclerotin through quinone tanning.

Our finding tyrosine (which contains a phenol radical), tyrosinase, and melanin-positive reaction in the SFGs suggests that sclerotin type protein is present in the eggshell and that quinone is derived from tyrosine. In *Pseudostylochus* sp., tyrosine and tyrosinase are detected only in parts 1 and 3 of SFGs and sclerotin only in the outer layer of the CSA, again indicating that parts 1 and 3 of SFGs form the outer layer of CSA as mentioned above. As tyrosine and tyrosinase are also detected in the shell = gland granules, sclerotin seems to be formed to some extent in EPGM. The wrinkled layer on the egg-plate surface in *P. multitentaculata* is especially positive to the melanin reaction, and must contain much sclerotin. Inasmuch as the inner egg-envelope formed by SGS is positive for melanin, some tyrosine and tyrosinase in the shell-gland granules appears to enter this layer. That is, tyrosine and tyrosinase in these granules seem to take part in the sclerotin formation in both the outer layer of the CSA and the EPGM.

The egg-plate soon after oviposition is white and soft, but it becomes yellowish, harder, and stronger as it ages. This change can be explained as follows. At the time of oviposition, tyrosine contained in SFGs and SGS becomes O-diphenol (dihydroxyphenylalanine, Dopa) under the action of tyrosinase and oxygen. As Dopa is colorless, the white color of the egg-plate soon after oviposition is due to the shell = gland granules surrounding the egg. The later-appearing yellowish color of the egg-plate is due to the enzymatic oxidation of Dopa to Dopa-quinone; the color peculiar to quinone appears on the eggshell and the EPGM. After the quinone combines with protein to form sclerotin, the whole egg-plate becomes more yellowish and harder. The fact that tyrosinase is not detected in the eggshell and EPGM shows that tyrosinase activity has been lost after the Dopa and Dopa-quinone forming reaction.

The inner and middle layers of the CSA seem to be a collagen-like scleroprotein (albuminoid) as it stains with aniline blue, lacks tryptophan, and is capable of swelling. But typical collagen fibers cannot be found in these layers by electron microscopy. The middle and outer layers of the BACS seem to consist of fibrous scleroprotein that contains basic amino acids and that easily swells. The fact that the eggshell is broken along this BACS at hatching shows that this area is more fragile than the sclerotin-containing CSA.

Although the EPGM contains sclerotin, other fibrous scleroproteins and other PAS-positive substances (except glycogen) are also present (Table 1). Its strong affinity for aniline blue in *P. multitentaculata* suggests the existence of collagen-like substances; however, aniline-blue affinity is not strong in *Pseudostylochus* sp. Some collagen fibers are characterized by striations with a 64 nm periodicity (Bucher, 1962). The bamboo-like fiber bundles

Table 1. Stainability and results of histochemical tests on the shell forming granules, yolk granules, shell-gland granules, eggshell and egg-plate gellatinoid matrix of *Pseudostylochus* sp.

Procedure	Stain	Each part of SFG 1	2	3	4	5	Yolk granules	Shell gland granules	Each layer of egg shell: CSA il	CSA ml	CSA ol	BACS ml	BACS ol	EPGM
Delafield's hematoxylin·eosin	Hematoxylin	○	○	▿	–	–	○	▿	–	–	○	○	○	○
	Eosin	▿	○	▿	–	±	○	□	–	–	○	○	○	▿
Azan	Azocarmin	○	□	○	–	–	▿	▿	–	–	○	▿	□	○
	Aniline blue	±	–	±	○	±	–	□	○	○	±	±	–	▿
	Orange G	○	–	–	○	±	○	±	–	±	○	○	–	–
A differential staining of semi-thin sections of epon-embedded tissues	Malachite green	▿	–	▿	–	–	±	▿	–	–	▿	–	–	±
	Toluidine blue	○	○	□	–	–	○	○	–	○	○	–	–	–
	Basic fuchsin	○	○	▿	○	–	○	○	○	±	–	±	○	○
PAS reaction		▿	○	○	–	–	○	○&±*	–	–	○	±	±	○
PAS after Salivary digestion		▿	○	○	–	–	○	○&±*	–	–	○	±	±	○
Ninhydrin-Schiff reaction		▿	○	▿	–	–	○	○&±*	–	–	▿	○	○	○
Bonhag's mercury bromphenol blue method		±	▿	±	○	–	▿	▿	±	○	±	○	▿	▿
Naphthol yellow S reaction		▿	▿	▿	±	±	▿	▿	○	○	▿	▿	▿	▿
Millon's tyrosine reaction		▿	–	○	–	–	±	○	–	–	▿	–	–	○
Fitzpatrick method for tyrosinase		○	–	▿	–	–	–	○	–	–	–	–	–	–
Masson's modified method for melanin		○	–	□	–	–	–	○	–	±	▿	±	–	±(○**)
Ferric ferricyanide method for melanin		○	±	▿	–	–	±	±&–*	–	–	○	±	±	–(○**)
Bruemmer, Carver & Thomas's modified method for indole		○	–	±	–	–	±	±	–	–	○	–	–	–

□, intensely stained; ▿, moderately stained; ○, slightly stained; ±, weak reaction; –, no stain.
* duality caused by two kinds of shell-gland granules.
** *Planocera multitentaculata*, wrinkled structure of egg-plate surface.
BACS, CSA, EPGM, il, ml, ol, SFG; see Figures 1, 2 and 3.

found in EPGM of *Pseudostylochus* sp., though striated, are unlike such collagen fibers in that they have narrower and irregular striations and have a different thickness and structure in cross section. But there is at least the similarity between the 0.3 μm unit-fiber length in the bamboo-like fiber bundles and the length (0.3 μm) of a collagen molecule.

Acknowledgments

The authors wish to express their gratitude to Professor Koichi Hiwatashi and Professor Kenzi Osanai of Tôhoku University for their valuable advice.

References

Boyer, B. C., 1972. Ultrastructural studies of differentiation in the oocyte of the Polyclad Turbellarian, *Prostheceraeus floridanus*. J. Morph. 136: 273–295.

Bucher, O., 1962. Histologie und Mikroskopische Anatomie des Menschen. Medizinischer Verlag, Hans Huber, Bern Stuttgart.

Bunke, D., 1972. Sklerotin-Komponenten in den Vitellocyten von *Microdalyellia fairchildi* (Turbellaria). Z. Zellforsch. 135: 383–398.

Bunke, D., 1981. Ultrastruktur-Untersuchungen an Vitellocyten von *Microdalyellia fairchildi* (Turbellaria, Neorhabdocoela). Zoomorphology 99: 71–86.

Bunke, D., 1982. Ultrastruktur-Untersuchungen zur Eischalenbildung bei *Microdalyellia fairchildi* (Turbellaria). Zoomorphology 101: 61–70.

Burton, P. R., 1963. A histochemical study of vitelline cells, egg capsules, and Mehlis' gland in the frog lung-fluke *Haematoloechus medioplexus*. J. Exp. Zool. 154: 247–257.

Domenici, L., L. Galleni & V. Gremigni, 1975. Electron microscopical and cytochemical study of egg-shell globules in *Notoplana alcinoi* (Turbellaria, Polycladida). J. Submicr. Cytol. 7: 239–247.

Gremigni, V. & L. Domenici, 1974. Electron microscopical and cytochemical study of vitelline cell in the fresh water Triclad *Dugesia lugubris* s. l. I. Origin and morphogenesis of cocoon-shell globules. Cell Tiss. Res. 150: 261–270.

Hofsten, N., 1912. Eischale und Dotterzellen bei Turbellarien und Trematoden. Zool. Anz. 39: 111–136.

Ishida, S., T. Gotoh & W. Teshirogi, 1981. Oogenesis and eggshell formation in polyclads. Rep. Fukaura Mar. Biol. Lab. No. 9: 32–48 (in Japanese).

Kato, K., 1940. On the development of some Japanese polyclads. Jap. J. Zool. 8: 537–573.

Kurotaki, M., 1972. A differential staining of semi-thin sections of epon-embedded tissues and a short review on the optical microscopy of plastic-embedded tissue. Acta Anatomica Nipponica 47: 237–250 (in Japanese with English summary).

Lang, A., 1884. Die Polycladen. Fauna u. Flora Golfes v. Neapel 11: 1–688.

Locke, M. & N. Krishnan, 1971. The distribution of phenoloxidases and polyphenols during cuticle formation. Tissue & Cell 3: 103–126.

Nollen, P. M., 1971. Digenetic trematodes: Quinone tanning system in eggshells. Experimental Parasitology 30: 64–72.

Nurse, F. R., 1950. Quinone tanning in the cocoon-shell of *Dendrocoelum lacteum*. Nature 165: 570.

Okamoto, K., M. Ueda, T. Maeda & A. Mizutani, 1965. Microscopical histochemistry (3rd Edition). Tokyo Igaku Syoin, Osaka (in Japanese).

Pryor, M. G. M., 1940. On the hardening of the ootheca of Blatta. Proc. Roy. Soc. London 128: 378–393.

Sano, Y., 1976. The method for histological research – Theory and Method. Nanzando, Tokyo (in Japanese).

Sato, T., 1968. A modified method for lead staining of thin section. J. Elect. Micro. 17: 158–159.

Smyth, J. D. & J. A. Clegg, 1959. Egg-shell formation in trematodes and cestoides. Experimental Parasitology 8: 286–323.

Takeuchi, T., Shimizu, N. & K. Ogawa (editor), 1967. Enzyme histochemistry. Asakura Syoten, Tokyo (in Japanese).

Vialli, M., 1934. Ricerche istochimiche sui granuli vitellini dei Policladi. Boll. Zool. 5: 21–23.

Wittrock, D. D., 1982. Structure and origin of the eggshell of *Quinqueserialis quinqueserialis* (Trematoda: Notocotylidae). Z. Parasitenkd 67: 37–44.

Yanagita, Y. & T. S. Yamamoto, 1981. Formation and chemical property of the cocoon in a freshwater planarian, *Dendrocoelopsis ezensis* Ichikawa et Okugawa (Turbellaria, Tricladida). Zool. Anz., Jena 207: 130–142.

Fine-structural characteristics of female and male germ cells in Proseriata Otoplanidae (Platyhelminthes)

Beate Sopott = Ehlers
Arbeitsstelle zur Erforschung der Mikrofauna des Meeresbodens der Akademie der Wissenschaften und der Literatur Mainz, II. Zoologisches Institut und Museum der Universität Göttingen, Berliner Strasse 28, D-3400 Göttingen, F.R.G.

Keywords: Platyhelminthes, ultrastructure, germocytes, vitellocytes, spermatozoa

Abstract

Fine-structural features of female germ cells differentiating within the germaria of *Otoplanella baltica* and *Notocaryoplanella glandulosa* are documented and compared with those of other free-living platyhelminths having ectolecithal eggs.

In these species, encompassed in the taxon Proseriata Lithophora, insemination of the germocytes occurs within the germaria. A sperm cell, that has penetrated a germocyte differs in special features from mature male germ cells found in the testes, in parts of the male genital system, or even in other regions of the organism. The hypothesis that dense bodies correspond to acrosomal material is supported.

Introduction

There is a large number of publications concerning the ultrastructure of male gametes in free-living platyhelminths (see Hendelberg, 1983a). In contrast, our knowledge of the fine structure of female gonads seems comparatively small, although some electron-microscopical (EM) studies of these organs have been carried out in recent years (*c.f. inter alies* Boyer, 1972; Bunke, 1981; Domenici & Gremigni, 1974; Gremigni, 1974, 1976, 1983; Gremigni & Banchetti, 1972; Gremigni & Domenici, 1975; Gremigni & Nigro, 1983; Oakley, 1982).

Within the taxon Seriata, the Tricladida (together with the Bothrioplanida) and the Proseriata (Lithophora and Unguiphora) are assumed to be sister groups (Sopott-Ehlers, 1985). While some publications discuss the fine structure of the female gonad of triclads, only two notices about the fine structure of this organ in proseriate species have been published until now (Lanfranchi in Gremigni, 1983; Gremigni & Nigro, 1984).

Material and methods

Specimens of *Otoplanella baltica* Ax, 1956, and *Notocaryoplanella glandulosa* (Ax, 1951) were extracted from sandy sediments taken from beaches of the island of Sylt (North Sea). The methods of preparation are described in detail in Ehlers & Ehlers (1977). Serial sections were examined using Zeiss EM 9 and Zeiss EM 10B electron microscopes.

Results

In the Seriata the germ- and yolk-cell producing parts of the female gonad are always separated. Furthermore, a characteristic feature of the Seriata is the follicular development and serial arrangement of the gonads. Follicles in two longitudinal rows form the vitellaria, while the germaria – in most cases – are developed as two compact organs.

Hydrobiologia 132, 137–144 (1986).

The germaria of the two otoplanid species studied differ: while *O. baltica* possesses one pair of compact organs, which are as a whole surrounded by a tunica, the germarium of *N. glandulosa* is a median complex of single follicles in which each germocyte is surrounded by a cellular tunica of its own. In both species and in the monocelidid *Archimonocelis oostendensis* Martens, 1981 (unpublished observations), the germaria are separated from the neighbouring tissue by a thin layer of extracellular matrix.

In both otoplanid species, germogonia as well as mature and already inseminated germocytes were found. Young germocytes or germogonia are characterized by a high number of free ribosomes and, as a result, a dark appearing cytoplasm. Aside from the ribosomes, the cytoplasm has a comparative paucity of organelles. Mitochondria and rough endoplasmic reticulum (ER) are to be found. In *O. baltica*, maculae adherentes were sometimes found between adjacent cells. The nuclei are round and show clearly developed nucleoli as well as synaptonemal complexes (Fig. 1A). Nuclear pores were also observed. The synaptonemal complexes only appeared in those cells in which electron-dense chromatin was scarcely condensed.

In germocytes of more advanced development, aggregations of centrioles (Fig. 1B) lie near the nucleus which assumes an increasingly lobed shape during maturation. The nucleolus appeared ring-shaped in some cases. Lipid droplets appeared in great quantity; and in restricted regions of the cytoplasm, membrane-bounded granules with a more or less electron-dense core were found (Fig. 1C). In mature germocytes, these granules (Fig. 1D) lie in a regular arrangement peripherally, directly beneath the cell membrane, the oolemma (Fig. 2A). I call these granules 'marginal granules'. Within the cytoplasm single microtubules were seen. Furthermore, rough ER and dictyosomes as well as a few multivesicular bodies are present.

Two different non-membrane bound electron-dense complexes were observed (Figs. 2B, C): small oval formations which were always associated with mitochondria and larger round ones. In the literature, these structures are called 'germinal plasm', 'chromatoid bodies', 'nuage', or 'nuclear extrusions' (for further information see Eddy, 1975). Electron-dense differentiations of this kind are considered to have passed from the nucleus through nuclear pores into the cytoplasm and are considered typical of germocytes.

The germinal vesicle was lobate. The nuclear envelope had numerous nuclear pores. Annulate lamellae (Figs. 2D, E), characteristic of oocytes, appeared directly next to the nuclear membrane in stacked parallel arrays with membranous components frequently interrupted by hexagonally arranged pores, structurally similar to those of the nuclear envelope. The subarchitecture of these pores (Fig. 2F, G) showed granular subunits, and with high magnification an electron-dense dot could be discerned in the center of some pores. Annulate lamellae are thought to play a role in nuclear-cytoplasmic exchange (*cf.* Kessel, 1983).

A large amount of glycogen and a strongly developed rough ER was observed within the vitellocytes of the otoplanid species studied. Furthermore, the cells contained lipid droplets and large osmiophilic droplets of yolk substance. Electron-dense material appeared within the vesicles of the dictyosomes. Larger vesicles contained material (presumably shell substance) deposited in a bowl-shaped manner (Fig. 2H). Some round granules have a substructure resembling the convolutions of a brain (Fig. 2I) (*cf.* Ehlers, 1985a).

The spermatozoa of the Proseriata have an elongated nucleus consisting of electron-dense fibrous material, mitochondria arranged in a row and so-called dense bodies. Two cilia with an axoneme of the 9×2+'1' pattern characteristic of the taxon Trepaxonemata Ehlers (Ehlers, 1985a, b) are interconnected with each other by a striated plate, the intercentriolar body, as already known from other

Fig. 1. Stages of germocyte maturation. A. *Otoplanella baltica.* Early prophase with synaptonemal complexes. B. *O. baltica.* Centriole aggregations near germinal vesicle. C. *Notocaryoplanella glandulosa.* Appearance of first marginal granules. D. *N. glandulosa.* Mature germocyte. Scale bars = 1 μm unless otherwise indicated. Abbreviations: al, annulate lamellae; cb, centriolar body; ci, cilium; ct, centrioles; db dense body; gv, germinal vesicle; lp, lipid droplet; mg, marginal granules; mi, mitochondrion, mt, microtubules; n, nucleus; ne, nuclear extrusions; nl, nucleolus; np, nuclear pores; pal, pores of annulate lamellae; rs, rootletlike structure; sc, synaptonemal complex; sg, shell secretion granule; tu, tunica; yg, yolk granule.

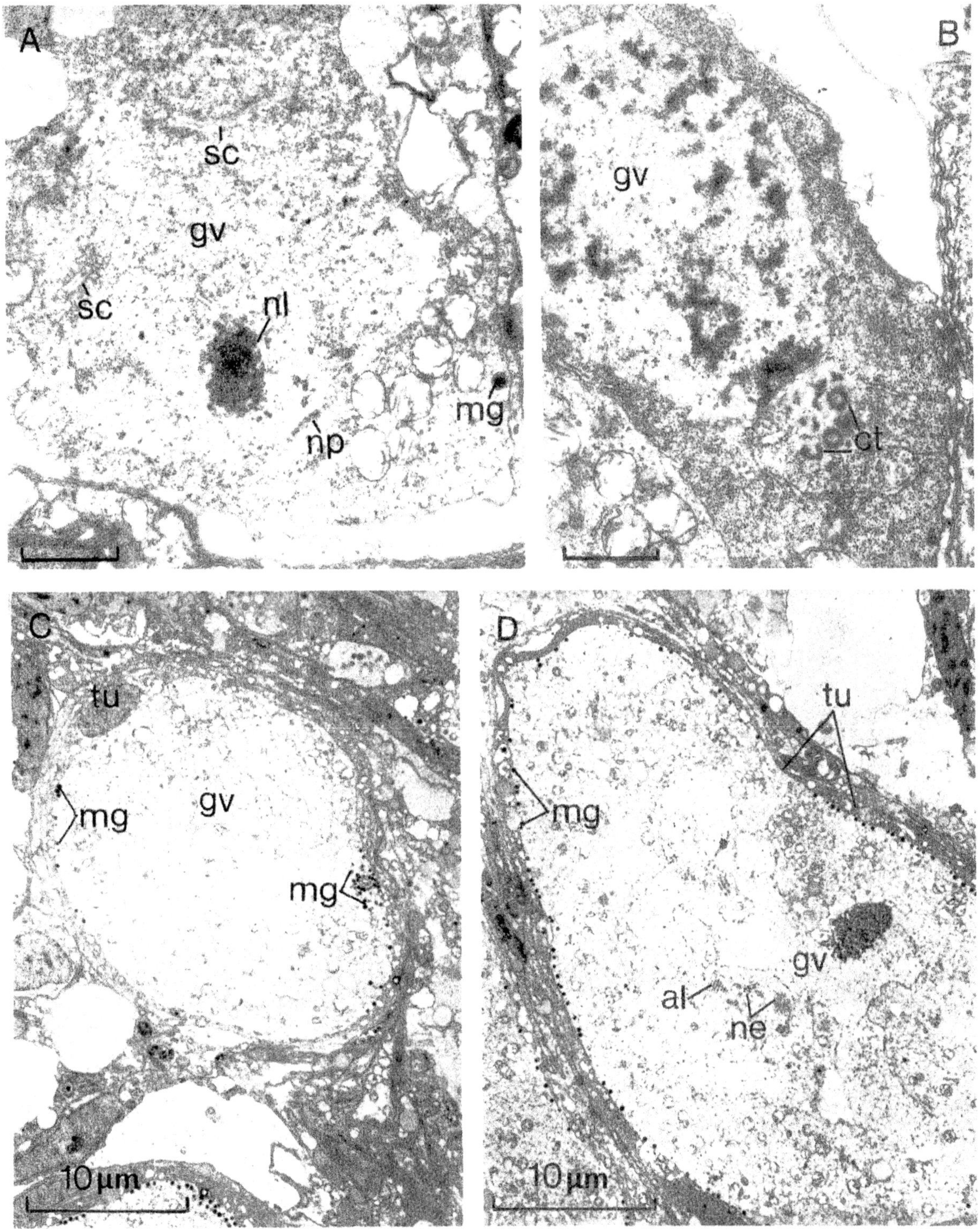
A
sc
gv
sc
nl
np
mg
B
gv
ct
C
tu
mg
gv
mg
10 µm
D
tu
mg
gv
al
ne
10 µm

freeliving and parasitic platyhelminths (Silveira & Porter, 1964; Kubo-Irie & Ishikawa, 1983; Franquinet & Lender, 1972; Williams, 1983; Burton, 1972; Rees, 1979; own unpublished observations on the lecithoepitheliate *Geocentrophora*). The cilia are anchored within the cytoplasm – at least in young spermatids – by rootlet-like differentiations (Fig. 3A). These structures run in the direction of the nucleus on one side and in the direction of the outer cell membrane on the other side. Both of the cilia are incorporated into the cytoplasm over a short distance proximally, and distally they stand free. Over a more or less long distance the nucleus is surrounded by mitochondria and dense bodies (Fig. 3B). Microtubules run directly beneath the cell membrane.

Spermatozoa of this morphology were found not only within the male organs, but also in other parts of the body, for instance between nerve cells within the capsule of the brain in *Carenscoilia bidentata* Sopott; that is to say, the spermatozoa in this region had the nucleus as well as mitochondria, cilia and dense bodies.

I have found inseminated germocytes within the germaria of the proseriate species studied. In the cytoplasm of these germocytes, the fibrous electron-dense material of the spermatozoal nucleus lay within a vacuole surrounded by nuclear envelope (Fig. 3C). Beyond this, residuals of the marginal microtubules of the sperm cell could be seen as well as mitochondria that had a smaller diameter than the mitochondria in the cytoplasm of non-inseminated oocytes. In one inseminated germocyte these mitochondria still lay in a row (Fig. 3D), indicating they were derived from a spermatozoon.

Furthermore, both flagella of the sperm cell (Figs. 3E, F) could be found within the cytoplasm of the germocyte. But dense bodies could not be seen at all. Otherwise the cytoplasm of the inseminated germocytes did not differ with regard to structure and organelles from that of non-inseminated germ cells; among other things the marginal granules were still in their previous shape and position.

Discussion

These observations on the ultrastructure of female gametes in different stages of maturation allow conclusions to be drawn on maturation, insemination, and the role of cell organelles.

The existence of numerous synaptonemal complexes in the nucleus indicates prophase of the first meiotic division. To date synaptonemal complexes have been observed in oocytes of not only polyclads (Boyer, 1972) but also triclads (Gremigni & Banchetti, 1972) and *Mesostoma ehrenbergi* (Oakley, 1982). The existence of these structures allows the conclusion that, in free-living platyhelminths, at least the first steps of meiosis happen within the ovaries or germaria.

Annulate lamellae, which occur most reguarly in germ cells, as well as chromatoid bodies (nuclear extrusions), which are considered as specially differentiated cytoplasmic structures in germ cells, have also been observed in differentiating cells of triclads (*cf. inter alies* Coward, 1974, 1979; Hori, 1982). The existence of these differentiations in gametes as well as in regenerating somatic tissue coupled with the occurrence of centriolar aggregations in germocytes indicate that germ cells are derived from totipotent stem cells.

Cortical granules are membrane-bounded bodies located in the cortical region of the mature oocyte. They form a layer at the egg surface after insemination and are thought to prevent polyspermy. Boyer (1972) and Gremigni & Domenici (1975) found in polyclads and triclads, respectively, that the structures considered to be cortical granules did not ex-

---

Fig. 2. N. glandulosa. A. Marginal granules and microtubules. B. Organelles in proximity of the germinal vesicle. C–G. Annulate lamellae and pores of annulate lamellae. D. Semi-oblique section through a stack of annulate lamellae (right). Arrows (left) point to granular substructure of pores. F and G. Substructure of pores of annulate lamellae. H–I. Part of vitellarium. Scale bars = 1 μm unless otherwise indicated. Abbreviations: see Fig. 1.

Fig. 3. A. *N. glandulosa.* Spermatid. B. *O. baltica.* Mature spermatozoa, part of head and middle piece. Arrow points to an axoneme. C–F. *N. glandulosa.* Inseminated germocyte. D. Mitochondria of sperm cell (mi_2) within the germocyte, mi_1 = mitochondrion of germocyte. E. Nucleus and cilia of sperm cell within germocyte. Scale bars = 1 μm. Abbreviations: see Fig. 1.

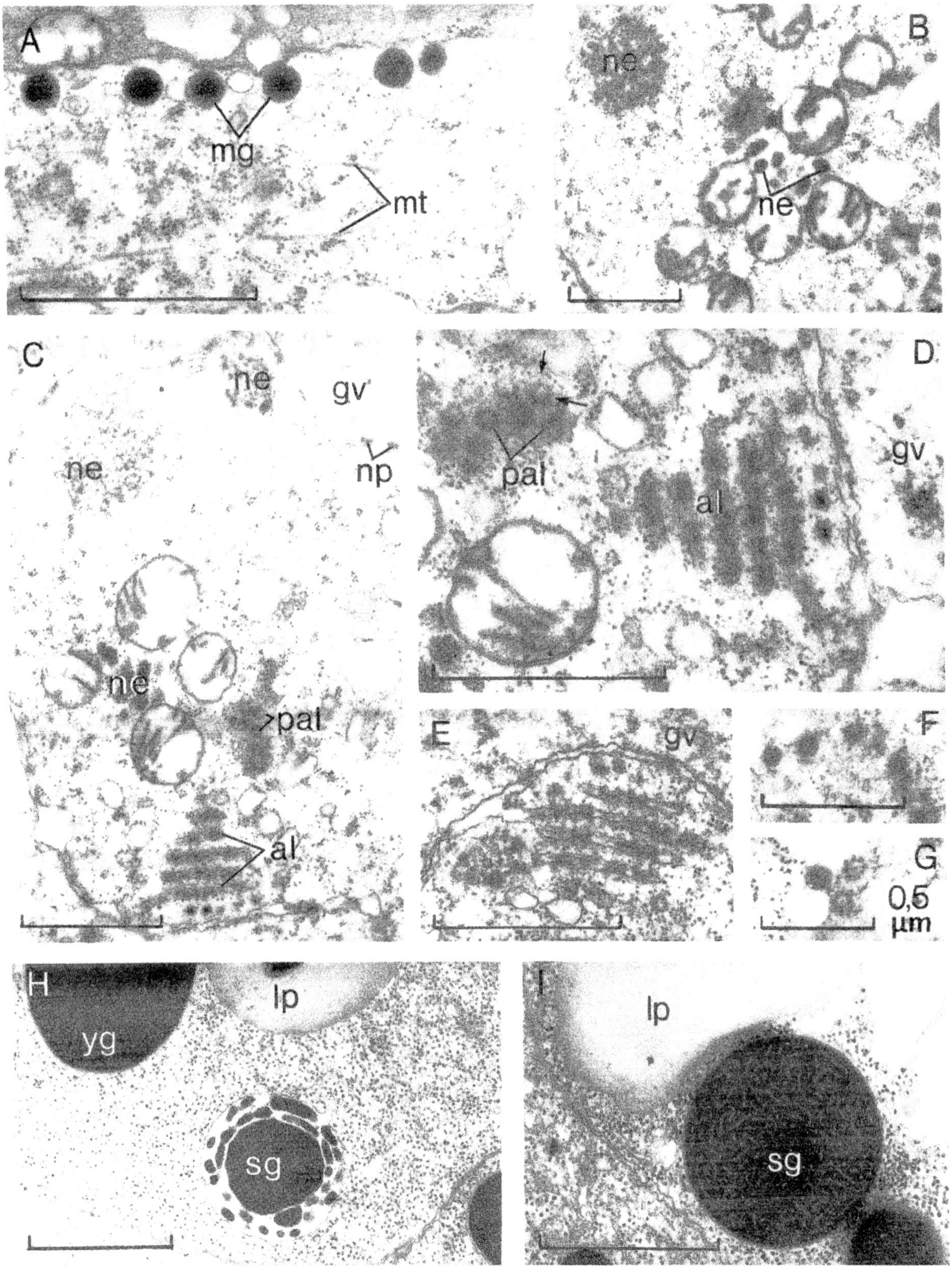
A
mg
mt
B
ne
ne
C
ne
gv
ne
np
ne
pal
al
D
pal
al
gv
E
gv
F
G
0,5
µm
H
lp
yg
sg
I
lp
sg

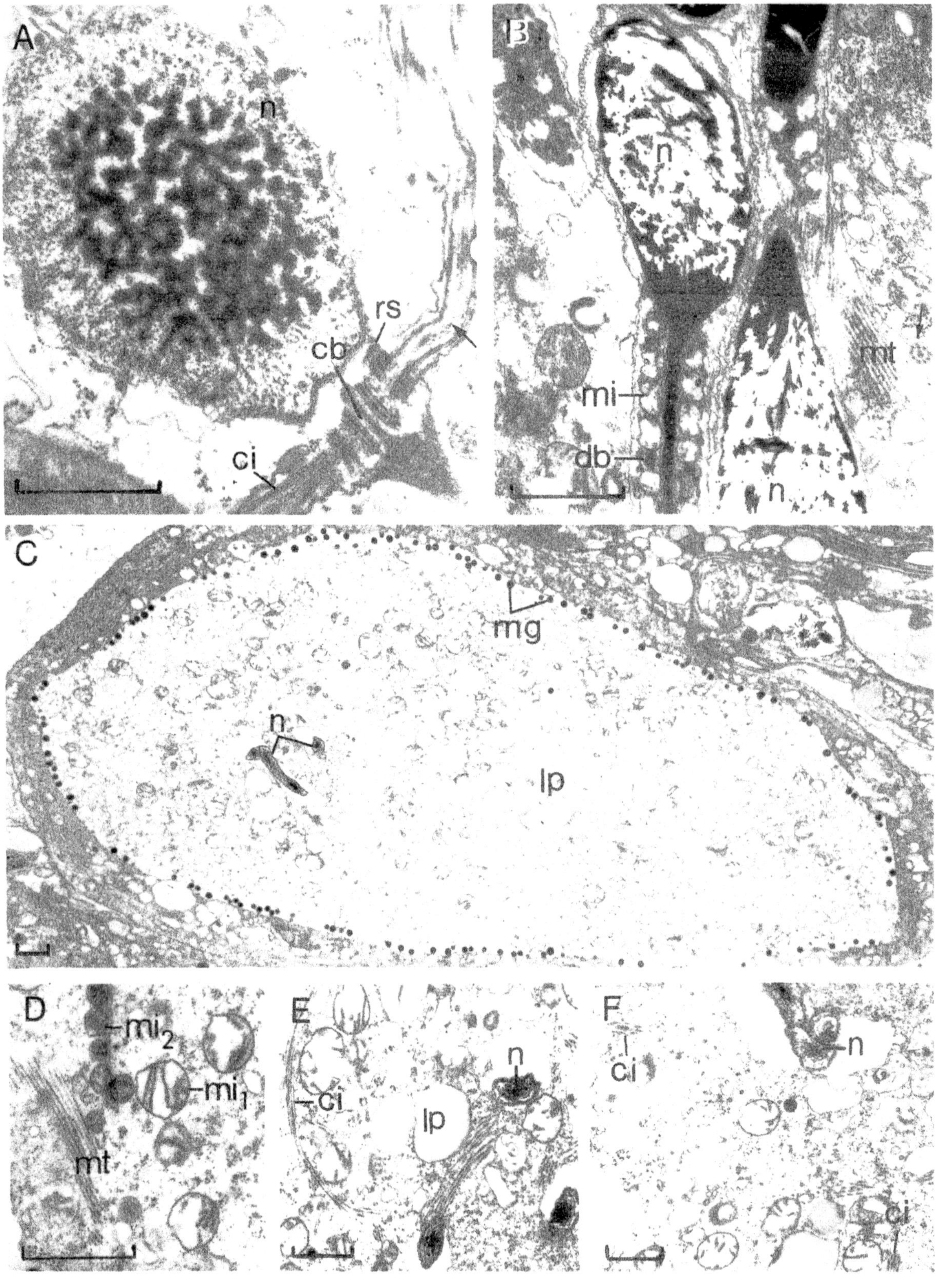
A
n
rs
cb
ci
B
n
mi
db
n
mt
C
mg
n
lp
D
mi2
mi1
mt
E
ci
n
lp
F
ci
n
ci

ist after fertilization. In the otoplanid species studied, the peripheral membrane-bounded granules could be found in all germocytes, whether inseminated or not. Because of this, it seems doubtful that these granules participate in forming a block to polyspermy. Therefore, I prefer to refer to them by the neutral term 'marginal granules'.

Sperm cells which had penetrated into germocytes have been observed by von Hofsten (1918) and by Marcus (1950) in *Otoplana intermedia* du Plessis and *Philosyrtis eumeca* Marcus. To my knowledge, EM observations on inseminated female gametes of platyhelminths other than the proseriates reported on here exist only for a triclad (Gremigni & Nigro, 1983), a prolecithophoran species (Rieger, 1981), two species of Digenea (Burton, 1967; Justine & Mattei, 1984), two species of Eucestoda (Swiderski, 1979; Mokhtar-Maamouri, 1980), and the prolecithophoran species *Pseudostomum quadrioculatum* (U. Ehlers, pers. commun.)

The spermatozoa of the Platyhelminthes lack an acrosome, which in many other metazoans enables the sperm cell to penetrate the oocyte. Hendelberg (1983b) supposes that the dense bodies which originate from the Golgi complex correspond to acrosomal material. The fact that in inseminated germocytes besides the nucleus of the sperm cell, mitochondria and both axonemata of the cilia can be found, but no trace of dense bodies, whereas spermatozoa, which have penetrated into other tissues still possess dense bodies, gives support to this hypothesis; that means the dense bodies could correspond to acrosomal material.

In the species studied, *N. glandulosa* and *O. baltica*, as well as in *Parotoplana macrostyla* Lanfranchi, 1978, the cytoplasm of the germocytes is devoid of inclusions that, on the basis of their ultrastructure, could be identified as shell or yolk material. Thus, germocytes of these representatives of the Proseriata Lithophora are alecithal. Not all taxa subsumed in the Neoophora have germocytes lacking yolk and shell substance, however. Within the Tricladida, some taxa have germocytes with yolky inclusions while other have alecithal germocytes (for references see Gremigni, 1983). Also in the mature germocytes of the prolecithophoran *Pseudostomum quadrioculatum*, for example, there are residual shell substance vesicles (Ehlers, 1985a) that are smaller than vesicles within vitellocytes and that have a similar appearance as those in vitellocytes of a dalyellioid studied by Bunke (1981).

In light of the present state of our knowledge, it appears germ plasm free of shell and yolk material is an apomorphic feature. Further investigations of the Proseriata are needed to show whether this feature can be a synapomorphy of the taxa Coelogynoporidae and Otoplanidae or whether it is only an autapomorphy of the Otoplanidae.

Acknowledgments

Financial support and support in EM-facilities was provided by the Akademie der Wissenschaften und der Literatur Mainz.

Thanks are due to Mrs. E. Hildenhagen-Brüggemann and Mrs. M. Frixe for technical assistence.

References

Boyer, B. C., 1972. Ultrastructural studies of differentiation in the oocyte of the polyclad turbellarian, Prostheceraeus floridanus. J. Morph. 136: 273–295.

Bunke, D., 1981. Ultrastruktur-Untersuchungen an Vitellocyten von Microdalyellia fairchildi. (Turbellaria, Neorhabdocoela). Zoomorph. 99: 71–86.

Burton, P. R., 1967. Fine structure of the reproductive system of a frog lung fluke. II. Penetration of the ovum by a spermatozoon. J. Parasitol. 53: 994–999.

Burton, P. R., 1972. Fine structure of the reproductive system of a frog lung fluke. III. The spermatozoon and its differentiation. J. Parasitol. 58: 68–83.

Coward, S. J., 1974. Chromatoid bodies in somatic cells of the planarian: observations on their behaviour during mitosis. Anat. Rec. 108: 533–546.

Coward, S. J., 1979. On the occurrence and significance of annulate lamellae in gastrodermal cells of regenerating planarians. Cell Biol. Int. Rep. 3: 101–106.

Domenici, L. & V. Gremigni, 1974. Electron microscopical and cytochemical study of vitelline cells in the fresh-water triclad Dugesia lugubris s.l. II. Origin and distribution of reserve materials. Cell Tissue Res. 152: 219–228.

Eddy, E. M., 1975. Germ plasm and the differentiation of the germ cell line. Int. Rev. Cytol. 43: 229–280.

Ehlers, U., 1985a. Das phylogenetische System der Plathelminthes. G. Fischer, Stuttgart, New York, 317 pp.

Ehlers, U., 1985b. Phylogenetic relationships within the Platyhelminthes. In S. Conway Morris, W. D. George, R. Gibson & H. M. Platt (eds), The origins and relationships of lower invertebrates. University Press, Oxford.

Ehlers, U. & B. Ehlers, 1977. Monociliary receptors in interstitial Proseriata and Neorhabdocoela (Turbellaria, Neoophora). Zoomorphologie 86: 197–222.

Franquinet, R. & T. Lender, 1972. Quelques aspects ultrastructuraux de la spermiogenèse chez Polycelis nigra (Planaires). Z. mikrosk. anat. Forsch. 86: 481–495.

Gremigni, V., 1974. The origin and cytodifferentiation of germ cells in the planarians. Boll. zool. ital. 41: 359–377.

Gremigni, V., 1976. Genesis and structure of the so-called 'Balbiani Body' or 'Yolk Nucleus' in the oocyte of Dugesia dorotocephala (Turbellaria, Tricladida). J. Morph. 149: 265–278.

Gremigni, V., 1983. Platyhelminthes-Turbellaria. In K. G. Adiyodi & R. G. Adiyodi (eds), Reproductive biology of invertebrates. Vol. I. Oogenesis, oviposition, and oosorption. Wiley & Sons, London: 67–105.

Gremigni, V. & M. Nigro, 1984. Ultrastructural study of oogenesis in Monocelis lineata (Turbellaria, Proseriata). Int. J. Invert. Reprod. Devel. 7: 105–118.

Gremigni, V. & R. Banchetti, 1972. Submicroscopic morphology of hyperplasic ovaries of ex-fissiparous individuals in Dugesia gonocephala s.l. Atti Acad. naz. Lincei 52: 539–543.

Gremigni, V. & L. Domenici, 1975. Genesis, composition, and fate of cortical granules in the eggs of Polycelis nigra (Turbellaria, Tricladida). J. Ultrastruct. Res. 50: 277–283.

Gremigni, V. & M. Nigro, 1983. An ultrastructural study of oogenesis in a marine triclad. Tissue & Cell 15: 405–415.

Hendelberg, J., 1983a. Platyhelminthes-Turbellaria. In K. G. Adiyodi & R. G. Adiyodi (eds), Reproductive biology of invertebrates. Vol. II. Spermatogenesis and sperm function. Wiley & Sons, London: 75–104.

Hendelberg, J., 1983b. Trends in the evolution of flatworm spermatozoa. In J. André (ed.), The sperm cell. Martinus Nijhoff, Dr. W. Junk Publ., London: 451–453.

Hofsten, N. von, 1918. Anatomie, Histologie und systematische Stellung von Otoplana intermedia Du Plessis. Zool. Bidr. Upps. 7: 1–74.

Hori, I., 1982. An ultrastructural study of the chromatoid body in planarian regenerative cells. J. Electron Microsc. 31: 63–72.

Justine, J.-L. & S. Mattei, 1984. Ultrastructural observations on the spermatozoon, oocyte and fertilization process in Gonapodasmius, a gonochoristic trematode (Trematoda: Digenea: Didymozoidae). Acta zool. 65: 171–177.

Kessel, R. G., 1983. The structure and function of annulate lamellae: porous cytoplasmic and internuclear membranes. Int. rev. Cytol. 82: 181–303.

Kubo-Irie, M. & M. Ishikawa, 1983. Spermiogenesis in the flatworm Notoplana japonica with special attention to the organization of an acrosome and flagella. Develop., Growth Differ. 25: 143–152.

Marcus, E., 1950. Turbellaria Brasileiros (8). Bol. Fac. Fil. Ciênc. Letr. Univ. São Paulo, Zoologia 15: 5–192.

Mokhtar-Maamouri, F., 1980. Particularités des processus de la fécondation chez Acanthobothrium filicolle Zschokke, 1888 (Cestoda: Tetraphyllidea, Onchobothriidae). Archs Inst. Pasteur Tunis 57: 191–205.

Oakley, H. A., 1982. Meiosis in Mesostoma ehrenbergii ehrenbergii (Turbellaria, Rhabdocoela). II. Synaptonemal complexes, chromosome pairing and disjunction in achiasmate oogenesis. Chromosoma 87: 133–147.

Rees, F. G., 1979. The ultrastructure of the spermatozoon and spermiogenesis in Cryptocotyle lingua (Digenea: Heterophyidae). Int. J. Parasitol. 9: 405–419.

Rieger, R. M., 1981. Morphology of the Turbellaria at the ultrastructural level. Hydrobiologia 84: 213–229.

Silveira, M. & K. R. Porter, 1964. The spermatozoids of flatworms and their microtubular systems. Protoplasma 59: 240–265.

Sopott-Ehlers, B., 1985. The phylogenetic relationships within the Seriata (Platyhelminthes). In S. Conway Morris, D. George, R. Gibson & H. M. Platt (eds), The origins and relationships of lower invertebrates. University Press, Oxford, in press.

Swiderski, Z., 1979. Fertilization in the cestode Hymenolepis diminuta (Cyclophyllidea, Hymenolepididae). Proc. Sixth Europ. Congr. Electron. Microsc., Jerusalem 1976: 311–312.

Williams, J. B., 1983. The genital system of Temnocephala. I. Ultrastructural features of the differentiating spermatid of Temnocephala novaezealandiae, including notes on a possible correlation between cellular autophagy and mitochondrial function. Aust. J. Zool. 31: 317–331.

Editor's note

Several terms in this article are applied in accordance with German rather than English usage. 'Insemination' is meant to refer to union of male and female gametes, irrespective of karyogamy; English-speaking readers may read 'fertilization' for this term. 'Germocyte' is meant to refer to female germ cells arising in germaria (as distinct from oocytes arising in overies). 'Cilia' is used to refer to what are traditionally known as flagella spermatozoa.

Ultrastructural features of oogenesis in some marine neoophoran turbellarians

V. Gremigni, M. Nigro & M. S. Settembrini
Istituto di Zoologia e Anatomia Comparata, Università di Pisa, via Volta n. 4, 56100 Pisa

Keywords: Turbellaria, neoophora, oogenesis, vitellogenesis, ultrastructure

Abstract

An ultrastructural study of oogenesis has been undertaken in some marine species of neoophoran turbellarians belonging to the Tricladida, Proseriata, and Prolecithophora. Among marine triclads, *Cercyra hastata* has oocytes with a remarkable amount of what appears to be autosynthetic proteinaceous yolk, while *Procerodes dohrni* and *P. lobata* have alecithal oocytes. Among the Proseriata, several species of the primitive family Monocelididae (subfamily Monocelidinae) have oocytes with a discrete amount of what appears to be autosynthetic yolk, while *Parotoplana macrostyla,* of the derived family Otoplanidae, has alecithal eggs. Finally preliminary observations on *Plagiostomum maculatum* (order Prolecithophora) suggest that oocytes have autosynthetic yolk globules.

These results support the hypothesis (previously formulated on the basis of similar ultrastructural investigations on freshwater triclads) that the presence of autosynthetic yolk in some neoophoran turbellarians can be interpreted as a primitive character inherited from an ancestor with archoophoran organization. This plesiomorphic character would still be maintained in some species while lost in others following differentiation of vitellaria which are characteristic of the Neoophora.

Introduction

Previous ultrastructural investigations of oogenesis in freshwater triclads yielded the unexpected finding that species belonging to the family Dugesiidae have oocytes provided with small amounts of autosynthetic, proteinaceous yolk (Gremigni, 1969a, 1969b, 1979). In contrast, species belonging to the families Planariidae and Dendrocoelidae have oocytes devoid of yolk and containing cortical granules (Gremigni & Domenici, 1975; Gremigni, 1979). Findings obtained on the dugesiids were unexpected since triclads are typical neoophoran turbellarians and, therefore, are provided with well-developed vitellaria containing specialized cells for the production and accumulation of reserve substances. Thus, eggs of triclads, like those of all other neoophorans, are commonly considered completely alecithal. The different features of oogenesis observed in different families of freshwater triclads led me (Gremigni, 1979, 1983) to hypothesize that the presence of intraoocytal autosynthetic yolk could represent a primitive character inherited from an archoophoran ancestor.

Therefore, we have extended the ultrastructural study of oogenesis to other species of neoophoran turbellarians. We first chose those belonging to groups that are closely related to freshwater triclads, i.e. marine triclads and proseriates, and then to a species belonging to another order of primitive neoophorans, the Prolecithophora. The aim of this research was to ascertain whether autosynthetic yolk is present in the oocytes of other neoophorans, and with this, to elucidate further the interesting but still controversial problems of turbellarian phylogeny.

Hydrobiologia 132, 145–150 (1986).

Materials and methods

Several specimens of the following species were collected from different marine habitats and were either immediately prepared for ultrastructural study or transferred and maintained in our laboratory where they were kept in glass aquaria containing filtered and standing seawater before being fixed.

Tricladida: *Cercya hastata, Procerodes dohrni, P. lobata.*

Proseriata: *Monocelis lineata, M. fusca, Archiloa petiti, Archilopsis unipunctata.*

Prolecithophora: *Plagiostomum maculatum.*

All specimens were prefixed with 3 – 5% glutaraldehyde buffered with 0.1 M phosphate or cacodylate buffer to which a few drops of $CaCl_2$ and 5 – 10% sucrose were added. After prolonged washings in the buffer, the specimens were fixed with 1 – 2% osmium tetroxide, dehydrated in ethanol and embedded in epoxy resin. One-μm-thick sections, obtained with glass knives, were stained with toluidine blue and methylene blue and observed with a light microscope. Thin sections were cut with a diamond knife, stained with uranyl acetate and lead citrate and observed with a Siemens Elmiskop 101 electron microscope.

Observations

Tricladida

Ovaries from all specimens were well delimited, paired and oblong, and contained several oocytes at different stages of maturation. Oogonia and very young oocytes were located at one pole of the ovary while the opposite pole was occupied by more mature oocytes.

Procerodes dohrni and *P. lobata.* No remarkable differences characterized oogenesis of these two closely related species. Differentiating oocytes had endoplasmic reticulum and Golgi complex from which small vesicles containing a dense material were elaborated (Fig. 1). Similar dense material was present in membrane-bounded inclusions about 1 μm in diameter which were most numerous at the periphery in a monolayer under the plasma membrane of mature oocyes (Fig. 2). These roundish inclusions with internal substructure had a similar morphology, diameter, and position as the cortical granules previously described in oocytes of freshwater triclads, granules that disappeared after fertilization (Gremigni & Domenici, 1975). No globules resembling proteinaceous yolk were observed in the ooplasm; some randomly distributed lipid droplets were present.

Cercyra hastata. Features similar to those described in *Procerodes* characterized early stages of oogenesis, during which nearly round and dense granules about 1 μm in diameter were produced. During the following stages of oogenesis, these granules appeared to be distributed in a monolayer under the oolemma and were interpreted as cortical granules (Gremigni & Nigro, 1983). In more mature oocytes, the rough endoplasmic reticulum had developed to form flattened cisternae or whorls from which several bodies containing a material of medium electron density were elaborated (Fig. 6). In mature stages, these bodies were 2 – 5 μm across, were surrounded by a double membrane (Fig. 5), and were randomly distributed in the ooplasm (Fig. 4). We presume these globules most probably represent proteinaceous yolk similar to those of freshwater dugesiids. We found no endocytotic activity at the oocyte surface and, therefore, suggest that yolk is only of endogenous origin.

Proseriata

The species studied from the family Monocelididae, *Monocelis lineata, M. fusca, Archiloa petiti,*

Figs. 1 – 8. (1) *Procerodes dohrni.* Golgi complex producing small vesicles that contain dense material. ×28 000. (2) *P. dohrni.* Mature oocyte with monolayer of dense granules under the oolemma. No yolk globules are visible. L.D. = lipid droplets. ×5600. (3) *P. dohrni.* Two presumed cortical granules. ×20 000. (4) *Cercyra hastata.* Light micrograph of ovary with oocytes containing presumed yolk globules. ×1 500. (5) *C. hastata.* Globule, presumably yolk, bounded by double membrane and containing finely granular material. ×25 000. (6) *C. hastata.* Cytoplasm of nearly mature oocyte with large amounts of granular endoplasmic reticulum and presumed yolk globules. ×4 300. (7) *Monocelis lineata.* Cytoplasm of nearly mature oocyte showing oblong presumed cortical granules (C.G.), a Golgi complex (G.C.) and several presumed yolk globules (Y.G.). ×19 000. (8) *M. lineata.* Golgi complex producing vesicles containing a medium-dense material similar to that of yolk globule (Y.G.). ×36 000.

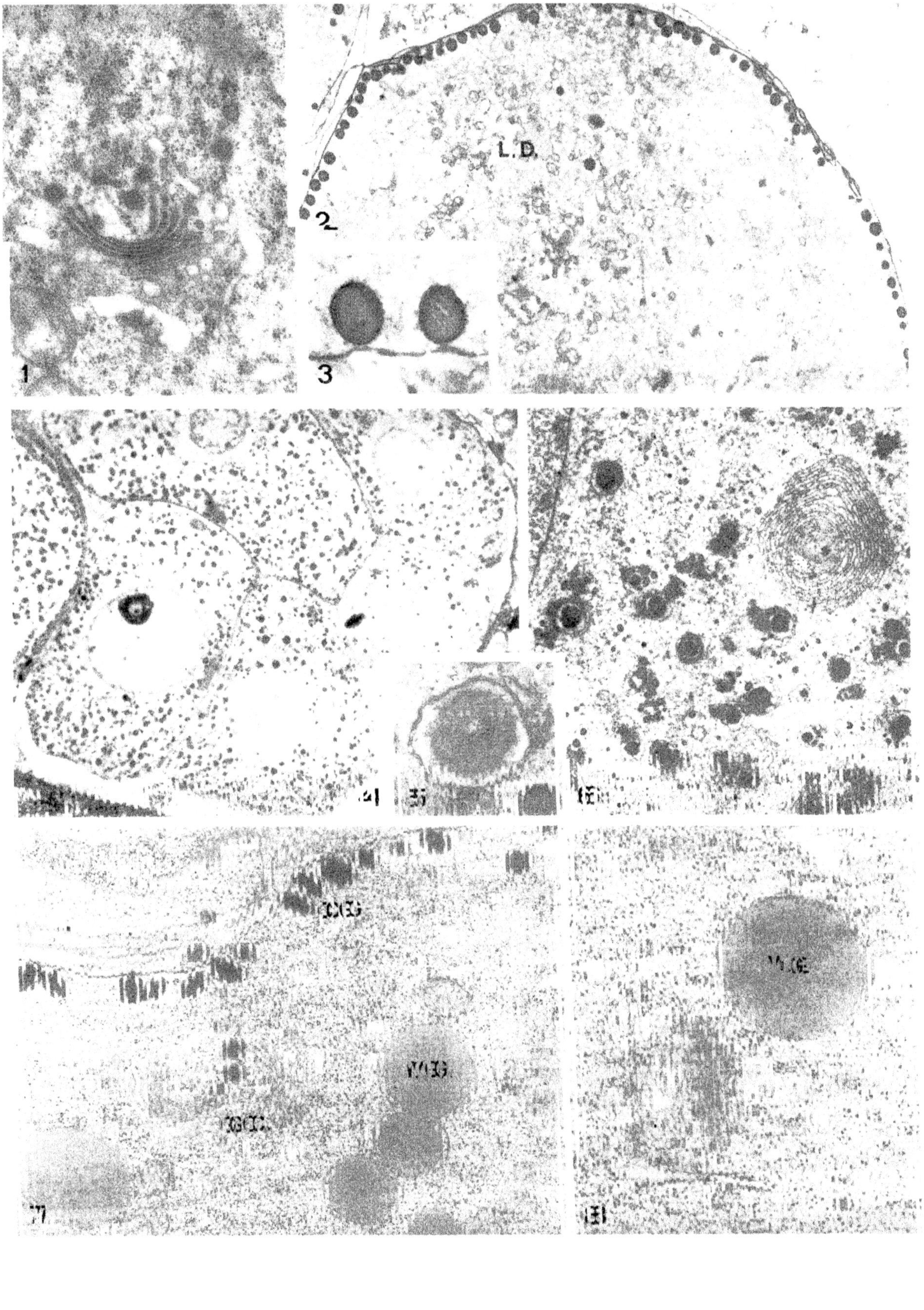
L.D.
2
3
1

Archilopsis unipunctata, showed similar features of oogenesis. The paired ovaries contained several oocytes at different stages of differentiation, each surrounded by thin and long cytoplasmic projections of follicle cells. In one of the better-studied species, *M. lineata* younger oocytes had well developed rough endoplasmic reticulum and Golgi complex. These organelles seemed mainly to be involved in the synthesis of vesicles containing a dense material which by fusion gave rise to oblong granules of about 1 μm that in mature oocytes were distributed in a monolayer under the oolemma (Fig. 7). They had a similar morphology, diameter and position as the 'marginal granules' described by Sopott-Ehlers (1985) in oocytes of otoplanid proseriates and were previously interpreted as being cortical granules (Gremigni & Nigro, 1984).

In more mature oocytes the endoplasmic reticulum and Golgi complex became involved in synthesis of a different type of vesicle which by fusion gave rise to larger inclusions (3–5 μm in diameter) containing a less-dense material than that of presumed cortical granules. These globules remained scattered in the ooplasm through the entire process of maturation, and we suggest they represent autosynthetic yolk (Fig. 8). We found no endocytotic activity at the oocyte surface and therefore assume yolk is exclusively of endogenous origin.

A similar course of cortical granule and yolk globule production characterized the oogenesis of the other members of the Monocelididae studied. However, in *Archilopsis unipunctata,* the second type of inclusion contained a very dense material (Figs. 9–10). In *Archiloa petiti,* what we presumed to be yolk globules were less numerous but reached a larger size (up to 7–8 μm) (Fig. 11).

Prolecithophora

Preliminary observations on the oogenesis of *Plagiostomum maculatum* demonstrated that the oocytes were not contained in well-delimited ovaries but rather were distributed in the parenchyma as isolated cells. Each oocyte seemed to be surrounded by a single follicle cell with which material interchanges may occur, but not apparently in the form of endocytosis. Differentiating oocytes showed a number of profiles of rough endoplasmic reticulum and Golgi complexes involved in the synthesis of small vesicles containing a dense material. This same material could be seen in larger membrane-bounded globules (2–4 μm in diameter), presumably derived from vesicle coalescence. In mature oocytes, they appeared less dense (Fig. 13) and remained randomly distributed in the ooplasm (Fig. 12). We presume they represent autosynthetic yolk. No granules were evident under the oolemma in either early or final stages of oogenesis.

Discussion

Tricladida. Marine triclads present features of oogenesis that resemble those previously observed in the closely related freshwater planarians. In fact, oocytes of *Cercyra hastata* contain a remarkable amount of presumably proteinaceous autosynthetic yolk as do freshwater triclads belonging to the family Dugesiidae. In contrast, oocytes of *Procerodes dohrni* and *P. lobata* apparently do not synthesize or accumulate proteinaceous yolk globules, and thus resemble freshwater species of the families Planariidae and Dendrocoelidae. Also, from a karyological point of view, the two marine genera show remarkable differences (see Galleni & Puccinelli, 1982; Galleni *et al.,* 1984). It, therefore, seems proper to suggest that further investigations are needed on these and other characters of marine triclads to better define a new subdivision for this group (see also Ball, 1977).

Proseriata. Present results obtained on some species of the family Monocelididae suggest that the genus *Monocelis* and other closely related genera of the subfamily Monocelidinae have oocytes that contain a certain amount of proteinaceous autosynthetic yolk. In contrast, the single observation made by Lanfranchi on *Parotoplana macrostyla* (see Gremigni, 1983) and those on other otoplanids made by Sopott-Ehlers (1985) suggest that members of the family Otoplanidae have alecithal oocytes. Further investigations on many other species from the families Monocelididae, Otoplanidae, and Coelogynoporidae are now in progress and are needed before ascertaining the significance of the presence of absence of yolk in proseriate oocytes. However, it is interesting that yolk seems to be present in oocytes of representatives of the family Monocelididae, commonly considered primitive within the order, while seemingly absent in repre-

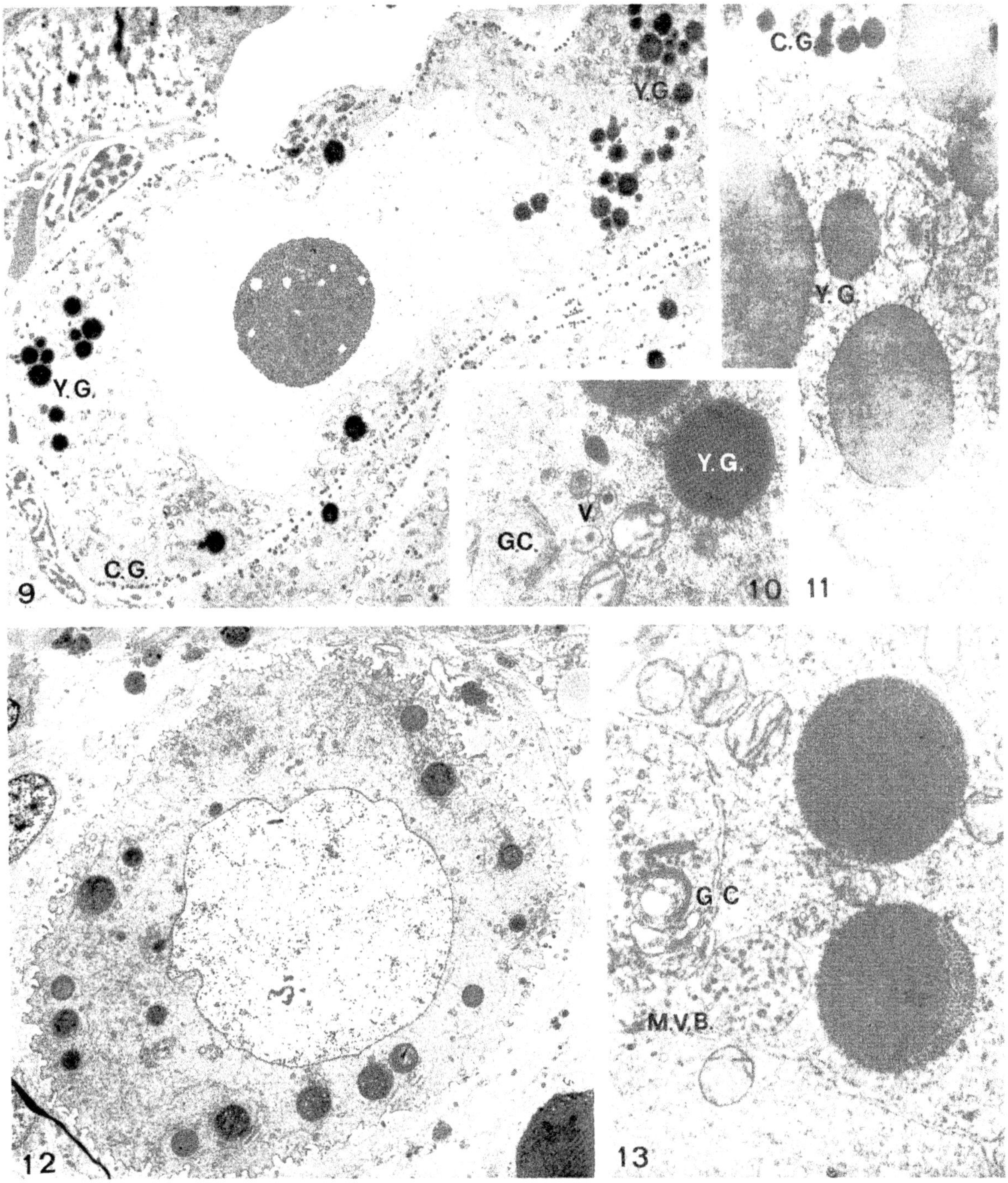

Figs. 9–13. (9) *Archilopsis unipunctata.* Maturing oocyte with small presumed cortical granules (C.G.) and dense yolk globules (Y.G.). ×3 400. (10) *A. unipunctata.* Golgi complex (G.C.), vesicles containing dense material (V.) and presumed yolk globules (Y.G.). ×18 000. (11) *Archiloa petiti.* Maturing oocyte. C.G. = cortical granules, Y.G. = yolk globules. ×25 000. (12) *Plagiostomum maculatum.* Oocyte containing several dense globules. ×2 500. (13) *P. maculatum.* Two dense globules, a Golgi complex (G.C.) and multivesicular bodies (M.V.B.) from a nearly mature oocyte. ×17 500.

sentatives of the family Otoplanidae, commonly considered to be more derived.

Prolecithophora. Preliminary results obtained from investigations on oogenesis in *Plagiostomum maculatum* suggest that oocytes of this species could contain autosynthetic yolk globules. At present, no other ultrastructural data on oogenesis of members of this order are available apart from those of Newton (1974) who did not find autosynthetic yolk globules in *Hydrolimax* but obtained some evidence of nutrient transfer from the accessory cell to the oocyte. Further investigations either on *Plagiostomum* or other prolecithophorans are needed to better understand this aspect of oogenesis in the family.

Conclusions

Ultrastructural investigations performed on oogenesis of neoophoran turbellarians have shown that intraoocytal autosynthetic yolk seems to be present in some and absent in other representatives. In particular, yolk is presumably present in the oocytes of *Cercyra,* which among marine triclads most probably has more plesiomorphic characters than *Procerodes,* a genus with oocytes devoid of yolk. A similar situation seems to characterize oocytes of Proseriata where some representatives of the primitive family Monocelididae have presumably endolecithal oocytes while representatives of the derived family Otoplanidae have presumably alecithal oocytes (see Gremigni, 1983; Sopott-Ehlers, 1985).

The prolecithophoran *Plagiostomum maculatum* also possibly has lecithal eggs. Further investigations are needed on this order.

Present data seem to support the previously formulated hypothesis that the presence of yolk in the oocytes of some neoophoran turbellarians is a primitive character inherited from one or more ancestors with archoophoran organization.

Acknowledgments

Most of the proseriates were given to us as prefixed or embedded specimens by Dr Els Martens and Dr Paul Martens of the Limburgs Universitair Centrum (Diepenbeek, Belgium) whom we thank.

References

Ball, I. R., 1977. On the phylogenetic classification of aquatic planarians. Acta zool. fenn. 154: 21–35.

Galleni, L. & I. Puccinelli, 1982. The chromosome complement of Cercyra hastata (Turbellaria: Tricladida) from the Adriatic Sea. Trans. am. microsc. Soc. 101: 347–352.

Galleni, L., M. Nigro & I. Puccinelli, 1984. The chromosome complement of Procerodes dohrni (Tricladida: Maricola). J. mar. biol. Ass. U.K. 64: 227–231.

Gremigni, V., 1969a. Ricerche istochimiche e ultrastrutturali sull'ovogenesi dei Tricladi. I: inclusi deutoplasmatici in Dugesia lugubris e Dugesia benazzii. Accad. naz. Lincei 47: 101–108.

Gremigni, V., 1979b. Origine del vitello negli ovociti di Dugesia lugubris e D. benazzii. Atti VII Congr. Soc. ital. Micr. elettr. Modena (Italy): 68–70.

Gremigni, V., 1979. An ultrastructural approach to planarian taxonomy. Syst. Zool. 28: 345–355.

Gremigni, V., 1983. Platyhelminthes-Turbellaria. In K. G. Adiyodi & R. G. Adiyodi (eds), Reproductive biology of invertebrates, vol. 1, Oogenesis, oviposition and oosorption. J. Wiley and Sons Ltd, Chichester: 67–107.

Gremigni, V. & L. Domenici, 1975. Genesis, composition and fate of cortical granules in the eggs of Polycelis nigra (Turbellaria, Tricladida). J. ultrastruct. Res. 50: 277–283.

Gremigni, V. & M. Nigro, 1983. An ultrastructural study of oogenesis in a marine triclad. Tissue & Cell 15: 405–415.

Gremigni, V. & M. Nigro, 1984. Ultrastructural study of oogenesis in Monocelis lineata (Turbellaria, Proseriata). Int. J. Invert. Reprod. Devel. 7: 105–118.

Newton, W. D., 1974. The accessory cell and yolk halo of the oocyte of the freshwater turbellarian Hydrolimax grisea (Platyhelminthes: Plagiostomidae). Biol. Bull. 147: 618–623.

Sopott-Ehlers, B., 1985. Fine-structural characteristics of female and male germ cells in Proseriata Otoplanidae (Platyhelminthes). This volume.

Ultrastructural investigations on the differentiation of genital hard structures in free-living platyhelminths and their phylogenetic significance

Jochen Brüggemann
II. Zoologisches Institut und Museum der Universität, Berliner Straße 28, D-3400 Göttingen, Fed. Rep. Germany

Keywords: Turbellaria, Platyhelminthes, ultrastructure, differentiation, copulatory organ, hard structure, phylogenetic system

Abstract

Ultrastructure and differentiation of penis stylets and stylet needles have been investigated in representatives of various groups of free-living platyhelminths, viz. the Acoela, Macrostomida, Typhloplanoida, Kalyptorhynchia, and Dalyellioida. In all these groups, the differentiation of such hard parts occurs intracellularly but in different ways in the different groups. The ultrastructure of the bursal mouth piece in an acoel platyhelminth is not comparable to the hard structures in male copulatory organs. The presence of penial copulatory organs having intracellular hard structures appears to be an autapomorphy of the Euplatyhelminthes. Several characters in the ultrastructure and development of these structures can be used as autapomorphies for various platyhelminth groups.

Introduction

Until now the ultrastructure of penis stylets and stylet needles of free-living platyhelminths has been investigated in only a few, taxonomically scattered representatives (Mainitz, 1977; Ehlers & Ehlers, 1980; Doe, 1982; Brüggemann, 1984). The differentiation of such hard elements has been described only for the proseriate *Carenscoilia bidentata* (Ehlers & Ehlers, 1980) and the unguiphore *Polystyliphora filum* (Brüggemann, 1984). In this presentation, ultrastructure and differentiation of genital hard structures in representatives of different groups of free-living platyhelminths are described.

Material and methods

Specimens of *Philocelis cellata* (Acoela), *Paromalostomum fusculum* and *P. proceracauda* (Macrostomida), *Ciliopharyngiella intermedia* (Typhloplanoida?), *Adenorhynchus balticus* (Typhloplanoida), *Promesostoma marmoratum, P. caligulatum* and *P. rostratum* (Typhloplanoida), *Marirhynchus longasaeta* (Kalyptorhynchia), and *Provortex psammophilus* and *P. tubiferus* (Dalyellioida) were extracted out of sandy sediments from various beaches of the North Sea island of Sylt. After light microscopical examination, the specimens were fixed in 2.5% glutaraldehyde buffered with 0.1 M sodium cacodylate buffer for 1.5 h, postfixed for 1.5 h in cacodylate-buffered osmium tetroxide (1%), dehydrated in an ethanol series, and embedded in Araldite. Sections were cut with a diamond knife on a Reichert OmU 3 ultramicrotome, stained with aequeous uranyl acetate and lead citrate (each 10 min), and examined in a Zeiss EM 10B electron microscope.

Results

The differentiation of the penis stylets and stylet needles in the species studied occurred partly in a successive, partly in a synchronous formation

Hydrobiologia 132, 151–156 (1986).

mode, but always intracellularly.

The male copulatory organ of the acoel *Philocelis cellata* contains a number of needle-shaped hard elements which are arranged in the form of a caudally turned, conical tube. In the specimens I studied, each needle differentiated in a separate needle-forming cell and consisted of a bundle of cross-striated microfibrils; they formed successively from the distal end to the base (Fig. 1). In some places, the microfibrils were fused as well as connected by a great number of microtubules. Even in the fully developed copulatory organ, the microfibrils remained without an envelope of electron-dense material.

The stylet apparatus of the macrostomids *Paromalostomum fusculum* and *P. proceracauda* consisted of a sperm-grain secretion tube, also called sperm tube, and a gland tube, whose walls were fused basally. Both tubes differentiated in a successive formation mode from the distal tip to the base. In a tube-shaped stylet-forming cell, a framework of microtubules was laid down and became enveloped by electron-dense material (Fig. 2). Thin-walled areas were constructed solidly, and eventually they appeared nearly homogeneous, whereas in thick-walled regions small cytoplasmic spaces remained. When the surfaces of the stylet walls were complete, the cytoplasm of the stylet-forming cell receded, so that the cell membrane was in direct contact with the hard structure. The differentiation of the shorter gland tube started later than the formation of the sperm-grain secretion tube, probably in a projection of the stylet-forming cell and in the same way as the sperm tube. The base of the hard structure was embedded in a finely fibrous intracellular matrix substance. Probably the stylets of the macrostomid genus *Macrostomum* are formed in the same way.

The penis stylet of *Ciliopharyngiella intermedia* (a species of uncertain placement in the phylogenetic system of the Platyhelminthes) differentiated successively, also. First, in the stylet-forming cell, a framework of cross-striated microfibrils was laid down. This became enveloped by electron-dense material, forming a so-called rough-form which had a relatively uneven surface (Fig. 3). The inner and outer surfaces of the stylet walls were smoothed by the addition of electron-dense material to the membranes of the stylet-forming cell, until the space between the membrane and the rough-form filled up. Only the stylet base remained unsmoothed; it became embedded in a fibrous intracellular matrix.

The proseriate *Polystyliphora filum* has numerous uniform prostatoid organs, which can be regarded as modified copulatory organs. Each of these prostatoid organs has a funnel-shaped stylet with a curved distal part and a flattened projection in the middle; funnel and projection differentiated together in a single cell. Both portions of the stylet exhibited a synchronous formation mode (Brüggemann, 1984). In an early phase of differentiation, a framework of microtubules was built in the basal part, which later became enveloped by electron-dense material. In a later phase, the space between the formed hard structure and the outer membrane of the stylet-forming cell partially filled up with electron-dense material. The distal part of the stylet was formed by addition of electron-dense material to the cell membrane.

Likewise, the mode of stylet formation in the typhloplanoid *Adenorhynchus balticus* was synchronous. Electron-dense material was added to the outer membrane of a funnel-shaped stylet-forming cell (Fig. 4). Though the wall thickness differed in places, all parts of the stylet were finished at the same time because of a varying rate of growth in wall thickness. The base of the stylet was embedded in a finely fibrous intracellular matrix.

The kalyptorhynch *Marirhynchus longasaeta* had a flattened stylet-forming cell which was coiled up to a funnel shape that sloped distally. The differentiation of the hard structure occurred in a synchronous mode and similar to what was seen in

Figs. 1–8. Genital hard structures of free-living platyhelminths in the process of differentiation: (1) Oblique section through a stylet needle of *Philocelis cellata*; (2) Oblique section through wall of sperm tube of *Paromalostomum proceracauda*; (3) Cross section through stylet of *Ciliopharyngiella intermedia*; (4) Cross section through stylet wall of *Adenorhynchus balticus*; (5) Oblique section through stylet wall of *Marirhynchus longasaeta*; (6) Cross section through stylet wall of *Promesostoma caligulatum*; (7) Cross section through stylet wall of *Provortex psammophilus*; (8) Oblique section through the bursal mouth piece of *Philocelis cellata*. (Scale in Figs. 1–8: 1 μm.)

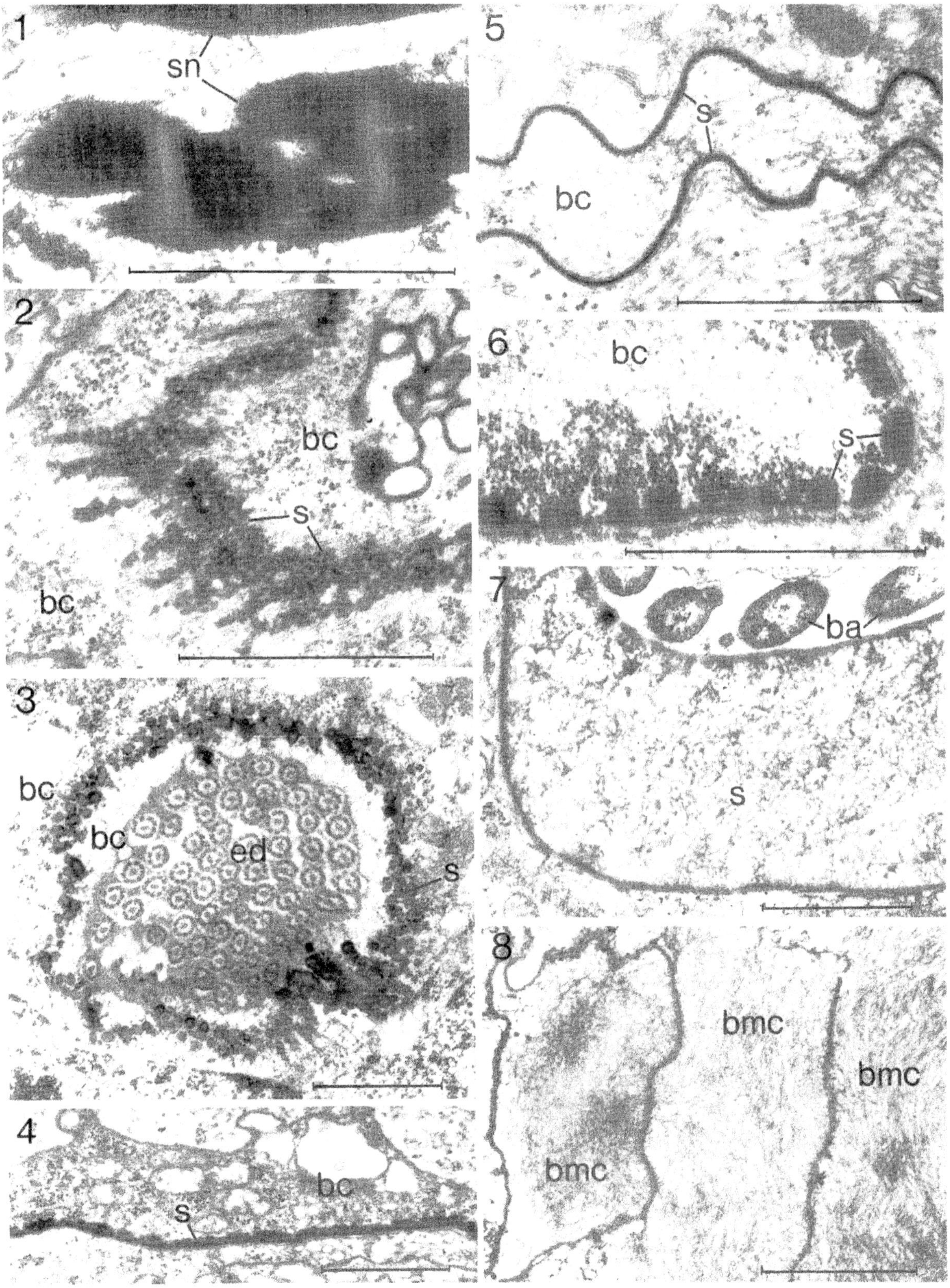
1
sn
2
bc
s
bc
3
bc
bc
ed
s
4
bc
s
5
s
bc
6
bc
s
7
ba
s
8
bmc
bmc
bmc

Adenorhynchus, but with a proportional growth in thickness of the walls. During the phase of differentiation, on the inner surfaces of the growing stylet walls, microtubules were arranged in parallel (Fig. 5). In the stylet base, however, at infoldings of the membrane of the stylet-forming cell many separate hard elements were built. Later, they were joined by electron-dense material and surrounded by a fibrillar intracellular matrix.

The long stylet tubes of the typhloplanoids *Promesostoma marmoratum, P. caligulatum,* and *P. rostratum* differentiated in a successive formation mode, beginning at the distal tip and proceeding basally. However, the differentiation of the stylet walls occurred as in the stylet base of *Marirhynchus* (Fig. 6). Many separate hard elements formed, and these were fused by electron-dense material.

The penis stylets of the dalyellioids *Provortex psammophilus* and *P. tubiferus* had a synchronous formation mode. First a thin even layer of electron-dense material was added to the inner and outer membrane of a tube-shaped stylet-forming cell except at the base. Then, the remaining spaces between these thin layers filled with moderately electron-dense material which partially condensed into fibers. Thin areas of the walls became filled up solidly, while thick parts of the stylet walls, especially distal ones, were more porous in structure, containing numerous cytoplasmic spaces (Fig. 7). The stylet base consisted only of the porous, moderately electron-dense material. A fibrous intracellular matrix was lacking at the base of the stylet. For further details concerning the formation of the stylets described above see also Brüggemann (1985a).

The bursal mouth piece, also called nozzle, of the acoel *Philocelis cellata* is not comparable to the stylets described above. The main components of the bursal mouth piece were formed by more than 26 flattened and ring-shaped, linked bursal-mouthpiece cells, arranged serially. Their central regions in the course of ontogeny became filled with a fibrous, moderately electron-dense substance, starting distally and proceeding proximally (Fig. 8). Later, moderately electron-dense amorphous material was added (Brüggemann, 1985b).

Phylogenetic significance

On the basis of the phylogenetic system of the Platyhelminthes proposed by Ehlers (1985a, b) and Ax (1984, 1985), and from my investigation of genital hard structures and their differentiation several hypotheses can be drawn. The bursal mouth piece of *Philocelis* is factored out because it probably is not homologous to penial hard structures.

One can assume that penial copulatory organs with intracellular hard structures are an autapomorphy of the taxon Euplatyhelminthes, perhaps even of the monophylum Platyhelminthes; hence the cirrus structures have developed secondarily and several times independently in different groups. Also, the occurrence of an unarmed penis, i.e. without hard structures, is at least an apomorph feature within the Rhabditophora, and perhaps also in the Acoela.

The fibrillar stylet needles are probably a plesiomorphy of the Acoela, i.e. structure and differentiation mode (microfibrils, successive formation mode) must have existed already in the common stem species of the Euplatyhelminthes; hence the absence of penial hard elements within the Nemertodermatida represents an autapomorphy of this taxon. Nevertheless, the possibility that penial hard structures have been developed convergently within the Acoela and the Rhabditophora cannot be absolutely excluded.

The Rhabditophora have probably assumed the successive formation mode and the differentiation of microfibrils as synapomorph features (like the sister group Acoelomorpha); however, the enveloping with electron-dense material and the embedding of the stylet base in a fibrous intracellular matrix are autapomorphies. The structure and formation mode of the stylet as it must have existed in the Rhabditophora stem species can be found again, presumably with little change, as a plesiomorphy in the macrostomid *Haplopharynx* sp. (see Doe, 1985), the proseriate *Carenscoilia bidentata* (see Ehlers & Ehlers, 1980), and probably in most stylet- or needle-bearing proseriates. On the other hand, the microtubular framework instead of microfibrils in the hard structures of the macrostomid genera *Macrostomum* (see Doe, 1982) and *Paromalostomum* represents an apomorphy within the Macrostomida.

Obviously, the stylets of the Rhabdocoela are

formed primarily by the synchronous secretion of electron-dense material onto both the inner and outer membranes of the stylet-forming cell. I designate this formation mode as holo-synchronous; it is varied within the Rhabdocoela. Also the successive stylet-formation of the *Promesostoma* species can be derived from the holo-synchronous formation; it represents an autapomorphy of the genus *Promesostoma* and is obviously connected with the extreme stylet length.

Distinct beginnings of a holo-synchronous formation can be found already in *Ciliopharyngiella intermedia* and *Polystyliphora filum*. The distal parts of the prostatoid hard structures of *Polystyliphora* and the stylets of the Rhabdocoela can be regarded as further developments of the intracellular smooth layer which appeared in *Ciliopharyngiella*. Regarding the structure and formation mode, the rough-form of the stylet of *Ciliopharyngiella* corresponds to the penial hard structure as it must have existed in the common stem species of the Rhabditophora.

The modified successive stylet formation mode of *Ciliopharyngiella* is more primitive than the stylet formation mode of *Polystyliphora*, which I designate as hemi-synchronous formation: the tips of the synchronously formed prostatoid hard structures correspond to the rhabdocoel stylet, whereas the basal parts are still formed from a rough-form with a tubular framework.

Because of the hemi-synchronous formation mode of the prostatoid hard structures of *Polystyliphora*, the Unguiphora, which comprises the Polystyliphoridae and Nematoplanidae, should not be grouped with the Seriata, but can be linked – with proviso – to a common stem species with the Rhabdocoela. With regard to the mode of stylet formation, the only (possible) autapomorphy of the Unguiphora is the lack of an intracellular matrix at the base of the hard structure. Nevertheless, the possibility that a synchronous formation mode has been developed convergently within the Seriata and the Rhabdocoela cannot be positively excluded.

On the basis of pharynx type or the gonads, the Ciliopharyngiellidae has until now either been relegated to the Proseriata or designated as primitive typhloplanoids. Because of their modified successive stylet formation mode, I place the Ciliopharyngiellidae as a sister group opposite the Unguiphora + Rhabdocoela. With respect to the mode of stylet formation, the Ciliopharyngiellidae does not show any autapomorphies but only synapomorphies with the sister group Unguiphora + Rhabdocoela.

Acknowledgments

I thank Prof. Dr. P. Ax and Dr. U. Ehlers for valuable advice and helpful criticism.

Abbreviations

ba	bacteria
bc	stylet-forming cell
bmc	bursal mouth piece cell
ed	ejaculatory duct
s	stylet
sn	stylet needle

References

Ax, P., 1984. Das phylogenetische System. Systematisierung der lebenden Natur aufgrund ihrer Phylogenese. G. Fischer, Stuttgart, New York, 349 pp.

Ax, P., 1985. The position of the Gnathostomulida and Platyhelminthes in the phylogenetic system of the Bilateria. In S. Conway Morris, D. G. George, R. Gibson & H. M. Platt (eds), The origins and relationships of lower invertebrates. University Press, Oxford: in press.

Brüggemann, J., 1984. Ultrastruktur und Differenzierung der prostatoiden Organe von Polystyliphora filum (Plathelminthes, Proseriata). Zoomorphology 104: 86–95.

Brüggemann, J., 1985a. Ultrastruktur und Bildungsweise penialer Hartstrukturen bei freilebenden Plathelminthen. Zoomorphology: 105: 143–189.

Brüggemann, J., 1985b. Ultrastructure and formation of the bursa mouth piece of Philocelis cellata (Plathelminthes, Acoela). Hydrobiologia, in press.

Doe, D. A., 1982. Ultrastructure of copulatory organs in Turbellaria. I. Macrostomum sp. and Microstomum sp. (Macrostomida). Zoomorphology 101: 39–59.

Doe, D. A., 1985. Ultrastructure of the copulatory stylet and accessory spines in Haplopharynx quadristimulus. This volume.

Ehlers, U., 1985a. Das phylogenetische System der Plathelminthes. Akademie der Wissenschaften und der Literatur Mainz & G. Fischer, Stuttgart. 317 pp.

Ehlers, U., 1985b. Phylogenetic relationships within the Platyhelminthes. In S. Conway Morris, D. G. George, R. Gibson & H. M. Platt (eds), The origins and relationships of lower invertebrates. University Press, Oxford: in press.

Ehlers, B. & U. Ehlers, 1980. Struktur und Differenzierung penialer Hartgebilde von *Carenscoilia bidentata* Sopott (Turbellaria, Proseriata). Zoomorphologie 95: 159–167.

Mainitz, M., 1977. The fine structure of the stylet apparatus in Gnathostomulida scleroperalia and its relationship to turbellarian stylets. Acta Zool. fenn. 154: 163–174.

Ultrastructure of the copulatory stylet and accessory spines in *Haplopharynx quadristimulus* (Turbellaria)

David A. Doe
Biology Department, Westfield State College, Westfield, MA 01086, USA

Keywords: Turbellaria, Haplopharyngida, ultrastructure, reproductive system, stylet

Abstract

The copulatory organ of *Haplopharynx quadristimulus* Ax, 1971 (Carolina form, Rieger, 1977) consists of a proximal prostatic vesicle and a distal stylet apparatus comprising a central tubular stylet and four to five peripheral accessory spines. By electron microscopy it could be seen that the stylet and spines were intracellular specializations. The copulatory organ can be interpreted as a specialization of an epithelial canal extending from the testes to the body wall. In the complex stylet apparatus, the epithelium was differentiated into six cell types. The stylet, which was formed in a matrix syncytium next to the prostatic vesicle, extended into the lumen of the stylet canal. The interior of the stylet apparatus contained one group of cells that had thick ciliary rootlets and another that had rootlet-like ribbons.

The cells that contain the rootlets enveloped bundles of longitudinally arranged muscles. The accessory spines were formed in cells which lay peripheral to the muscle bundles. The spines, stylet, rootlet-like ribbons, and rootlets had similar patterns of periodic cross striations. The similarity in striation patterns suggests that the accessory spines and stylet are composed of modified ciliary rootlets.

Introduction

Ultrastructural investigations have shown that the cuticles and cuticular derivatives (e.g. spines, teeth, hooks, and copulatory structures) found in many lower bilateria are analogous and not homologous as previously thought. True cuticles are found in the Annelida and most aschelminth phyla (see Rieger & Rieger, 1977). False cuticles are found in the Platyhelminthes, Rotifera, and Acanthocephala (see Lee, 1972; Rieger & Doe, 1975; Rieger & Rieger, 1977).

Within the Platyhelminthes, the 'cuticular' spines and hooks of the parasitic trematodes and cestodes are intracellular specializations (Lee, 1972), while 'cuticular' differentiations in the Turbellaria are either derivatives of the basement membrane or intracellular specializations. Basement-membrane derivatives include the proboscis hooks in the Kalyptorhynchia-Schizorhynchia (Rieger & Doe, 1975; Doe, 1976) and the copulatory cirrus in the proseriate *Archilopsis* (Martens & Schockaert, 1981). Intracellular specializations include the proboscis hooks in the Kalyptorhynchia-Eukalyptorhynchia (Doe, 1976); copulatory stylets in the orders Acoela (Mainitz, 1977), Macrostomida (Doe, 1977; 1982) and Proseriata (Lanfranchi, 1978; Ehlers & Ehlers, 1980); and the stylet in the prostatoid organ of the proseriate *Polystyliphora filum* (Brüggemann, 1984).

This paper presents the ultrastructure of the stylet and accessory spines in another turbellarian order, the Haplopharyngida. The data on the fine structure of the other cell types in the copulatory organ and other parts of the male reproductive system (sperm ducts, seminal vesicles, seminal ducts and male canal) is published separately.

Hydrobiologia 132, 157–163 (1986).

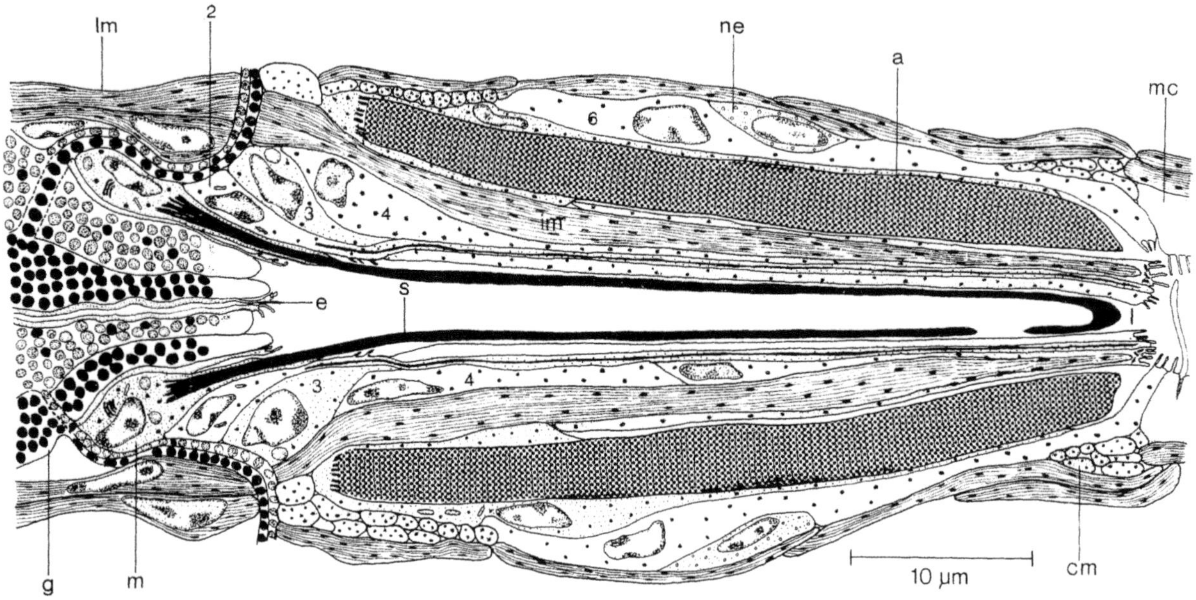

Fig. 1. Sagittal reconstruction of stylet apparatus of copulatory organ in *Haplopharynx quadristimulus*. Slightly schematic.

Material and methods

Specimens of *Haplopharynx quadristimulus* Ax, 1971 (Carolina form, Rieger, 1977) were collected from a sediment depth of 15 – 30 cm at the HTL of the high energy beach at Bogue Bank and from the upper 5 cm of sand at the HTL of a low energy sand flat at the White Oak River Inlet, Swansboro, North Carolina. For a description of extraction and fixation procedures see Doe (1982).

Observations

The copulatory organ consists of a proximal prostatic vesicle and a distal stylet apparatus. The stylet apparatus in the specimens studied contained a central, tubular stylet and four to five peripheral accessory spines (Figs. 1 & 5). A pair of wing-like, bow-shaped lateral extensions were located in the distal half of each spine. The stylet apparatus appeared to be a specialization of an epithelial canal extending from the testes to the body wall. This epithelium was differentiated into several cell types which appeared as concentric layers in a transverse section through the apparatus (Fig. 5).

The stylet was a specialization within a matrix syncytium, which was the cell type closest to the prostatic vesicle (Figs. 1 & 2). The tubular syncytium was thickest (3 μm) at the proximal end, which contained most of the cytoplasmic components. Distally, the syncytial thickness decreased to 0.5 μm as the outer and inner syncytial membranes closed toward the stylet (Figs. 1 & 5). A thin rim of cytoplasm remained between the inner membrane and the inner surface of the stylet. The outer membrane adhered more closely to the stylet, but thin areas of cytoplasm were also observed along this surface. The proximal region of the syncytium contained several elongate and often lobate nuclei. Also present were stylet material, dictyosomes, ribosomes, mitochondria, electron-dense granules, 400 – 600-nm-diameter vesicles, and filaments.

The stylet mainly consisted of electron-opaque material which, in the proximal one-half of the stylet, appeared to be fibrillar in nature (Fig. 8). 250 – 300-nm-thick ribbons of this material overlapped and were fused together distally to form a tube with 450 – 500-nm-thick walls (Figs. 2, 3 & 5). Proximally, the stylet material was electron-opaque but distally, where the ribbons overlap, each had an electron-opaque core and a 50-nm-thick electron-dense rim. Further distally, where the stylet wall fused into a single piece, the electron-dense outer

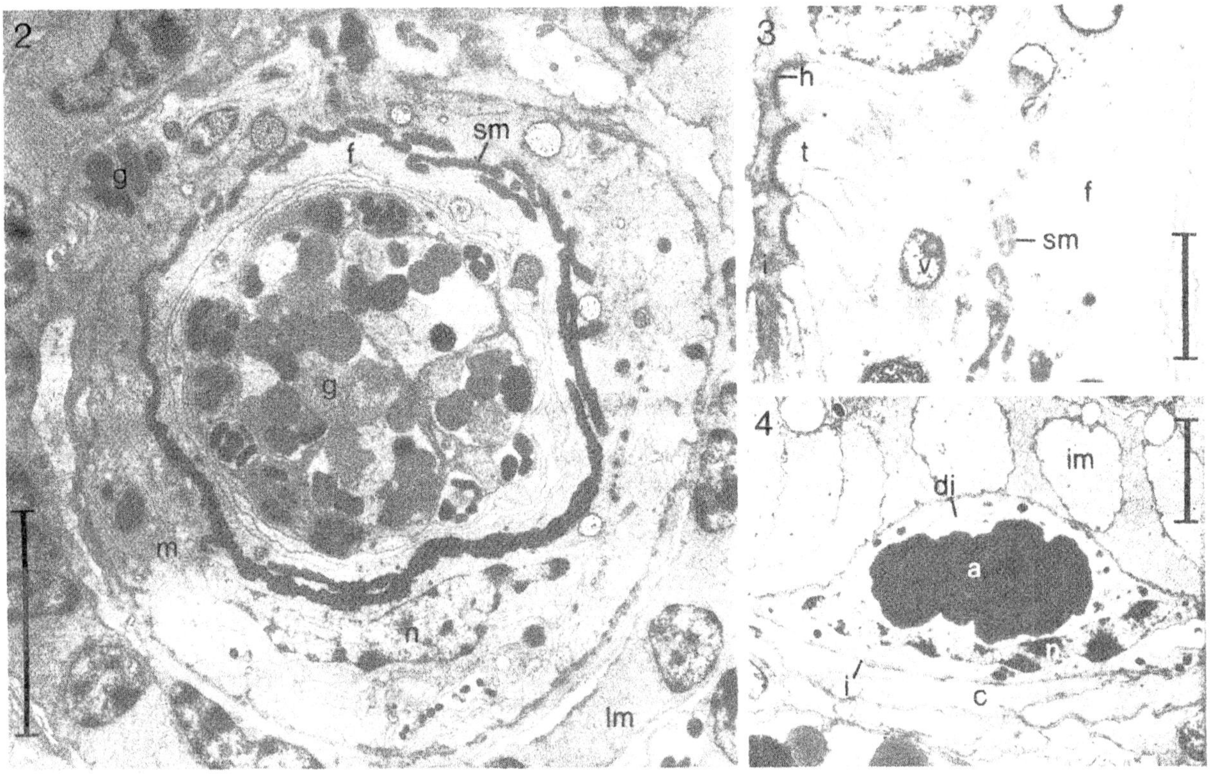

Figs 2–4. (2) Slightly oblique transverse section through proximal region of stylet apparatus. Note intracellular nature of stylet material, filaments, prostatic gland cell necks. Scale: 5 μm; (3) Fine structure of proximal region of matrix syncytium. Note the hemidesmosomes on the basal cell membrane, tonofilaments, filament bundles, and stylet material. Scale: 1 μm; (4) Transverse section through proximal region of spine cell. Note spine material, nucleus, dictyosomes, electron-dense granules, and position of the cell between musclelayers. Scale: 2 μm.

and inner surfaces were approximately 100 nm thick (Fig. 5). The stylet wall thickness decreased to approximately 300 nm at the level of the subterminal stylet opening. The tip of the stylet was a 500 nm thick swelling (Fig. 6).

The fibrillar material in the proximal region of the stylet displayed a pattern of periodic cross striations (Fig. 8) with a periodicity of approximately 71 nm, consisting of alternating dense and less dense bands. The striations were obscured in the electron-opaque material in the distal region of the stylet. The electron-dense outer and inner surface of the stylet did not appear striated.

Large areas of the proximal region of the matrix syncytium were filled with non-striated filaments, which appeared to be tonofilaments (Figs. 2 & 3). Filaments were located in the cytoplasm proximal to the stylet and also along the inner and outer surfaces of the circular stylet wall. The filaments in the cytoplasm on the inside of the stylet wall were arranged in a 0.5–0.75-nm-thick and 2.0–2.5-μm-long circular bundle. The filaments along the outer surface of the stylet formed a less discrete bundle as they intermingled with ribbons of stylet material and other organelles. Some of the filaments along both stylet surfaces were attached to the stylet material. In the proximal region of the matrix syncytium, filaments extended toward the basal cell membrane and attached to large (600 nm long) hemidesmosomes (Fig. 3). The filaments in the cytoplasm and the filamentous attachments to hemidesmosomes in the stylet therefore linked (or anchored) the stylet to the intercellular matrix.

Prostatic gland cell necks, which connect the gland cell bodies located outside of the copulatory organ to the prostatic vesicle, passed between the basal cell membrane of the syncytium and the intercellular matrix (Fig. 2).

The stylet apparatus contained five additional epithelial cell types (designated types 2–6, see Fig. 1). These cell types are numbered according to their position from the center of the stylet apparatus outward. One of these formed the accessory spines (type 5). In addition, there were internal longitudinal muscles and neurons. The fine structure of these cells is presented in a separate publication and they will be mentioned here only as they relate to the matrix syncytium and spine cells. The type 2 cells formed the proximal end of the stylet canal (Fig. 2). These cells were joined to the matrix syncytium by spot desmosomes. A short distance distally the membranes of the two cell types separated, forming the stylet canal lumen. The type 2 cells also contained striated rootlet-like ribbons that had a periodicity pattern of approximately 68 nm (Fig. 7). Cell type 3 appeared to be a syncytium which proximally surrounded the type 2 cells. A short distance distally the syncytium passed between type 2 cells to line the remainder of the canal lumen.

Cells of cell type 4 occupied part of the interior of the stylet apparatus (Fig. 5). In the area of the spines, cytoplasmic extensions of type 4 cells were located between the spine cells and the internal muscles. Proximally, the membranes of the type 4 cells and the spine cells were joined together by zonulae adhaerentes and septate junctions. However, distally the membranes separated to form an intercellular space which continued to the distal end of the stylet apparatus. At the level of the stylet tip, the type 4 cells lay between the spines (Fig. 6). Here the cells were filled with numerous thick, irregularly arranged ciliary rootlets which had a striation pattern similar to that of the stylet and type 2 cell ribbons but with a periodicity of approximately 66 nm (Fig. 6). Cilia were not associated with them.

The accessory spines were intracellular specializations within four of five cells (cell type 5) whose shape paralleled that of the spines (Figs. 1, 4 & 5). The cells were bordered by different cell types along their length. Proximally they were situated between the circular muscles of the stylet apparatus wall and the internal longitudinal muscles (Fig. 4). Distally, on the inward surface, type 4 cells were interposed between the spine cells and the internal muscles (Fig. 5). On the peripheral surface, type 6 cells were interposed between the spine cells and the apparatus wall musculature.

Along the proximal part of the main axis, each cell was circular to ovoid in cross-section (up to 2.2 μm $\times$ 3.5 μm) (Fig. 4). Here, each cell was attached to the surrounding intercellular matrix by hemidesmosomes. The elongate nucleus lay peripheral to the spine. Distally, the cells extended laterally so that the leading edges of adjacent spine cells overlapped, forming a complete ring of cells around the internal muscles (Fig. 5). These cytoplasmic extensions, which distally contained the spine lateral extensions, maintained their contact with each other until just proximal to the tips of the spines. At the distal end of the stylet apparatus, the spine cells anchored to extensions of the intercellular matrix by hemidesmosomes.

The spines mainly consisted of electron-opaque material that exhibited a striation pattern similar to that observed in the stylet, in type 2 cell rootlet-like ribbons, and in type 4 cell rootlets but with a periodicity of approximately 65 nm (Fig. 9). As in the case of the stylet, the spine material at the proximal end was arranged in ribbons that were not completely fused together. Here the spines were ovoid in transverse section (up to 2×6 μm) with a very irregular surface (Fig. 4). Distally, the main axis became almost circular in transverse section (up to 3.5×4.0 μm). In the region of the lateral extensions, the spines appeared T-shaped in transverse section (Fig. 5). The thin lateral extensions spread out from the main axis of each spine reaching a maximum width of 11 μm, edge to edge. Each lateral extension tapered in thickness from approximately 0.75 μm near the stalk to 200 nm at the edge. In the region of the extensions, the spines had a rim of electron-dense material similar to that of the stylet. The rim, which was 100–300 nm thick, was absent at the lateral edges of the lateral extensions. Distally, the spines maintained the T-shape as they tapered to the tip.

Each spine was separated from the spine cell membrane by cytoplasm over its entire length except for the inner surface (toward the internal muscles) in the region of the lateral extensions (Fig. 5). Here, the membrane adhered tightly to the spine. The spine cells also contained organelles and inclusions similar to those in the other cell types including small electron-dense ovoid–to–sausage-shaped granules (up to 300 nm wide and 400 nm long).

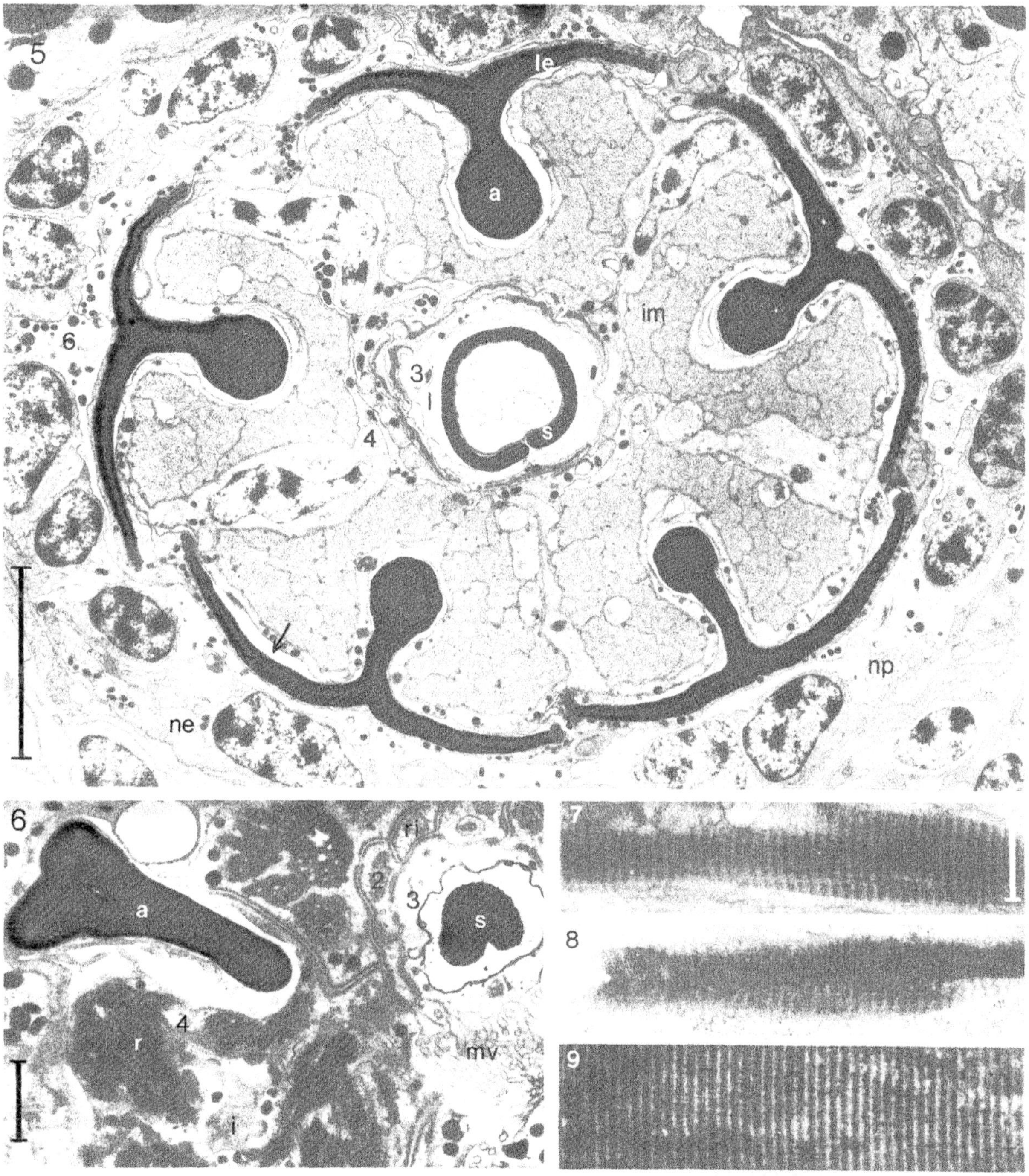

Figs 5–9. (5) Slightly oblique transverse section through stylet apparatus at the level of the stylet opening. Note the main axis and lateral extensions of the spines, internal muscle bundles, intercellular space (arrow) between spine cell and type 4 cell. Numbers specify different cell types. Scale: 5 μm; (6) Slightly oblique transverse section through stylet apparatus at level of stylet tip. Note solid stylet tip, spine tip, cell-type-4 rootlets, cell-type-2 rootlet-like ribbons. Scale: 1 μm; (7) Sagittal section through cell-type-2 rootlet-like ribbon. Scale: 0.5 μm; (8) Sagittal section through proximal region of stylet. Scale same as in Fig. 7; (9) Sagittal section through proximal region of accessory spine. Scale same as in Fig. 7.

Discussion

Fine-structural analysis of the copulatory stylet and accessory spines in *H. quadristimulus* identifies these structures as intracellular specializations. These results are consistent with the intracellular nature of stylets in Acoela (Mainitz, 1977), Macrostomida (Doe, 1982), and Proseriata (Ehlers & Ehlers, 1980; Lanfranchi, 1978; Brüggemann, 1984).

As in other turbellarian intracellular stylets, there are prominent dictyosomes, ribosomes, and electron-dense material in the proximal region of the matrix syncytium and of the spine cells (see Ehlers & Ehlers, 1980; Doe, 1982; Brüggemann, 1984). My use of adult specimens of *H. quadristimulus* precludes any developmental information. However, the similar patterns of striations in the stylet, spines, type 2 cell rootlet-like ribbons and the type 4 cell rootlets suggests a common morphological base for all four structures. Ehlers & Ehlers (1980) observed cross-striated microfibrils in the stylet of *Carenscoilia bidentata* (Proseriata) and Westheide (1979) stated that the similarities in striation patterns in intracellular stylet rods and ciliary rootlets in *Microphthalamus* c. f. *similis* suggest the possibility that the rods are very large rootlets. Striations were not observed in the stylets of *Macrostomum* sp. and *Microstomum* sp. (Doe, 1982), but they may have been obscured by electron-dense material in the adult specimens. Brüggemann (1984) has shown that the stylet of the prostatoid organ in *Polystyliphora filum* is built on a framework of microtubules.

Although a connection to basal bodies and cilia may be obligate to justify using the term 'ciliary rootlet' (Flood, 1975), rootlet-like structures have been seen in non-ciliated cells (Nagano, 1962; Sakaguchi, 1965; Doe, 1982). The striation periodicities of 65 – 71 nm in the rootlets, rootlet-like ribbons, stylet, and spines fall within the 12 – 90 nm range frequently cited for ciliary rootlets and (extracellular) collagen (see Flood, 1975; Olson & Rattner, 1975; Larson & Dingle, 1981). The differences in the periodicities in the stylet, spines, rootlets, and ribbons may not be significant and may only represent variability in the striations of the structural protein (see Simpson & Dingle, 1971).

Mechanical support of cilia is one of the proposed functions of ciliary rootlets (Pitelka, 1974). However, Flood (1975) showed that in larval amphioxus, the ciliary anchoring function is secondary to the mechanical support of the tail fin. Here, extraordinarily large and thick rootlets provide a rigid but elastic resistance against bending forces (Flood, 1975). This finding supports the possibility that the stylet rods in *Microphthalmus* and the spines (and rootlet-like ribbons) in *H. quadristimulus* are modified rootlets providing mechanical support. The formation of a tubular stylet from rootlets involves more complicated morphological changes in the arrangement of the rootlet proteins and in the shape of the stylet-forming cell or syncytium (see Stephens, 1975; Larson & Dingle, 1981 for descriptions of rootlet proteins).

Large bundles of intracellular filaments associated with stylets have not been observed previously. However, hemidesmosomes and tonofilaments have been seen closely associated with stylets in *C. bidentata* (Ehlers & Ehlers, 1980), *M.* c.f. *similis* (Westheide, 1979), *Microstomum* sp. and *Macrostomum* sp. (Doe, 1982), and with intracellular proboscis hooks in *Gnathorhynchus* sp. (Doe, 1976). In all cases the hemidesmosomes attach the epithelial basal cell membrane to the basement membrane (intercellular matrix) and the tonofilaments are attached directly to the stylet or hook material. These serve to anchor the hard parts to supporting musculature. The filament bundle inside the stylet in *H. quadristimulus* may provide additional support for the stylet.

The results support the view (Rieger & Doe, 1975) that, within the Platyhelminthes, supportive and other hard structures are developed as derivatives of the basement membrane or as elaborations of intracellular components in the outer part of an epithelium (modifications of microvilli, microtubules, and possibly ciliary rootlets).

Acknowledgments

I thank Dr. Seth Tyler and Dr. Julian Smith for their critical reading of the manuscript. I thank Renee Gilman for technical assistance, Bryan Kantor for a preliminary drawing of Fig. 1, and Fran Curran and Kim Mills for typing the manuscript. This study was supported by NSF grant #BSR 82-14376.

Abbreviations

a – accessory spine
c – circular muscle
di – dictyosome
e – ejaculatory duct
f – filaments
g – prostatic gland cell neck
h – hemidesmosome
i – intercellular matrix
im – internal muscle
l – stylet – canal lumen
le – spine lateral extension
lm – longitudinal muscle
m – matrix syncytium
mc – male canal
mv – microvilli
n – nucleus
ne – neuron
np – nerve process
r – rootlet
ri – rootlet-like ribbon
s – stylet
sm – stylet material
t – tonofilaments
v – vesicle
2 – cell type 2
3 – cell type 3
4 – cell type 4
6 – cell type 6

References

Brüggemann, J., 1984. Ultrastruktur und Differenzierung den prostatioden Organe von Polystyliphora filum (Platyhelminthes, Proseriata). Zoomorphology 104: 86–95.

Doe, D. A., 1976. The proboscis hooks in Karkinorhynchidae and Gnathorhynchidae (Turbellaria, Kalyptorhynchia) as basement membrane or intracellular specializations. Zool. Scr. 5: 105–115.

Doe, D. A., 1977. Fine structure of 'cuticular' structures in Platyhelminthes. Am. Zool. 17: 970.

Doe, D. A., 1982. Ultrastructure of copulatory organs in Turbellaria. I. Macrostomum sp. and Microstomum sp. (Macrostomida). Zoomorphology 101: 39–59.

Ehlers, B. & U. Ehlers, 1980. Struktur und Differenzierung penialer Hartgebilde von Carenscoilia bidentata Sopott (Turbellaria, Proseriata). Zoomorphologie 95: 159–167.

Flood, P. R., 1975. Ciliary rootlet-fibres as tail fin-rays in larval amphioxus (Branchiostoma lanceolatum, Pallas). J. Ultrastruct. Res. 51: 218–225.

Larson, D. E. & A. D. Dingle, 1981. Isolation, ultrastructure, and protein composition of the flagellar rootlet of Naegleria gruberi. J. Cell. Biol. 89: 424–432.

Lanfranchi, A., 1978. Morphology and taxonomy of two new otoplanids (Turbellaria, Proseriata) from the Ligurian Sea. Zool. Scr. 7: 249–254.

Lee, D., 1972. The structure of the helminth cuticle. Adv. Parasitol. 10: 347–377.

Mainitz, M., 1977. The fine structure of the stylet apparatus in Gnathostomulida Scleroperalia and its relationship to turbellarian stylets. In T. G. Karling and M. Meinander (eds), The A. Luther Symposium on Turbellaria. Acta. Zool. fenn. 154: 163–174.

Martens, E. E. & E. Schockaert, 1981. Observations on the ultrastructure of the copulatory organ of Archilopsis unipunctata (Fabricius, 1826) (Proseriata, Monocelididae). Hydrobiologia 84: 277–285.

Nagano, T., 1962. An electron microscopic observation on the cross-striated fibrils occurring in the human spermatocyte. Z. Zellforsch. mikrosk. Anat. 58: 214–218.

Olson, G. E. & J. B. Rattner, 1975. Observations on the substructure of ciliary rootlets. J. Ultrastruct. Res. 51: 409–417.

Pitelka, D. R., 1974. Basal bodies and root structures. In M. A. Sleigh (ed.), Cilia and Flagella. Academic Press, Inc., New York: 437–469.

Rieger, G. E. & R. M. Rieger, 1977. Comparative fine structure study of the gastrotrich cuticle and aspects of cuticular evolution within the Aschelminthes. Z. Zool. Syst. Evolut.-forsch. 15: 81–124.

Rieger, R. M. & D. A. Doe, 1975. The proboscis armature of Turbellaria-Kalyptorhynchia, a derivative of the basement lamina. Zool. Scr. 4: 25–32.

Sakaguchi, H., 1965. Pericentriolar filamentous bodies. J. Ultrastruct. Res. 12: 13–21.

Simpson, P. A. & A. D. Dingle, 1971. Variable periodicity in the rhizoplast of Naegleria flagellates. J. Cell Biol. 51: 323–328.

Stephens, R. E., 1975. The basal apparatus. Mass isolation from the molluscan ciliated gill epithelium and a preliminary characterization of striated rootlets. J. Cell Biol. 64: 408–420.

Westheide, W., 1979. Ultrastruktur der Genitalorgane interstitieller Polychaeten II. Mannliche Kopulationsorgane mit intrazellularen Stilettstaben in einer Microphthalmus-Art. Zool. Scr. 8: 111–118.

Comparative ultrastructure of copulatory organs having a stylet in the Proseriata (Turbellaria)

Els E. Martens
Dept. SBM, Limburgs Universitair Centrum, B-3610 Diepenbeek, Belgium

Keywords: Turbellaria, Proseriata, ultrastructure, copulatory organ, hard structures

Abstract

The ultrastructure of male copulatory organs having a stylet has been studied in some genera of the Proseriata.

Within the Monocelididae there was a variety of stylet-like hard structures. The stylet in *Monocelis fusca* was a differentiation of the basement membrane of the epithelium lining a penis-like muscular papilla. The penis papilla in *Ectocotyla* consisted of circular muscles surrounded by a thickened basement membrane and an epithelium. *Archilopsis* sp. and *Archilina* sp. with a duplex copulatory bulb, had a stylet within a spiny cirrus. The stylet in *Archilopsis* sp. was a cylindrical muscular protrusion with a thickened basement lamina that lined the cirrus lumen. The stylet structure in *Archilina* sp. was composed of four long spines which were derivatives of the basement membrane. In *Ectocotyla multitesticulata* and *Dupliminona corsicana,* the accessory prostatoid organ was provided with a hook-shaped stylet that was differentiated in the basement membrane and of which the material was continuous with the fibrous matrix between the muscles of the prostatic bulb. The stylet and needles in the *Archimonocelis* species were intracellular differentiations. The copulatory organ in *Carenscoilia biforamen* consisted of a tubiform stylet and four needles, all of which were also intracellular specializations.

I consider copulatory hard structures in the Turbellaria to be taxonomically significant in terms of structure, differentiation, and location (whether subcellular, in the basement membrane, or intracellular).

Introduction

With advancing knowledge on the ultrastructure of the Turbellaria new characteristics become available which can validly be applied in phylogenetic systematics. Previous studies on the ultrastructure of the copulatory organ have revealed that the hard parts are intracellular specializations in the Acoela (Mainitz, 1977), Macrostomida (Doe, 1982), Haplopharyngida (Doe, 1985), and Proseriata (Lanfranchi, 1978; Ehlers & Ehlers, 1980; Brüggemann, 1984; E. Martens & Schockaert, 1984), or derivatives of the basement membrane in the Proseriata (E. Martens & Schockaert, 1981; E. Martens, 1984b).

In this paper I present observations on the ultrastructure of stylet-bearing copulatory organs in several genera of the Proseriata. Such a comparative study may reveal new characteristics that clarify relationships within the Proseriata and, eventually, within the Turbellaria.

Material and methods

Specimens of *Monocelis fusca* were collected in St. Andrews (Canada) from shells of barnacles, and specimens of *Ectocotyla multitesticulata* were collected from the gills of crabs, *Chionoecetes oplio,* from Newfoundland (Canada).

Hydrobiologia 132, 165–173 (1986).

Archilina sp., *Archimonocelis* sp., and *Dupliminona corsicana* were collected in the Bay of Calvi (Corsica). *Archimonocelis oostendensis, Carenscoilia biforamen,* and *Archilopsis* sp. were found in sand samples from the low tide level of the beach of Mariakerke and in the Zwin-estuary in the North of Belgium.

The animals were extracted by decantation with a $MgCl_2$-solution isotonic to seawater, and fixed in 2% glutaraldehyde in 0.1 M phosphate buffer with 10% sucrose (pH 7.3) at 4 °C for 2–3 h. After rinsing in the same buffer they were postfixed in 1%-phosphate-buffered osmium tetroxide at 4 °C for 2 h, dehydrated in a graded acetone series, and embedded in Araldite. Semithin and ultrathin sections were cut with a Reichert OMU 3 ultramicrotome. Ultrathin sections were mounted on pioloform-coated copper grids, stained with 2% aqueous uranyl acetate and 1.2% aq. lead citrate, and examined in a Philips EM 400. The semithin sections were stained with toluidine blue and examined by light microscopy.

Observations

1. The stylet in Monocelis fusca *(Oersted, 1845) (Monocelididae)*

Under the light microscope the copulatory organ of *M. fusca* appeared to consist of a single ovoid vesicle surrounded by several muscle layers (Fig. 1). The proximal part of the vesicle appeared to function as seminal vesicle, and in its distal part prostate glands entered the vesicle from the left and the right side. A simple straight, conical stylet, 70–85 μm long, was inserted in the distal end of this vesicle.

By electron microscopy it could be seen that the proximal part of the stylet consisted of circular muscles embedded in a fine fibrous matrix and covered on both sides by an epithelium (Fig. 2). The electron dense matrix was similar to and continuous with the basement membrane. Distally, the muscles and epithelium were lacking, and only the electron dense material formed the stylet here (Figs 3–4).

The epithelia covering the luminal side and the outer side of the stylet could be followed to the edge of the muscles.

The vesicle itself was lined in its proximal part by a flattened epithelium with microvilli and cilia with a long rootlet parallel to the cell membrane. Lateral cell membranes were highly convoluted and connected to each other by a zonula adherens and a deeper septate desmosome. In the distal part of the vesicle, the epithelium was high, and its entire extent here was pierced by numerous cell necks of the prostate glands.

2. The 'penis papilla' in Ectocotyla multitesticulata *Fleming & Burt, 1978*

The copulatory organ consisted of a simple vesicle similar to that in *Monocelis fusca* but without any prostate glands. It was surrounded by a thick layer of muscles, of which the inner layer of circular muscles continued into a tubular eversible penis papilla, and a part of the longitudinal muscle layer extended to the atrial wall (Fig. 5).

In its proximal part, the seminal vesicle was lined by a flattened ciliated epithelium. Distally, the epithelial cells had long branching expansions around the inverted penis papilla. In the penis papilla, the basement membrane under this epithelium became very thick and electron-dense. It consisted of fine fibrous material and was continuous with the matrix in which the muscles were embedded. The penis papilla was covered on its luminal side by flattened epithelial cells which were connected to each other by a zonula adherens and a deeper septate desmosome and to the basement membrane by hemidesmosomes (Fig. 6). In the epithelial-cell cytoplasm several microfibrils, mitochondria, and granules of different electron densities could be seen, but no nuclei were found.

3. The stylet within a cirrus in Archilopsis *sp.*

This new species resembles *Archilopsis unipunctata* (Fabricius, 1826) in many respects, but in highly squeezed individuals a cylindrical stylet could be seen within the cirrus. A more detailed description of the species will be given later.

By electron microscopy the stylet could be seen to be double-walled. The stylet was basically a tube-shaped cirrus spine. Only in the proximal half of the stylet, was the luminal side lined with an epithelium (Fig. 12). The nuclei of this epithelium were situated far inside the copulatory bulb at the beginning of the ejaculatory duct. The rather thin basement membrane consists of fine fibrous material and is continuous with that of the ejaculatory duct. More distally, this basement membrane formed the lining on the luminal side. The outer side of the stylet consisted only of a basement membrane, thicker here than on the inner side, and with a trilamellar structure as found in the spines (see also Martens, 1984b). The basement membranes of the inner side and of the outer side were continuous at the tip of the stylet. The outer basement membrane was also continuous with the layer that formed the spines of the surrounding cirrus (Fig. 13). The stylet wall, between the two basement membranes, consisted of processes of the muscle cells situated at the base of the stylet.

The ultrastructure of the other parts of the copulatory organ was comparable with that in *A. unipunctata* (see Martens & Schockaert, 1981). In this new species, the prostate glands discharge their contents into two 'prostatic ducts' which enter the ejaculatory duct, midway between the seminal vesicle and the cirrus.

4. The 'stylet' within a cirrus in Archilina *sp.*

This species from the Mediterranean has an anatomy very similar to that of *Archilina endostyla* (Ax, 1959). The stylet within the cirrus is, however, much shorter than in *A. endostyla,* and in a highly squeezed specimen it appeared to consist of four long spines. A description of the species will be given later.

By electron microscopy this 'stylet' could be seen to consist of four spines that had the same ultrastructure as the smaller surrounding spines of the cirrus: extensions of the sarcoplasm of underlying muscles covered by a trilamellar basement membrane without a covering epithelium (Figs. 14, 15). The four spines were closely apposed to form a tube.

In this species, the prostate glands send their necks separately through the epithelium of the seminal duct between the seminal vesicle and the cirrus. No 'prostatic ducts' have been seen.

5. The stylet of the prostatoid organ in Ectocotyla multitesticulata *Fleming & Burt, 1978 and* Dupliminona corsicana *Martens, 1984*

The accessory prostatoid organ consisted of a vesicle filled with secretory granules surrounded by a thick layer of muscles and with a stylet extending into a small atrium (Figs. 7, 10). The vesicle was lined by a basement membrane composed of fine fibrous material.

The stylet had the shape of a conical gutter, broad at its base and with a sharp, curved end (Figs. 8, 9, 11). Its lumen narrowed from the base towards the tip of the stylet where it was almost closed, the two margins nearly touching each other (Figs. 9, 11). The wall of the stylet was thinnest in its proximal part and thickened gradually towards the tip.

In *Ectocotyla* the stylet was composed of a thin outer layer of homogenous material and irregular electron-dense granules which adhered to the outer layer where their contents seemed to spread (Fig. 9). Distally, the granules became more densely packed and formed a solid electron-dense matrix in the tip of the stylet. At the base of the stylet the material was continuous with the fibrous matrix between the muscles in which strands of electron-dense granules could be seen. The outer layer of the stylet was continuous with the basement membrane surrounding the prostatoid vesicle.

In *Dupliminona,* the stylet consisted of coarse fibrillar dense material with an outer rim of homogeneous electron density (Fig. 11). Towards the tip, the fibrous matrix coalesced into a more homogeneous material. This fibrous matrix was continuous with the matrix between the muscles at the base of the stylet, while the outer rim was continuous with the basement membrane surrounding the vesicle.

6. The stylet and needles in Archimonocelis

Two species of *Archimonocelis* were studied: *A. oostendensis* Martens & Schockaert, 1981, and a new species from the Mediterranean that will be described later. In both species the copulatory organ consisted of a stylet surrounded by long needles (fourteen fine plus two thicker needles in *A. oostendensis*) and an accessory organ within the same muscular sheath (Fig. 16). This accessory organ consisted of a bundle of glands surrounded by a separate girdle of needles (eleven fine needles plus three thicker needles) in *A. oostendensis,* while in the other species the nine needles encircled the ac-

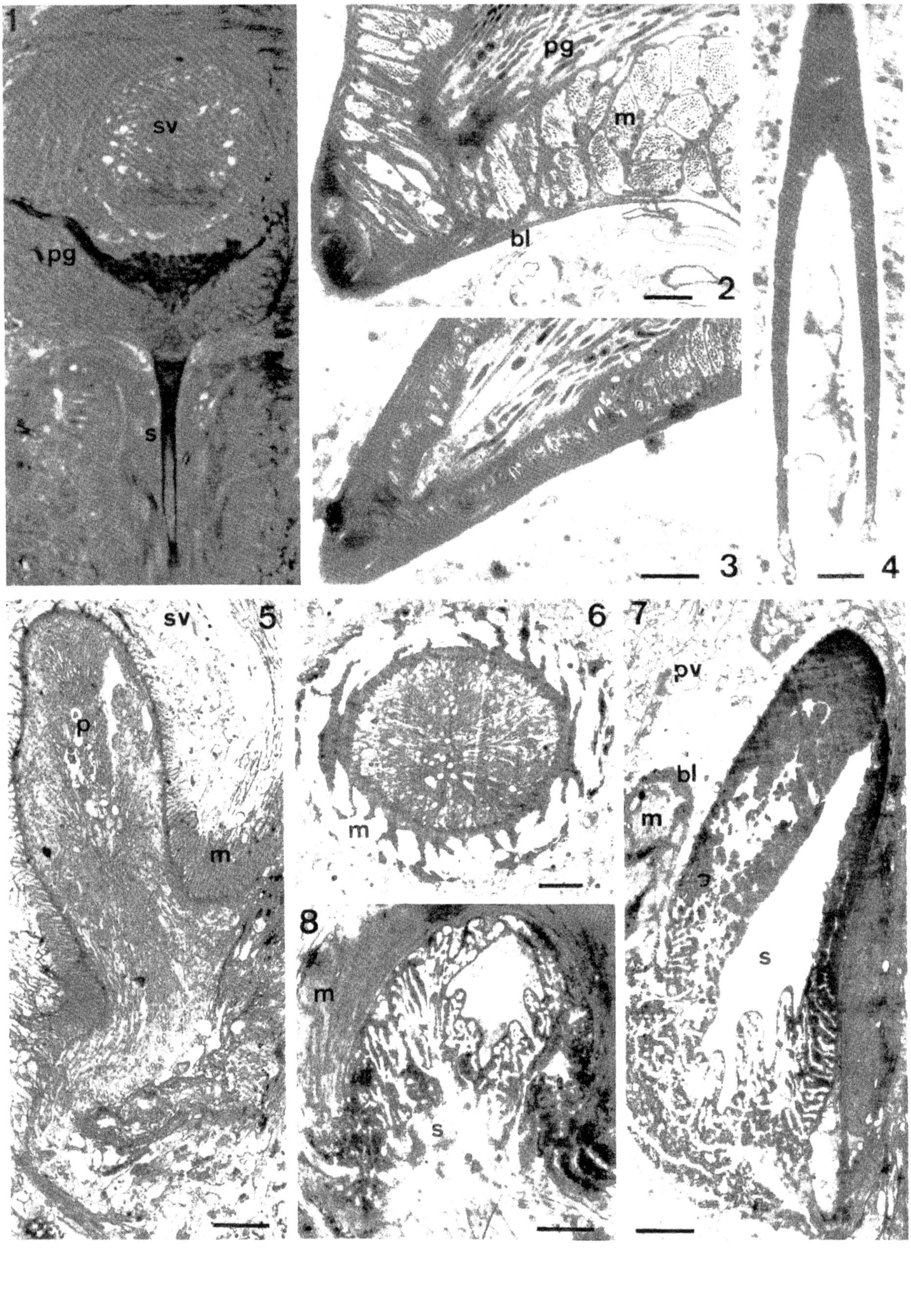
1
sv
pg
s
pg
m
bl
2
3
4
sv
5
p
m
6
m
7
pv
bl
m
s
8
m
s

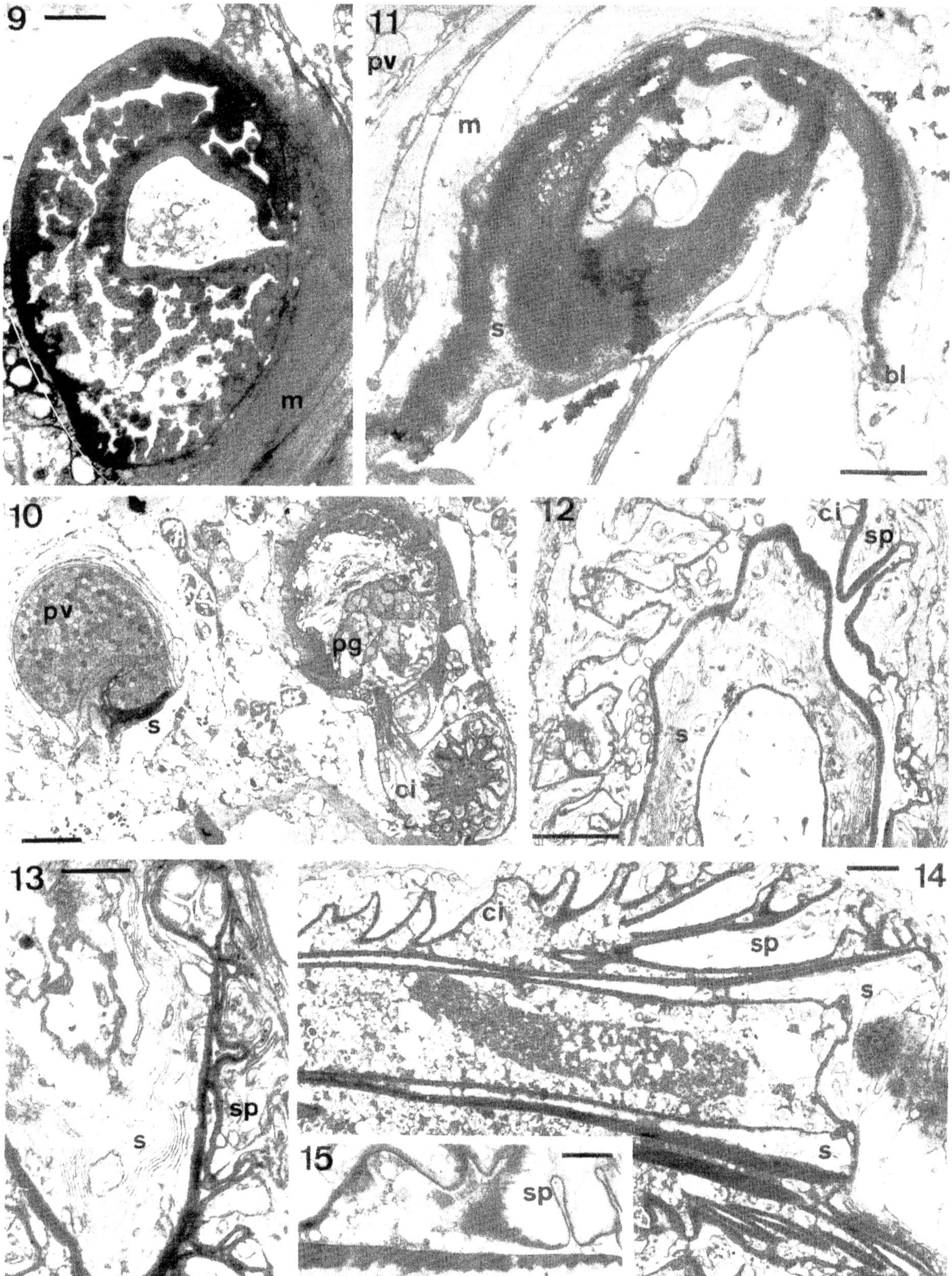
9
m
11
pv
m
s
bl
10
pv
s
pg
ci
12
ci
sp
s
13
sp
s
14
ci
sp
s
s
15
sp
s

cessory glands as well as the stylet. The stylet material consists of a coarse fibrous electron-dense matrix surrounded by a more homogeneous less-dense layer. The stylet base, however, was composed of a more flocculent material accumulated in the cytoplasm of the (probably single) matrix cell which was attached to the basement membrane by several hemidesmosomes.

The stylet had a more or less cylindrical form and a subterminal opening. Cross sections indicated that the stylet was a closed gutter-shaped structure with margins overlapping at the caudal side of the stylet. The opening was situated rostrally, so that from the opening towards the tip, the stylet consisted of two overlapping halves (Figs. 17, 18).

The needles surrounding the stylet and the accessory glands were also intracellular differentiations. Each needle was formed in its own matrix cell and consisted of a core of cross-banded fibrils embedded in a coarse fibrous material and an outer more homogeneous layer (Fig. 19). The periodicity of the cross-banding was about 73 nm. At the base, the material was more flocculent and not delimited by a membrane from the surrounding cytoplasm. The cytoplasm of a matrix cell contained a large nucleus, extensive smooth endoplasmic reticulum, ribosomes, mitochondria, Golgi complexes, and electron-dense granules.

7. The stylet apparatus in Carenscoilia biforamen *Sopott, 1972*

The stylet apparatus in *Carenscoilia bidentata* Sopott, 1972, has been studied by Ehlers & Ehlers (1980) using a subadult specimen.

Observations on adult and subadult specimens (Martens, 1984a) of *C. biforamen* confirmed the findings of Ehlers & Ehlers. As in the subadult, the stylet and needles in the adult consisted of a core of cross-banded fibrils, with a periodicity of about 73 nm, embedded in a moderately electron-dense material which was more flocculent and dense towards the base of the stylet (Figs. 20, 21, 22). Cross sections indicated that the stylet was a closed tube with an opening under the slightly curved tip.

Whether the stylet apparatus is formed in a syncytial matrix, as described by Ehlers & Ehlers (1980) or in a single cell (as is probably the case in *Archimonocelis*) could not be ascertained.

Discussion

In earlier studies, we have shown that spines in the cirrus of proseriates may have two separate origins: in the Monocelididae (see Martens & Schockaert, 1981; E. Martens, 1984b) they are differentiations of the basement membrane; in *Cirrifera* (Coelogynoporidae) they are intracellular structures (Martens & Schockaert, 1984). The present study

Figs 1–4. Monocelis fusca: (1) Photograph of section through copulatory bulb and stylet; (2) Electron micrograph of section through proximal part of stylet showing the epithelial lining on both sides of the stylet. Scale bar = 1 μm; (3) Section through the stylet more distally. Scale bar = 1 μm; (4) Section through the distal tip of the stylet at its opening. Scale bar = 0.5 μm.

Figs 5–9. Ectocotyla multitesticulata: (5) Sagittal section through copulatory bulb showing the penis papilla lined on both sides by an epithelium. Scale bar = 5 μm; (6) Cross section through the penis papilla. Scale bar = 2 μm; (7) Sagittal section through the accessory prostatoid vesicle and stylet. The stylet is gutter-shaped and has a subterminal opening. Scale bar = 5 μm; (8) Cross section through the stylet at its base. Scale bar = 5 μm; (9) Cross section through the stylet near its distal tip. Scale bar = 1 μm.

Figs 10–11. Dupliminŏna corsicana: (10) Sagittal section showing the copulatory organ with spiny cirrus and accessory organ with stylet. Scale bar = 5 μm; (11) Section through the stylet at its base. The stylet is gutter-shaped. Scale bar = 1 μm.

Figs 12–13. Archilopsis sp.: (12) Sagittal section showing part of the stylet within the spiny cirrus. The inner epithelium of the stylet disappears more distally. Scale bar = 2 μm; (13) Section through the stylet near its base. Scale bar = 1 μm.

Figs 14–15. Archilina sp.: (14) Sagittal section through the copulatory bulb showing stylet-spines within the spiny cirrus. Scale bar = 2 μm; (15) Higher manification of the basal lamina forming stylet and cirrus spines. Scale bar = 0.5 μm.
Abbreviations:
ao: accessory organ, bl: basement membrane, ci: cirrus, m: muscles, mc: matrix cell, n: needle, p: penis papilla, pg: prostate glands, pv: prostatic vesicle, s: stylet, sp: spines, sv: seminal vesicle, sz: spermatozoa.

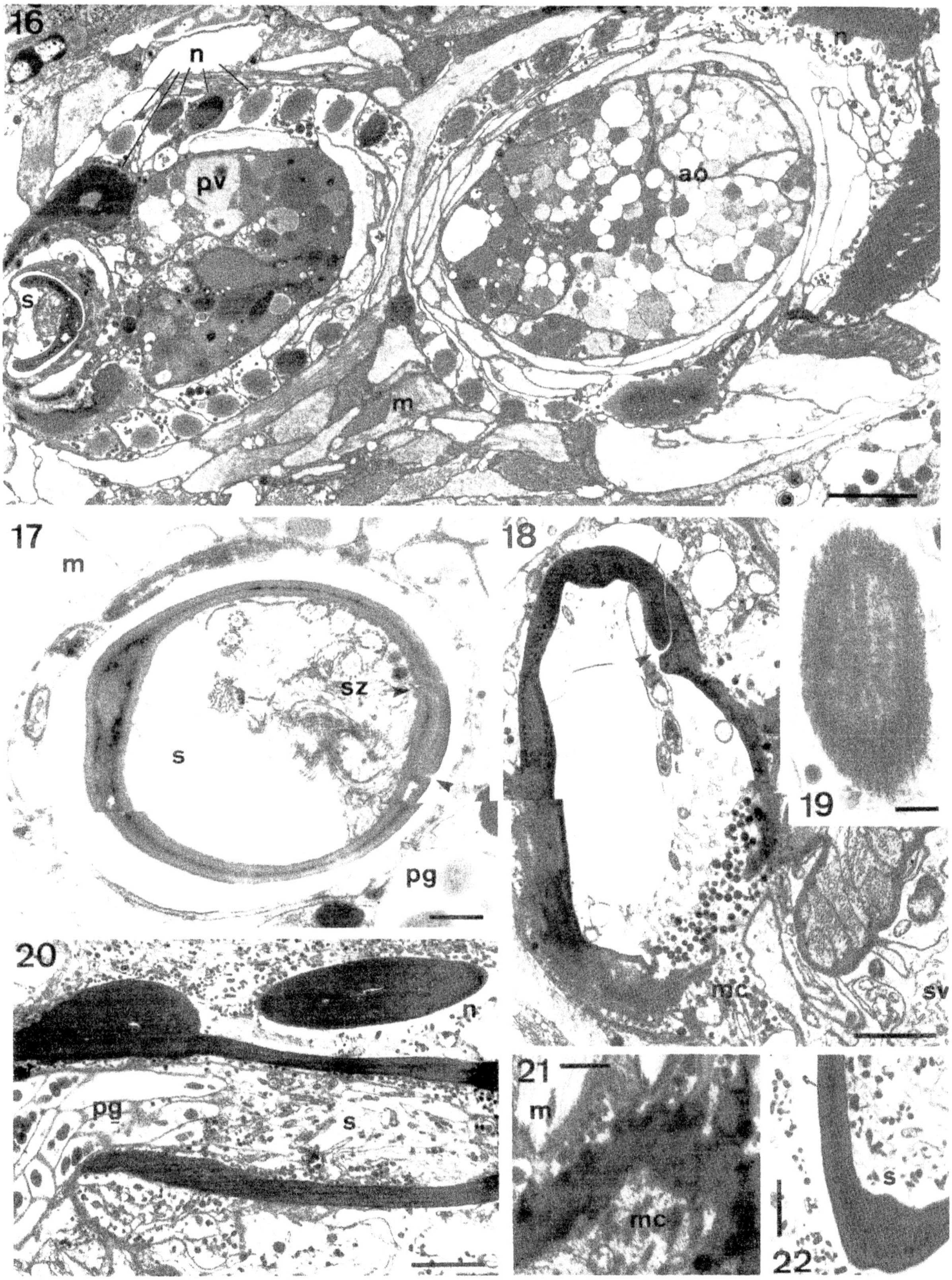
16
n
pv
ao
s
m
17
m
sz
s
pg
18
mc
sv
19
20
n
pg
s
21
m
mc
22
s

shows that the same applies to stylet structures: again only in the Monocelididae (exclusive of *Archimonocelis*) are stylets derivatives of the basement membrane, while those in other proseriates are intracellular formations: in *Archimonocelis* (present paper), in the Coelogynoporidae (Ehlers & Ehlers, 1980; present paper), in the Otoplanidae (Lanfranchi, 1978; Ehlers & Ehlers, 1980), and in the Polystyliphoridae (Brüggeman, 1984).

As pointed out before (Martens, 1984b), the origin of hard parts in the copulatory organ of the Monocelididae from basement membrane is an autapomorphy for this family, and so *Archimonocelis* can no longer be considered as a monocelid. The intracellular origin of hard parts is the plesiomorphous situation for the Proseriata by the principle of parsimony and by out-group occurrence in Acoela (Mainitz, 1977), Macrostomida (Doe, 1982), Haplopharyngida (Doe, 1985), and in Typhloplanoida, Kalyptorhynchia, and Dalyellioidea (Brüggeman, 1984: 95).

Some other ultrastructural features of the copulatory organ may be of taxonomic importance. The fine structure of the prostatoid organ in *Ectocotyla* and in *Dupliminona* appears identical, which is strong evidence for the homology of this organ within the Minoninae – a homology that has been questioned by P. Martens (1983). This accessory organ may thus be a synapomorphy for all genera that possess it and hence is supportive of the monophyletic origin of the Minoninae.

For the time being, only for the genus *Archilopsis* is it known that the prostate glands discharge their contents into the seminal duct via a pair of 'prostatic ducts'. This character may be considered as diagnostic for this genus.

The localization and structure of the needles in *Archimonocelis* are very similar to those of the spines in the cirrus of *Cirrifera* (Martens & Schockaert, 1984). Each needle, like each cirrus spine, is enclosed (at its base) by a single cell which is characterised by a large nucleus and numerous dense granules. The stylet in *Archimonocelis, Carenscoilia,* and the Otoplanidae is basically the same. Rather surprising is the different form of the intracellular stylets: a closed tube in *Carenscoilia* and *Polystyliphora* (Brüggeman, 1984), gutter-shaped in *Archimonocelis,* and composed of two pieces fitting into each other in the Otoplanidae (Lanfranchi, 1978). Whether this represents an evolutionary line (from long spines to a stylet or the reverse) must be considered in a larger context (see also Lanfranchi, 1978).

In subadult individuals of *Carenscoilia* (Ehlers & Ehlers, 1980; E. Martens, 1984a) and Otoplanidae (E. Martens, in preparation), the core of cross-banded fibrils arises first and then becomes enveloped by electron-dense material. In differentiating spines in the cirrus of *Cirrifera* (Martens, 1984a) and stylets of the prostatoid organs in *Polystyliphora filum* (Brüggeman, 1984), the framework forming the core consists of a bundle of microtubules. The core of the cross-banded collagen-like fibrils in the needles of *Archimonocelis* and in the stylet of *Carenscoilia* may perhaps be related to the length of the hard structures. Bundles of thicker fibers give a more solid framework for long structures.

It may be of some interest to note that the stylet of *Monocelis fusca,* the stylet in the cirrus of *Archilopsis* sp., and the penis papilla in *Ectocotyla* are basically of the same construction: double-walled with an epithelial covering on the luminal side and on the outer side but not more distally and with the basement membrane (especially of the outer side) usually thickened. Though the penis papilla may be

Figs 16–17. Archimonocelis oostendensis: (16) Cross section through copulatory bulb showing the stylet surrounded by needles and the accessory prostatoid organ with a separate circle of needles. Scale bar = 2 μm; (17) Cross section through the stylet, formed by a gutter-shaped structure with overlapping margins (arrow). Scale bar = 0.5 μm.

Figs 18–19. Archimonocelis sp.: (18) Sagittal section through the stylet at its base, which lies free in the cytoplasm of the matrix cell. Scale bar = 1 μm; (19) Section through a needle which consists of cross-banded fibrils surrounded by a layer of coarse fibrous material. Scale bar = 0.2 μm.

Figs 20–22. Carenscoilia biforamen (adult specimen): (20) Sagittal section through the stylet structure and one of the needles. Scale bar = 2 μm; (21) Cross section through the stylet wall at its base. Scale bar = 0.5 μm; (22) Cross section through the stylet wall showing the cross-banded fibrils surrounded by a more homogeneous layer. Scale bar = 1 μm. Abbreviations: see Figs 1–15.

inverted into the copulatory organ (seminal vesicle) at rest and everted when used (as we suppose) it is not comparable to a cirrus in duplex organs.

Acknowledgments

I am indebted to Prof. E. R. Schockaert for valuable advice and criticism of the manuscript and to Miss E. Plaum and H. Zurings for technical assistance. I thank Prof. Burt and his staff for help and hospitality from the Department of Biology, University of New Brunswick, Canada, where a part of this work was carried out.

References

Brüggeman, J., 1984. Ultrastruktur und Differenzierung der prostatoiden Organe von *Polystyliliphora filum* (Plathelminthes, Proseriata). Zoomorphology 104: 86–95.

Doe, D. A., 1982. Ultrastructure of copulatory organs in Turbellaria. I. *Macrostomum* sp. and *Microstomum* sp. (Macrostomida). Zoomorphology 101: 39–59.

Doe, D. A., 1985. Ultrastructure of the male reproductive system in *Haplopharynx quadristimulus* Ax, 1971. I. Copulatory stylet and accessory spines. This volume.

Ehlers, B. & U. Ehlers, 1980. Struktur und Differenzierung penialer Hartgebilde von *Carenscoilia bidentata* Sopott, 1972 (Turbellaria, Proseriata). Zoomorphologie 95: 159–168.

Lanfranchi, A., 1978. Morphology and taxonomy of two new otoplanids (Turbellaria, Proseriata) form the Ligurian Sea. Zool. Scr. 7: 249–254.

Mainitz, M., 1977. The fine structure of the stylet apparatus in Gnathostomulida scleroperalia and its relationship to turbellarian stylets. In T. G. Karling & M. Meinander (eds.). The A. Luther Symposium on Turbellaria. Acta zool. fenn. 154: 163–174.

Martens, E. E., 1984a. Differentation in the copulatory organ of Turbellaria (Platyhelminthes). In W. Engels (ed.). Advances in invertebrate reproduction 3. Elsevier, Amsterdam: 611.

Martens, E. E., 1984b. Ultrastructure of the spines in the copulatory organ of some Monocelididae (Turbellaria, Proseriata). Zoomorphology: 104: 261–265.

Martens, E. E. & E. R. Schockaert, 1981. Observations on the ultrastructure of the copulatory organ of *Archilopsis unipunctata* (Fabricius, 1826) (Proseriata, Monocelididae). In E. R. Schockaert & I. R. Ball (eds.). The biology of the Turbellaria. Hydrobiologia 84: 277–285.

Martens, E. E. & E. R. Schockaert, 1984. Studies on the ultrastructure of the genital organs in Proseriata (Turbellaria). I. *Cirrifera aculeata* (Ax, 1951) (Coelogynoporidae). Zool. Scr.: 14: 81–90.

Martens, P. M., 1983. Three new species of Minoninae (Turbellaria, Proseriata, Monocelididae) from the North Sea, with remarks on the taxonomy of the subfamily. Zool. Scr. 12: 153–160.

Development of the reproductive apparatus of the land planarian *Rhynchodemus sylvaticus* (Turbellaria: Tricladida) and its significance for classification in the genus

Robert E. Ogren
Biology Department, Wilkes College, Wilkes-Barre, PA 18766, U.S.A.

Keywords: Turbellaria, land planarians, copulatory organ development, systematics

Abstract

Seven specimens of *Rhynchodemus sylvaticus* (Leidy) collected from a variety of localities in the US and having variously developed copulatory organs are believed to represent stages in the development of the copulatory apparatus. Four specimens were juveniles with under-developed male components, one specimen had a well-developed female atrium and small male component, and two specimens were mature with a male organ twice the size of the female part. In early stages, the male component had sperm ducts, seminal vesicle, and narrow atrium; more mature stages had a considerable elongated atrium with thick folds in its muscularized wall, a massive muscular bulbus; and a sigmoid ejaculatory duct opening into the large bulbar cavity. Morphological features of mature male copulatory organs in all species of the genus *Rhynchodemus* are basically similar whereas external body features (color and number of dorsal stripes) of these same species differ.

Introduction

The genus *Rhynchodemus* is distinguished from other genera of the Rhynchodemidae in such major anatomical features as the presence of a single pair of large eyes well anterior, arrangement of longitudinal subepithelial musculature in bundles, and the rounded elongate body shape. The simple copulatory apparatus has taxonomically significant features such as the absence of a penis papilla and genito-intestinal connection. In searching for characters to distinguish species of *Rhynchodemus*, investigators have turned to the copulatory apparatus. Because specimens found in nature differ in degree of development, knowledge of developmental changes in the reproductive complex can be useful in determining degree of sexual maturity and in ensuring that differences due to developmental stage are not assumed to be taxonomically significant.

Materials and methods

Routine histological methods were used to prepare serially sectioned specimens stained with hematoxylin and eosin. Serial diagrams were made of the copulatory apparatus, and a single composite profile was reconstructed from these diagrams. The seven specimens selected for this study were collected over the past 32 years from various localities (Table 1). The external appearance considered typical of the genus was exemplified by RW-1, the least developed specimen. It had a body color of pale chocolate brown with darker pigmentation at the anterior third and two brown longitudinal stripes that join as a dark pigmented spot over the pharynx and continue posteriorly somewhat united. Darker pigmentation with indistinct lines was found in specimen RK-83.

Hydrobiologia 132, 175–180 (1986).

Table 1. Specimens.

Specimen & Data	Perserved body length	Distance between genital pore and mouth	Sexuality
RW-1, New Glarus Woods, Roadside, Wisconsin; August 10, 1952 (Ogren, 1955)	6.0	0.7	rudimentary
RY-3, Urbana, Illinois, July 1952 (Ogren, 1955)	9.0	1.4	more advanced
RK-83, Kingston, PA. May 14, 1983 (Ogren, 1983)	11.0	2.0	juvenile
RHY-57, Carlisle, PA. Sept, 1957 (Ogren, 1981)	(15 mm alive, moving) 11.0	1.8	juvenile
RC-3, Chautauqua, Co., N.Y.; June, 1954 (Ogren, 1957)	16.0	2.1	formed egg capsule
RHY-81, Chautauqua Co., N.Y.; July 11, 1981; (Ogren, 1983)	(19.mm alive, moving) 14.0	2.8	mature male organ
RHY-80, Kingstoon, PA. June 12, 1980; (Ogren, 1983)	(18 mm alive, moving) 14.0	3.0	mature male organ

Description of developmental stages

RW-1 had the least developed condition; (Fig. 1): the female part was larger than the male part (Fig. 1). Ovovitelline ducts were narrow, joining to form a common chamber lined by a cuboidal epithelium and not enclosed by glands. The narrow glandular duct was lined with columnar epithelium penetrated by abundant eosinophilous glands. The vagina had a small cavity lined with ciliated columnar epithelium and enclosed by weak musculature with single layers of inner circular and outer longitudinal fibers distributed loosely beneath the epithelium. The male organ was undifferentiated with a short narrow male atrium and weak musculature and without a seminal vesicle. The sperm ducts were incomplete; testes were absent.

RY-3 was larger than RW-1 but was still immature (Fig. 2). The female organ was elongate and broad, with some epithelial folds. The male organ was rudimentary, with a narrow atrium, weak musculature, and a small seminal vesicle receiving the sperm ducts.

RK-83 represents a still larger juvenile condition (Fig. 3). The voluminous vagina had extensive epithelial folds with well-developed musculature; the glandular duct was long. The male organ had a seminal vesicle and an elongated tubular atrium about 50% the length of the vagina, with small epithelial folds and weak musculature. RHY-57 was similar to RK-83.

RC-3 had produced an egg capsule, and its female part was considered mature (see Ogren, 1955). Restudy of this material resulted in a new interpre-

Figs 1–6. (Figs. 1–5: Reconstructed profiles of copulatory organs in *Rhynchodemus sylvaticus*; to same scale; based on serial sections. Scale line applies to all except Fig. 5): (1) Early juvenile (RW-1), without seminal vesicle; (2) Juvenile condition, slightly more advanced (RY-3); (3) Juveniles condition, more advanced (RK-83); (4) Female-mature; specimen had formed egg capsule (RC-3); (5) Male organ shows beginning of maturation. Profile redrawn from Hyman, 1943, Fig. 7). Drawn proportionally to scale of other Figures; (6) Male-mature condition (RHY-80 and RHY-81) redrawn from Ogren (1983, Fig. 2).
Abbreviations used in Figures: BM bulbus musculature; CC common chamber of joined ovovitelline ducts; CE cilitated epithelium of creeping sole; CG shell or eosinophilous cement glands; CM circular muscle layer; ED ejaculatory duct, ES eosinophile granule secretions in male organ; GP gonopore, GD glandular duct (Drusengang), LM longitudinal muscle layer, MA male atrium, OD ovovitelline duct, SV seminal vesicle, SD sperm duct or vas deferens, VG vagina or female atrium.

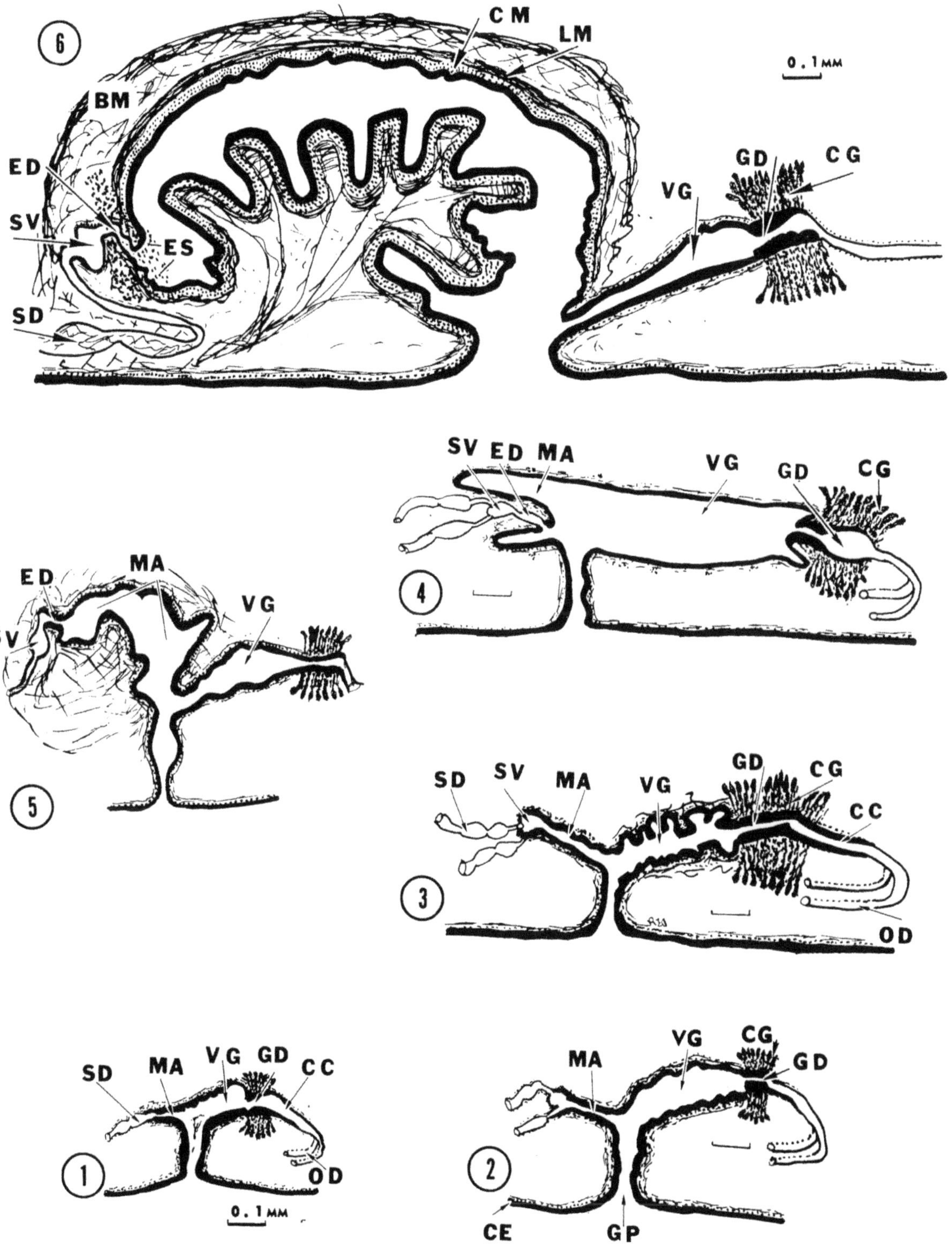
6
CM
LM
0.1MM
BM
ED
SV
ES
SD
VG
GD
CG
SV ED MA
VG
GD
CG
4
ED
MA
VG
SV
5
SD
SV
MA
VG
GD
CG
CC
3
OD
SD
MA
VG
GD
CC
OD
1
0.1MM
MA
VG
CG
GD
2
CE
GP

tation of copulatory-organ structure (Fig. 4). The female atrium, much larger than in the preceeding three specimens, had a long glandular duct obscured by secretions, and the inflated common chamber was enclosed by glands. The male organ was small, lacking heavy musculature and epithelial folds. The reduced ejaculatory duct opened into the chamber by a short fold which is not interpreted as a penis papilla.

Specimens RHY-80 and RHY-81 were male-mature (Fig. 6). The female organ in these was large and well-developed, with an elongated atrium but few epithelial folds. The male organ, described in detail elsewhere (Ogren, 1983), was twice the size of the female. The atrial walls and floor had great epithelial folds bulging into the chamber. The musculature consisted of thick sub-epithelial layers, a retractor system for ventral folds, and a massive bulbus enclosed by a fibrous tunic. A fold of tissue separated male and female chambers. The ejaculatory duct was long, sigmoid, and connected the seminal vesicle with the bulbar cavity. Abundant granular secretion from remote glands filled the atrium. This massive copulatory organ produced a distinct localized swelling of the body behind the pharynx.

The distance between genital pore and mouth was greater in more sexually mature specimens, in those with a more developed copulatory apparatus, and in longer-bodied specimens (Table 1). Large individuals about 16 mm in length were male-mature. Such individuals may have a visible swelling behind the pharynx near the gonopore.

Discussion

The presumed sequence of developing copulatory organs described above supplements the description of this organ by von Graff (1896, 1899). In *Rhynchodemus sylvaticus*, the male part matures later than the female part and has massive folds rather than a penis papilla. Comparison of available specimens shows progressive changes toward the mature condition. Female structures differentiate first, while the male remains undeveloped (Fig. 1). The female atrium enlarges, acquires extensive epithelial folds, heavier musculature, and an increase in glands along the glandular duct (Figs. 2, 3), with further enlargement during egg capsule formation (Fig. 4). The early male atrium is short, narrow, undifferentiated, and without a seminal vesicle (Fig. 1). It subsequently elongates (Fig. 2), and weak musculature and the seminal vesicle appear, the latter receiving the sperm ducts. Significant enlargement and muscularization accompany further maturation (Figs 3, 5, 6).

A nearly mature condition is depicted in Fig. 5, reproduced from Hyman (1943). The female organ is well-developed with spacious vagina and narrow glandular duct. The male organ is equal in length to the female part, epithelial folds are prominent, a strong muscular bulbus is present, an epithelial fold separates male and female chambers, and a short ejaculatory duct connects the seminal vesicle with the bulbar cavity.

The following features are acquired gradually and serve as criteria for male maturity: 1) an epithelial fold separates male from female chambers; 2) massive epithelial folds arise from the atrial floor and walls; 3) eosinophilous secretions appear in the epithelium and lumen of the ejaculatory duct and bulbar cavity; 4) subepithelial muscle layers thicken and multiply; 5) a massive muscular bulbus, enclosed by a fibrous tunic appears; 6) the prominent sigmoid ejaculatory duct connects the seminal vesicle with the large bulbar cavity; and 7) the male organ becomes twice the length of the female part.

If one arranges known underdeveloped and juvenile conditions in species of *Rhynchodemus* according to degree of maturity, the following groupings emerge. 1) Male organ largely absent, female organ present: *R. piptus* Marcus, 1952. 2) Male organ rudimentary with small atrium and seminal vesicle, female organs well-developed: *R. americanus* Hyman, 1943; *R. ijimai* Kaburaki, 1922; *R. misus* du Bois-Reymond Marcus, 1955; *R. pellucidus* Graff, after Heinzel, 1929, and Marcus, 1952; *R. scius* du Bois-Reymond Marcus, 1955; *R. sylvaticus* (Leidy, 1851) after Ogren, 1955, 1957. 3) Male organ moderately developed about equal to length of well-developed female organ: *R. aripensis* Prudhoe, 1949a; *R. blainvillei* Graff, after Heinzel, 1929; *R. oahuensis* Hyman, 1939; *R. putzei* Graff, after Prudhoe, 1949b; *R. sylvaticus* (Leidy) (= *bilineatus*) after Hyman, 1954, Ball and Reynoldson, 1981. For the above species the male-mature condition is known only for *R. putzei*, *R. scius* and *R. sylvaticus*.

Knowledge of the copulatory apparatus in *Rhynchodemus* can be applied to classification. Both juvenile and mature conditions have characters that allow one to distinguish the subfamily Rhynchodeminae from the Microplaninae and that allow one to distinguish *Rhynchodemus* from *Dolichoplana* or *Platydemus*. Diagnosis at the species level, however, depends on knowledge of external features as well as of copulatory-organ structure. Characters of juvenile conditions, while helpful, are not sufficient since only the female part is developed in such conditions. On the other hand, the taxonomic utility of the male-mature condition of the copulatory organ may also be limited as there are uncertainties in using differences and similarities that are found. Apparent differences can be due to mechanical distortions, level of development or function, or differences in their diagramatic representation by different authors. Nevertheless, unique characters have been identified – for example in *R. nematopsis* de Beauchamp, 1930, which has a massive elongate bulbus many times the length of the female part and a long narrow exit canal, features not found in other species, and in *R. scius* which has a type of false penis papilla in a late juvenile condition.

An additional problem is that fundamentally similar male-mature copulatory organs are found in species having very different external features in terms of color and number of dorsal stripes (du Bois-Reymond Marcus, 1955). Morphologically similar (but not identical) male organs are found, for example in *R. bromelicola* de Beauchamp, 1912 (with one median stripe), in *R. schuberti* du Bois-Reymond Marcus, 1955, and *R. sylvaticus* (Leidy) (both with two stripes), in *R. ochroleucus* Graff and *R. schmardaei* Graff (with three stripes), and in *R. hectori* Graff (with four stripes). Thus, although it is difficult to recognize unique features in mature copulatory organs of these species, they can be distinguished on the basis of external appearance, in conjunction with morphology of the mature copulatory apparatus and other histological features.

In conclusion, accurate diagrams or photographs of external features as well as accurate reconstructions of the mature copulatory apparatus are essential for species description. And knowledge of development can be useful in recognizing juvenile features of the copulatory apparatus.

Acknowledgments

The assistance of Wilkes College grants-in-aid for research, the cooperation of the Eugene Shedden Farley Library staff, the resources of previous National Science Foundation grants, consultations, in Summer of 1983, with Dr. Ian R. Ball, Institute of Taxonomic Zoology, University of Amsterdam, Netherlands, and Nederlandse organisatie voor zuiver-wetenschappelijk onderzoek, are gratefully appreciated and hereby acknowledged.

References

Ball, I. & Reynoldson, T. B., 1981. British Planarians, Synopsis of British Fauna No. 19, Cambridge University Press, London, 141 pp.

de Beauchamp, P. M., 1912 (1913). Planaires de Bromeliacees de Costa Rica (2 note). Arch. Zool. exp. gen., 51 (Notes et Revue no 2): 41–52.

de Beauchamp, 1930. Triclades Terricoles. Arch. Hydrobiol. Suppl. 8: 72–91.

du Bois-Reymond Marcus, E., 1955. On Turbellaria and Polygordius from the Brazilian coast. Bol. Fac. fil. Cien. Letr., Zool. 20: 19–65.

Heinzel, L., 1929. Zur Kenntnis der Rhynchodemiden. Zool. Jb. Syst. Ökol. Geogr. Tiere 56: 425–462.

Hyman, L. H., 1939. Land planarians from Hawaiian Islands. Arch. Zool. exp. gen. 80: (Notes et Revue, No. 3): 116–124.

Hyman, L. H., 1943. Endemic and exotic land planarians in the United States with a discussion of necessary changes in the Rhynchodemidae. Am. Mus. Novitates 1241: 1–21.

Hyman, L. H., 1954. Some land planarians of the United States and Europe, with remarks on nomenclature. Am. Mus. Novitates 1667: 1–21.

Kaburaki, T., 1922. On the terrestrial planarians from Japanese territories. J. Coll. Sci. imp. Univ. Tokyo 44: 1–54.

Leidy, J., 1851. Corrections and additions to former papers on helminthology published in the proceedings of the academy. Proc. Acad. nat. Sci., Philad. 5: 289.

Marcus, E., 1952. Turbellaria Brasileiros (10). Bol. Fac. Fil. Cien. Letr., Zool. 17: 5–187.

Marcus, E., 1953. Turbellaria Tricladida. In Exploration du Parc National de l'Umpemba, Mission de Witte, Brussels, Institute des Parcs nationaux du Congo Belge, Fasc. 21, 62 pp.

Ogren, R. E., 1955. Ecological observations on the occurrence of Rhynchodemus, a terrestrial turbellarian. Trans. am. microsc. Soc. 74: 54–60.

Ogren, R. E., 1957. Developmental observations, the egg capsule and sexual maturity of the land planarian Rhynchodemus sylvaticus. Trans. am. microsc. Soc. 76: 45–52.

Ogren, R. E., 1981. Land planarians in Pennsylvania. Proc. Pa Acad. Sci. 55: 52–56.

Ogren, R. E., 1983. Reproductive complex and sexual maturity of the land planarian Rhynchodemus sylvaticus (Leidy) in

the United States. Proc. Pa Acad. Sci. 57: 69–73.
Prudhoe, S., 1949a. Some roundworms and flatworms from the West Indes and Surinam. IV. Land planarians. J. linn. Soc., Zool. 41: 420–433.
Prudhoe, S., 1949b. Some land planarians from the Marquesas Islands. Occ. Pap. Bernice P. Bishop Mus. 19(13): 247–255.
von Graff, L., 1896. Ueber die Morphologie des Geschlechtsapparates der Landplanarien. Verh. dt zool. Ges. 6: 75–93.
von Graff, 1899. Monographie der Turbellarien. II. Tricladida Terricola. Vol. I, Leipzig.

Cell migration and differentiation during wound healing and regeneration in *Microstomum lineare* (Turbellaria)

Irmeli Palmberg
Institute of Biology, Åbo Akademi, 20500 Åbo, Finland
Mailing address: *Faculty of Education, Åbo Akademi, Kyrkoesplanaden 12–14, 65100 Vasa, Finland*

Keywords: Turbellaria, wound healing, regeneration, [^{3}H]T-autoradiography, differentiation, Macrostomida

Abstract

Using transmission electron microscopy and serial sections with light-microscopic autoradiography, I have investigated the ultrastructure of wound healing, the distribution of cells preparing for proliferation, and the fates of cells labelled with exogenous tritiated thymidine ([^{3}H]T) in *Microstomum lineare* undergoing wound healing and regeneration. Immediately after decapitation the open wound was reduced to a minimum by strong contraction of circular muscle fibers. The wound epidermis was cellular, consisting of thin parts of epidermal cells from the epidermis around the wound. These epidermal cells maintained close adhesive contact with one another through zonulae adherentes and septate junctions. No proliferating cells were found in the old epidermis. The only cells taking up [^{3}H]T were mesenchymal and gastrodermal neoblasts which proliferated and migrated towards the surface. The final epidermis was formed by conjunction of the wound epidermis and newly differentiated epidermal cells. Regeneration in *Microstomum*, in contrast to that of planarians, occurs mainly by morphallaxis, without the formation of a regeneration blastema, but also through continuous cell proliferation, migration, and differentiation.

Introduction

Regeneration in turbellarian flatworms involves a combination of morphallactic and epimorphic events (Hyman, 1951). Early histological investigations, mostly on planarian regeneration (see review by Brøndsted, 1969), attempted to identify and describe the cells participating in wound healing and blastema formation. More recently, the ultrastructure of cells participating in regeneration (Pedersen, 1972; Morita & Best, 1974; Hay & Coward, 1975), the process of wound closure and wound healing (Hauser, 1971; Spiegelman & Dudley, 1973; Pedersen, 1976; Chandebois, 1980) and the role of cell autolysis and deletion (Bowen *et al.*, 1982) have been examined through various microscopical, cytochemical, and biochemical techniques. Experiments dealing with mitotic counts (Baguña, 1976) and with cytophotometric analysis of total DNA content (Gremigni *et al.*, 1980) or measurements of DNA, RNA, and proteins by ^{32}P (Martelly *et al.*, 1981) have also been applied to regenerating planarians. There have been no investigations performed, however, with the technique of tritiated-thymidine labelling ([^{3}H]T) for detection of the distribution and subsequent fate of proliferative cells during regeneration, largely because this technique is not applicable to triclads (see e.g. Hay & Coward, 1975). Moreover, very little work has been done on other (non-triclad) turbellarians with high regeneration power.

The present investigation, performed with the microturbellarian *Microstomum lineare*, combines light and electron microscopic studies with [^{3}H]T-autoradiography of regenerating specimens. These methods, supplemented with observations on living regenerating specimens, reveal new information about the process of wound healing and regeneration in turbellarians.

Hydrobiologia 132, 181–188 (1986).

Material and methods

Specimens of *M. lineare* (Müller, 1774) (Order Macrostomida) were caught in fresh water at Teijo, Bjärnå (SW Finland). Approximately 100 worms were decapited just posterior to the eyespots and exposed for 30 min to tritiated thymidine ([6-^{3}H] thymidine, spec. act. 5 Ci/mmol) in a sterilized medium of a concentration of 90 μCi/ml and washed thoroughly in sterilized water before fixation. Intact worms, serving as controls, were labelled simultaneously. Specimens were fixed in 2.5% glutaraldehyde and 1% OsO_4 in collidine buffer (Palmberg *et al.*, 1980) at intervals of 1/2, 1, 3, 6, 12, 25 h of regeneration time. Subsequent steps in light microscopic autoradiography followed the method described by Palmberg & Reuter (1983). Ultrathin sections were cut from the same blocks as those used for LM autoradiography.

Depending on the various contraction states of the unanaesthetized worms, the exact level of amputation was difficult to define, resulting in minor differences in the constitution of the amputated piece and the cut surface. The chief difference was the presence or absence of the caecum, the apical part of the intestine, which in a contracted state is cut off with the head piece.

Results

Control tissues

Ultrastructural characteristics of the intact epidermis of *M. lineare* conformed in general to descriptions of the epidermis of Macrostomida as presented by Bedini & Papi (1974), Rieger (1981), and Tyler (1984). The free surface of the epithelial cells bears cilia intermingled with microvilli. The cytoplasm is subdivided into a thin apical portion and a more extensive basal one by a terminal web. Cytoplasmic organelles, except the ciliary basal bodies, ultrarhabdites, and some vacuoles, are confined to the basal region. The basal part of the epidermal cells lies, without any basement membrane, on the subepidermal muscle cells which are surrounded by fine granular or fibrous extracellular matrix.

Two main types of mesenchymal cells could be discerned. The first type, corresponding to the planarian fixed parenchymal cell, appeared together with the extracellular matrix as a loose net-like tissue. The second cell type is an undifferentiated basophilic cell like stem cells (Baguña, 1981) or neoblasts (Pedersen, 1972) of planarians. Such neoblasts occur sparsely in the highly differentiated, intact head piece but are more abundant in the postpharyngeal part. These cells, as well as neoblast-like cells in the gastrodermal tissue, were the only cells incorporating [^{3}H]T. In addition, cells possessing a bundle of centrioles in their cytoplasm, but otherwise similar to neoblasts, were found in the mesenchyme.

The nervous tissue (Reuter *et al.*, 1980) and epidermal eyespots (Palmberg *et al.*, 1980) have been described previously. In this investigation, I used the appearance of eyespots as a criterion of completed regeneration.

Initial wound healing (0–1 h)

Immediately after decapitation, the open wound was reduced to a minimum by strong contraction of circular muscle fibers. Cell fragments, broken membranes in the form of fibrillar strands, released secretory material, and unidentifiable cell debris were observed at the surface of the wound. At this stage, it was difficult to distinguish between cells participating in wound healing and necrotic cells. One hour later, more stretching of the marginal regions of the epidermal cells adjacent to the wound was apparent, and a thin wound epidermis with widely spaced cilia and microvilli covered the whole wound area. Cells near the wound were ovoid with extreme elongations and cytoplasmic processes extending towards the regenerating surface (Fig. 1). Neither these cells nor the wound epidermis were labelled 1/2 h after treatment with [^{3}H]T. In fact, there were only a few labelled cells in the prepharyngeal region at this time. An abundance of labelled cells, on the other hand, were found behind the pharynx in the mesenchyme and gastrodermis (Fig. 2).

Three h after decapitation

Two h later, the wound borders were still uneven, but no protruding cells were observed on the surface. The wound epidermis was cellular, consisting of thinly stretched epidermal cells. They maintain a close adhesive contact with one another through

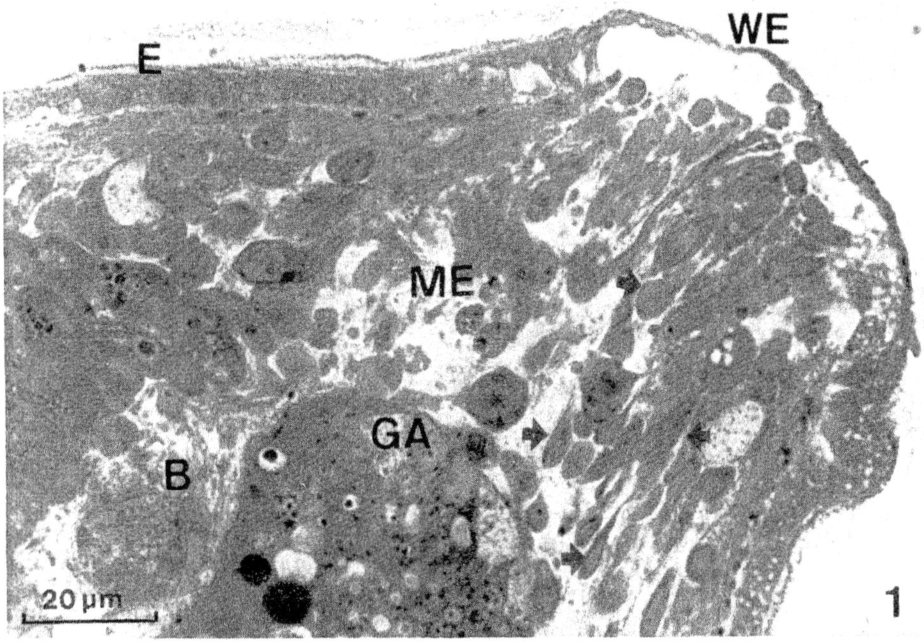

Fig. 1. M. lineare. Wound epidermis in 1-h regenerate. Arrows indicate elongate cells, possibly migrating. Abbreviations: B–brain; E–epidermis; GA–gastrodermis; ME–mesenchyme; MU–muscle; N–nucleus; PH–pharynx; WB–wound bulge; WE–wound epidermis.

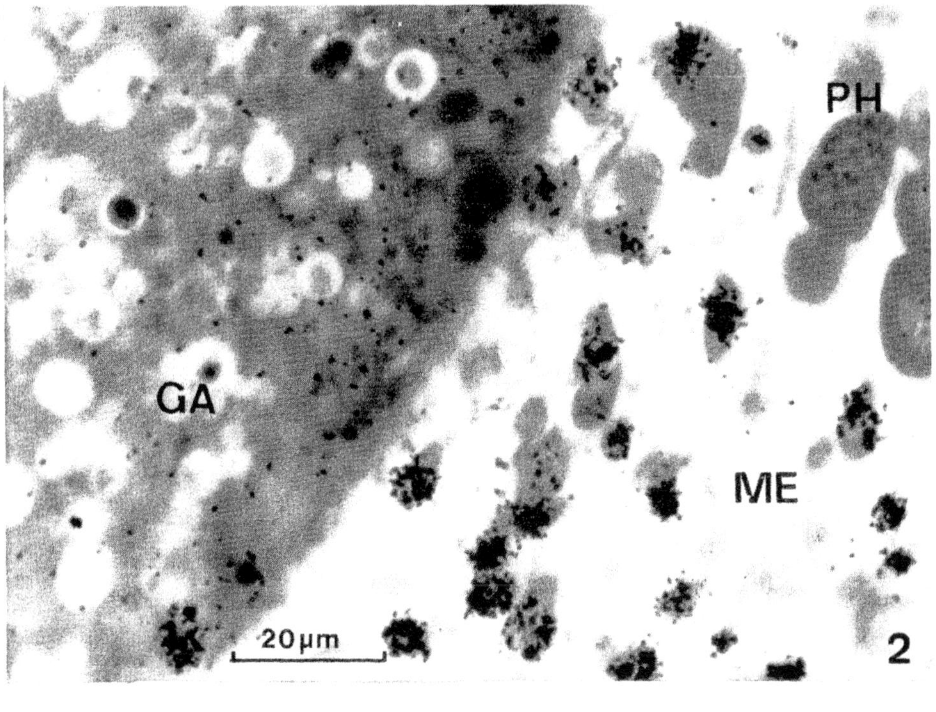

Fig. 2. Labelling in mesenchyme and gastrodermis of postpharyngeal region at 1 h regeneration time. Abbreviations see Fig. 1.

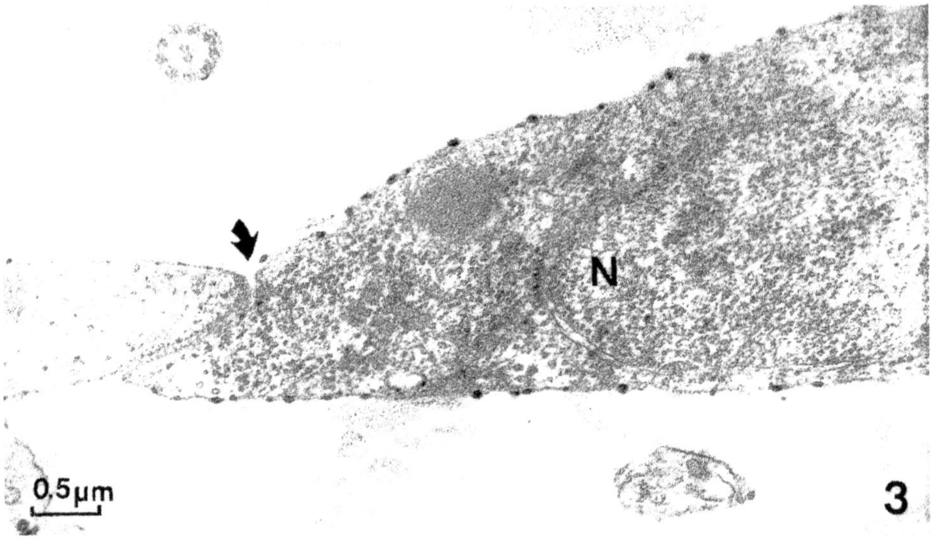

Fig. 3. Septate junction (arrow) between stretched epidermal cells at 3 h regeneration time. Abbreviations see Fig. 1.

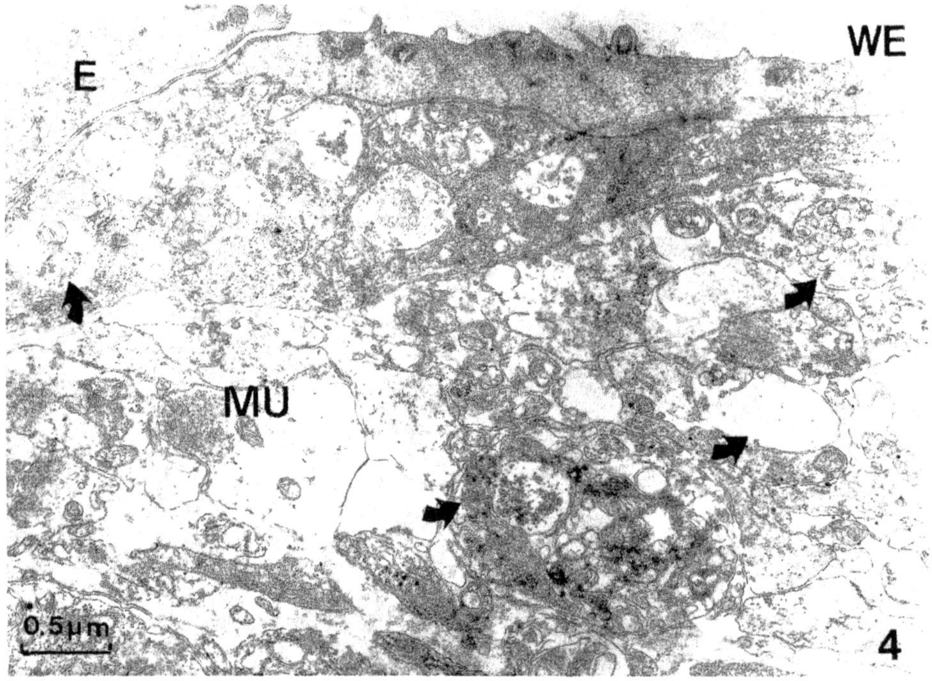

Fig. 4. Elongated epidermal cell at wound at 3 h regeneration time. Arrows indicate autophagosomes. Abbreviations see Fig. 1.

zonulae adherentes and septate junctions (Fig. 3). The cytoplasm of the wound epidermis contained few cell organelles. The nuclei of these cells were elongated and oriented with their long axes parallel to the wound surface. Only the epidermal cells adjacent to the wound exhibited changes in form and pattern. Cell autolysis and deletion near the wound borders appeared most commonly around 3 h regeneration time. Cell organelles – e.g. Golgi bodies, endoplasmic reticulum, and mitochondria – showed signs of degeneration. Several autophagosomes were observed in the elongated epidermal cells and in the adjacent subepidermal muscle layers (Fig. 4). These muscle cells probably became necrotic. Gland processes containing secretory granules and their cell bodies, as well as cells with bundles of centrioles and vacuoles in their cytoplasm were found between the wound epidermis and mesenchymal tissues where the extracellular space was enlarged. No aggregation of cells like that of a planarian blastema existed near the wound. In addition, there was no sign of labelling in the wound epidermis or in the old epidermal cells, although labelling was clearly discernible in the mesenchyme around and posterior to the pharynx.

Six h after decapitation

An interesting morphological formation, which I designate a 'wound bulge' (Fig. 5), characterizes the wound surface at this stage. This formation, located only on one side of the wound, consisted of a dilated space between the thin, but still continuous, wound epidermis and the regenerating tissues in front of the pharynx. Several kinds of cells undergoing lysis appeared in this space. Adjacent to the space, on the other hand, lay cells of uncertain identity at varying stages of differentiation. Several of the cells containing bundles of centrioles were found in the epidermis near the wound bulge. These new epidermal cells lacked cilia and microvilli, but the centrioles were located immediately beneath the plasmalemma and served as basal bodies of developing cilia.

The brain, if only injured by amputation, underwent a complete regeneration. Labelled cells appeared around the neuropile. No aggregation of cells occurred at this time of regeneration, rather the cells were loosely situated in large intercellular spaces as in intact worms.

Twelve to twenty-five h after decapitation

In some cases, the wound bulge could still be seen 12 h after decapitation, but usually the remainder of the degenerated wound epidermis had fallen off. The new epidermis formed a continuous layer consisting of newly differentiated epidermal cells intermixed with old, reorganized epidermal cells. The newly differentiated cells could be distinguished from old ones by their less-dense cytoplasm and extensive Golgi bodies. Numerous cilia and microvilli occurred on the surface, and terminal web appeared between the zonulae adherentes of the new epidermal cells. About 10 h

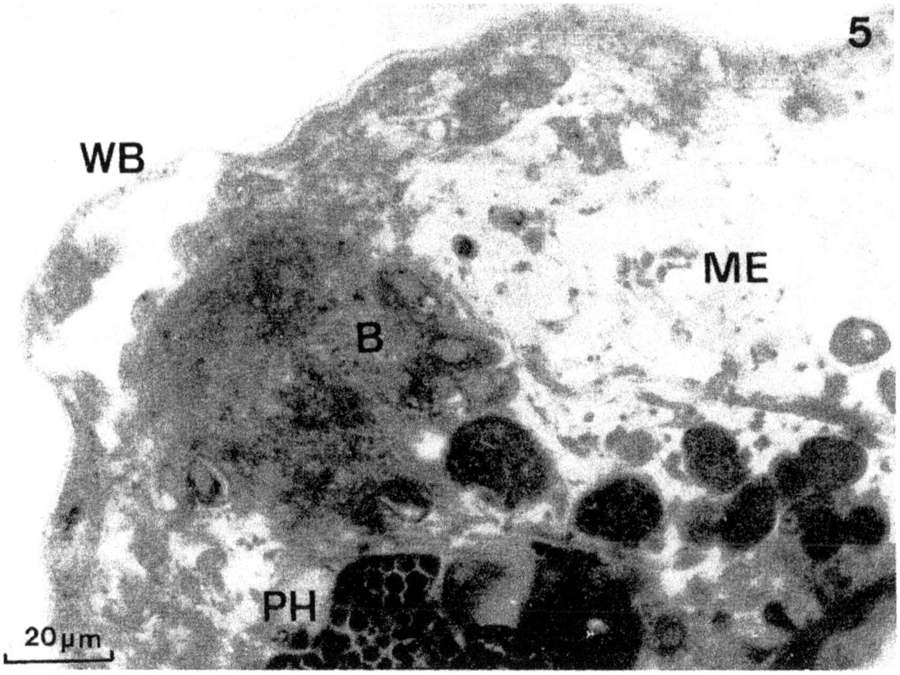

Fig. 5. Wound bulge at 6 h regeneration time. Abbreviations see Fig. 1.

later, gland and sensory processes formed junctional complexes with the epidermal cells and became incorporated into the epidermis. Characteristic features of differentiating muscle cells were recognized. Labelling, as well as dense extranuclear chromatin aggregates, was observed in subepidermal muscle layers at 25 h.

Eye differentiation was initiated simultaneously with the differentiation of muscles. The pigment of the eyespots became visible at 22 h. Developing pigment cells had immature pigment granules in the vicinity of the Golgi apparatus as well as several microtubules and many vacuoles. They were, however, unlabelled. Differentiated photoreceptor cells were not found at this time. Nor could differentiating photoreceptor cells be identified with certainty due to their confusing similarity to epidermal cells containing centrioles.

Discussion and conclusions

Good regeneration capacity is closely related to asexual reproduction and decreases with the appearance of sexuality (Hyman, 1951; Brønsted, 1969). Wound healing and regeneration of the head piece are very rapid processes in *M. lineare*, taking 25 h on an average. In other turbellarians, time periods varying from 6 h to 3 days have been reported for wound closure (Pedersen, 1972). For regeneration of the head, reported time periods range from 60 h in *Catenula* (Moraczewski, 1977) up to 22 days in planarians (Teshirogi, 1985). The size of the wound and the extent of the wound closure are, however, extremely variable and dependent on many factors affecting the whole regeneration process (Pedersen, 1972).

Wound epidermis

The initial wound closure involves mostly passive stretching of the pre-existing epidermal cells as a consequence of the strong contraction of the muscle layers. While active cell migration must be involved in the rapid (within 1 h) completing of the wound epidermis, such migration does not play as large a role as reported in planarians by Hori (1978). The wound epidermis serves as a shelter for the developing tissues, degenerates gradually as the new surface closes around it, and eventually falls off. Before full degeneration and falling off, it forms a structure, the wound bulge, which may serve as a site for cell autolysis. Phagocytosis of cell fragments and cell debris, as well as cell autolysis in wound healing of other turbellarians, result in products assumed to be recycled (Bowen *et al.*, 1982). A wound bulge has not been reported, however, in other regenerating turbellarians. In *Catenula*, a space filled with fluid has been observed between the intestine and the wound epidermis as a result of lysis (Moraczewski, 1977), and it could be homologous to the wound bulge. In addition, the regeneration of *Catenula* and *Microstomum* are similar insofar as both lack a regeneration blastema which in planarians (Pedersen, 1972) is formed immediately under the wound epidermis. The final epidermis is built under the wound epidermis by a process of reorganization of old tissues and by migration of new cells from the mesenchyme.

Migration

The aim of the present investigation was to follow the fates of cells labelled immediately after the amputation, i.e., their probable migration and differentiation during the regeneration of the head. Only cells preparing for proliferation at the moment of the decapitation and during the following 1/2 h were followed; variations in spatial and temporal labelling during regeneration will be the subject of another investigation.

In *M. lineare*, it is evident that neoblasts migrate from the postpharyngeal parts of the body to participate in regeneration of the head. In planarians, on the other hand, pre-existing local neoblasts have been observed to proliferate to a degree sufficient to form the regeneration blastema (Baguña, 1976); but also cell migration from other parts of the body and independent of local proliferation has been reported (Gabriel, 1970; Banchetti & Gremigni, 1973; Gremigni & Miceli, 1980). Furthermore, the possibility of both local proliferation and migration taking place has been recognized (Pedersen, 1972; Slack, 1980; Gremigni *et al.*, 1982). Migration in *M. lineare* is indicated by an increase in labelled cells (neoblasts) in the prepharyngeal region during regeneration and by the appearance of cytoplasmic processes extending towards the wound as is reported for planarian regeneration by Betchaku (1967), Morita & Best (1974), and Pederson (1976). The dilation of the intercellular spaces also suggests migration (*cf.* Spiegelman & Dudley, 1973).

Differentiation

The only cells capable of proliferation in intact as well as in regenerating *M. lineare* are the mesenchymal and gastrodermal neoblasts. The notion of neoblast has been widely discussed (see review by Pedersen, 1972). In a recent paper, Baguña (1981) considered neoblasts to be the stem cell of most differentiated cells. The same conclusion can be drawn in the present study. Continuously proliferating neoblasts, independent of any growth zones or axial gradients (Palmberg & Reuter, 1983), replenish the population of stem cells available for regeneration and asexual reproduction. Thus dedifferentiation is not necessary. Only a fraction of the population of proliferative cells could be followed by the methods I used; therefore, only two cell types were observed to differentiate from neoblasts, namely nerve cells in the brain and body-wall muscles. Neither those cells already differentiating nor those undifferentiated cells not dividing before differentiation were labelled, however. This, together with morphallactic processes, may explain the absence of labelled cells anterior to the pharynx during the first hours of regeneration. If a great number of undifferentiated cells are in the G2 state of the cell cycle, as supposed by Baguña (1976), it would mean that this cell population, too, is unlabelled. This can be controlled in future experiments, for example, by analysis of mitosis. Results of ultrastructural autoradiography and further pulse-labelling experiments will, I expect, provide more evidence for the totipotency of neoblasts.

References

Baguñà, J., 1976. Mitosis in the intact and regenerating planarian Dugesia mediterranea n. sp. II. Mitotic studies during regeneration and a possible mechanism of blastema formation. J. exp. Zool. 195: 65–80.

Baguñà, J., 1981. Planarian neoblasts. Nature 290: 14–15.

Banchetti, R. & V. Gremigni, 1973. Indirect evidence for neoblast migration and for gametogonia dedifferentiation in exfissiparous specimens of Dugesia gonocephala s.l. Accad. naz. Lincei 55: 107–115.

Bedini, C. & F. Papi, 1974. Fine structure of the turbellarian epidermis. In N. W. Riser & M. P. Morese (eds), Biology of the Turbellaria. McGraw-Hill, N.Y.: 108–147.

Betchaku, T., 1967. Isolation of planarian neoblasts and their behavior in vitro with some aspects of the mechanism of the formation of regeneration blastema. J. exp. Zool. 164: 407–434.

Bowen, I. D., J. E. den Hollander & G. H. J. Lewis, 1982. Cell death and acid phosphatase activity in the regenerating planarian Polycelis tenuis Iijima. Differentiation 21: 160–167.

Brønsted, H. V., 1969. Planarian regeneration. Pergamon Press, London.

Chandebois, R., 1980. The dynamics of wound closure and its role in the programming of planarian regeneration. II – Distalization. Develop. Growth Differ. 22: 693–704.

Gabriel, A., 1970. Étude morphologique et evolution biochemnique des néoblastes au cours des premières phases de la régénération des planaires d'eau douce. Ann. Embryol. Morphogen. 3: 49–69.

Gremigni, V. & C. Miceli, 1980. Cytophotometric evidence for cell 'transdifferentiation' in planarian regeneration. Wilhelm Roux Arch. dev. Biol. 188: 107–113.

Gremigni, V., C. Miceli & E. Picano, 1980. On the role of germ cells in planarian regeneration: II. Cytophotometric analysis of the nuclear Feulgen – DNA content in cells of regenerated somatic tissues. J. Embryol. exp. Morph. 55: 65–76.

Gremigni, V., M. Nigro & I. Puccinelli, 1982. Evidence of male germ cell redifferentiation into female germ cells in planarian regeneration. J. Embryol. exp. Morph. 70: 29–36.

Hauser, J., 1971. Non-cellular regeneration processes in the integument of the flatworm Geoplana abundans. Oncology 25: 258–268.

Hay, E. D. & S. J. Coward, 1975. Fine structure studies on the Dugesia. I. Nature of the 'neoblast' and other cell types in noninjured worms. J. ultrastruct. Res. 50: 1–21.

Hori, I., 1978. Possible role of rhabdite-forming cells in cellular succession of the planarian epidermis. J. electron Microsc. 27: 89–102.

Hyman, L. H., 1951. The Invertebrates, Vol. II. Platyhelminthes and Rhynchocoela. McGraw-Hill, N.Y., 551 pp.

Martelly, I., C. Rey & A. Le Moigne, 1981. Planarian regeneration: DNA metabolism in adults. Int. J. invert. Reprod. 4: 107–121.

Moraczewski, J., 1977. Asexual reproduction and regeneration of Catenula (Turbellaria, Archoophora). Zoomorphologie 88: 65–80.

Morita, M. & J. B. Best, 1974. Electron microscopic studies of planarian regeneration. II. Changes in epidermis during regeneration. J. exp. Zool. 187: 345–374.

Palmberg, I. & M. Reuter, 1983. Asexual reproduction in Microstomum lineare (Turbellaria). I. An autoradiographic and ultrastructural study. Int. J. invert. Reprod. 6: 197–206.

Palmberg, I., M. Reuter & M. Wikgren, 1980. Ultrastructure of epidermal eyespots of Microstomum lineare (Turbellaria, Macrostomida). Cell Tissue Res. 210: 21–32.

Pedersen, K. J., 1972. Studies on regeneration blastemas of the planarian Dugesia tigrina with special reference to differentiation of muscle-connective tissue filament system. Wilhelm Roux Arch. devl Biol. 169: 134–169.

Pedersen, K. J., 1976. Scanning electron microscopical observations on epidermal wound healing in the planarian Dugesia tigrina. Wilhelm Roux Arch. devl Biol. 179: 251–273.

Reuter, M., M. Wikgren & I. Palmberg, 1980. The nervous system of Microstomum lineare (Turbellaria, Macrostomida) I. A fluorescence and electron microscopic study. Cell Tissue Res. 211: 31–40.

Rieger, R. M., 1981. Morphology of the Turbellaria at the ultrastructural level. In E. R. Schockaert & I. R. Ball (eds), The Biology of the Turbellaria, Hydrobiologia 84: 213–229.

Slack, J. M. W., 1980. The source of cells for regeneration. Nature 286: 760.

Spiegelman, M. & P. L. Dudley, 1973. Morphological stages of regeneration in the planarian Dugesia tigrina: A light and electron microscopic study. J. Morph. 139: 155–184.

Teshirogi, W., 1985. On the origin of neoblasts in freshwater planarians. This volume.

Tyler, S., 1984. Platyhelminthes Turbellaria. In J. Bereiter-Hahn, A. G. Matoltsy & K. S. Richards (eds), Biology of the integument. Springer-Verlag, Berlin, New York: 112–131.

Physiological regeneration of the digestive parenchyma in *Convoluta convoluta* and *Oxyposthia praedator* (Turbellaria, Acoela)

I. M. Drobysheva
Laboratory of Evolutionary Morphology, Zoological Institute, Academy of Sciences, Leningrad, 199034, USSR

Keywords: Turbellaria, Acoela, histology, cell proliferation

Abstract

Autoradiography has been applied to two acoel turbellarians, *Convoluta convoluta* and *Oxyposthia praedator*, to determine the distribution and fate of proliferative cells. In *C. convoluta*, mitotic figures and nuclei that labelled with [^{3}H]thymidine could be observed in the peripheral parenchyma but not in the middle zone of the central parenchyma. The time required for regeneration of physiologically competent digestive cells was about 10–15 days. In *O. praedator*, mitotic figures (in metaphase and telophase) were observed in the peripheral parenchyma while none were found in the epidermis either in untreated animals or after treatment with colchicine. Mitotic figures were found only rarely in the central parenchyma and only in its marginal zone. Autoradiographs of *O. praedator* demonstrated [^{3}H]thymidine incorporation into both the nuclei and the cytoplasm of peripheral parenchymal cells. In the central parenchyma, no nuclei with primary labelling were observed. The digestive parenchyma of the acoels is regarded as a unique histological system involving both specialized cells of the central parenchyma and stem cells located in the peripheral parenchyma.

Introduction

An elucidation of the peculiarities of the organization of the acoel digestive parenchyma is of interest in terms of the comparative study of the digestive system in lower metazoans and is important for determining the phylogenetic status of the Acoela among other turbellarian groups. Complete study of any tissue system should include an analysis of three of its qualities: structure, function, and proliferation. The last item involves proliferation of cells at the expense of which tissue growth occurs (or has occurred in early ontogeny) and renewal of tissues for maintenance. The question of to what extent the central parenchyma of acoel turbellarians may be regarded as a homologue of the endodermal digestive epithelium of other animals, including other turbellarians, cannot be solved completely only on the basis of study of static tissue structure; the sources for regeneration of digestive cells must also be identified and located. Thus Smith (1981), in establishing the central syncytium and wrapping cells as an 'anatomical unit' in *Convoluta* sp., performed an excellent morphological analysis without assessing the proliferative potentialities of the digestive parenchyma, in particular of its wrapping cells. Therefore, we cannot exclude the possibility that the wrapping cells may originate from undifferentiated (though not yet identified) cells of the peripheral parenchyma.

Thymidine autoradiography, which allows labelling and tracing the fate of proliferating cells, is a tool I have applied recently to analysis of the digestive parenchyma in the Acoela (Drobysheva, 1976, 1978, 1982, 1983). The data obtained have been used previously for phylogenetic discussion by Ivanov and Mamkaev (Ivanov & Mamkaev, 1977; Mamkaev, 1979).

On the basis of the results I present here, I compare the digestive parenchyma of the Acoela with

Hydrobiologia 132, 189–193 (1986).

intestinal epithelia of other metazoans in terms of cell replacement.

Materials and methods

Morphological and autoradiographic studies were made of two acoel turbellarians: *Convoluta convoluta* (Abildgaard 1806) Graff, 1902, from the White Sea and *Oxyposthia praedator* A. Ivanov, 1952, from the Sea of Japan. Tritiated thymidine in doses of 0.2 and 0.1 MBq ml^{-1} and colchicine at 10^{-5} g ml^{-1} were used. For autoradiographic analysis, specimens of *O. praedator* were placed in the isotope for 2–3 h and then were fixed immediately. Specimens of *C. convoluta* were dosed with [^{3}H]thymidine over 10 d, 3 h daily and fixed in groups, some after the first incubation, others after every 5 d through the 37th day after the beginning of the experiment (see Drobysheva, 1983). Autoradiographs were prepared from sectioned animals.

To compare the peripheral parenchyma with the central parenchyma, an index of labelled nuclei in the post-oral part of the body was calculated for 1 000 cells in each region. Statistical comparisons were made with the White nonparametric criterion (a = 0.05). The mitotic index in *C. convoluta* was determined for 5 000 cells, and in *O. praedator* for 1 500–3 000 cells of the peripheral parenchyma of each animal. To estimate the distribution of mitotic figures in the oral and post-oral regions of *O. praedator*, all cells were examined in every third cross-section (6 μm section thickness).

Results and discussion

C. convoluta and *O. praedator*, like other acoels, have their parenchyma characteristically divided into peripheral and central regions (von Graff, 1891). The peripheral parenchyma comprises various types of cells; and in the central parenchyma, one type of strongly ramifying cell predominates (Pedersen, 1964; Mamkaev, 1967; Markosova, 1982). It is expedient to distinguish in the central parenchyma a marginal (along the border with the peripheral parenchyma) and a middle region.

In *C. convoluta*, the cell composition of the marginal zone of the central parenchyma was relatively heterogeneous; the middle region consisted only of digestive elements. The chromatin in the nuclei of individual digestive cells was not stained by methyl green, gallocyanin, or by the Feulgen method. Some wrinkled, partly damaged nuclei appeared degraded.

Some mitotic figures and cells with [^{3}H]thymidine-labelled nuclei were observed in the peripheral parenchyma throughout the body. Among five individuals, the average mitotic index in the peripheral parenchyma was 0.37%, ranging from 0.30% to 0.40% (see Fig. 1). In the autoradiographic experiment, the proportion of peripheral parenchymal nuclei with primary labelling after the initial 3-h incubation (LNI – labelled nuclei index) ranged from 6.7% to 10.3% and averaged 8.4% (Table 1). In the central parenchyma, neither mitotic figures nor nuclei with primary labelling were present in the middle zone; the LNI varied from 0.2% to 3.3% (1.5% average), all of the labelled cells being limited to the marginal zone. On the fifth day of the experiment, the LNI of the central parenchyma (1.6%) was also due entirely to labelled nuclei in its marginal zone. The statistically significant differ-

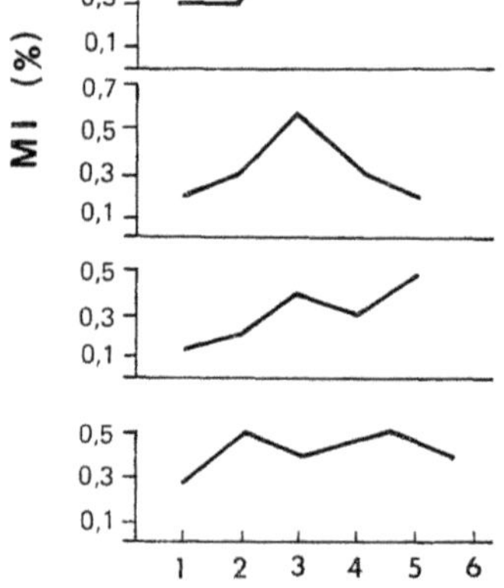

Fig. 1. Distribution of mitotic figures along the body of *C. convoluta* in the oral and post-oral zones (each plot is data for one animal; five animals total). Abscissa: zones along length of animal's body; ordinate: mitotic index (%).

ence between the LNI of the peripheral parenchyma and that of the central parenchyma on the first and fifth days of the experiment disappeared by the tenth day (Table 1, Fig. 2). Over the same time period, an absolute increase relative to the initial value was observed in the LNI of the central parenchyma. By the tenth day, some labelled digestive cells were found in the middle zone of the central parenchyma. Later (in 20, 25, and 30 days), the central parenchyma LNI in the majority of animals surpassed that of the peripheral parenchyma both in its average value and in individual indices (Fig. 2). But even the greatest average difference (observed on days 20 and 25) was not statistically significant. As the differentiated peripheral parenchymal elements – which developed from labelled stem cells – accumulated, the average value of the peripheral parenchymal LNI became higher than that of the central parenchyma (Fig. 2).

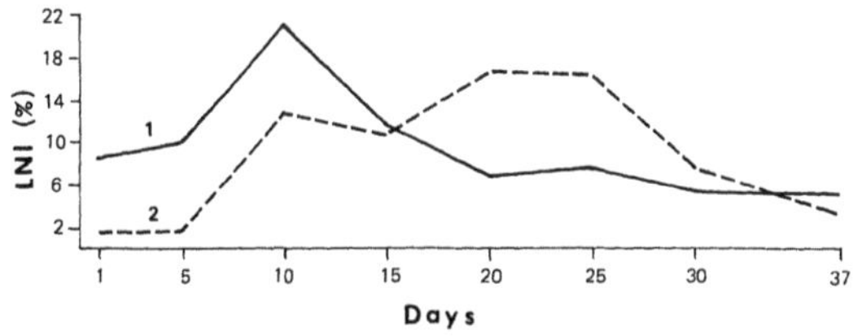

Fig. 2. Changes in LNI (labelled-nuclei index) in peripheral (1) and central (2) parenchyma of *C. convoluta* under conditions of 10 daily 3-h incubations in [^{3}H]thymidine solution, followed by maintenance in media without label. Abscissa: time between first [^{3}H]thymidine incubation and histological fixation (in days); ordinate: LNI (%). (See also Table 1 for statistical evaluation.)

In general, the intensity of labelling of nuclei in digestive cells was appreciably lower than in peripheral parenchymal cells. The presence of strongly labelled digestive elements indicates that some cells migrating into the central parenchyma had not undergone any division after the S phase; that is, these cells were polyploid. By the end of the experiment, the number of silver grains over nuclei of the digestive cells had decreased from that observed at the start. But in the peripheral parenchyma, a small number of heavily labelled cells remained. This may indicate that the proliferative cell populations are more heterogeneous in terms of cell-cycle duration.

In untreated specimens of *O. praedator* as well as in specimens after treatment with colchicine, mitotic figures were observed everywhere in the peripheral parenchyma. Proliferative cells in both the oral and post-oral regions were distributed rather randomly (Fig. 3). Mitotic figures occurred also along the border between the peripheral and central parenchyma but were infrequent in the marginal zone of the central parenchyma. No mitotic figures were detected in mature digestive cells nor in the epidermis. The one mitotic figure observed in the middle zone of the central parenchyma (Fig. 3c) was situated along the course of a dorsoventral muscle fiber. The mitotic index of the peripheral parenchyma averaged 0.23% (0.19%, 0.21%, 0.30%). After the three-hour treatment with colchicine, the mitotic index increased to an average of 0.46% (0.42%, 0.45%, 0.52%). (Mitotic figures occurring on the border between the peripheral and central parenchyma were not included in calculating these indices.)

Table 1. Labelled-nuclei index values (LNI) in parenchyma of *C. convoluta.*

Time from beginning of experiment (days)	Number of animals	LNI, %, $\bar{x} \pm s_x$	
		Peripheral parenchyma	Central parenchyma
1	5	8.4 ± 0.6	1.5 ± 0.5*
5	5	10.2 ± 1.9	1.6 ± 0.3*
10	6	21.2 ± 2.6	13.0 ± 3.7
15	5	11.6 ± 0.9	11.1 ± 2.4
20	5	7.1 ± 1.7	16.9 ± 4.6
25	6	7.8 ± 1.6	16.8 ± 6.2
30	5	5.3 ± 0.6	7.5 ± 2.5
37	5	5.1 ± 0.7	3.3 ± 0.6

* Significantly differs from value for peripheral parenchyma ($P < 0.05$).

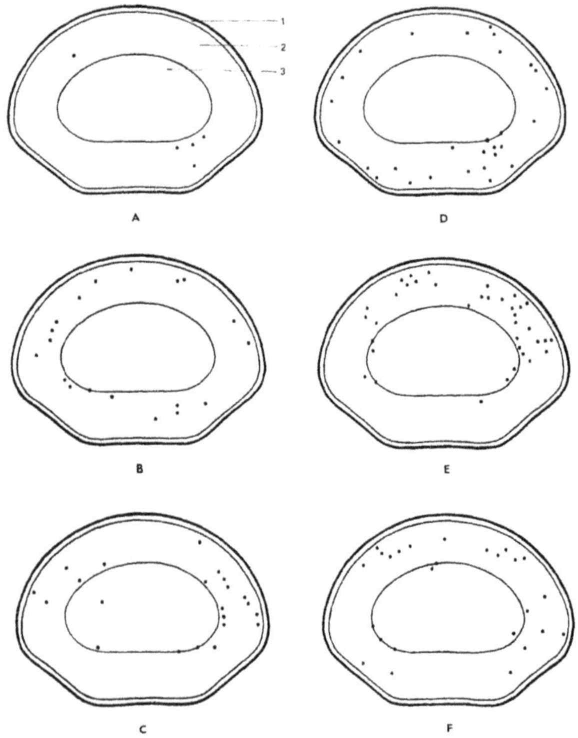

Fig. 3. Distribution of mitotic figures in cross sections of the oral and post-oral regions of *O. praedator* in untreated state (a, b, c) and following 3-h treatment in colchicine (d, e, f). 1 – epithelium; 2 – peripheral parenchyma; 3 – central parenchyma. Each dot represents one mitosis. Data is from a) 2 600 peripheral parenchymal cells (p.p.), 2 870 central parenchymal cells (c.p.), 1 730 epidermal cells (e.); b) 8 580 p.p., 3 100 c.p., 1 030 e.; c) 6 070 p.p., 3 000 c.p., 1 400 e.; d) 5 980 p.p., 1 190 c.p., 910 e.; e) 4 710 p.p., 810 c.p., 570 e.

Nuclei with primary labelling in *O. praedator* were found in the peripheral parenchyma and in a number of cases along its border with the central parenchyma. Some nuclei with primary labelling were observed in the marginal zone of the central parenchyma. No nuclei in the middle zone of the central parenchyma were labelled. Some labelled nuclei were also encountered among the body-wall muscle fibers. Nuclei with one end in the circular muscle layer and the other in the epidermis were also labelled. Autoradiographs of the oral and post-oral zones of five animals had only one labelled epidermal cell. There was considerable variation in the intensity of labelling over the peripheral parenchyma. As a rule, the most heavily labelled nuclei seemed to be concentrated around the mouth.

In *O. praedator*, [^{3}H]thymidine is incorporated not only in the nuclei, but also in the cytoplasm of individual cells in the peripheral parenchyma. It appears that there are more cells with labelled cytoplasm in the ventral part of the peripheral parenchyma than in the dorsal part.

Pedersen (1964), in his ultrastructural study of *C. convoluta*, pointed out two possible sources for digestive elements: 1) a caudal cluster of glandular cells adjoining the central parenchyma, and 2) free connective tissue elements similar to neoblasts of planarians and located mainly in the peripheral parenchyma. I have confirmed the second possibility with an autoradiographic study (Drobysheva, 1983). An electron microscopic study of *O. praedator* revealed undifferentiated cells only within the peripheral parenchyma and having no apparent morphological relation to the central syncytium (Mamkaev & Markosova, 1979). Since the material available does not allow one to distinguish between undifferentiated cells of the digestive parenchyma and undifferentiated cells of other systems, the question of the 'anatomical unit' of the central parenchyma with respect to stem cells is still open (cf. Smith, 1981). I am inclined to believe that the absence of a distinct morphological border (e.g., in the form of basal lamina) between the central and peripheral parenchyma even when the latter is strongly developed as it is in *C. convoluta* and *O. praedator*, shows the undifferentiated cells are elements of the peripheral parenchyma, irrespective of whatever cells they may adjoin.

These findings lead to the conclusion that the central parenchyma of *C. convoluta* and *O. praedator* is a specialized area of the parenchyma in which mature digestive cells do not divide. The experimental data also indicate that the migration of cells in *C. convoluta* from the peripheral to the central parenchyma is one-way. The time between origin of the cells and their migration to the central parenchyma is variable, ranging in the individuals I studied between 10 and 15 days. Thus, the physiological regeneration of digestive cells in *C. convoluta* occurs much more slowly than in the intestinal epithelium of mammals (cf. Zavarzin, 1967). It is comparable, however, to the renewal rate in digestive tissues in amphibians and invertebrates (see Lanskaya, 1974; Punin, 1978; Yoshikoshi, 1980; Obukhova, 1982). The central parenchyma of acoel turbellarians appears to be most comparable, therefore, to the crests of intestinal epithelial folds of other metazoans: both consist of specialized digestive cells that lose their proliferative ability in the course of differentiation.

However, the populations of stem cells in both the vertebrate and invertebrate animals mentioned are part of the gut epithelial layer. They are often isolated in crypts or in bases of folds, whereas in acoel turbellarians the stem cells of the digestive parenchyma are not spatially distinct from the peripheral part of the phagocytoblast. Stem cells are distributed diffusely and are possibly totipotent. The integumentary epithelium is also evidently restored at the expense of stem cells from the peripheral parenchyma.

My results indicate that the digestive parenchyma of acoels, in spite of its morphological peculiarities, is built on the same organizational principles as the intestinal epithelium of higher multicellular animals. In the course of phylogenetic formation of digestive tissues, the functional features geared to cell-population kinetics at early stages of evolution (as exemplified by the Acoela) appear to outstrip the spatial shaping of the system – i.e., the epithelialization seen in digestive systems of higher organisms.

References

Drobysheva, I. M., 1976. Avtoradiografischeskoe issledovanie organizatsii parenkhimy beskishechnoj turbellarii Convoluta convoluta (Abildgaard). In: Evolyutsionnaya morfologiya bespozvonochnykh zhivotnykh. Leningrad: 11–12.

Drobysheva, I. M., 1978. Issledovanie organizatsii parenkhimy beskishechnoj turbellarii Convoluta convoluta (Abildgaard) s pomoshch'yu timidinovoj avtoradiografii. In: Morfologiya, sistematika i evolyutsiya zhivotnykh. Leningrad: 59–60.

Drobysheva, I. M., 1982. Osobennosti fiziologicheskoj regeneratsii pishchevaritel'noj parenkhimy Convoluta convoluta (Turbellaria, Acoela). Dokl. Akad. Nauk SSSR 268: 717–719.

Drobysheva, I. M., 1983. Avtoradiografischeskoe issledovanie pishchevaritel'noj parenkhimy Convoluta convoluta. Tsitologiya 25: 1270–1277.

Ivanov, A. V. & Mamkaev, Y. V., 1977. Über die Struktur des Digestionsparenchyma bei Turbellaria Acoela. Acta Zool. Fenn. 154: 59–61.

Lanskaya, A. B., 1974. Avtoradiografischeskoe issledovanie kinetiki kletochnykh populyatsij v epitelii tonkoj kishki i pecheni travyanoj lyagushki. Tsitologiya 16: 329–333.

Mamkaev, Y. V., 1967. Ocherki po morfologii beskishechnykh turbellarij. In: Trudy Zool. inst. AN SSSR. Leningrad 44: 26–108.

Mamkaev, Y. V., 1979. O gistologischeskoj organizatsii pishchevaritel'noj sistemy turbellarij. In: Trudy Zool. inst. AN SSSR. Leningrad 84: 13–24.

Mamkaev, Y. V., Markosova, T. G., 1979. Elektronnomikroskopicheskoe issledovanie parenkhimy predstavitelej beskishechnykh turbellarij. In: Trudy Zool. inst. AN SSSR. Leningrad 84: 7–12.

Markosova, T. G., 1982. Ultrastrukturnye osobennosti pishchevartel'noj parenkhimy beskishechnoj turbellyarii Oxyposthia praedator. Dokl. Akad. Nauk SSSR 267: 950–952.

Obukhova, E. V., 1982. Izuchenie kinetiki kletochnykh populyatsij epiteliya pishchevaritel'noj zhelezy golozhabernogo mollyuska Dendronotus arborescens. Tsitologiya 24: 1117–1184.

Pedersen, K. J., 1964. The cellular organisation of Convoluta convoluta, an acoel turbellarian: a cytological, histochemical and fine structural study. Z. Zellforsch. Mikrosk. Anat. 64: 655–687.

Punin, M. Yu., 1978. Avtorradiograficheskoe issledovanie organizatsii konechnykh otdelov pishchevaritel'noj zhelezy dvukh vidov plastinchatozhabernykh mollyuskov. Tsitologiya 20: 58–65.

Smith, J. P. S., 1981. Fine structural observations on the central parenchyma in Convoluta sp. Hydrobiologia 84: 259–265.

Yoshikoshi, K., 1980. On the structure and function or the alimentary canal of Trigriopus japonicus (Copepoda, Harpacticoida). II. Cellular renewal in the mid-gut epithelium. Bull. Jap. Soc. Sci. Fish. 46: 705–710.

Von Graff, L., 1891. Die Organisation der Turbellaria Acoela. W. Engellmann, Leipzig, 90 pp.

Zavarzin, A. A., 1967. Sintez DNK i kinetika kletochnykh populyatsij v ontogeneze mlekopitayushchikh. Izd. Nauka, Leningrad, 196 pp.

Asexual reproduction, regeneration, and somatic embryogenesis in the planarian *Dugesia tigrina* (Turbellaria)

Elena B. Krichinskaya
Department of Embryology, State University of Leningrad, Leningrad 199164, USSR

Keywords: regeneration, morphogenesis, asexual reproduction, somatic embryogenesis, hypermorphosis

Abstract

In reviewing recent research published in Russian on regeneration and asexual reproduction, the following morphogenetic processes in the planarian *Dugesia tigrina* are considered: 1) regeneration of lost parts of the body; 2) regeneration of the whole worm from fragments of the body, either by normal regeneration when the inital polarity of the fragment is retained or by somatic embryogenesis when one or more new axes of polarity arise; 3) somatic embryogenesis, or development of individuals from somatic cells; 4) hypermorphosis, or the presence of more than the usual number of organs or body parts, a process that can be interpreted in terms of somatic embryogenesis; and 5) asexual reproduction. Some morphological, biochemical, and physiological studies of the 'division zone' in *D. tigrina* demonstrate peculiarities of a local breakdown of integrative functions, a breakdown which in turn causes division of the individual to take place at this zone; timing of division is controlled by the organism as an integrated whole.

Introduction

Evolution of regenerative processes and evolution of asexual reproduction have occurred conjointly in the animal kingdom, and, as a result, the mechanisms of these processes are closely linked with the level and character of organismal differentiation and of integration (Tokin, 1959; Korotkova & Tokin, 1979). At Leningrad University (Department of Embryology), comparative studies of the correlation between these processes in a number of animal groups have been carried out for many years. Because this research is published in Russian, it is not readily accessible to Western scientists. I review here those studies that pertain particularly to turbellarians, specifically to the planarian *Dugesia tigrina,* in an effort to remedy that situation.

Planarians are known for their high regenerative capacity. Different planarian species, however, have various degrees of regenerative capacity. Moreover, closely related species and even different ecological populations of one and the same species have different capacities (Teshirogi & Kawakatsu, 1967; Teshirogi *et al.*, 1978).

Among the phenomena of 'restoration morphogenesis' in worms, four categories can be distinguished: 1) regeneration of lost parts, 2) regeneration of the whole organism from fragments of the body, 3) somatic embryogenesis, and 4) hypermorphosis.

Somatic embryogenesis, a concept introduced by Tokin (1959), is the development of new individuals from somatic cells. Involved in this phenomenon are destructive processes, dedifferentiation of cell masses, and loss of initial polarity. In contrast, regeneration requires retention of initial polarity and a limited degree of destructive processes.

Asexual reproduction of present-day multicellular animals is mostly of secondary origin, probably arising as a consequence of various restoration phenomena, including regeneration and somatic embryogenesis (Ivanova-Kazas, 1977). Modes of

Hydrobiologia 132, 195–200 (1986).

asexual reproduction are closely correlated with the organization of those animals having it and they have developed on the basis of mechanisms of regeneration or somatic embryonogenesis; this is why quite a variety of types of asexual reproduction is found in different animal groups.

The research reported in the following sections was conducted on an asexual race of the North-American planarian *Dugesia tigrina* (Girard). This planarian has a rather high regenerative capacity: a whole organism can develop from a fragment excised from any level of the body. It can also reproduce asexually by a kind of architomy in which transverse division occurs in the postpharyngeal region of the body.

Restoration morphogenesis

I. *Regeneration of lost body parts* in *D. tigrina* occurs by way of epimorphosis together with some morphallaxis. The anterior-posterior physiological gradient (in the broad sense of the concept), which is characteristic of planarians, is undoubtedly one of the factors controlling this morphogenetic processes. The morphological course of this regeneration in *D. tigrina* has been described by a number of authors (see e.g. Curtis & Schulze, 1934; Isayeva, 1972; Spiegelman & Dudley, 1973). A subject of dispute in planarian regeneration is the origin of regenerative cells (Krichinskaya, 1980). Only by elucidating the nature of neoblasts – their origin in ontogeny and the character of their differentiation – can the conflicting opinions be reconciled.

II. *Restoration of the whole worm form a fragment of the body* can proceed in two ways depending on polarity of the fragment. When the initial polarity of the body is retained, the head region is restored at the anterior end of the fragment and the tail at the posterior end. In many respects, the morphogenetic processes occur similarly to those in regeneration of lost parts but with a greater share of morphallactic phenomena. The second way occurs when there is a change in the direction of the main polarity axis or when a second or several new polarity axes arise; such changes are evidence of somatic embryogenesis.

III. *Somatic embryogenesis* from a fragment can be observed in two special situations: a) when the fragment is very short such that the gradient of the physiological activity disappears (the fragment becomes 'apolar' and one or more new polar axes arise) and b) when the tissue systems lose integration for any of a number of reasons (such 'tissue disintegration' includes cell separation, disturbance of the cell cycle, loss of coordination between tissues, etc.).

In experimental conditions under the effect of a number of substances (such as alkaloids, hormonoids, chelators, neuropharmacological agents, and carcinogens) multipolar and bipolar forms arise (Krichinskaya & Voronkov, 1979). All these substances, as a rule, prevent regeneration and stimulate asexual reproduction in planarians. The question of by what mechanism the substances affect the course of restoration remains open. My own data and those found in the literature indicate that all these substances interfere with tissue integration in one way or another.

Low concentrations of colchicine (concentrations too low to block mitosis) and trilon B (to a lesser extent), when acting on fragments formed from two transverse cuts and amounting to 1/8–1/10 of the body, cause various teratogenic effects and induce the development of multipolar forms. Repeated removal of the regenerative blastemas enhances the disorganizing effect of colchicine: about 46% of the fragments showed somatic embryogenesis (Fig. 1a, b; see also Krichinskaya & Efimova, 1978). Short exposure of small fragments to colchamin leads to formation of a great number (44.3%) of bipolar forms (Krichinskaya & Korsakova, 1980).

In the case of colchicine-induced somatic embryogenesis, the time to achieve complete restoration is significantly extended (Krichinskaya & Voronkov, 1979). We could not detect any external changes of importance: after the wound heals, the fragment takes the form of a low pyramid, and no regeneration blastema is formed (Fig. 1c). The fragment remains in such a state for 5–7 days, a period we call the latent period, and then transforms rapidly. Some processes channeling the restoration of the fragment into somatic embryogenesis appear to occur in the latent period.

If one compares the processes of regeneration in fragments excised from the prepharyngeal region with those of fragments excised from the 'division zone', it is evident that under the influence of colchicine and colchamin, the change in polarity char-

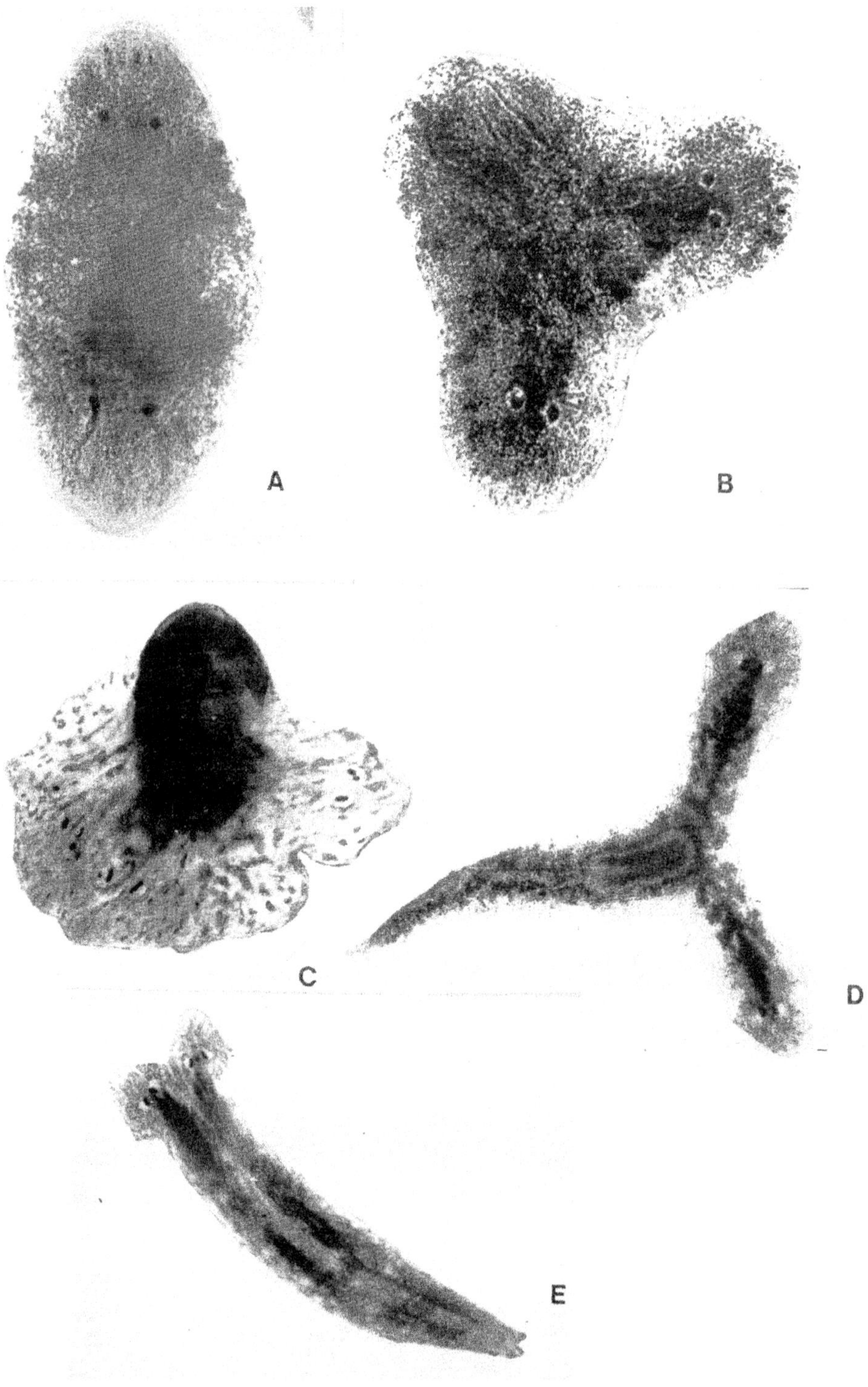

Fig. 1. Incomplete somatic embryogenesis in *Dugesia tigrina:* (a, b) regenerate after repeated removal of the regenerative blastema and under the influence of colchicine; (c) regenerate from body fragment (from other than the 'division zone') exposed to colchicine; (d) regenerate from fragment excised from the 'division zone'; (e) regenerate showing hypermorphosis.

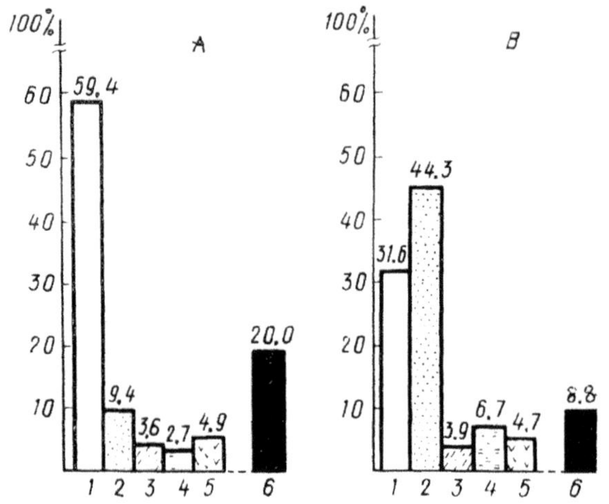

Fig. 2. Morphogenetic processes in prepharyngeal fragments (A) and in fragments excised from the 'division zone' (B) following treatment with colchicine: (1) normal regenerates; (2) double-headed regenerates; (3) regenerates with outgrowths; (4) regenerates with abnormal eyes; (5) forms showing only limited regeneration; (6) mortality.

acteristic of somatic embryogenesis arises more frequently in those fragments from the division zone (Fig. 2; see also Krichinskaya & Korsakova, 1980). The resulting bi- and multipolar forms represent incomplete somatic embryogenesis (Fig. 1d). If such forms survive over extended periods and divide, two or more worms are formed from each.

IV. *Hypermorphosis* is the development of a larger than normal number of organs or parts of the body, such as multiple pharynges, heads, and tails. It is seen in planarians from both natural and laboratory-cultivated populations as well as in experimental conditions. The causes of hypermorphosis in planarians are, according to different authors, variously attributed to modifications of the polarity or of the physiological gradient, hyperplasia of the nervous system, or traumatization – that is, to some mechanical disorganizing factor (Krichinskaya, 1978).

The mechanism of this phenomenon is still far from being clear. Hypermorphosis can well be interpreted in terms of the somatic embryogenesis theory: any local breakdown in integration of the organism's tissues, in particular that resulting from mechanical or other injury, may precipitate somatic embryogenesis. The process, however, is of an incomplete type, as the disorganization is local, not involving the whole organism (Fig. 1e). The number of organ-augmentation phenomena developed in an individual passes to daughter zooids in asexual reproduction (in *D. tigrina*) but is not passed on to progeny of sexual reproduction [e.g. in *D. lugubris*; hence these changes do not influence the animal's genetic material (Krichinskaya, 1978)]. There are, however, some examples among planarians of a genetically fixed polypharyngy arising on the basis of asexual reproduction (Krichinskaya, 1978; Ivanov & Mamkaev, 1973).

Asexual reproduction

Characteristic features of the division zone in D. tigrina

To comprehend mechanisms of asexual reproduction – in particular, to elucidate the stimulus for division – it is necessary to study the 'division zone', that is, the region of the planarian body where it is divided into zooids under asexual reproduction. The division zone is not morphologically distinct from other parts of the body, but as division always occurs here one can assume that it is physiologically distinct.

Support for this view is found in the fact, for example, that a number of planarian species have elevated metabolic activity (with respect to the anterior-posterior decreasing gradient of physiological activity) in the region of the division zone (Child, 1929; Coward & Flickinger, 1965). Tokin (1959) suggested that the division zone of planarians is distinguished from other parts of the body by some local breakdown in integration caused by endogenous factors. Such breakdown may produce the restoration processes peculiar to the division zone (Isayeva, 1968) and bring about somatic embryogenesis in fragments excised from the division zone.

1. Morphological investigation of the division zone
The site of the division zone in *D. tigrina* is determined by proportional relations of body parts and is a function of the worm's age (Chapayeve, pers. commun.). In young individuals this is more pronounced: division starts when the length of the anterior part of the body (including the pharynx) and that of the postpharyngeal region become equal. The division plane lies at a level between the first

and second thirds of the postpharyngeal region. Older planarians may simply fragment (consecutive separation of two daughter zooids).

Some authors believe that both the capacity for asexual reproduction and the capacity to regenerate lost parts are determined by the distribution of neoblasts, and they cite in support of this an anterior-posterior gradient in the number of neoblasts, even in an asexual race of *D. tigrina* (e.g. Curtis & Schulze, 1984). However, their counting of neoblasts has been irrespective of the stage of the asexual-reproductive cycle. By following neoblast distribution through various stages of asexual reproduction (from one division to the next) we (Krichinskaya & Lenicque, 1968; Krichinskaya & Martynova, 1974) found neoblasts to be distributed in an *increasing* gradient, i.e. opposite that of sexually reproducing species. Stagno & Grasso (1965) noted the accumulation of neoblasts in the posterior part of the body just before division. We also observed an increase in neoblast number of the incipient division zone and regard it as a mechanism for preparation for division.

2. Biochemical investigation of the division zone

The concentration of RNA per mg wet weight is lowest in the head region of *D. tigrina,* higher in the prepharyngeal and postpharyngeal regions, and highest in the pharyngeal region (Krichinskaya & Avenirova, 1970). Such distribution corresponds to that of the neoblasts to a large extent. Neoblasts are rich in RNA, and it is in these cells that a great part of the planarian RNA is found.

Degree of labelling by [^{14}C]uracil does not correspond to the distribution of RNA (Krichinskaya & Avenirova, 1970). Labelling is highest in the head region, is lower in the prepharyngeal region, and especially lower in the pharyngeal and tail regions, and is intermediate in the division zone. Similar results were obtained by Coward & Flickinger (1965).

3. Physiological investigation of the division zone

Following Kanatani (1960), who demonstrated that weak concentrations of such substances as colchicine and heparin cause an acceleration of division in *Dugesia gonocephala,* we found that colchicine at concentrations sufficiently low to not arrest mitotic division (as monitored on histological sections) stimulated division in *D. tigrina* (see Krichinskaya, 1971). If, as Tokin (1959) suggests, a local breakdown in integration occurs before division, it is appropriate to assume this process is enhanced by the disorganizing effect of colchicine. Evidence for such a disorganizing effect by colchicine is the formation in these experiments of a great number of abnormal worms from pieces taken from worms exposed briefly to colchicine (Fig. 1c); the largest number of abnormalities was formed from the division zone (Krichinskaya, 1971).

In a number of cases, when specimens were cut transversely into pieces 1/8–1/12 of the body size, abnormal forms arose from fragments taken from the division zone (Krichinskaya, 1971). Such abnormal forms had several heads and tails, various outgrowths, or anomalies in eye structure. In the course of asexual reproduction, such abnormal forms arose from areas corresponding to the division zone of worms which had only recently undergone division or were about to divide (Krichinskaya & Lenicque, 1968).

A number of workers have demonstrated a greater resistance to altering effects in fragments taken from the division zone as compared to fragments excised from the prepharyngeal region (McWhinni, 1955; Krichinskaya & Korsakova, 1981). This observation reveals another physiological peculiarity of the division zone.

Taking into consideration the integrating role the nervous system plays, we (Krichinskaya *et al.,* 1974) studied the effect extirpation of the cerebral ganglion has on the asexual reproductive cycle. Two hundred twenty planarians were decapitated and the head regenerate was removed every third day for a total of 10 decapitations. After the third decapitation, the planarians ceased to divide; by 8–14 d after the decapitations were stopped, the ability to divide was restored, that is, the suppression of asexual reproduction was reversible.

Conclusions

On the grounds of the experiments described above, I conclude that the division zone has special characteristic features indicative of some local loss of integrating control that causes division of the individual at this particular zone; the regularity of division, i.e. regulation of the asexual reproductive cycle, appears to be controlled by the organism as an integrated whole.

References

Child, C. M., 1929. The physiological gradients. Protoplasma 5: 447–476.

Coward, S. F. & K. A. Flickinger, 1965. Axial patterns of protein and nucleic acid synthesis in intact and regenerating planaria. Growth 29: 151–169.

Curtis, W. C. & L. M. Schulze, 1934. The contrasting powers of regeneration in Planaria and Procotila. J. Morph. 95: 474–512.

Isaeva, V. V., 1968. Osobennosti regeneratzionnikh yavleny v zone delenia u razmnozhajushikhsya bespolym putyom planariy. Biol. nauki 4: 7–10.

Isaeva, V. V., 1972. Vossyanovitelnye yavlenya u nekotorykh Turbellariy. V. sb.: Bespoloye razmnozhenie, somaticheskiy embriogenez i regenereatsia. L.: 233–267.

Ivanov, A. B. & Yu. V. Mamkaev, 1973. Resnichnie chervy (Turbellaria). Ich proiskhozhdenie i evolutsiya. Phylogeneticheskye ocherky. Ed. 'Nauka', L. Leningrad. 220 pp.

Ivanova-Kazas, O. M., 1979. Bespoloye razmnozhenie zhivotnikh. Izd. Leningradskogo Universiteta, Leningrad, 238 pp.

Kanatani, H., 1960. Studies on fission in the planarian Dugesia dorotocephala. 2. Effects of colchicine and demecolcine on the occurrence of fission. J. Fac. Sci. Univ. Tokyo, Sec. 1V, 9: 60–66.

Korotkova, G. P. & B. P. Tokin, 1979. O zakonomernostyakh evolucii vosstanovitelikh morfogenesov. Biol. nauki 11: 5–18.

Krichinskaya, E. B., 1971. Osobennosti zony delenya planarii Dugesia tigrina. Biol. nauki 4: 4–10.

Krichinskaya, E. B., 1978. O tak nazivaemihk gipermorfozakh u planarii. Biol. nauki 1: 67–73.

Krichinskaya, E. B., 1980. Kletochnie istochniki regeneracii u planarii. Sovremennie predstavlenya o neoblastakh. Arkhiv anat., gistol. i embriol. 79: 102–109.

Krichinskaya, E. B. & E. L. Avenirova, 1970. Isuchenie raspredelenya i obnovlenya RNK vdol'tela planarii Dugesia tigrina (Girard). Ontogenez 1: 526–529.

Krichinskaya, E. B. & G. V. Efimova, 1978. Vosstanovlenie tzelogo chervya iz nebolshogo fragmenta tela planarii Dugesia tigrina posle mnogokratnogo udalenya regeneratov. Ontogenez 9: 510–513.

Krichinskaya, E. B. & P. Lenicque, 1968. Distribution des néoblastes le long du corps de la planaire Dugesia tigrina et pouvoir de développement de différents morceau isolé de celui-ci. Acta. zool. 50: 69–76.

Krichinskaya, E. B. & I. G. Malikova, 1969. Vosstanovitelnye processy pri bespolum razmnozhenii i regeneracii planarii Dugesia tigrina. Arkhiv. anat., gistol. i embriol. 7: 48–51.

Krichinskaya, E. B. & M. G. Martinova, 1974. Raspredelenie neoblastov i mitozov v khode bespologo razmnohzenya planarii Dugesia tigrina (Girard). Ontogenez 5: 349–355.

Krichinskaya, E. B. & L. Yu. Korsakova, 1981. Vliyanie kolkhamina na vosstanovitelnye prozessy vo fragmentakh tela planarii Dugesia tigrina (Girard). Biol. nauki 9: 64–68.

Krichinskaya, E. B. & A. R. Voronkov, 1979. Formoobrazovatelnye processy u ekstirpirovannikh iz tela planarii Dugesia tigrina fragmentov posle vozdeistvia kolkhicinom, trilonom B i khloristym litiem. Biol. nauki 12: 57–59.

Krichinskaya, E. B., A. E. Koroleva & N. O. Ermolina, 1974. Vliyanie decapitacii na bespoloe razmnozhenie planarii Dugesia tigrina (Girard). V sb.: Morfogeneticheskie processy pri raznikh tipakh razmnozhenya i v khode regulyacii. Izd. Leningradskogo Universiteta, L.: 83–88.

McWhinni, M., 1955. The effect of colchicine on regeneration reconstitutional development in Dugesia dorotocephala. Biol. Bull. 108: 54–65.

Spiegelman, M. & P. L. Dudley, 1973. Morphological stages of regeneration in the planarian Dugesia tigrina. A light and electron microscopic study. J. Morph. 139: 155–184.

Stagni, A. & M. Grasso, 1965. Osservazioni preliminery sulla schizogenesi di Dugesia tigrina. Linc. Rend. Sc. fis. mat. e nat. 38: 9005.

Teshirogi, W. & M. Kawakatsu, 1967. Comparative studies on regenerative capacities of some freshwater planarians from Japan, Europe and North America. Sci. Rep. Hirosaki Univ. 1I4: 23–36.

Teshirogi, W., S. Ishida & H. Yamazaki, 1978. Regeneration capacities of transverse pieces in two species of freshwater planarians Dendrocoelopsis locata and Polycelis sapporo from Amori Prefecture, especially compared with those of same species from Hokkaido. Dobutsugaku zasshi, Zool. Mag. 87: 3.

Tokin, B. P., 1959. Regeneracya i somaticheskii embriogenez. L. 268 pp.

Positional information and gonadal differentiation in the planarian *Dugesia lugubris** (Turbellaria)

E. Ghirardelli, A. Di Marcotullio & F. Persi
Department of Biology, University of Trieste, Trieste, Italy

Keywords: Turbellaria, Dugesia, gonad, transplantation, regeneration, positional information

Abstract

The planarian *Dugesia lugubris* is a balanced hermaphrodite, meaning that male genetic factors are in equilibrium with female factors. Differentiation of the gonads is controlled by the region in which they develop. According to the classical theory of germ cell formation, these cells stem from neoblasts that are induced to differentiate by factors specific to the gonadal regions, factors presumably due to gradients formed by neurosecretory activity of the cephalic ganglia and longitudinal nerve cords. A more recently proposed theory holds that germ cells in regenerates originate not from neoblasts but from dedifferentiated cells and that characteristics of the gonadal regions are determined by direct interactions of cells here. Results of our experiments with homo- and autoplastic grafst support the classical theory. Prepharyngeal portions grafted onto posterior body portions retained their ability to maintain or induce development of ovaries. Postpharyngeal portions grafted onto anterior portions produced only testes even though the brain developed normally in these regenerates. Under these experimental conditions, gonad regeneration took longer than it does in normal regeneration (i.e., that in which body regions are not displaced).

Introduction

The determination and differentiation of germ cells in hermaphroditic animals has been studied most intensively in freshwater hydroids and planarians (Vannini, 1979). Freshwater hydroids usually reproduce asexually by budding; but when sexual reproduction is initiated, testes appear in the upper parts of the gastric column below the hypostome and ovaries arise in the lower part. This condition, a form of 'balanced hermaphroditism,' is particularly characteristic of *Pelmatohydra braueri.* It appears that such a condition is a product of a polygenic system, one in which the number of masculinizing and feminizing genetic factors are in equilibrium, but in which other factors can intervene to displace the equilibrium locally. In *P. braueri,* sexual differentiation is regulated by 'positional information' under the control of the nerve net in the hypostome. Neurosecretory cells here produce low-molecular-weight polypeptides that diffuse along an oral-aboral gradient which in turn is responsible for the morphogenetic polarity. As a consequence, the hypostomal region is considered the inductive center for differentiation of the gonads (Stagni, 1966).

In the same way, the planarian *Dugesia lugubris* is a balanced hermaphrodite, with the two ovaries lying ventral and immediately posterior to the cephalic ganglia and the numerous testis follicles lying dorsal along the length of the longitudinal nerve cords. And also in the same way, differentiation of the germ cells in planarians could be through positional information under control of the cephalic ganglia and longitudinal cords, whose

* Translated, from the French, by Marianne Klauser and Seth Tyler.

influence on anterio-posterior polarity has been described by Sperry *et al.* (1973) and Kishida & Kurabuchi (1978). We provide here evidence for the validity of the positional-information hypothesis and weigh this against another more recent hypothesis of cell determination, Chandebois's (1984) 'cellular sociology' hypothesis.

Materials and methods

Our observations were made on specimens of *D. lugubris* collected from the karst lake Doberdo near Trieste, Italy, and from the vicinity of Paris. (The latter gave more consistent results.) Homo- and autoplastic grafts were performed essentially according to the technique of Sugino (1952). Cuts were made at any of three levels of the body, essentially dividing the worms into four portions, viz. head, prepharyngeal, pharyngeal (midbody), and tail portions. In order to augment the contact surface between grafts, the amputations were made with a V-shaped cut. Samples in which the grafting was at least in part successful were fixed one to five months after the operation in Carnoy's fluid (chloroform, absolute alcohol, and acetic acid in volumetric ratio of 3:6:1). Reconstruction of the organs was done from 5-μm sections stained with hematoxylin and eosin.

Results

Homoplastic grafts

About 60 grafts were attempted in these experiments, but in only 10 of these was there any fusion of graft pieces. These can be arranged into three groups. In the first group, a prepharyngeal portion of one specimen was grafted into the postpharyngeal region of another, but since the tail piece did not fuse to the postpharyngeal portion, the regenerate consisted only of a prepharyngeal portion grafted onto the posterior edge of the pharyngeal region of another worm (Fig. 1, upper row). After 30 days, the original form of the body had been reconstituted. These regenerates had four ovaries, two anterior and two posterior, but testes and copulatory apparatus were missing (Fig. 1, upper right).

In the second group (three examples), a prepharyngeal portion from one worm fused with the tail portion of another (Fig. 1, lower half). One specimen, in which the head had regenerated in an irregular fashion, had neither gonads nor a copulatory apparatus after 135 days in culture. In a second specimen, the head had regenerated normally by one month after the operation, and by three months after the operation the organism had well developed ovaries. The testes were lacking at this

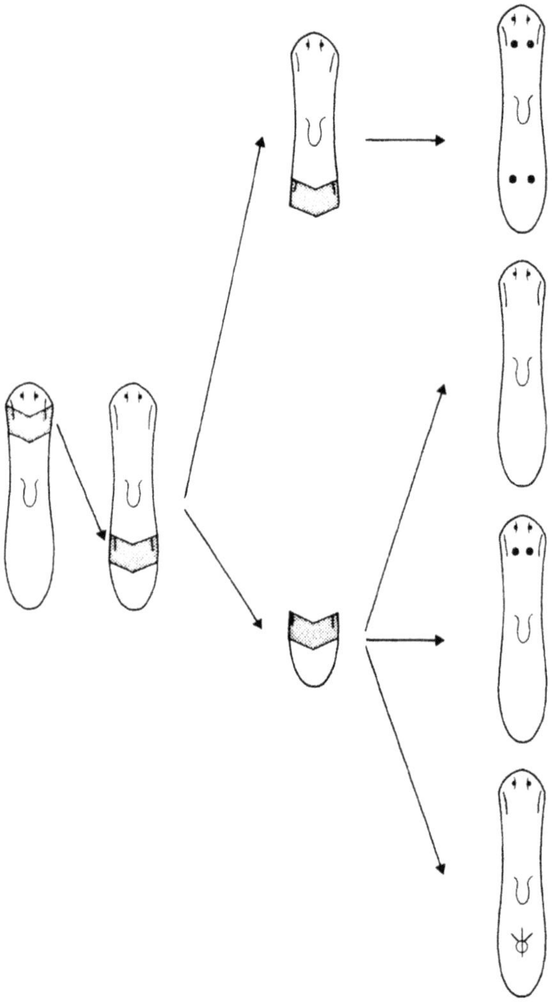

Fig. 1. Homoplastic grafts: anterior portion of the prepharyngeal region was removed and grafted into the postpharyngeal region of another worm. Failure of grafted edges to hold produced regenerates that essentially had either prepharyngeal portion grafted to posterior end of pharyngeal region (upper path) or prepharyngeal region and tail grafted together (lower paths). Results, after regeneration, are at right. Filled circles: ovaries. Spoked circle: copulatory apparatus. (After Ghirardelli *et al.*, 1981.)

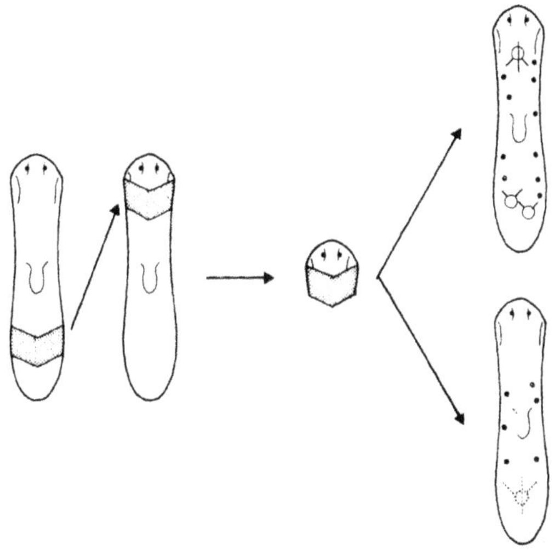

Fig. 2. Homoplastic grafts: postpharyngeal region was grafted in place of prepharyngeal region of another worm. Failure of graft to hold produced regenerate that essentially had postpharyngeal region grafted onto head portion. Results, after regeneration, at right. Small filled circles: testes. Spoked circle: rudimentary or complete copulatory apparatus. Dotted spoked circle: regeneration copulatory apparatus. (After Ghirardelli *et al.*, 1981.)

stage but a node of neoblasts was clearly visible in the area where the copulatory organ usually develops. The third specimen, fixed 145 days after the operation, had no gonads but a copulatory apparatus was present (Fig. 1, lower right); this probably was simply the copulatory apparatus of the recipient and had not regressed, but it is also possible that it is regenerating in the absence of testes.

In the six specimens of the third group, only the head of one worm and the postpharyngeal portion of another fused (Fig. 2). In five of these specimens fixed between 95 and 175 days after the operation, we observed an intercalary type of regeneration and some testes in regression in the prepharyngeal region. Nevertheless, in the same region as well as in the postpharyngeal region, there were testes follicles in the process of reorganizing, and nearly completely developed copulatory apparatuses (Fig. 2, lower right). In the sixth specimen of this group (Fig. 2, upper right), a completely differentiated copulatory apparatus could be seen in the prepharyngeal region 107 days after the operation, and as before this was probably a transplanted copulatory apparatus. Two other rudimentary copulatory apparatuses had been regenerated in the postpharyngeal region. The ovaries were absent and the testes regressing, with their spermatogonia dispersed in the parenchyma, giving the impression that they were dedifferentiating. There appeared to be a mass migration of neoblasts from the ventral side of the body to the dorsal side where other testes were regenerating.

Autoplastic grafts

Only four specimens out of 40 autoplastic grafting operations produced successful grafts. In two of these (Fig. 3, upper row), head, pharyngeal, and

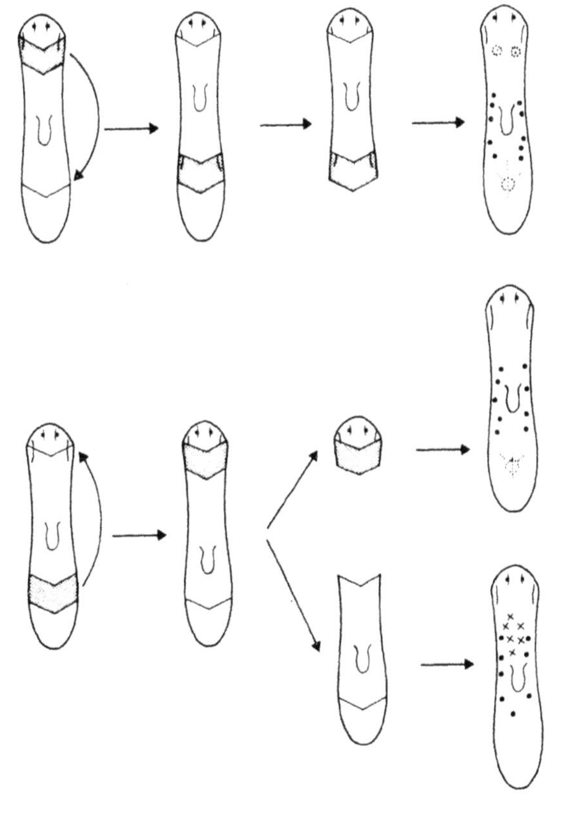

Fig. 3. Autoplastic grafts. Upper course: Prepharyngeal portion grafted into postpharyngeal region. The grafts produced regenerates that had head, pharyngeal, and prepharyngeal portions fused. Lower course: Postpharyngeal portion grafted into prepharyngeal region. Failure of some graft edges to hold produced either regenerates with head portion and postpharyngeal portion grafted together or with tail and pharyngeal portions grafted together. Results, after regeneration, at right. Small filled circles: testes. Dotted circles: incompletely differentiated ovaries. Dotted spoked circles: regenerating copulatory apparatus. Crosses: dispersed oocytes. (After Ghirardelli *et al.*, 1981.)

prepharyngeal portions fused, intercalary regeneration took place, and, by 109 days after the operation, ovaries, testes, and copulatory apparatus were almost completely regenerated. In a third specimen, a head fused with a postpharyngeal portion; and, by 109 days after the operation, there were no ovaries in the regenerate, but the testes and copulatory apparatus were in the process of regenerating (Fig. 3, center). The fourth specimen, in which only the pharyngeal portion and tail fused, produced, after 109 days in culture, an irregular distribution of testicular follicles in the middle of the body and some dispersed oocytes at the level where testes would normally appear (Fig. 3, lower row).

Discussion

We interpret our results according to the classical view that the head is dominant and so regulates the maturation of the gonads, especially the testes. In decapitated specimens, the testes regress within one week (Grasso, 1963; Ghirardelli & Gordini, 1964; Ghirardelli, 1965; Ghirardelli & Di Marcotullio, 1976; Ghirardelli *et al.*, 1981) and do not regenerate again if one impedes regeneration of the head (Ghirardelli & Brandi, 1963).

Differentiation of the germ cells could be regulated by neurosecretory products of the cephalic ganglia and nerve cords. In effect, Grasso & Quaglia (1971) observed neurosecretory cells in the proximity of male and female gonads as well as release of neurosecretory granules and transport of these granules along axons. Neurosecretory products may also induce sexuality in asexual planarians (Benazzi & Grasso, 1977).

The presence of neurosecretory cells and variations in their numbers and activity during regeneration could be profitably studied as was first shown by Lender & Klein (1961). Franquinet & Le Moigne (1979) found that the concentration of serotonin increases at the beginning of regeneration and that of dopamine increases at the end of morphogenesis. The binding of serotonin and dopamine to specific receptors on the cell membrane causes activation of the enzyme adenylcyclase which catalyzes the synthesis of cAMP. As a consequence, the concentration of RNA in the activated neoblasts increases (Ghirardelli, 1959).

Undoubtedly the various regions of the planarian body have different properties affecting regeneration. Teshirogi (1963) found that two pieces obtained by transverse sectioning of either the pre- or the postpharyngeal regions of the planarian *Bdellocephala brunnea* could reciprocally inhibit regeneration if they were grafted together in inverted polarities. Our experiments show that the interaction of grafts cut from more or less distantly separated levels along the body retard re-establishment of the normal gradient. Alteration of the anterior-posterior gradient was evident, for example, in the distribution of the reproductive organs in all of the autoplastic grafts (Fig. 3). And regeneration of gonads is slower in grafts than it is following simple amputation (i.e., without displacement of the respective body parts) probably because of an 'inertia' the tissues must overcome to change their original determination. After simple amputation, the time necessary for complete regeneration of gonads is less than three weeks.

Chandebois (1984) also studied the consequences of homo- and autoplastic grafts in the planarian *D. lugubris*, in particular intercalary regeneration. Her conclusions were contrary to classical hypotheses, especially those relating to the concept of axial gradients and to the dominance of the head, as well as that of positional information as proposed by Wolpert (1971). According to Chandebois & Faber (1983) each cell possesses individual characteristics governed by many sources of information arising from the various types of cells in its vicinity; and for this reason cells take part simultaneously in several systems that interact mutually at various levels of the body ('cellular sociology'). When two pieces taken from more or less distantly separated regions of the body are grafted together, reconstruction should take place by morphallaxis. As this occurs, a given level is transformed into a certain number of other levels of reduced dimensions to replace those levels missing, all at the expense of one of the grafted pieces as it is gradually remodeled. Interference between the various regulatory systems may create a kind of *vis inertiae* that impedes the immediate conversion of one level into another (Chandebois, 1984); and such interference might also explain the persistence of the individuality of transplanted pieces. However, we feel this persistence is better explained as being due to the influence of the head and of the cephalocaudal gradient the results of our experiments suggest.

The regeneration blastema, according to the classical concepts of Wolff and his school (Lender, 1965; Stephan-Dubois, 1965), is formed by neoblasts. However, results of several authors raise questions of this theory. Chandebois (1965), for example, thought that neoblasts play no role in regeneration, claiming rather the blastema originates from parenchymal cells. Coward (1969), in reference to the opinion of other authors, admits that the blastema originates from the dedifferentiation of other cells followed by their multiplication, migration, and redifferentiation in the blastema. One of the more demonstrative cases of such dedifferentiation is that described by Gremigni & Puccinelli (1977), in which spermatogonia reportedly dedifferentiated and then redifferentiated to give rise not only to other male germ cells but also to oocytes and somatic cells. Such dedifferentiation does not, however, necessarily exclude the possible existence of cells that maintain an embryonic character, i.e. neoblasts.

Whatever the origin of the blastemal cells in gonadal regeneration of *D. lugubris,* our experiments show that positional information plays an important role in their determination and differentiation. In any case, more research on the origin of the blastema and on the regulatory role of a longitudinal gradient are needed. The classical concepts are no longer entirely valid, but the newer hypotheses require confirmation.

Conclusions

Our results indicate that the differentiation of germ cells in the planarian *D. lugubris,* a balanced hermaphrodite, is controlled by 'positional information.' Conclusive evidence of this, however, will require more experiments than we were able to accomplish. These results are best explained by assuming that the prepharyngeal region has a feminizing effect while more posterior regions have a masculinizing effect, all depending on a cephalocaudal gradient. This gradient is apparently sustained by the action of neurosecretory products whose distribution varies qualitatively and quantitatively along the length of the longitudinal nerve cords.

Grafting disrupts the continuity of the gradient, but the grafts maintain their original characteristics for some time despite the manipulation of the tissue. The interaction between grafts cut from more or less distantly separated regions of the body retards the re-establishment of the normal gradient and gonadal regeneration, perhaps because the tissues resist changing their original determination.

Acknowledgments

We thank Prof. T. Lender for providing specimens from the environs of Paris, and we thank Marianne Klauser and Seth Tyler for critically reviewing and translating the original French draft of this manuscript. Financial support of this research was provided by the Consiglio Nazionale delle Ricerche and the Ministero Pubblica Istruzione.

References

Benazzi, M. & M. Grasso, 1977. Comparative research on the sexualization of fissiparous planarians treated with substances contained in sexual planarians. Monit. zool. ital. 11: 9–19.

Chandebois, R., 1965. Cell transformation systems in planarians. In V. Kiortsis & H. A. L. Trampusch (eds), Regeneration in animals and related problems. North-Holland Publ. Co., Amsterdam: 131–142.

Chandebois, R., 1984. Intercalary regeneration and level interactions in the fresh-water planarian *Dugesia lugubris.* I. The anteroposterior system. Roux's Arch. dev. Biol. 193: 149–157.

Chandebois, R. & J. Faber, 1983. Automation in animal development. Monogr. Dev. Biol. vol. 16, Karger, Basel. 204 pp.

Coward, S. I., 1969. Regeneration in planarians: some unsolved problems and questions. J. biol. Psychol. 11(2): 15–19.

Franquinet, R. & A. Le Moigne, 1979. Relation entre les variations des taux de sérotonine et l'AMP cyclique au cours de la régénération d'une planaire (Polycelis tenuis). Biol. Cell. Fr. 34: 71–76.

Ghirardelli, E., 1959. Gli acidi nucleici nella rigenerazione di dischetti isolati dal corpo di *Dugesia lugubris.* Acta Embryol. Morphol. exp. 2: 320.

Ghirardelli, E., 1965. Differentiation of the germ cells and regeneration of the gonads in planarians. In V. Kiortsis & H. A. L. Trampusch (eds), Regeneration in animals and related problems. North-Holland Publ. Co., Amsterdam: 176–184.

Ghirardelli, E. & L. Brandi, 1963. L'azione del cervello sulla rigenerazione delle gonadi di *Dugesia lugubris.* Rc Accad. naz. Lincei 35: 121–125.

Ghirardelli, E. & S. Gordini, 1964. Influenza della regione cefalica sulla rigenerazione delle gonadi in esemplari bicefali di *Dugesia lugubris.* Rc Accad. naz. Lincei 37: 92–96.

Ghirardelli, E. & A. Di Marcotullio, 1976. Influenza del territorio nella rigenerazione delle gonadi di *Dugesia lugubris.* Atti Accad. Sci. Bologna 3: 93–112.

Ghirardelli, E., A. Di Marcotullio & F. Persi, 1981. 'Informazione di posizione' e differenziamento delle gonadi nella planaria *Dugesia lugubris.* Atti Accad. Sci. Bologna 8: 79–96.

Grasso, M., 1963. L'organizzazione dell'apparato genitale ermafrodita in esemplari rigenerati ed in esemplari bicefali di *Dugesia lugubris.* Rc Accad. naz. Lincei 35: 101–104.

Grasso, M. & A. Quaglia, 1971. Studies on neurosecretion in planarians. Neurosecretory fibres near the testes and ovaries of *Polycelis nigra.* J. Submicr. Cytol. 3: 171–180.

Gremigni, V. & I. Puccinelli, 1977. A contribution to the problem of the origin of the blastema cells in planarians: a caryological and ultrastructural investigation. J. exp. Zool. 199: 57–71.

Kishida, Y. & S. Kurabuchi, 1978. The role of nervous system in the planarian regeneration. I, Regeneration of body fragments deprived of ventral cords. Annot. zool. Japon. 51: 90–99.

Lender, T., 1965. La régénération des planaires. In V. Kiortsis & H. A. L. Trampusch (eds), Regeneration in animals and related problems. North-Holland Publ. Co., Amsterdam: 95–111.

Lender, T. & N. Klein, 1961. Mise en évidence de cellules sécrétrices dans le cerveau de la planaire *Polycelis nigra.* Variation de leur nombre au cours de la régénération postérieure. C. r. Acad. Sci. Fr. 253: 331–333.

Sperry, P. J., K. D. Ansevin & F. K. Tittel, 1973. Inductive role of the nerve cord in regeneration of isolated postpharyngeal body sections of *Dugesia dorotocephala.* J. exp. Zool. 186: 159–174.

Stagni, A., 1966. Some aspects of sexual polymorphism in *Chlorohydra viridissima.* J. exp. Zool. 163: 87–92.

Stéphan-Dubois, F., 1965. Les néoblastes dans la régénération chez les planaires. In V. Kiortsis & H. A. L. Trampusch (eds), Regeneration in animals and related problems. North-Holland Publ. Co., Amsterdam: 112–130.

Sugino, H., 1952. Recombination experiments of small pieces in *Dugesia gonocephala* (Dugès). Annot. zool. Japon. 25: 198–208.

Teshirogi, W., 1963. Transplantation experiments of two short pieces of a freshwater planarian *Bdellocephala brunnea.* Japan. J. Zool. 14: 21–48.

Vannini, E., 1979. Alcuni modelli di differenziamento sessuale in animali pluricellulari. Seminario sulla evoluzione biologica. Accad. Naz. Lincei, Centro interdisciplinare scienze, 48: 87–149.

Wolpert L., 1971. Positional information and pattern formation. Curr. Topics dev. Biol. 6: 183–224.

On the origin of neoblasts in freshwater planarians (Turbellaria)

Wataru Teshirogi
Department of Biology, Faculty of Science, Hirosaki University, Hirosaki 036, Japan

Keywords: Turbellaria regeneration, neoblast, intestine, sexuality induction, dedifferentiation, cell culture

Abstract

Experiments on 1) regeneration of the cave-adapted planarian, *Sphalloplana zeschi,* 2) induction of sexuality in an asexual strain of *Dugesia japonica japonica* by feeding, and 3) culture of dissociated planarian cells, show that neoblasts originate from intestinal cells, i.e. phagocytic cells and granular clubs.

Introduction

It is generally agreed that neoblasts in freshwater planarians play the chief role in blastema formation. Concerning the origin of neoblasts, there are two main theories: the embryonic-stock-cell theory (Curtis, 1936; Curtis & Schulze, 1934; Wolff & Dubois, 1947; Brønsted, 1955; Lender, 1962; Baguña & Romero, 1981), and the dedifferentiated-cell theory (Bartsch, 1923; Weigand, 1930; Woodruff & Burnett, 1965; Rose & Shostak, 1968; Hay, 1968; Coward, 1969).

In the present study, results of three experiments on the regeneration of the cave-adapted planarian *Sphalloplana zeschi,* the induction of sexuality in an asexual strain *Dugesia japonica japonica,* and the culture of dissociated cells indicate that neoblasts originate from intestinal cells, i.e. phagocytic cells and granular clubs.

Materials and methods

Planarians used in the regeneration experiment were mature individuals of *Sphalloplana zeschi* Mitchell, 1968, and asexual individuals of *Dugesia dorotocephala* (Woodworth, 1897). *S. zeschi,* a cave-adapted eyeless planarian (Fig. 1A), was collected in Zesch Ranch Cave, Mason County, Texas. *D. dorotocephala* was collected from a stream at Buffalo Springs Lake, Lubbock County, Texas. Individuals 18–20 mm in length were used of both species. The planarians used in the sexuality induction experiment were *Dugesia japonica japonica* Ichikawa et Kawakatsu, 1964, *Bdellocephala brunnea* Ijima et Kaburaki, 1916, and *Polycelis sapporo* (Ijima et Kaburaki, 1916). An asexual strain *D. j. japonica* was collected from a stream of Mitaki Hot Spring, Sendai, Miyagi Prefecture, and worms of a sexual strain of the same species from a stream of Sakunami Hot Spring, Miyagimachi, Miyagi Prefecture (Fig. 1B, C). *B. brunnea* was collected from a spring in *Hirosaki City,* Aomori Prefecture and *P. sapporo* from a stream in Tashirotai Heights, Aomori City, Aomori Prefecture. *B. brunnea* was also used in the culture of dissociated cells.

Regeneration experiment. Before the experiment, the individuals of *S. zeschi* and *D. dorotocephala* were kept fasting for at least a week. They were decapitated, and the remaining bodies were cut transversely into five short pieces of about the same length. In each experiment, 60 animals (360 pieces) were used. To observe regeneration, these pieces were cultured in Petri dishes containing about 200 ml culture water (dechlorinated and aerated for 1–2 days), at 20 °C room temperature without feeding.

Hydrobiologia 132, 207–216 (1986).

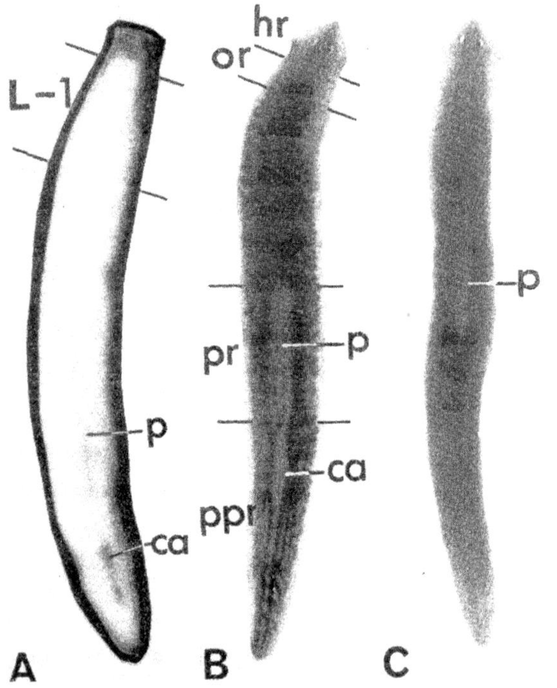

Fig. 1. *Sphalloplana zeschi* (A) and *Dugesia japonica japonica* (B and C). B, sexual race from Sakunami, showing body regions used as food. C, asexual race from Mitaki. ca: copulatory apparatus. hr: head region. L-1: level-1 piece. or: ovary region. p: pharynx. ppr: post-pharyngeal region. pr: pharyngeal region.

Induction of sexuality by feeding. Asexual individuals of *D. j. japonica* of 18 – 20 mm in length were chosen for these experiments. They were starved for seven days and then treated in one of three ways. In one, the animals were fed frozen whole bodies of sexually matured *B. brunnea* or *P. sapporo* twice a week for eight weeks at 12 – 13 °C. In a second treatment, the animals were fed one of four parts cut from the bodies of sexual individuals of *D. j. japonica,* i.e., the head region containing the brain, the ovary region containing the ovaries and testes, the pharyngeal region with testes, and the post-pharyngeal region containing testes and copulatory apparatus (Fig.1 B). The third treatment were controls in which whole bodies of asexual *D. j. japonica* or chicken liver were given as food.

Culture of dissociated cells. Individuals of *B. brunnea* were rinsed several times in autoclaved culture water with added streptomycin (200 unit/ml) and penicillin (200 unit/ml) and maintained two or three days in the same culture water. They were then directly irradiated with ultraviolet rays from about 30 cm distance for 5 min, and the pre-pharyngeal region of the worm immediately extirpated and cut into 0.1 – 0.5-mm pieces which were treated in the refrigerator at 10 – 12 °C for 24 – 48 hours in Eagle's MEM with amino acids and vitamins at five times the concentration of the amino acids usually used for mammalian tissue (hereinafter called E.A.V. five-times Medium treatment). They were then removed into a Kooper dish with normal E.A.V. and observed.

Histology. Regenerating pieces and planarians with induced sexuality were fixed with Zenker's or Bouin's fluid. Paraffin-embedded material was sectioned and stained with Delafield's hematoxylin and eosin. Chromosome analysis was made in the regenerating pieces by the air-drying method (Teshirogi & Ishida, 1981).

Results

1. Regeneration of Sphalloplana zeschi

The head-frequency curve decreased gradually from the anterior to the posterior levels. The cut pieces of this species remained unchanged for the first seven days after cutting. The speed of regeneration was remarkably delayed compared with the epigean planarian, *D. dorotocephala* (Fig. 2). In contrast, the frequency of tail formation decreased from the posterior towards the anterior levels, even in the later stage of regeneration. Apparently the tail-formation frequency curve became the reverse of the head-frequency curve.

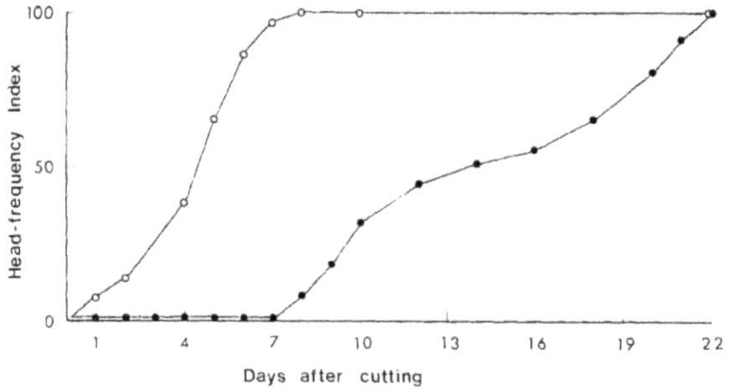

Fig. 2. Comparison of head-regeneration speed between *S. zeschi* and *D. dorotocephala.* Pieces in region adjoining head region used (level-1 piece in Fig. 1). ○, *D. dorotocephala.* ●, *S. zeschi.*

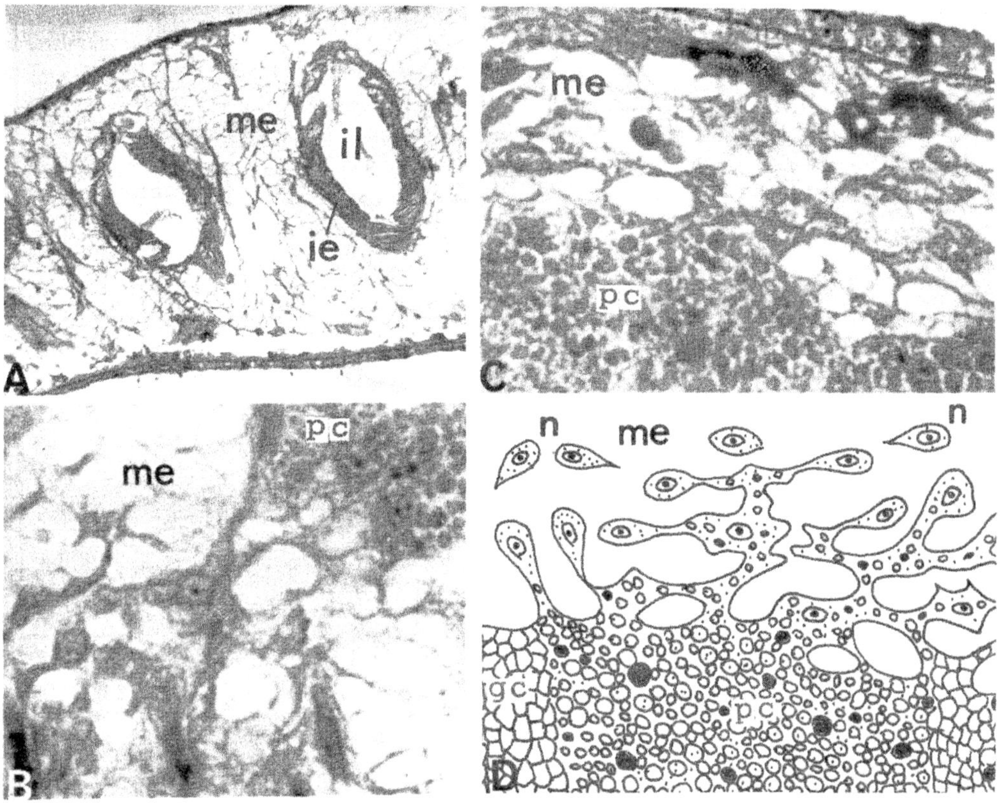

Fig. 3. Regenerating pieces of *S. zeschi.* A, Longitudinal section of intact worm; few neoblasts are seen in mesenchyme. B, Five days after cutting. Intestinal epithelium begins to extend into mesenchyme. C, Eight days after cutting. Amoeboid phagocytic cells migrating actively into mesenchyme. D, Diagrammatic representation of intestinal cells transforming into neoblasts. gc: granular club. ie: intestinal epithelium. il: intestinal lumen. me: mesenchyme. n: neoblast. pc: phagocytic cell.

Histological observations

Few neoblasts were evident in intact *S. zeschi* as compared with *D. dorotocephala* (Fig. 3A). However, in the course of time after cutting, the number of neoblasts increased rapidly. During this time, intestinal epithelium transformed to a sponge-like or amoeboid structure and then dissociated into phagocytic cells or granular clubs, which had relatively little cytoplasm around the nucleus. Soon after dissociation from intestinal epithelium, the liberated cells contained food vacuoles common to phagocytic cells and varied in size and shape, but in time they became oval. Some of them appeared to be transforming into neoblasts forming the first pseudopodia (Fig. 3B-D).

2. Induction of sexuality by feeding in asexual D. j. japonica

When an asexual strain of *D. j. japonica* was fed on whole bodies of sexual worms of *B. brunnea* or *P. sapporo,* sexuality was induced (Table 1). When fed on regions cut from sexual worms of the same species, the worms became mature only with feeding on ovary, pharyngeal, and post-pharyngeal regions (Fig. 4). But in worms given the head region, only an incomplete ovary was induced, and no other genital organs were formed. In worms given asexual strain *D. j. japonica* or chicken liver, sexuality was not induced.

The chromosome number of the sexual strain *D. j. japonica* from Sakunami was $2n=16$, $n=8$. In the asexual strain from Mitaki two karyotypes were found: individuals with $3n=24$ and a mixoploid with neoblasts having $2n=16$ and $3n=24$ (Fig. 5). In the mixoploid the number of neoblasts of $3n=24$ was remarkably low. Sexuality was induced only in mixoploid individuals, not in triploid ones.

A sexuality-induced individual deposited one cocoon, 2 mm in diameter, and after 35 days ten juve-

Table 1. Induction of sexuality in an asexual strain of *Dugesia japonica japonica* fed on mature planarians.

	Food			Induction of sexuality — Months after feeding 1	2	3	4	5	6	Non-induction	Total
Experiment	Sexual worm	*Bdellocephala brunnea*	(1)	0	0	0	0	2	5	5	10
		Polycelis sapporo	(2)	0	0	0	0	1	4	6	10
	Sexual strain *D. j. japonica*	Head region	(3)	0	0	0	0	0	4*	14	18
		Ovary region	(4)	0	0	5	8	9	12	6	18
		Pharyngeal region	(5)	0	1	7	7	11	13+1**	4	18
		Post-pharyngeal region	(6)	0	1	4	7	10	12+1**	5	18
Control	Asexual strain	*D. j. japonica*	(7)	0	0	0	0	0	0	10	10
	Chicken liver		(8)	0	0	0	0	0	0	10	10

(1), (2) and (7), whole body.
(3)~(6), Each region is indicated in Fig. 1, B.
* with one or two imperfect ovaries.
** with two copulatory apparatuses.

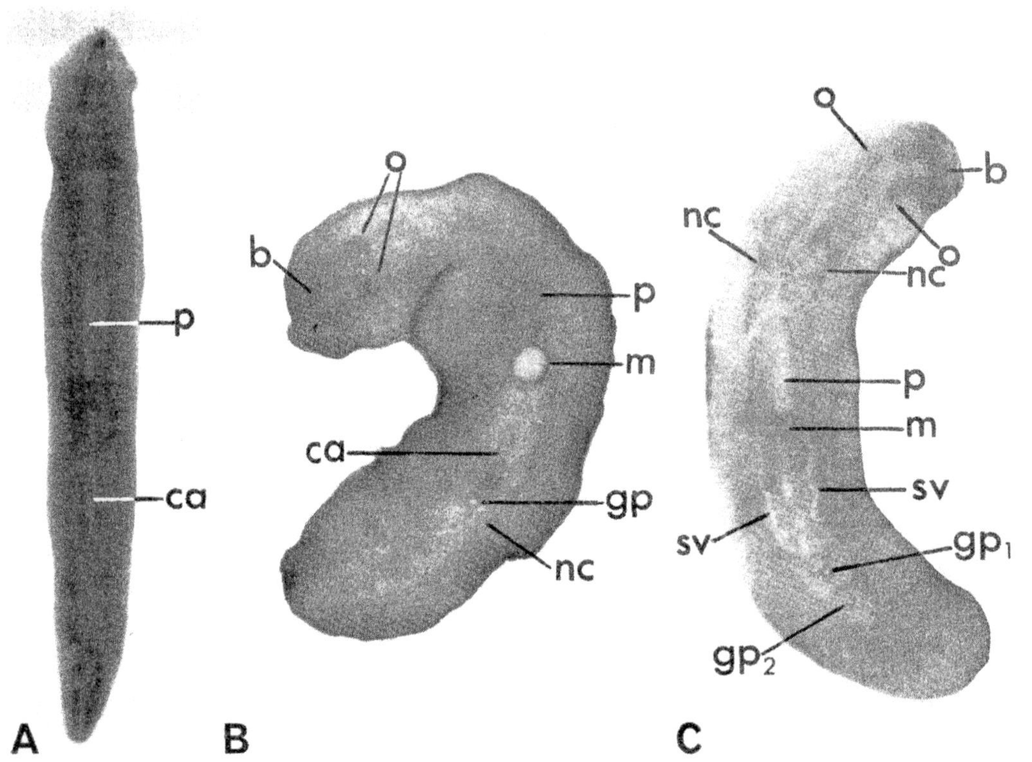

Fig. 4. Individuals from asexual race *D. j. japonica* with sexuality induced. A, dorsal view. B, ventral view. C, individual with two copulatory apparatuses, ventral view. b: brain. ca: copulatory apparatus. gp: genital pore. m: mouth. nc: nerve cord. o: ovary. p: pharynx. sv: spermiducal vesicles.

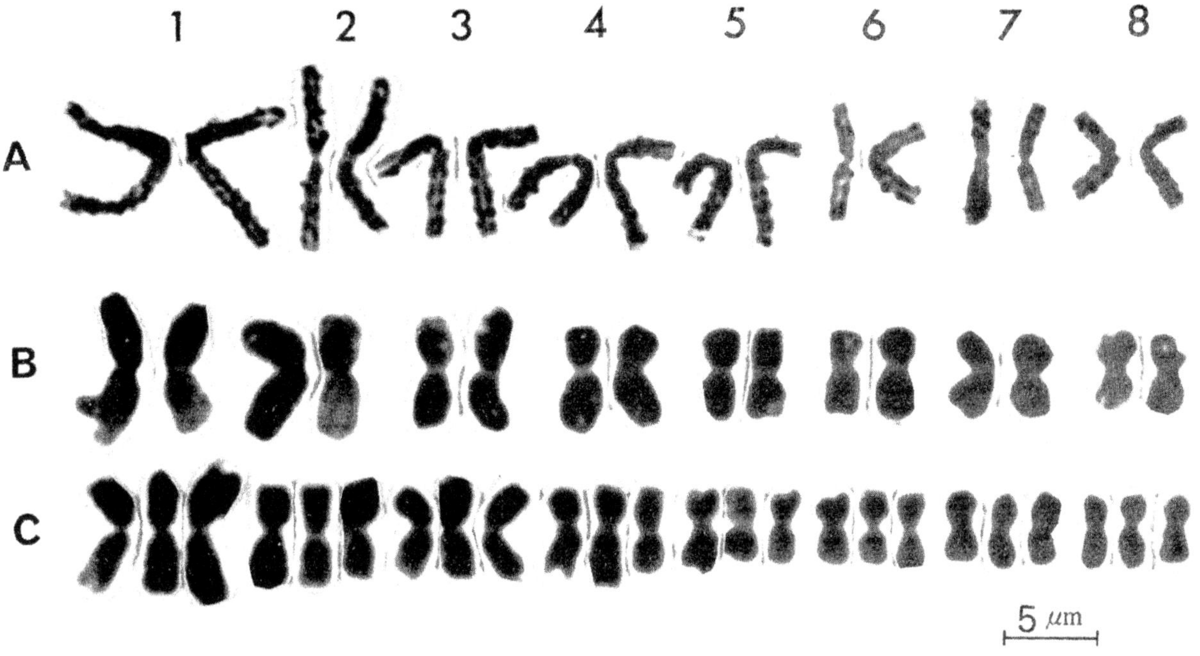

Fig. 5. Karyotypes of neoblasts in regenerating pieces of *D. j. japonica.* Chromosomes arranged according to length. A, sexual race from Sakunami (2n=16). B and C, asexual race from Mitaki, single specimen; mixoploid of 2n=14 (B) and 3n=24 (C).

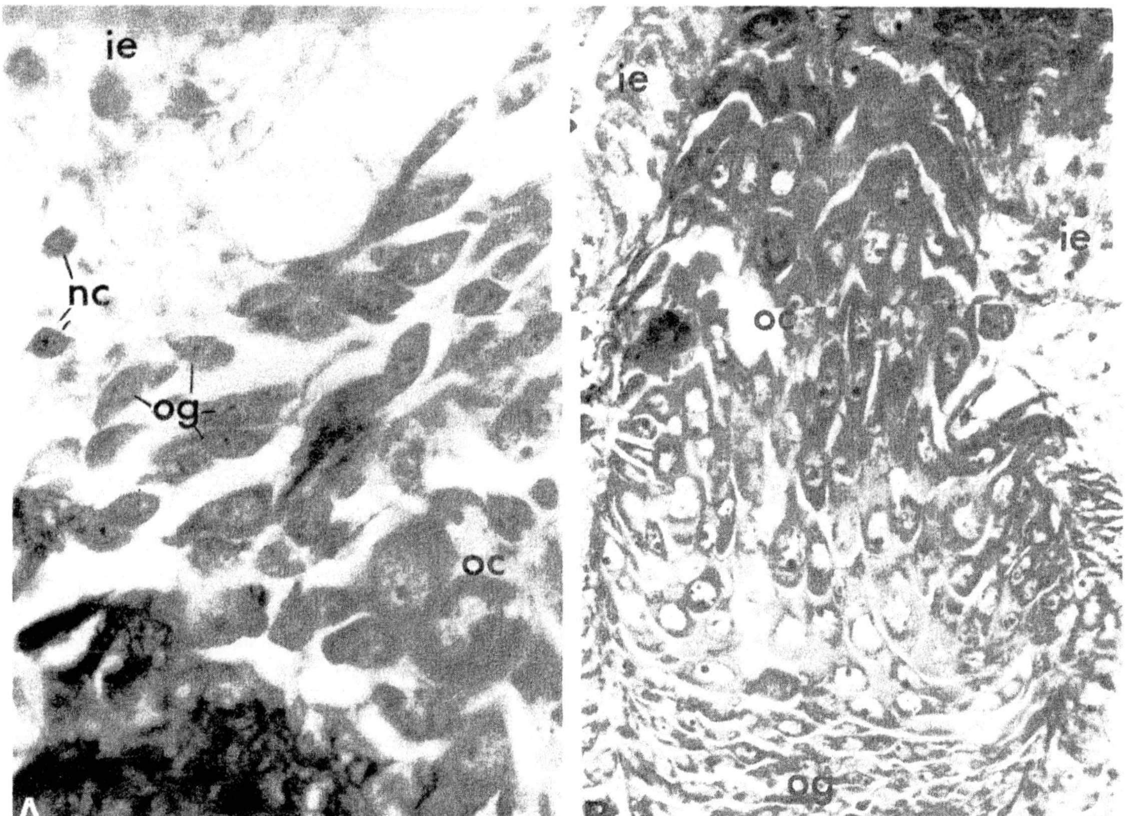

Fig. 6. Ovary induced in previously asexual *D. japonica japonica.* Differentiating ovarial region. B. Completed ovary. ie: intestinal epithelium. nc: neoblast-like cell. oc: oocyte. og: oogonia.

nile worms hatched from it. By about six months later, they had grown to 18-mm length and were sexually mature.

Ovary formation

Figure 6 shows ovary formation in individuals of induced sexuality. As in the case of regeneration of *S. zeschi,* phagocytic cells and granular clubs dissociated from the intestinal epithelium with a small amount of cytoplasm around the nucleus. The dissociated cells became neoblast-like cells and transformed into oogonia, which developed into oocytes (Fig. 6A). The induced ovary was remarkably large, touching the epidermis dorsally and with oogonia extending in many transverse layers ventrally (Fig. 6B).

Testis formation

Figure 7 shows the process of testis differentiation. At the initial stage, cells at the tips of intestinal branches differentiated and transformed into a group of neoblast- or spermatogonium-like cells. Such primitive testes and intestinal branches ran together without boundary (Fig. 7A). At the more advanced stage, a boundary began to be formed between the intestine and the induced testes. In the complete testis, spermatogonia, spermatocytes, spermatids, and sperm were observed. This observation suggests that the intestinal cells dedifferentiated into neoblasts and then redifferentiated into spermatogonia, resulting in testis formation.

Formation of copulatory apparatus

The copulatory apparatus appeared first as an assembly of numerous neoblasts. It could be recognized in an early form one month after feeding and was completed in five or six months (Table 1). The completed copulatory apparatus was morphologically almost the same as that of the sexual strain of *D. j. japonica,* and the penis showed the dorsal-ventral asymmetric structure peculiar to this species.

Differentiation of yolk gland

Yolk glands differentiated in asexual *D. j. japonica* when they were fed mature planarians (Fig. 9, cf. Fig. 8). Three months after feeding, granular clubs began to transform into yolk gland cells, and dense granules in them came to resemble, morphologically, the yolk globules of yolk gland cells. After about

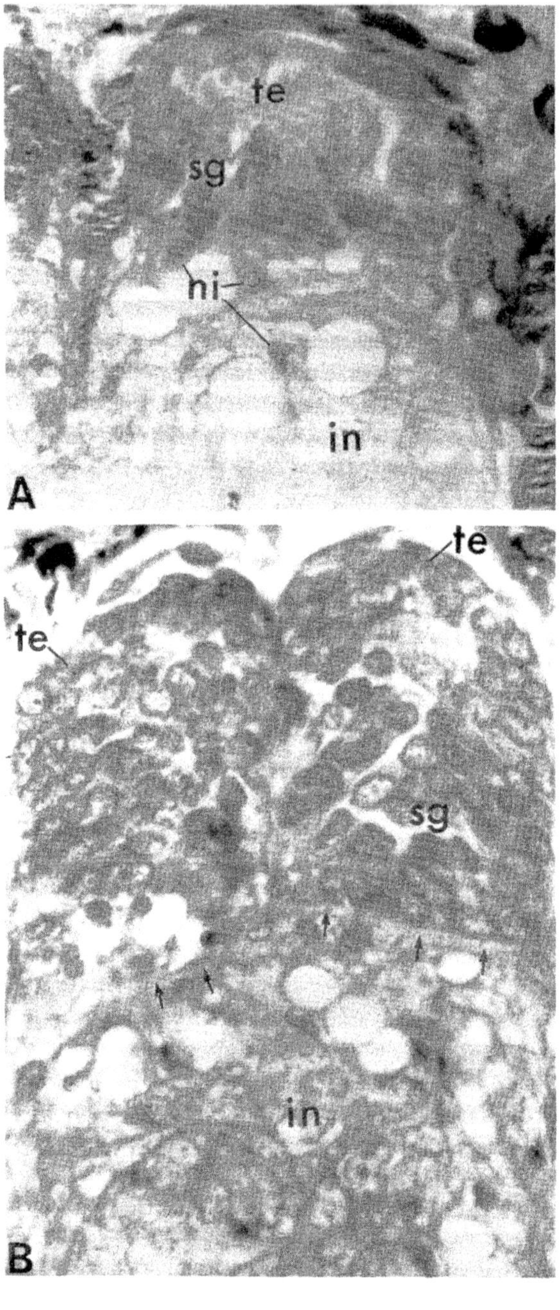

Fig. 7. Testis induced in previously asexual *D. j. japonica.* A, Two months after feeding. Tip of the intestinal branch showing neoblast- or spermatogonium-like cells. There is no boundary between the testis and the intestine. B, Three months. Boundary is forming between the intestine and testes (arrows). in: intestine. ni: nucleus of intestinal cells. sg: spermatogonium. te: testis.

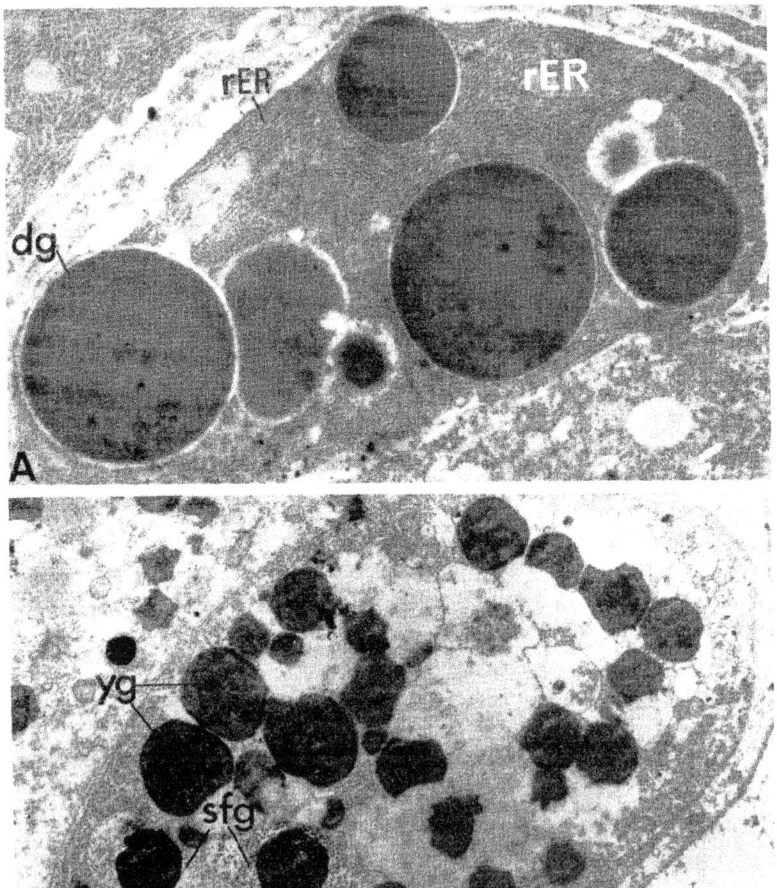

Fig. 8. TEM of granular club (A) and yolk gland cell (B) in *D. j. japonica.* dg: dense granule, rER: rough endoplasmic reticulum, sfg: shell forming granule, yg: yolk globule.

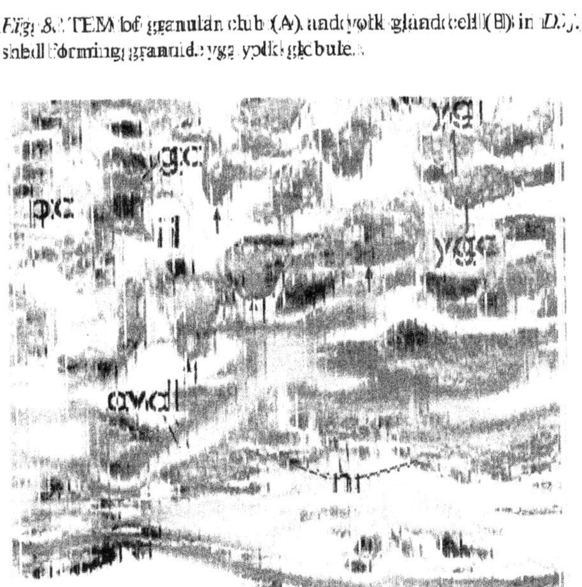

Fig. 9. Yolk gland cells in previously asexual *D. j. japonica* fed on sexual individuals of *B. brunnea.* Three months after feeding. Arrows indicate beginning of transformation into yolk gland cell. gc: granular club, il: intestinal lumen, nr: neoblast, ovd: ovovitelline duct, pc: phagocytic cell, yg: yolk globule, ygc: yolk gland cell.

four months, the granular clubs were transformed completely into yolk gland cells, and the yolk gland was nearly entirely formed. Within four or five months, the yolk glands were completely connected with ovovitelline ducts.

3. *Culture of isolated cells*

About 24 hours after removal into EAW normal

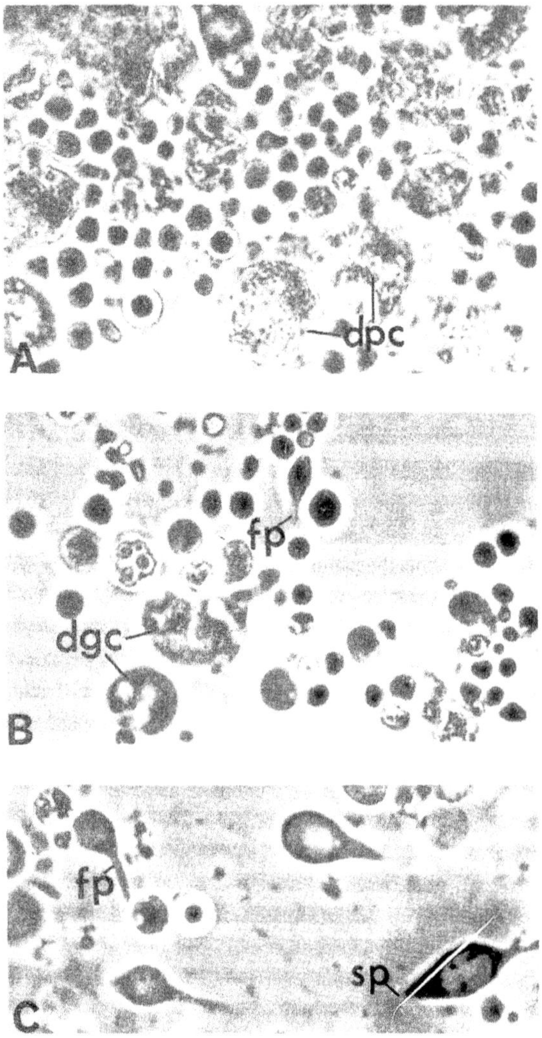

Fig. 10. Culture of dissociated planarian cells. A, Twenty-four hours after treatment with E.A.V. five times Medium. Small cut pieces disintegrated. B, Two days after treatment. Neoblast-like cells with first pseudopodium. C, Three days after treatment. Many neoblasts with first and second pseudopodia. fp: first pseudopodium. dgc: disintegrated granular club. dpc: disintegrated phagocytic cell. sp: second pseudopodium.

Medium, cultured prepharyngeal pieces of *B. brunnea* dissociated, and neoblast-like cells, which were spherical or oval in shape and of varied size (Fig. 10A), arose from the intestinal epithelium. After 24–48 hours, some of them had a pseudopodium (Fig. 10B). Hours later, these cells formed second pseudopodia (Fig. 10C).

Discussion

The intestinal epithelium of freshwater planarians consists of two different types of cells: phagocytic cells and granular clubs. Previously, several investigators (Lang, 1881; Arnold, 1909) regarded the granular clubs as gland cells because of their morphological characteristics, especially the presence of specific proteinaceous granules thought to contain precursors of digestive enzymes. Other investigators (Willier *et al.,* 1925; Jennings, 1957; Ishii, 1965) considered them to be protein reserve cells, and this view is more widely accepted.

1. Origin of neoblasts in regeneration of Sphalloplana zeschi

The head-frequency curve of this species decreases from anterior levels posteriorly, as in the Sapporo population of *Dendrocoelopsis lactea* (Teshirogi *et al.,* 1977). The regeneration speed of *S. zeschi* is much slower than that of the epigean planarian, *D. dorotocephala,* for in the former no regeneration blastema was found a week after cutting, while in the latter, almost 100 percent head formation was attained a week after cutting. The difference in the regeneration speed seems to be due to the number of neoblasts distributed in the intact worms; many neoblasts are distributed in *D. dorotocephala* and other epigean planarians (Teshirogi, 1962), while few neoblasts are distributed in *S. zeschi.* In *S. zeschi,* in the course of days after cutting, the number of neoblasts increased in parallel with regeneration. It appeared that most phagocytic cells and granular clubs of the intestinal epithelium dissociated into the mesenchyme and transformed into neoblasts. Thus, in the regeneration of *S. zeschi,* the embryonic-stock theory on the origin of neoblasts is refuted.

Juvenile *B. brunnea,* immediately before or after hatching, have a high regenerative capacity and regenerate within two days (Teshirogi, 1955). In this case, it is highly probably that the neoblasts originate from embryonic stock cells. In the intact adult worms of this species many neoblasts are always widely distributed and play the principal role in formation of a regeneration blastema. However, these neoblasts may not originate from embryonic cells but from the intestinal cells, from which they have transformed. The same conclusion was reached by Woodruff & Burnett (1965) who reported that in *D. tigrina*, granular clubs transform into neoblasts by dedifferentiation, and by Rose & Shostak (1968) who reported a similar transformation from phagocytic cells in *Phagocata gracilis*.

2. Origin of neoblasts in induction of sexuality

Using agamic fissiparous specimens of *D. gonocephala*, Grasso & Benazzi (1973) succeeded in inducing hermaphroditic genital organs by feeding crushed tissues of sexually mature *Polycelis nigra*. They proposed that the sexuality-inducing factor(s) was of neurosecretory origin. Sakurai (1981) induced sexuality in an asexual strain of *D. j. japonica* by feeding it frozen sexual worms of the same species or of *B. brunnea*. He stated that the sexualizing factor(s) necessary for complete maturation of the gonads was not distributed homogeneously throughout the body of sexual worms for the head piece lacked that factor(s). In the present experiments, sexuality was induced when the worms were fed body pieces other than the head region; while in worms fed the head region only, incomplete ovaries were rarely induced, as reported by Sakurai (1981). The testes must have an important role in inducing sexuality, because they are present in the ovarian, the pharyngeal, and the copulatory apparatus regions. It seems that the sexualizing factors are not species-specific, as shown by Benazzi & Grasso (1977) and Sakurai (1981).

It appears that the neoblasts forming ovaries and testes originate from intestinal cells. Histological observation reveals that phagocytic cells and granular clubs dissociate from the intestinal epithelium to transform into neoblasts, and granular clubs transform directly into yolk gland cells. As mentioned above, granular clubs are protein reserve cells (Willier *et al.*, 1925; and others). Yolk cells contain many shell-forming granules and yolk granules (Fig. 8B). According to Yanagita & Yamamoto (1981), shell-forming granules consist of many kinds of amino acids. It is well known that yolk granules consist of many kinds of proteins, phospholipids, and neutral fats (Yamamoto, 1968). It appears that the shell-forming granules and yolk granules in the induced yolk gland cells originate from dense granules in the granular clubs, but further study on this relationship is expected in the future.

Most of the neoblasts that assemble at the initial stage of copulatory apparatus formation may originate from the intestinal cells, but the present study could not affirm that.

It is noteworthy that asexual-strain individuals with induced sexuality laid eggs and that the progeny grew to be sexual individuals.

3. Observation of dissociated cells

The cut pieces of planarians treated with E.A.V. at high amino-acid concentration began to deform about two days after treatment, and many neoblast-like cells appeared. Disintegration of intestinal cells in these was remarkable (Fig. 10). The neoblast-like cells formed one and sometimes a second pseudopodium and had a small cytoplasm-to-nucleus size ratio as is characteristic of neoblasts (Betchaku, 1967). These findings support the theory that neoblasts originate from intestinal cells.

Acknowledgments

I express my thanks to Professor Dr. R. W. Mitchell of Texas Tech University for help in providing room and materials. The present study was partly supported by a Grant-Aid for Scientific Research from the Japanese Ministry of Education, Science and Culture (1979).

References

Arnold, G., 1909. Intra-cellular and general digestive processes in Planariae. Quart. J. micr. Sci. 54: 207. (cited from Willier *et al.*, 1925.)

Baguña, J. & R. Romero, 1981. Quantitative analysis of cell types during growth, degrowth and regeneration in the planarians *Dugesia mediterranea* and *Dugesia tigrina*. Hydrobiologia 84: 181–194.

Bartsch, O., 1923. Die Histiogenese der Planarienregenerate. Arch. mikr. Anat. 99: 187–221.

Benazzi, M. & M. Grasso, 1977. Comparative research on the sexualization of fissiparous planarian treated with substances contained in sexual planarians. Monit. zool. ital. 11: 9–19.

Betchaku, T., 1967. Isolation of planarian neoblasts and their behavior in vitro with some aspects of the mechanism of the formation of regeneration blastema. J. exp. Zool. 164: 407–434.

Brønsted, H. V., 1955. Planarian regeneration. Biol. Rev. 30: 65–126.

Coward, S. J., 1969. Regeneration in planarians: some unresolved questions and problems. J. biol. Psychol. 11: 15–19.
Curtis, W. C., 1936. Effects of X rays and radium upon regeneration. Biological Effects of Radiation 1: 411–458.
Curtis, W. C. & L. M. Schulze, 1934. Studies upon regeneration. I. The contrasting powers of regeneration in planaria and procotyla. J. Morph. 55: 477–512.
Grasso, M. & M. Benazzi, 1973. Genetic and physiologic control of fissioning and sexuality in planarians. J. Embryol. exp. Morph. 30: 317–328.
Hay, E. D., 1968. Dedifferentiation and metaplasia in vertebrate and invertebrate regeneration. In H. Ursprung (ed.), The Stability of the Differentiated State. Springer-Verlag, Berlin, N.Y.: 85–108.
Ishii, S., 1965. Electron microscopic observations on the planarian tissue II. The intestine. Fukushima J. med. Sci. 12: 67–87.
Jennings, J. B., 1957. Studies on feeding, digestion, and food storage in free-living flatworms (Platyhelminthes: Turbellaria). Biol. Bull. 112: 63–80.
Lang, A., 1881. Der Bau von Gunda segmentata und die Verwandtschaft der Plathelminthen mit Colenteraten und Hirudineen. Mich. Zool. Stat. Neapel 3: 187. (cited from Willier *et al.,* 1925.)
Lender, T., 1962. Factors in morphogenesis of regenerating freshwater planaria. In M. Abercrombie & J. Brachet (eds), Advances in Morphogenesis 2. Academic Press, New York: 305–331.
Rose, C. & S. Shostak, 1968. The transformation of gastrodermal cells to neoblasts in regenerating *Phagocata gracilis.* Exp. Cell Res. 50: 553–561.
Sakurai, T., 1981. Sexual induction by feeding in an asexual strain of the freshwater planarian, *Dugesia japonica japonica.* Annot. zool. Jap. 54: 103–112.
Teshirogi, W., 1955. On the regeneration of a turbellarian, *Bdellocephala brunnea,* especially its frequencies of head, tail, pharynx and genital organ formation. Zool. Mag. 64: 294–299. (In Jap. With Eng. summ.)
Teshirogi, W., 1962. Dynamic morphological change and timetable during the regeneration of the turbellarian *Bdellocephala brunnea.* Sci. Rep. Hirosaki Univ. 9: 21–48.
Teshirogi, W., S. Ishida & H. Yamazaki, 1977. Regenerative capacities of transverse pieces in the two species of freshwater planarian, *Dendrocoelopsis lactea* and *Polycelis sapporo.* Sci. Rep. Hirosaki Univ. 24: 55–72.
Teshirogi, W. & S. Ishida, 1981. Studies on the speciation of Japanese freshwater planarian *Polycelis auriculata* based on the analysis of its karyotypes and constitutive proteins. Hydrobiologia 84: 69–77.
Weigand, K., 1930. Regeneration bei Planarien und Clavilina unter dem Einfluss von Radiumstrahlen. Z. wiss. Zool. 136: 255–318.
Willier, B. H., L. H. Hyman & S. A. Rifenburgh, 1925. A histochemical study of intracellular digestion in triclad flatworms. J. Morph. Physiol. 40: 299–340.
Wolff, E. & F. Dubois, 1947. Sur une méthode d'irradiation localisée permettant de mettre en évidence la migration des cellules de régénération chez les Planaires. C. r. Soc. Biol., Paris 141: 903–906.
Woodruff, L. & A. L. Burnett, 1965. The origin of the blastema cells in *Dugesia tigrina.* Exp. Cell Res. 38: 295–305.
Yamamoto, M. 1968. Oogenesis. In J. Ishida (ed.), Developmental Biochemistry. Syôkabo, Tokyo: 22–41. (In Japanese.)
Yanagita, Y. & T. S. Yamamoto, 1981. Formation and chemical property of the cocoon in a freshwater planarian, *Dendrocoelopsis ezensis* Ichikawa et Okugawa (Turbellaria, Tricladida). Zool. Anz. 207: 130–142.

Ultrastructural observations on gastrodermal regeneration in the planarian *Dugesia japonica* (Turbellaria)

Isao Hori
Department of Biology, Kanazawa Medical University, Uchinada-machi, Ishikawa-ken 920-02, Japan

Keywords: Turbellaria, *Dugesia japonica*, regeneration, gastrodermis, ultrastructure

Abstract

The earliest detectable change during regeneration of the gastrodermis in *Dugesia japonica* was an aggregation of regenerative cells underneath the gastrodermis remaining at the wound margin. The gastrodermal cells in experimental regenerates retained some of their original characters and presented no indication of cell dedifferentiation. The regenerative cells came into contact with the basal surface of gastrodermal cells, forming stratified cell layers. Differentiation of these cells into gastrodermal cells was initiated by the development of synthetic organelles within their cytoplasm. These differentiating cells gave rise to two different types of gastrodermal cells, namely phagocytic cells and sphere cells. In later stages, there was an apparent movement of differentiated gastrodermal cells towards the parenchyma.

Introduction

Several authors have postulated that most regenerative cells forming the planarian blastema are derived from dedifferentiation of gastrodermal cells (Woodruff & Burnett, 1965; Rose & Shostak, 1968; Coward, 1979; Teshirogi, 1985). By morphological criteria, however, this view of the origin of undifferentiated cells cannot be accepted.

The planarian gastrodermis, though complex in its structure, is composed of only two types of cells, namely phagocytic cells and sphere cells. Thus we can follow comparatively easily how gastrodermis is regenerated. The precise mechanism of gastrodermal regeneration, however, has not yet been established. For example, the extent to which regenerative cells are responsible for reconstitution of this organ has not been made clear.

The definite markers with which we can identify types of cells before any specific inclusions are produced in them are their synthetic organelles, including the nuclear envelope which determines into what kind of cells they will develop (Hori, 1983a). Although several reasonable suggestions have been offered by Lentz (1967) and Spiegelman & Dudley (1973), there is still not enough information on the hypothesis of multipotentiality of planarian regenerative cells to support it.

The present study was designed to examine the relation between regenerating gastrodermis and regenerative cells involved in the process of wound healing.

Material and methods

The triclad flatworm *Dugesia japonica* was employed. After starvation for seven days, anterior parts of worms transected at the prepharyngeal region were allowed to regenerate in tap-water for one to fourteen days. Some of the samples were placed in a solution containing colchicine (M/2000) for two days immediately after transection. All regenerates were carefully trimmed and fixed in 2.5% glutaraldehyde in 0.1 M cacodylate buffer (pH 7.4) for 60 min. Subsequently, they were

Hydrobiologia 132, 217–222 (1986).

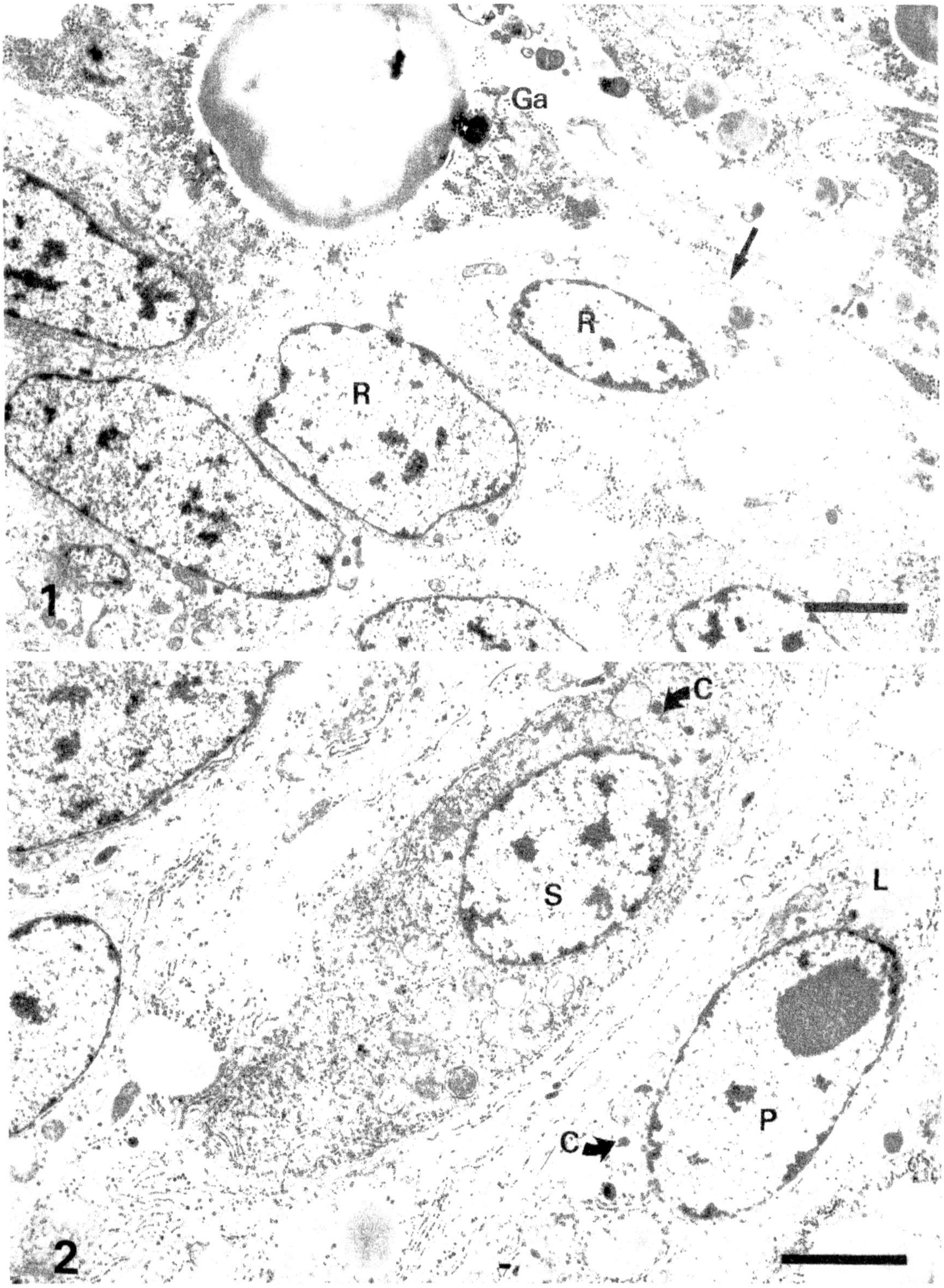

Figs 1–2. (1) Boundary region at the wound margin, 2 days post-wounding. Treated with colchicine (M/2000). Arrow: regenerative cell (R) closely associated with gastrodermis (Ga). Scale=2 μm; (2) Stratified cell layer of regenerating gastrodermis, 4 days post-wounding. Phagocytic cell. S–Sphere cell, L–Lipid droplet. C–Chromatoid body. Scale=2 μm.

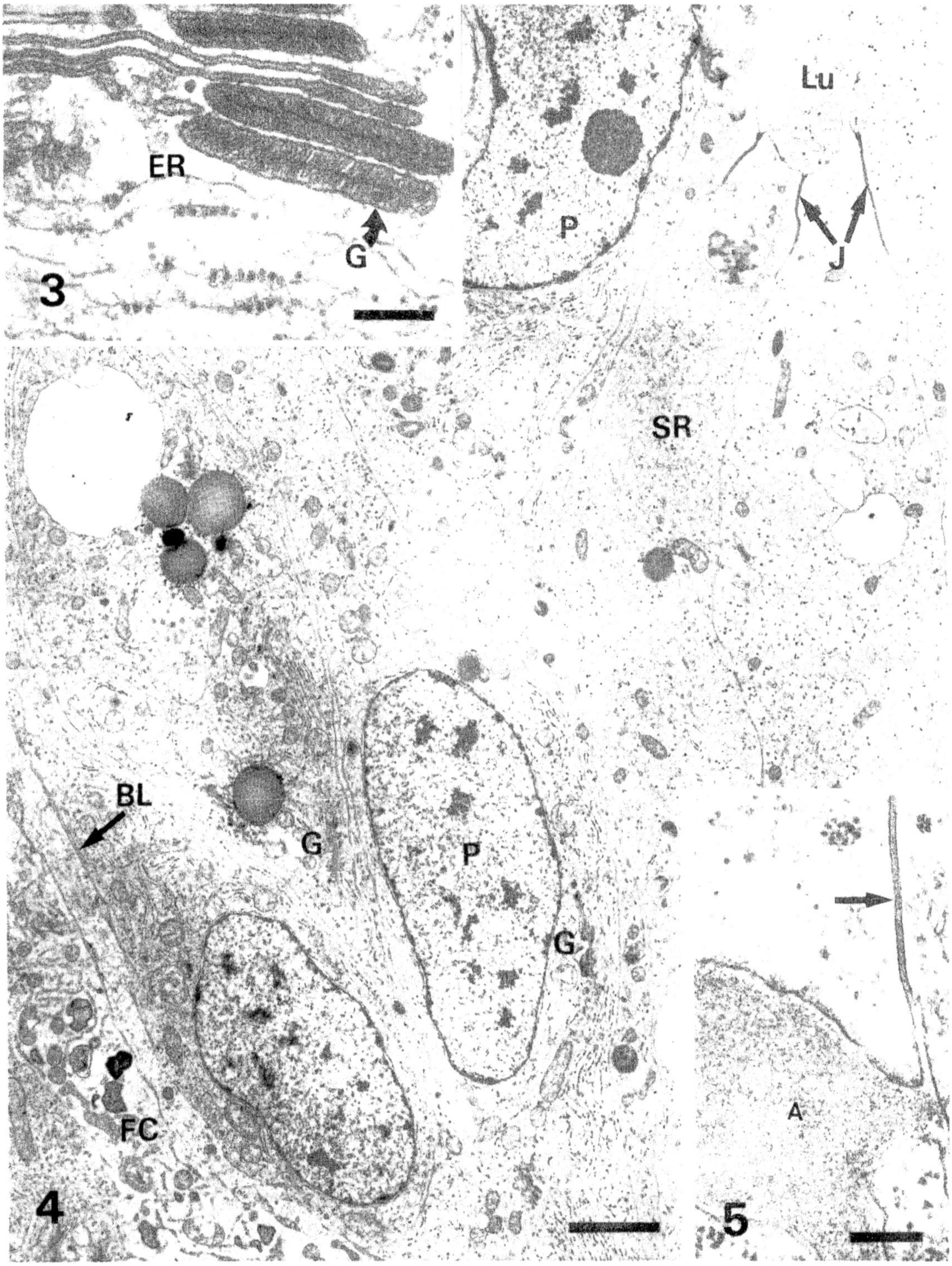

Figs. 3–5. (3) Within the cytoplasm of the differentiating phagocytic cell, 7 days post-wounding. ER–RER containing fibrous materials. G–Golgi lamella containing feather-like structures. Scale=0.2 μm; (4) Low magnification view of the regenerated gastrodermis, 7 days post-wounding. P–Phagocytic cell. SR–SER vesicles. Lu–Intestinal lumen. J–Septate junctions. FC–Fixed parenchyma cell. BL–Basal lamina. G–Golgi complex. Scale=2 μm; (5) Basal portion of gastrodermis, 7 days post-wounding. Arrow indicates a gap junction. A–Amorphous precursor of basal lamina. Scale=0.2 μm.

washed in buffer and post-fixed in 2% osmium tetroxide in 0.1 M cacodylate buffer (pH 7.4) for 90 min. Tissues were dehydrated in a graded series of alcohols and embedded in Epon 812. Thin sections were cut using a diamond knife and stained with uranyl acetate and lead citrate. Micrographs were taken on a Hitachi H-500 electron microscope at 100 kV.

Results

In the initial stage, shortly after completion of wound closure, numerous regenerative cells, which appeared more or less elongated, were seen gathering close to the basal surface of the gastrodermis at the wound margin. These cells could be observed even in colchicine-treated worms (Fig. 1). Although most gastrodermal cells in this region seemed structurally intact, a close examination revealed some degenerative features in their cytoplasm. Cytological aspects indicating dedifferentiation, however, were not apparent in any tissue cells in the vicinity of wounded areas.

In the subsequent stage, these regenerative cells began to differentiate and form stratified cell layers as though lining the pre-existing gastrodermis. Two to four days after wounding there was a rapid increase in the number of differentiating phagocytic cells. This predominant cell type was characterized by numerous lipid droplets within cytoplasm of low electron-density (Fig. 2). Besides the phagocytic cells with earlier stages of differentiation, there was another type of cell with dense cytoplasm. Most of this cytoplasm comprised a large populaton of free and membrane-bound ribosomes (Fig. 2), and lipid inclusions were absent. It is presumed from their appearance that these cells were a juvenile form of the sphere cell. In both cell types conspicuous chromatoid bodies were present (Fig. 2).

Development of the regenerating gastrodermis markedly progressed by five to seven days. During this period the homogeneity of the gastrodermal cell became particularly obvious (Fig. 4). The se-

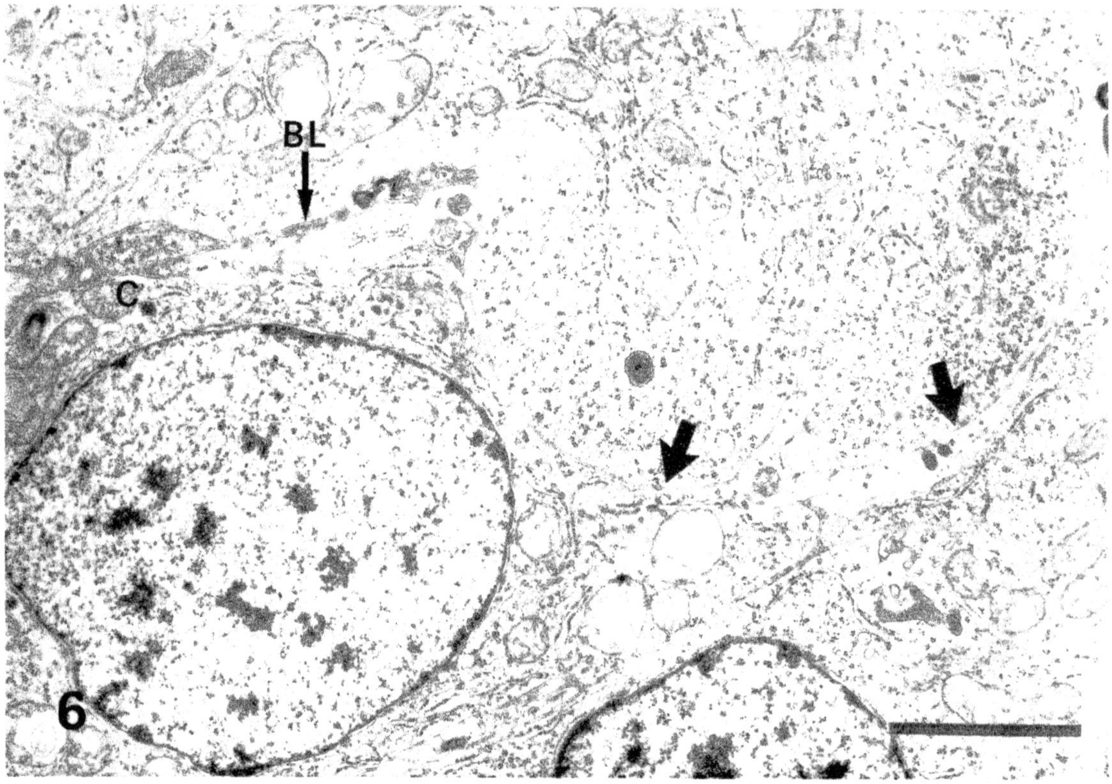

Fig. 6. Leading edge of developing gastrodermis, 7 days post-wounding. A direct contact between this edge and regenerative cells are marked (arrows). BL. New basal lamina. C. Chromatid body. Scale = 2 μm.

quence of differentiation of individual phagocytic cells appeared to begin with the accumulation of fibrous contents in cisternae of the RER (Fig. 3). The most noticeable feature of these cells was the development of Golgi lamellae that included specific 'feather-like structures' of high electron density (Figs 3 & 4). Some of the phagocytic cells acquired ultrastructural characters of fully differentiated states, with a polarity of intracellular inclusions such that SER vesicles accumulated in the apical cytoplasm. The apical lateral borders were joined by septate junctions while the basolateral borders were joined by minute gap junctions as is characteristic of the mature gastrodermis (Figs 4 & 5). Chromatoid bodies became obscure as cell differentiation advanced.

Precursors of basal lamina were noticeable as amorphous deposits just beneath the well differentiated gastrodermal cells (Fig. 5). A continuous sheet of the basal lamina was seen intervening between newly differentiated fixed parenchymal cells and gastrodermal cells (Fig. 4).

It is noteworthy that basal cytoplasmic portions of the new gastrodermal cells tended to project into the parenchyma in later stages (Fig. 6). Often several regenerative cells came in contact with the leading edge of these cytoplasmic protrusions, apparently contributing to the development of gastrodermis. The interspace at the boundary region was devoid of formed basal lamina which was more distinct at proximal regions (Fig. 6). The new basal lamina was necessarily supported by underlying parenchymal cells.

Discussion

The general organization of gastrodermal cells located at the wound margin did not appear to be markedly affected by wounding while most cells at the wound surface were completely destroyed. It is true that we cannot rule out completely the possibility that planarian regenerative cells are derived from dedifferentiated cells of other tissues during the initial stage of regeneration as mentioned by several authors (Hay, 1968; Coward *et al.*, 1976). But in order to prove that such process may occur in the gastrodermis remaining at the wound margin, the following evidence, at least, must be obtained: a) presence of characteristic inclusions that are degenerating by autophagocytosis, b) reappearance of prominent chromatoid bodies, c) migration of some gastrodermal cells toward the wounded area through the basal lamina and d) mitotic figures of these cells in the boundary regions. Such evidence was not obtained in the present study.

Free regenerative cells rapidly gathered around the wounded gastrodermis and then replaced the degenerative cells at the boundary area. Some of these cells came into direct contact with the remaining gastrodermal cells because of the loss of integrity of the basal lamina caused by wounding. Even in the case of colchicine-treated worms, numerous regenerative cells were seen accumulating around the gastrodermis. This suggests that mitosis of undifferentiated cells may not be responsible for the initial aggregation of regenerative cells after wounding.

The fate of planarian regenerative cells is determined in earlier stages, of course, as pointed out by Quick & Johnson (1977) but actually each cell type cannot be identified until these cells display their own characteristic features (Hori, 1983b). There have been no previous descriptions of the peculiar structure (the feather-like structure) in individual Golgi lamellae of differentiating phagocytic cells. This feather-like content was observed in no other tissue cells, thus providing a marker with which we could identify phagocytic cells even in early stages of differentiation. During gastrodermal regeneration, the cells begin to differentiate simultaneously at the wound margin so that most of the cells reconstituting the new gastrodermis appear similar to each other; these cells represent transitory forms between undifferentiated cells and mature gastrodermal cells.

It is still unknown to what extent cell movement occurs during the development of the new gastrodermis. The continued fusion of regenerative cells with the newly formed gastrodermis and cell movement in this process may be simultaneous in later stages. This continuation may account for the highly branched morphology characteristic of the intestine in triclads. Betchaku (1967) has observed the behaviour of planarian tissue *in vitro* and depicted the gastrodermal cells as having an ability to move actively.

Sufficient resolution was not achieved with regard to the differentiation of sphere cells, but the present data suggest that they also come from regenerative cells.

Acknowledgments

The author would like to thank Miss N. Hasegawa for her helpful assistance.

References

Betchaku, T., 1967. Isolation of planarian neoblasts and their behavior in vitro with some aspects of the mechanism of the formation of regeneration blastema. J. exp. Zool. 164: 407–434.

Coward, S. J., 1979. On the occurrence and significance of annulate lamella in gastrodermal cells of regenerating planarians. Cell Biol. Int. Rep. 3: 101–106.

Coward, S. J., C. E. Burnett & B. L. Hazlehurst, 1976. Lysosomes and lysosomal enzyme activity in the regenerating planarian: Evidence in support of dedifferentiation. J. exp. Zool. 189: 133–146.

Hay, E. D., 1968. Fine structure and origin of regenerative cells in planarians. Anat. Rec. 160: 363.

Hori, I., 1983a. Ultrastructural variation of rough endoplasmic reticulum – relation to the types of planarian regenerative cells. J. submicrosc. Cytol. 15: 975–989.

Hori, I., 1983b. Differentiation of myoblasts in the regenerating planarian Dugesia japonica. Cell Differentiation 12: 155–163.

Lentz, T. L., 1967. Rhabdite formation in planaria: The role of microtubules. J. ultrastruct. Res. 17: 114–126.

Quick, D. C. & R. G. Johnson, 1977. Gap junctions and rhombic particle arrays in planaria. J. ultrastruct. Res. 60: 348–361.

Rose, C. & S. Shostak, 1968. The transformation of gastrodermal cells to neoblasts in regenerating Phagocata gracilis (Leidy). Exp. Cell Res. 50: 553–561.

Spiegelman, M. & P. L. Dudley, 1973. Morphological studies of regeneration in the planarian Dugesia tigrina: a light and electron microscopic study. J. Morph. 139: 155–184.

Teshirogi, W., 1985. On the origin of neoblasts in freshwater planarians. This volume.

Woodruff, L. S. & A. L. Burnett, 1965. The origin of the blastemal cells in Dugesia tigrina. Exp. Cell Res. 38: 295–305.

Protein phosphorylation and the role of Ca^{2+} in planarian turbellarian regeneration

J. Moraczewski[1], I. Martelly & R. Franquinet
Laboratoire de Biologie Animale, Université Paris Val de Marne, Avenue du Général de Gaulle, Créteil 94010, Cedex France
[1]*Permanent address: Institute of Zoology, University of Warsaw, 00-927, Warszawa, Poland.*

Keywords: DNA synthesis, protein kinases, phosphorylation, regeneration, calcium, calmodulin

Abstract

During planarian regeneration, relationships in timing were established between variations of calcium and calmodulin contents on one hand, and protein phosphorylation and stimulation of DNA and RNA synthesis on the other. Special attention was paid to changes in histone phosphorylation in regenerating fragments. Using *in vitro* experiments on dissociated planarian cells, we demonstrated causal relationships between these events. In particular, the key role of Ca^{2+} in the activation of protein kinases and in the initiation of DNA synthesis was emphasized.

Introduction

Several aspects of the stimulation of DNA and RNA synthesis in regenerating planarians have been studied in our laboratory. Neuromediators, such as serotonin and dopamine, have been shown to stimulate DNA and RNA synthesis through activation of adenylate cyclase systems (Franquinet, 1979, 1981; Franquinet *et al.*, 1978, 1981; Franquinet & Martelly, 1981; Martelly *et al.*, 1981a) and the resultant cellular activation (Moraczewski, 1981). We also showed that variation in Ca^{2+} level was related to rates of nucleic acid synthesis in both regenerating planarians and planarian cell cultures (Martelly *et al.*, 1983; Martelly, 1984a, 1984b). Furthermore, variation in the activities of Ca^{2+}- and cAMP-protein kinases (PK) was discovered during the first 24 h of regeneration (Moraczewski *et al.*, 1983).

In this study we compare the variations in total calcium and calmodulin contents and the variation of endogenous phosphorylation in regenerating fragments of the planarians *Polycelis tenuis* and *Dugesia gonocephala*. Temporal relationships could be detected between these events and initiation of DNA and RNA synthesis after cutting. With *in vitro* experiments on dissociated planarian cells, we have attempted to demonstrate some causal relationships between them, and have found that Ca^{2+} plays a role in the activation of PK and in the initiation of nucleic acid synthesis. Special attention was paid to nuclear proteins, particularly to the histones.

Endogenous protein phosphorylation *in vivo*

Variation in endogenous protein phosphorylation could be interpreted in light of our previous results concerning the changes in PK-cAMP and PK-Ca^{2+} activity during regeneration (Fig. 1a) (Moraczewski *et al.*, 1983). As regards the significance of the *in vivo* protein phosphorylation with [^{32}P]orthophosphate (Fig. 1b), the high level at 1 h might result from an increase in PK-Ca^{2+} activity mostly. Phosphorylation dropped at 2 h, perhaps because of decline in PK-Ca^{2+}. The peak of phosphorylation at 3 h might result chiefly from PK-cAMP activity, as the PK-Ca^{2+} level was low at that time whereas cAMP-PK increased. Similar-

Hydrobiologia 132, 223–227 (1986).

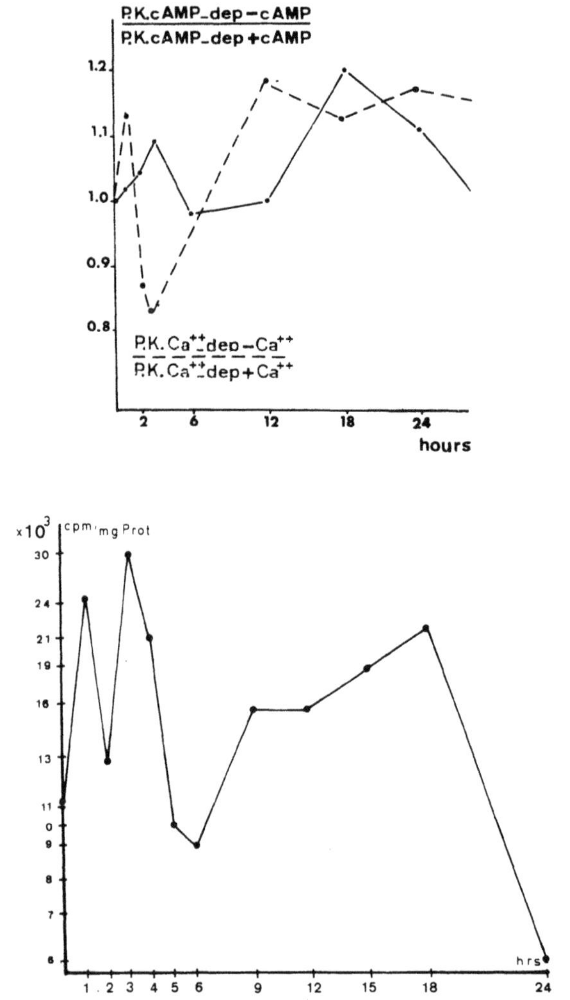

Fig. 1a and b. Endogenous protein phosphorylation during the first 24 h of regeneration.
(a) Variations in activity of protein kinases. Each point represents the ratio between activity of enzyme alone to that with stimulation substance (cAMP or Ca^{2+}, respectively) present. (● – – – ●) Ca^{2+}-dep. PK, (●——●) cAMP-dep. PK.
(b) ^{32}P incorporation by regenerates into proteins phosphorylated *in vivo*.

ly, the low phosphorylation rate observed at 6 h might be related to the fact that both PK activities were low at that moment. The subsequent increase of total endogenous protein phosphorylation at 9–12 h and 15–18 h might be caused principally by PK-Ca^{2+} and PK-cAMP activation, respectively. These increases in protein phosphorylations are related in time to peaks of DNA synthesis (12 h) and RNA synthesis (18 h) (Martelly *et al.*, 1981).

Ca^{2+} and calmodulin *in vivo*

Changes in total Ca^{2+} and calmodulin contents (Fig. 2), and in Ca^{2+} uptake (Fig. 3) could be compared to the pattern of variation of protein phosphorylation during planarian regeneration. In intact animals, we usually found 34 nM Ca^{2+} per planarian. This content decreased drastically (by 57%) within the first minute of cutting (time 0), remained at that low level during the next 6 h (cicatrisation period) and then rose significantly to 25 nM at 12 h of regeneration. Ca^{2+} content fell to time 0 level at 18 h and then rose again.

The effects of calcium are generally mediated by Ca^{2+}-binding proteins, the most widely distributed being calmodulin. We detected calmodulin in planarian tissues, but its concentration was much lower than in vertebrates (Le Peuch *et al.*, 1979; Thomasset *et al.*, 1981). We subsequently attempted to determine whether the increase in total Ca^{2+} observed after Ca^{2+} depletion corresponded to an increase in intracellular Ca^{2+} (Fig. 3). Accordingly, Ca^{2+} uptake in regenerating fragments was measured between 0 and 3 h and between 9.5 and

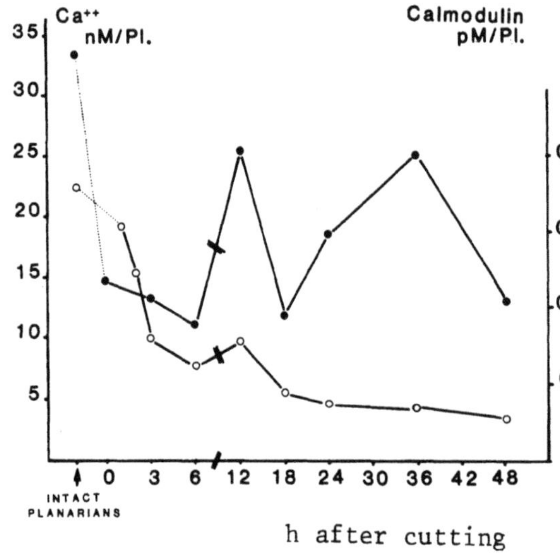

Fig. 2. Variation in total calcium and calmodulin content during the first 48 h of regeneration. (●), Ca^{2+} content; each point is the mean ± SEM for five to six determinations, pooled material from 20 animals was used for each determination. (○), calmodulin content; each point represents the mean of two determinations; pooled material from 100 animals was used for each determination.

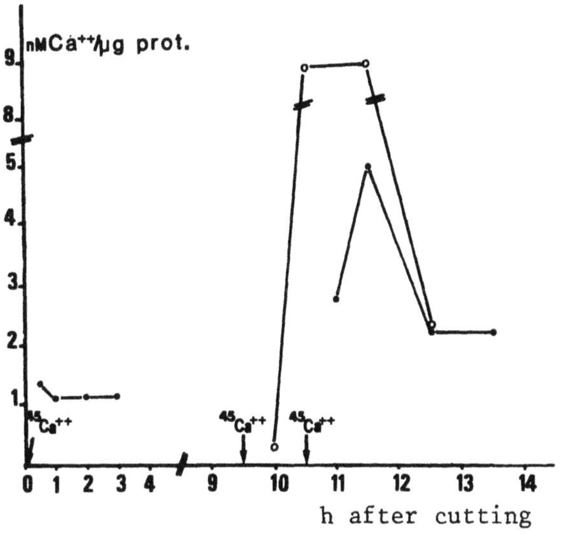

Fig. 3 Ca^{2+} uptake during regeneration. Planarians grown in tap water were cut and transferred to a Ca^{2+}-free buffered solution containing 10 µCi/ml $^{45}Ca^{2+}$, either just after cutting or 9.5 or 10.5 h later. After different periods of incubation in $^{45}Ca^{2+}$, the fragments were washed and dissociated into cells in Ca^{2+}-free medium containing 1 mM $LaCl_3$. Ca^{2+} uptake was then estimated from measurements of $^{45}Ca^{2+}$ contents taking into account the isotopic dilution.

13.5 h after cutting. During the first 3 h of regeneration, Ca^{2+} uptake remained low and unchanged. In contrast, when $^{45}Ca^{2+}$ was added at 9.5 h, Ca^{2+} uptake rose rapidly. The intracellular $^{45}Ca^{2+}$ radioactivity was maximal between 10.5 and 11.5 h, and subsequently decreased. When $^{45}Ca^{2+}$ was supplied at 10.5 h, there was a transitory increase in Ca^{2+} uptake which peaked at 11.5 h, declined until 12.5 h, and then remained unchanged during the next hours. These experiments showed that the increase in total Ca^{2+} content observed at 12 h in animals regenerating in tap water (Fig. 2) was due, at least partially, to the transitory increase in Ca^{2+} uptake which occurred between 10.5 and 12.5 h of regeneration.

These results suggest the need for a Ca^{2+} surge in order to trigger DNA synthesis. Indeed, DNA synthesis reached a first peak at 12 h simultaneously with increase in Ca^{2+} uptake and in Ca^{2+} and calmodulin contents. This Ca^{2+} might, in particular, be necessary for the phosphorylation of proteins involved in initiation of DNA synthesis since increase in PK-Ca^{2+} was observed too at 12 h.

Changes in histone phosphorylation *in vivo*

Phosphorylation of H_1 histones increased 35 fold from 0 to 24 h of regeneration (Fig. 4). It gradually rose from the moment of cutting, and the pattern of phosphorylation during regeneration displayed three peaks at hours 0–3, 6–12 and after 15 h of regeneration. Phosphorylation of H_3 histones increased five fold, chiefly from hours 12 to 21. H_{2b} histones phosphorylation increased four fold between hours 18 and 24. Phosphorylation of H_{2a} histones started 3 h after sectioning and rose seven-fold, showing two peaks at 12–15 h and 18–21 h. The pattern of H_4 phosphorylation was biphasic with the first maximum between hours 6 and 12 (12-fold rise) and the second between hours 18 and 24 (8-fold rise). Thus, time relationships were observed between increases in PK activities, histone phosphorylation, and initiation of nucleic acid synthesis.

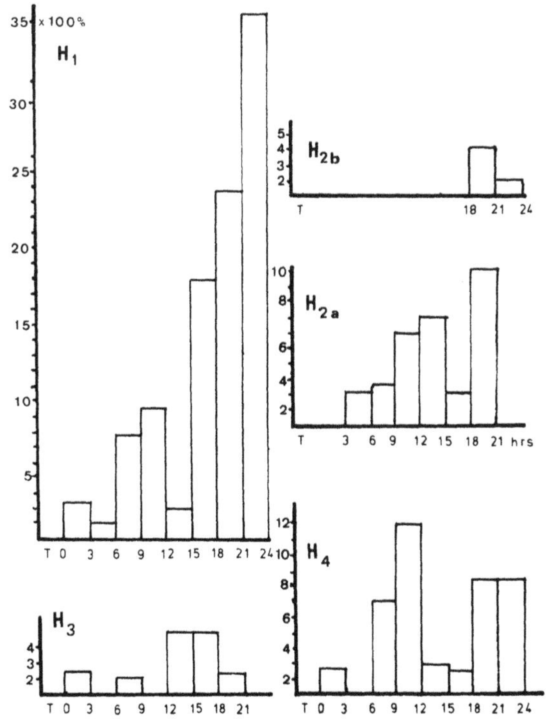

Fig. 4. Histone phosphorylation rates during regeneration. In each group, the phosphorylation level in controls is equated to 1, which corresponds to 1220 cpm/mg for H_1, 300 for H_3, 870 for H_{2b}, 220 for H_{2a}, and 190 for H_4. The width of each bar represents the 3 h duration of each incorporation. Only significant values are shown ($P < 0.05$ Student test).

Ca^{2+} dependence of DNA synthesis and protein phosporylation *in vitro*

The preceding results suggest there is a need for a calcium surge in the medium in order to activate protein phosphorylation and DNA synthesis. In dissociated planarian cells, the rate of DNA synthesis depended on the Ca^{2+} concentration in the medium (Fig. 5): 10^{-6} M Ca^{2+} was the minimum that allowed DNA synthesis *in vitro*. In the absence of Ca^{2+}, DNA synthesis gradually decreased during the first 24 h after dissociation. This might reflect a progressive arrest of cell growth late in the G_1 phase, because, as in the other system, Ca^{2+} depletion prevented cell growth (Whitfield *et al.*, 1980). Protein phosphorylation was shown to be very low in Ca^{2+}-free medium (Fig. 6). It rose significantly when cultures were supplemented with 10^{-9} M Ca^{2+} and reached a plateau in medium containing 10^{-6} M Ca^{2+}. The level of phosphorylation then was nearly 100 times the level recorded in calcium-free medium.

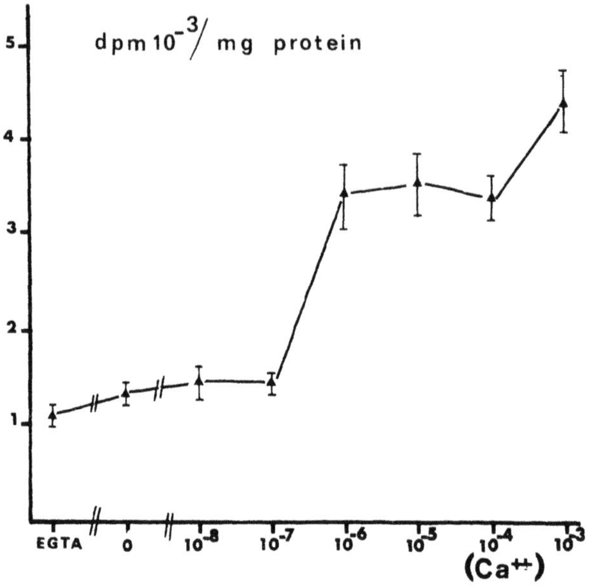

Fig. 5. Effect of increasing Ca^{2+} concentration in the culture medium on DNA synthesis by dissociated cells. Planarian cells were dissociated and grown for 24 h in media to which either 1.5 mM EGTA or different concentrations of $CaCl_2$ were added. DNA labeling was measured after a 2 h pulse with ^{32}P (10 μCi/ml). Each point is the mean ± SEM of the values for four cultures.

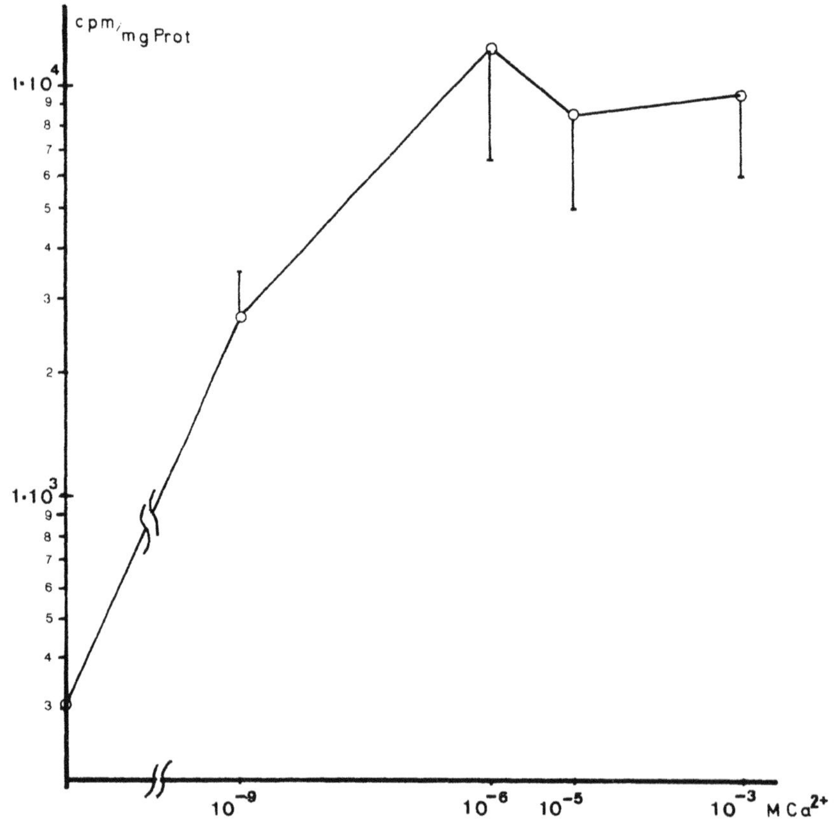

Fig. 6. Dose-dependent effect of Ca^{2+} concentration on protein phosphorylation in cell culture (bars represent SEM, n = 8).

Conclusion

These results indicate that Ca^{2+} plays a key role in planarian regeneration, in particular in stimulating DNA synthesis. A critical Ca^{2+} threshold of about 10^{-6} M was necessary for stimulating DNA synthesis and for protein phosphorylation in dissociated planarian cells and regenerating fragments. After the calcium depletion induced by cutting, increases in Ca^{2+} and calmodulin contents occurred at 12 h regeneration, and these allow stimulation of DNA synthesis. Protein phosphorylation might be necessary to allow this synthesis. In particular, histones, the phosphorylation of which are known to be linked to the cell cycle, displayed changes which could be related to stimulation of nucleic acid synthesis.

The suddenness of the onset and the high level of nucleic acid synthesis suggest that the cells involved in the regeneration process are synchronized during their first cycle following cutting. One of the factors determining the cell synchronisation might be variations in calcium levels.

References

Franquinet, R., 1979. Rôle de la sérotonine et des catécholamines dans la régénération de la planaire Polycelis tenuis. J. Embryol. exp. Morph. 51: 85–95.

Franquinet, R., 1981. Synthèse d'ADN dans les cellules de planaires cultivées in vitro. Rôle de la sérotonine. Biol. Cell. 40: 41–46.

Franquinet, R., A. Le Moigne & J. Hanoune, 1978. The adenylate cyclase system of planarian Polycelis tenuis. Activation by serotonin and guanine nucleotides. Biochim. Biophys. Acta 539: 88–97.

Franquinet, R. & I. Martelly, 1981. Effects of serotonin and catecholamines on RNA synthesis in planarians; in vitro and in vivo studies. Cell. Differ. 10: 201–209.

Le Peuch, C. J., C. Ferraz, M. P. Walsh & J. G. Demaille, 1979. Calcium and cyclic nucleotide dependent regulatory mechanisms during development of chick embryo skeletal muscle. Biochemistry 18: 5267–5273.

Martelly, I., 1984a. Planarian regeneration: effects of the external calcium concentration on total calcium, hormonal contents and DNA synthesis. Comp. Biochem. Physiol. 78A: 329–333.

Martelly, I., 1984b. Calcium thresholds in the activation of DNA and RNA synthesis in cultured planarian cells: relationship with hormonal and DB cAMP effects. Cell. Differ. 15: 25–36.

Martelly, I., R. Franquinet & A. Le Moigne, 1981. Relationship between variations of cAMP, neuromediators and the stimulation of nucleic acid synthesis during planarian (Polycelis tenuis) regeneration. Hydrobiologia 84: 195–201.

Moraczewski, J., 1981. Cell activation during regeneration of planarians. Hydrobiologia 84, 203–207.

Moraczewski, J., R. Franquinet & A. Le Moigne, 1983. Characterization and activity during planarian regeneration of two types of protein kinases: cyclic AMP-dependent and calcium-dependent, cyclic AMP-independent. Biol. Cell. 47: 171–178.

Thomasset, M., A. Molla, O. Parkes & J. G. Demaille, 1981. Intestinal calmodulin and calcium-binding protein differ in their distribution, and in the effect of vitamin D steroids on their concentration. FEBS Lett. 127: 13–17.

Whitfield, J. F., A. L. Boynton, J. P. Mac Manus, R. H. Rixon, M. Sikorska, B. Tsang, P. R. Walker & S. H. H. Swierenga, 1980. The roles of calcium and cyclic AMP in cell proliferation. Ann. N.Y. Acad. Sci. 339: 216–240.

Biology of long slender land planarians (Turbellaria) in Tokyo and environs

Naoya Makino & Yasuko Shirasawa
Department of Biology, Tokyo Medical College, 6-1-1 Shinjuku, Shinjuku-ku, Tokyo, Japan

Keywords: Turbellaria, Bipalium, fragmentation, reproduction, identification

Abstract

The common short-bodied species of *Bipalium* does not fragment, but individuals of two newly discovered long-bodied species – *B. nobile* Kawakatsu & Makino, 1982, and *B. multilineatum* Makino & Shirasawa, 1983 – do regularly fission, usually behind the mouth or genital pore. Some experimental regenerates of these species form rings by adhesion of the anterior with the posterior cut surface. We found two other forms of *Bipalium*, perhaps representing a further two species, in Hino City, Tokyo, in 1983; and we have preliminarily arranged the forms of *Bipalium* known in the region into four groups distinguished on the basis of body coloring, position of the mouth, and structure of the copulatory organ.

Introduction

In contrast to the common *Bipalium fuscatum,* which we have found does not reproduce asexually by fragmentation, the newly discovered *B. nobile* Kawakatsu & Makino, 1982, (Fig. 1) and *B. multilineatum* propagate regularly in gardens (private garden of N.M.) and in the laboratory by fragmentation. Identification of species of *Bipalium* is difficult, however, and has been complicated especially by the discovery of several other forms of the genus, perhaps new species. We have devised methods for identification based on the natural coloration, relative position of the mouth and genital pore, and structure of the copulatory organ. Our preliminary work permits arrangement of the known forms into four groups.

Material and methods

Two long slender species were employed for the study of fragmentation: *Bipalium nobile* and *B. multilineatum*. Both were collected in the senior author's garden, and the latter was collected also in other areas of Tokyo. The worms were maintained at 18 °C in glass dishes in controlled-temperature chambers.

B. fuscatum, B. fuscolineatum, and *B. hilgendorfi* were collected in Hino City, Tokyo. Unnamed forms of *Bipalium* used include two worms with creamy body color and a strong medio-dorsal brown stripe and 10 worms having grey-green color with three dorsal stripes, both of which forms were collected in the neighborhood of the senior author's residence. An unidentified species of small worms having three dorsal stripes was obtained from the campus of Hirosaki University, Aomori Pref.

Specimens were anaesthetized by carbon dioxide (Makino & Seo, 1983) and fixed in Bouin's fluid. For histological study, Paraffin-embedded specimens were sectioned at 7–9 μm and stained with Mayer's haematoxylin and eosin.

Hydrobiologia 132, 229–232 (1986).

Results and discussion

Asexual and sexual reproduction in Bipalium nobile

Of the species of *Bipalium* we have raised in controlled-temperature chambers, *B. nobile* and *B. multilineatum* propagated asexually, while short-bodied species of the genus did not. Hyman (1951) reported that *B. kewense* Moseley propagated asexually; Morgan (1900) and Hauser *et al.* (1976) also studied regeneration in *Bipalium.*

The number of fragmentation pieces in *B. nobile* appeared to be directly related to body length. Individuals broke posterior to the mouth into two to four pieces of approximately equal size. If the initial fragment was small, as from a very small worm, no further divisions occured; when the fragment was large, it fragmented further in a single step into equal-sized secondary pieces (Fig. 5). After three or four weeks, each piece regenerated a complete worm. The process in *B. multilineatum* differed in that all pieces (5–8 in number) formed simultaneously (Fig. 5) rather than in two or three steps. Fragmentation in both species was restricted to the region behind the mouth and was possibly initiated by starvation. Well-fed individuals did not undergo fragmentation; and, indeed, a portion cut from such an individual simply flexed so that the anterior and posterior ends met, forming a ring, which died after a few weeks. Comparable results were obtained with the polyclad *Stylochus ijimai* (see Yamao & Makino, 1954).

The process of copulation in *B. nobile* was observed in the senior author's garden the evening of 28 June, 1984, over the period from 2120 to 2310 (Fig. 3). One of the individuals was brought into the laboratory for further observation. It weighed 7.31 g, and 18 days after copulation it laid a cocoon 8 by 7 mm weighing 280 mg (Fig. 4). Two off-

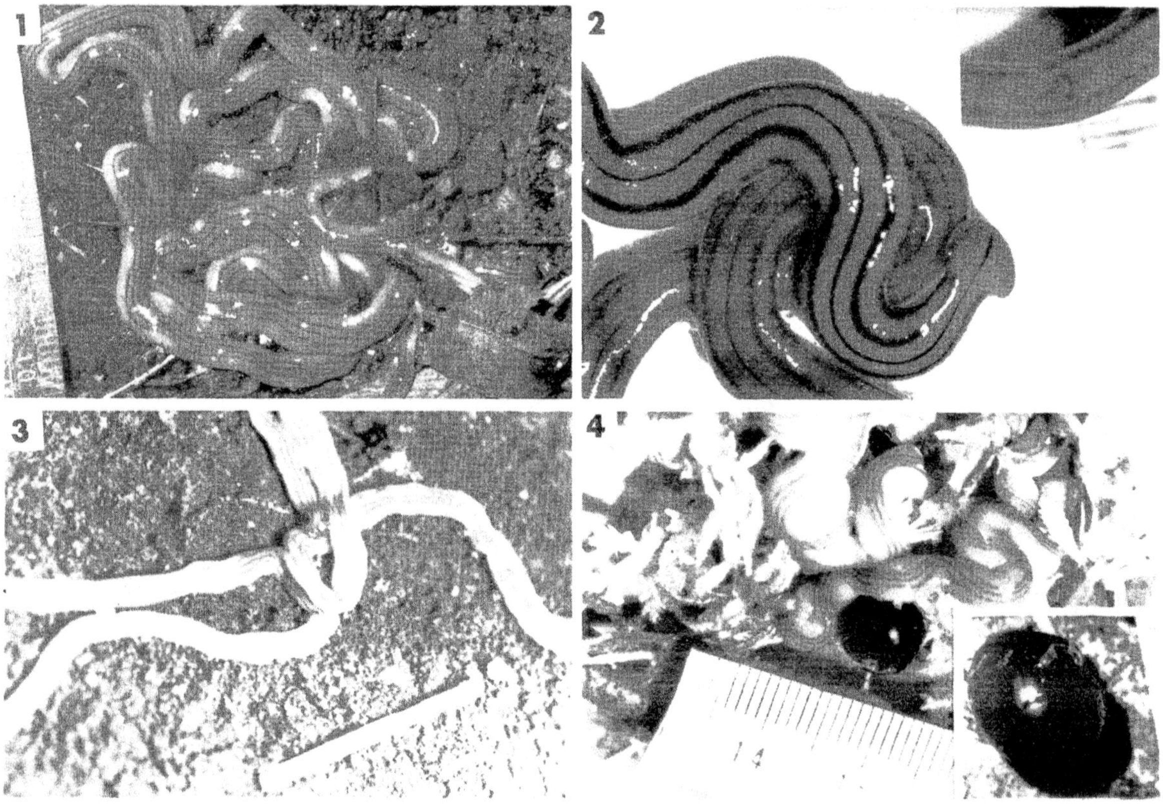

Fig. 1–4. (1) *Bipalium nobile*, a long slender yellowish-brown species with three dorsal longitudinal stripes; (2) *Bipalium multilineatum*, a long somber-green species with five dorsal and three ventral stripes, with a melanin ring around the mouth, and lacking a genital pore; (3) *B. nobile, in copula*; (4) Cocoon of *B. nobile.*

Table 1. Relative position of mouth and genital pore in species of Bipalium.

Species	Specimen No.	Body Weight	A/L	(A + B)/L
B. nobile	No. 404	3.700 g	0.23	0.29
	No. 405	3.230	0.20	0.26
B. multilineatum	No. 410	0.252	0.43	-
	No. 533	0.457	0.44	-
B. sp. A*	No. 602	0.253	0.43	0.63
	No. 603	0.153	0.45	0.64
B. sp. B**	No. 613	0.251	0.47	0.59
	No. 649	0.313	0.47	0.59
B. hilgendorfi	No. 633	0.123	0.52	0.71
	No. 696	0.199	0.52	0.70
B. fuscatum	No. 474	1.431	0.52	0.62
	No. 475	1.677	0.52	0.67
B. fuscolineatum	No. 650	0.783	0.59	0.67
	No. 651	0.684	0.57	0.67

A – distance between head and mouth.
B – distance between mouth and genital pore.
L – total body length.
* milky colored with a brown stripe.
** grey-green with three black stripes.

5

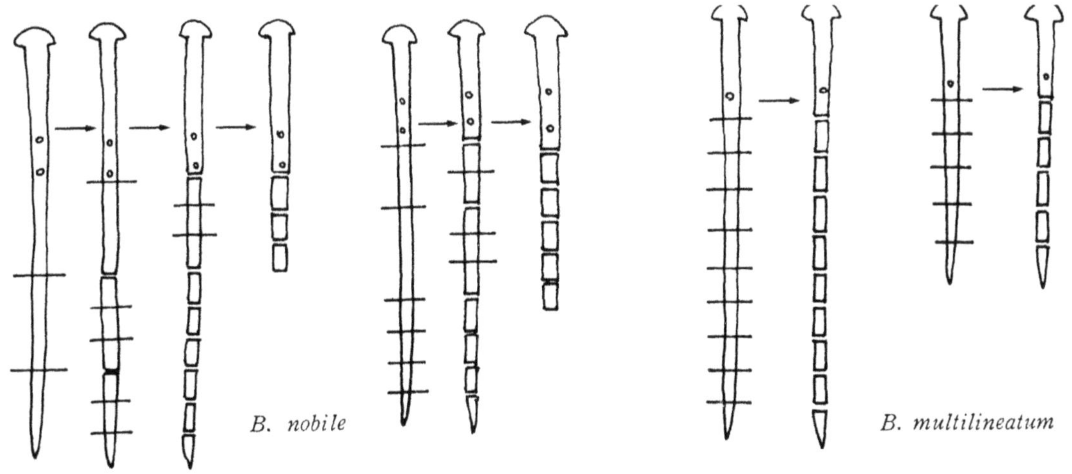

6

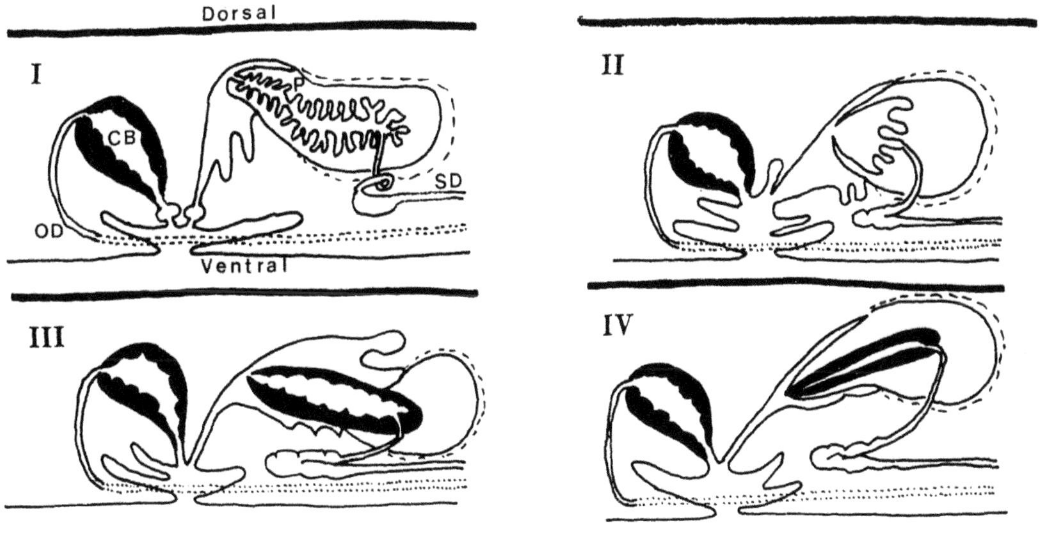

Fig. 5–6. (5) Fragmentation in *B. nobile* and *B. multilineatum*; (6) The four types of copulatory organ in *Bipalium*. CB – copulatory bursa; OD – oviduct; P – penis; SD – seminal duct.

spring emerged from the cocoon on 22 August (38 d after oviposition). A second cocoon deposited on 14 August yielded one juvenile on 13 September (30 d after oviposition).

Taxonomic considerations

Species of *Bipalium* are cryptic and only a few specimens are usually collected at a time. We recommend the routine use of color photography to document distribution of pigment and shape of the head. The positions of the mouth and genital pore relative to total body length has proved to be significant in identifying species (Table 1).

Morphology of the copulatory organ appears especially important inasmuch as some individuals having similar external morphologies have different copulatory-organ structure. We have identified four general types of copulatory organ in our material (see also Kaburaki, 1922):

Type I. Penis lumen strongly plicate (Fig. 6, I) as in *B. kewense, B. nobile*, and perhaps *B. multilineatum*.

Type II. Penis globular as in *B. fuscatum* and *B. fuscolineatum* (Fig. 6, II).

Type III. Penis lumen broad and slightly folded as in *B. hilgendorfi* and in a cream-colored, unidentified species from the senior author's garden (Fig. 6, III).

Type IV. Penis lumen narrow (Fig. 6, IV) as in what appears to be a new species from Hirosaki (Aomori Pref.) and Hino City, Tokyo.

An illustrated guide to the Japanese species of *Bipalium* based on these types is in preparation (Makino & Shirasawa, in prep.).

References

Hauser, J., M. G. Friedrich, & M. I. Diaz, 1976. Morphallactic processes in the regeneration of Bipalium kewense Moseley. Ber. Nat. Med. Ver. Innsbruck 63: 165–175.

Hyman, L. H., 1951. The Invertebrates. II. Platyhelminthes and Rhynchocoela. McGraw-Hill, New York: 550 pp.

Kaburaki, T., 1922. On the terrestrial planarians from Japanese territories. J. Coll. Sci. Tokyo Imp. Univ. 44: 1–44.

Makino, N. & N. Seo, 1983. About a simple anesthetic technique and the effect of gaseous carbon dioxide on slugs. Bull. Tokyo med. Coll. 9: 63–68. [In Japanese; Engl. summary]

Makino, N., & Y. Shirasawa, 1980. On the study of sexual individuals in land planaria (Bipalium kewense Mosely?). Bull. Tokyo med. Coll. 6: 87–98. [In Japanese; Engl. summary]

Morgan, T. H., 1900. Regeneration in Bipalium. Arch. Entwmech Org. 9: 563–568.

Yamao, Y., & N. Makino, 1954. Studies on regeneration of a seaplanarian. J. Coll. Art Sci., Chiba University, 1: 259–262.

Karyology of a marine population of *Gyratrix hermaphroditus* (Turbellaria, Rhabdocoela) and chromosomal evolution in this species complex

J.-P. L'Hardy
Laboratoire de Biologie Animale, Université du Maine, Route de Laval, B.P. 535, 72017 Le Mans-Cedex, France

Keywords: Turbellaria, Rhabdocoela, karyology, aneuploidy, chromosomal evolution

Abstract

Karyology and reproductive biology of a marine population of the species complex *Gyratrix hermaphroditus*, from Roscoff (Brittany, France), have been investigated. A diploid complement of six chromosomes was determined from spermatogonial mitotic figures. One chromosome pair is metacentric, the second is intermediate between meta- and submetacentric, and the third is subtelocentric.

In this population, regular meiosis occurs in both female and male germ lines, and the animals reproduce only by means of amphimictic eggs. Certain specimens of the population showed the elimination of one of the three bivalents during the first meiotic division in spermatogenesis. It seems that such animals produce normal and aneuploid sperm simultaneously; the aneuploid sperm are not capable of fertilization.

The Roscoff population differs in its karyotype (2n = 6) from freshwater populations, which are either diploid (2n = 4) or polyploid (3n = 6, 4n = 8). These results suggest that aneuploidy played a role in the differentiation of freshwater populations from an originally marine species complex.

Introduction

Gyratrix hermaphroditus is distributed world wide (except for being absent in Australia) in marine, brackish, and fresh waters. The morphologically polytypic nature of this taxon was shown by Meixner (1929), Kromhout (1943), Marcus (1946), and Evdonin (1971); and Reuter (1961) identified the existence of a number of karyologically distinct 'races' on the Finnish coast. Heitkamp (1978) studied the populations of some fifteen ponds in Lower Saxonia (Germany) and identified five reproductively isolated biotypes. He concluded that *G. hermaphroditus* is a species complex, as had been suggested by de Beauchamp (1961).

In continuation of research on the biology of littoral rhabdocoel turbellarians from Brittany coasts (L'Hardy, 1972), I have studied a marine population of *G. hermaphroditus* to 1) determine its karyotype and 2) compare its reproductive cytology with that of forms already known.

Material and methods

All animals used in this study were collected from the Ile Verte Channel in the vicinity of the Roscoff Biological Station (North-Finistere, Brittany, France).

Specimens of *Gyratrix hermaphroditus* were found in coarse sands. They were sorted from the sediment under a dissecting microscope without use of anesthesia and put in culture with small live copepods (*Paronychocamptus nanus*, found near the Biological Station) as food. Both water and food were renewed in the cultures every second day. Such cultures can be maintained at least two months at laboratory temperature (16–17 °C). The egg masses produced were collected and isolated.

The analysis of both mitosis and meiosis were done on squashes stained with a 2% solution of orcein in a mixture of equal parts lactic and propionic acids (Sharma & Sharma, 1965: 124). Stages of spermatogenesis were studied on whole animals

Hydrobiologia 132, 233–238 (1986).

that had been submerged ten minutes in the staining solution and then squashed between slide and coverglass. Mitotic figures in the male germ line were easily analyzed, though the figures were sparce because of the small size of the animals. More difficult was the study of oocyte maturation because this takes place in the sclerotinized egg capsule. A preliminary fixation of the egg (in 4% glutaraldehyde solution in sea water) for two or three hours was necessary before staining and squashing.

The diploid number of chromosomes was determined from study of numerous spermatogonial mitoses. Chromosomes of five metaphase plates of different specimens were measured and the relative length (length of chromosome × 100/total length of haploid genome) and the centrometric index (length of short arm × 100/length of the whole chromosome) calculated. The chromosomes were classified according to the nomenclature of Levan *et al.* (1964).

Results

Observations of biology

On the Brittany coast, as in other marine environments (Ax, 1951; Karling, 1963; Hoxold, 1979), *G. hermaphroditus* was found in intertidal sands. It was particularly abundant at the mid-tide level (*Fucus serratus* zone), in pools or outflows, and extended down to the low-tide zone. It was present in coarse sands and at times in gravels more or less filled with mineral particles, but never in fine clean sands.

Mature specimens were found all year round in favorable habitats but their numbers fluctuated markedly. They were scarce from November to April and abundant from May to October, the period of sexual reproduction. Under controlled laboratory conditions, reproduction took place at temperatures warmer than 10 °C; this correlates with what takes place in nature.

During copulation, which was preceded by a brief pursuit, the posterior parts of the partners were apposed as reported by Hallez (1873) and Heitkamp (1978). As Hallez (*op. cit.*) mentioned, 'the stylet itself plays no part in copulation, but the stylet sheath produces a continual backward and forward movement'.

The ectolecithal egg masses consisted each of a single ovum and about one hundred vitelline cells which accumulate, before laying, in the uterus. The eggshell is formed from sclerotin precursors contained in the vitellocyte cytoplasm, and it hardens gradually by quinone tanning (L'Hardy, 1979). The egg is fastened to a substratum (sand particle or side of culture jar) by means of a uterine glandular secretion which forms the stalk of the eggshell.

In the Roscoff population of *G. hermaphroditus*, the ovoid egg capsule was more or less swollen and of variable size (140 μm to 275 μm in length). The stalk measured 30–60 μm long as does that of German maritime populations (Hoxhold, 1971). The eggshell showed an apical valve similar to that depicted by Meixner (1923: 200, Fig. 3); this opened during hatching.

Spermatogonial division and karyotype

The single testis in *G. hermaphroditus* is situated dorsally in the middle of the body. As in other turbellarians, the differentiating male germ cells cluster into cytophores (rosettes). Each cytophore arises from the multiplication of one initial germ cell. Spermatogonia were found mostly in the anterior part of the testis and the cytophores were apparently gradually forced into the posterior part as they matured.

Spermatogonial mitoses showed a diploid set of 2n = 6 chromosomes (Fig. 1, Table I). One pair of

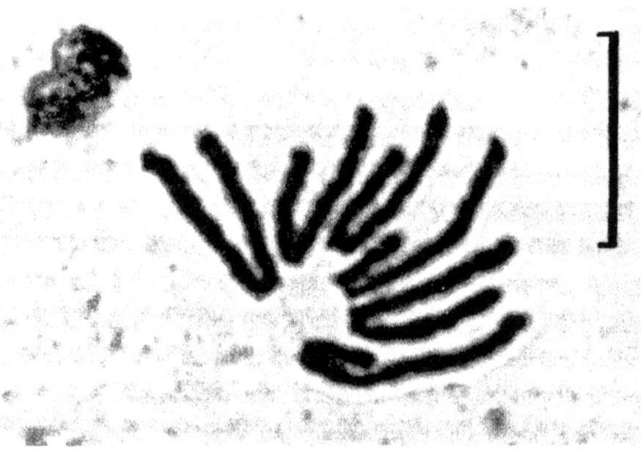

Fig. 1. Spermatogonial mitotic premetaphase in Roscoff population of *Gyratrix hermaphroditus* complex. Lacto-propionic orcein squash preparation. Bar: 10 μm.

Table 1. Relative length and centromeric index of the three chromosome pairs in Roscoff population of *Gyratrix hermaphroditus* complex. Means and standard deviations from five spermatogonial mitotic metaphase plates.

Chromosome number	Relative length	Centromeric index
1	37.28 ± 1.70	47.37 ± 0.93
2	31.56 ± 1.03	37.34 ± 2.98
3	31.16 ± 2.51	24.15 ± 0.59

chromosomes is metacentric, the second is intermediate between meta- and submetacentric, and the third is subtelocentric (Fig. 2).

Meiosis in the male germ line

During leptotene, the nucleus of a primary spermatocyte measured approximately 20 μm in diameter. At diplotene, the three bivalents could be seen, reaching a length of 20–28 μm each. During pachytene (Fig. 3), they became notably thicker and shorter so that they measured no more than

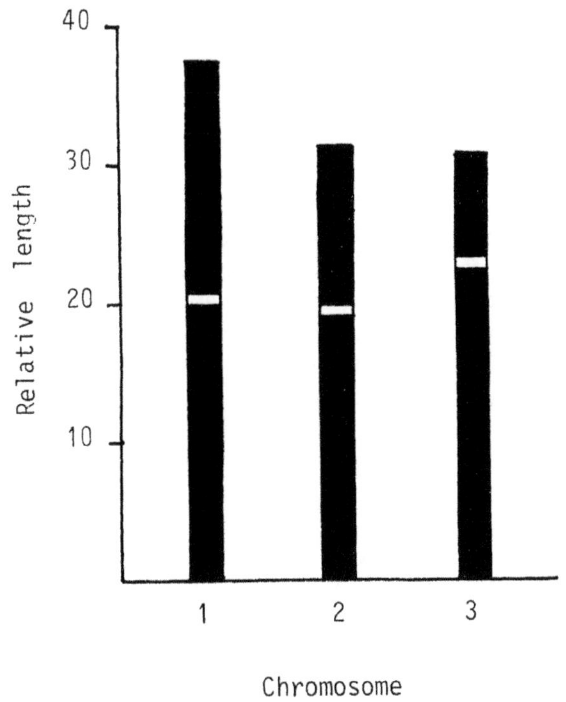

Fig. 2. Idiogram of the karyotype in the Roscoff population of *Gyratrix hermaphroditus* complex.

20 μm in length and 2–3 μm in width at first metaphase (Figs. 4, 5). At this stage, two ring-shaped bivalents had a chiasma on each arm. The third subtelocentric bivalent was often rod-shaped with one terminal chiasma on the long arm, though in some instances it had two chiasmata (one on each arm) and showed a delayed segregation at the beginning of anaphase I (Fig. 6). During the second maturation division, the chromosomes measured only 10 μm long and 1 μm thick.

It must be noted, however, that certain specimens of the Roscoff population had both normal and abnormal first metaphase plates (Fig. 7), the latter containing only two ring-shaped bivalents. In all cases observed the subtelocentric bivalent was eliminated (see Figs. 8, 9). However, both anaphase and telophase of this first meiotic division appeared to be normal. The existence of spermatids of different size indicate that the aneuploid cells continue to mature. Because there is no pycnosis nor cellular degeneration in the course of spermatogenesis, it seems likely that aneuploid secondary spermatocytes produce sperm.

Fertilization and oocyte maturation

Meixner (1924) specifies that fertilization can take place in the ovary of *G. hermaphroditus.* As Hallez did (1879: Pl. 11), I observed maturation and polar-body extrusion taking place inside the egg capsule. I surmise that in the Roscoff population, fertilization takes place either in the uterus before formation of the egg shell or in the genital atrium as it does in *Mesostoma* (see Bresslau, 1904) and in *Paravortex* (see Hallez, 1909; Patterson, 1912; Ball, 1916).

Female gametes that had just been released from the ovary were primary oocytes, 60–80 μm in diameter, in late prophase I. Such oocytes that were fertilized contained only a single male pronucleus. Sperm nuclei in oocytes undergoing metaphase of meiosis I or in later stages no longer had the thread-like shape characteristic of spermatozoa, but were globular masses, 8 μm in diameter, of heterogeneous content. In oocytes undergoing the second meiotic division, the male pronucleus was larger, 15 μm in diameter, and in oocytes in later stages of meiosis it reached 20–30 μm in diameter.

Oocytes in metaphase I showed three thick bivalents, with one or two chiasmata each (Fig. 10). In

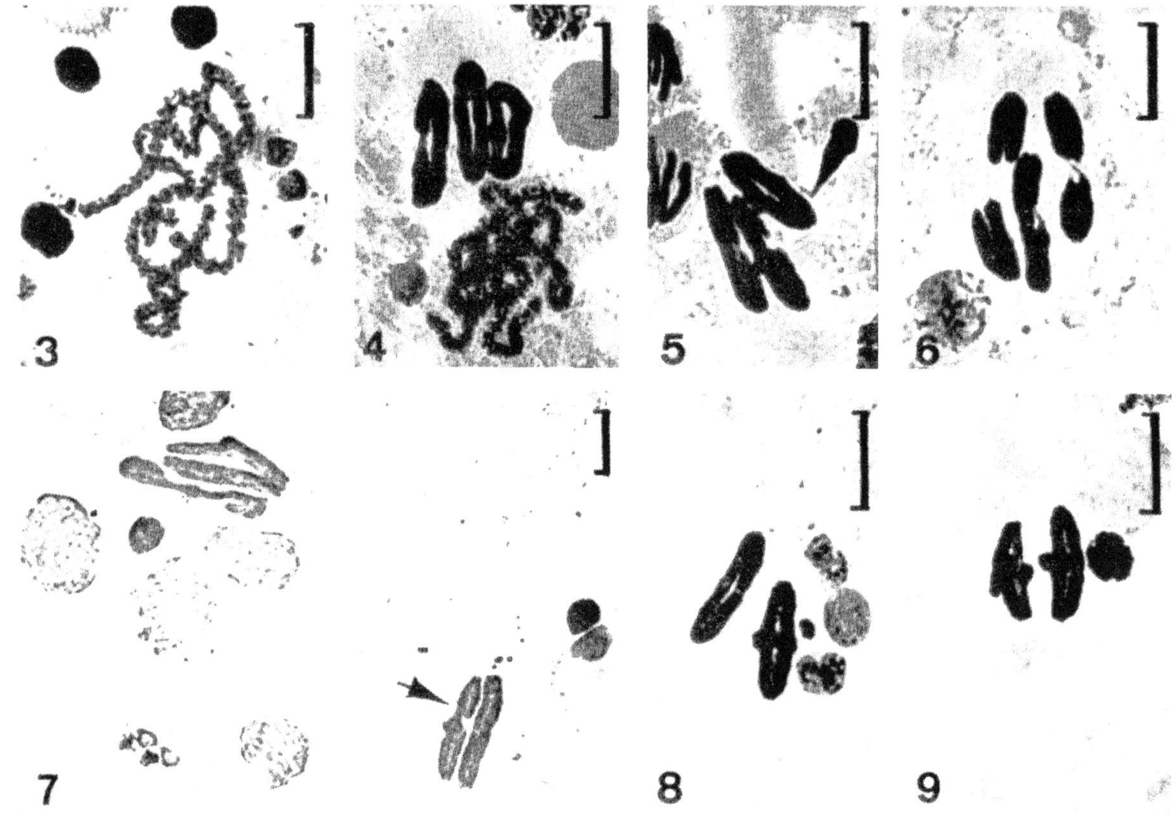

Figs 3–9. Spermatogenetic meiosis in Roscoff population of *Gyratrix hermaphroditus* complex: (3) pachytene stage; (4–5) metaphase 1; (6) anaphase 1; (7) simultaneous normal and aneuploid (arrow) metaphases 1; (8–9) aneuploid metaphases 1 in two different specimens. Lacto-propionic orcein squash preparations. Bar: 10 μm.

telophase I, the first polar body was eliminated as an elongated cortical cytoplasmic bud (Fig. 12). In oocytes that had eliminated the second polar body, the female pronucleus was less lobed than in previous stages and enlarged to a 30 μm diameter. The three chromosomes were readily evident in both male and female pronuclei before fusion.

Discussion

The Roscoff population of *G. hermaphroditus*, by virtue of its diploid complement of six chromosomes, is distinct from other species of the species complex (cf. Ruebush, 1938; Reuter, 1961; Heitkamp, 1978). Indeed, the most common karyotype of the brackish- and freshwater populations is 2n = 4 [Ruebush (*op. cit.*); Reuter's (*op. cit.*) H and Heitkamp's (*op. cit.*) populations 1, 2, 4, and 5]. Rarer is the somatic complement of eight chromosomes, found in Reuter's (*op. cit.*) population K and Heitkamp's (*op. cit.*) population 3. One could explain these populations as having originated as autotetraploids from diploid freshwater populations with 2n = 4 karyotypes.

Reuter's (*op. cit.*) population L deserves a more detailed examination since its complement of six chromosomes is the same as that of the Roscoff population. The two are different, however, in that Reuter (*op. cit.*) found three long metacentric chromosomes along with three short submetacentric ones in the Finnish population, whereas I found three distinct pairs in specimens from the Brittany coast. The L population must be considered to be a triploid derivative of the diploid freshwater populations as Reuter (*op. cit.*) suggested. In addition,

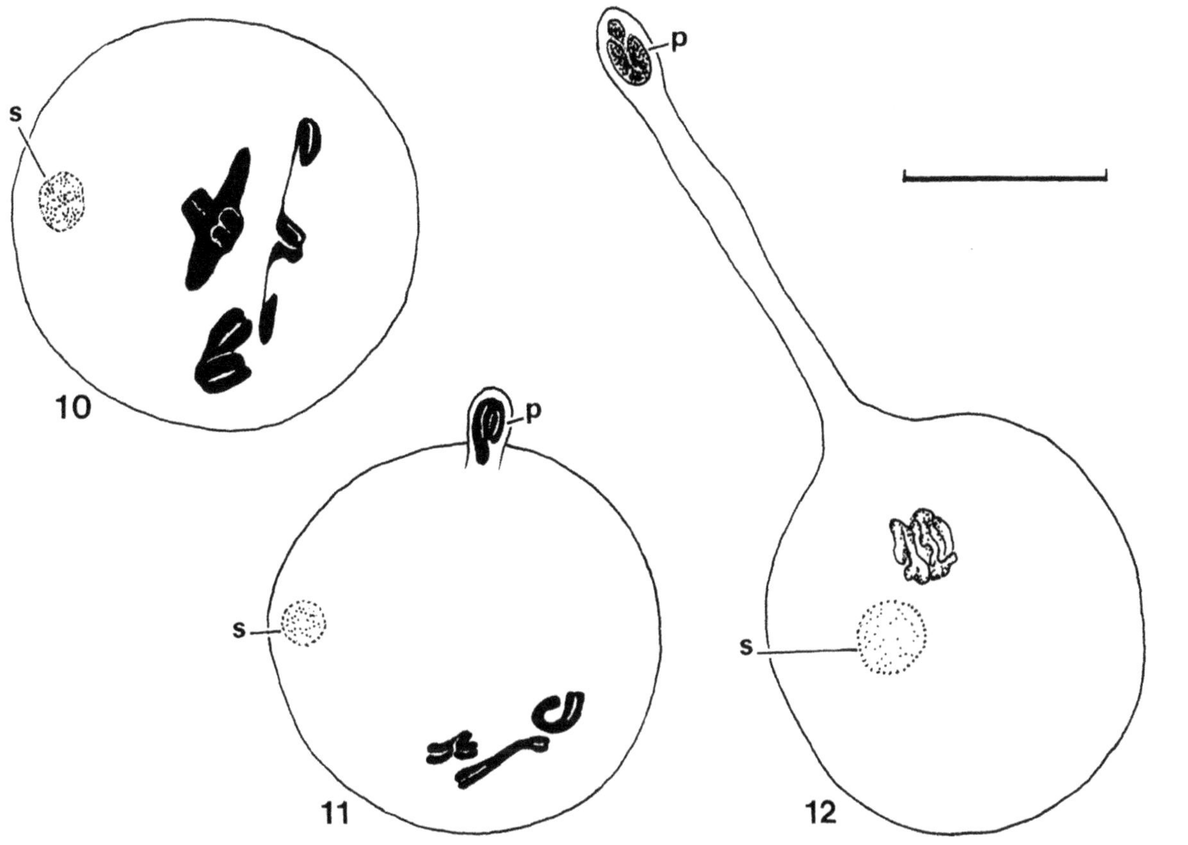

Figs 10–12. Oocyte maturation in Roscoff population of *Gyratrix hermaphroditus* complex: (10) metaphase 1. (11) anaphase 1. (12) telophase 1. Lacto-propionic orcein squash preparations. Bar: 50 μm. p: first polar body, s: sperm nucleus.

this Finnish L population is parthenogenetic, lacking spermatogenesis, whereas the Roscoff population appears to be obligatorily amphimictic, with gametogenesis in both germ lines.

Like all kalyptorhynchs (Turbellaria Kalyptorhynchia), *G. hermaphroditus* is evidently of marine origin. Meixner (1943) and Kolasa (1977) analyzed the kind of evolutionary mechanisms that allowed certain species to colonize fresh water. If this universally accepted hypothesis is applicable to *G. hermaphroditus*, one must consider that all freshwater forms have originated from marine populations of this complex. However, the only marine population actually known has a somatic set of six chromosomes, whereas the freshwater diploid populations have only four chromosomes. In the marine Roscoff population, the first metacentric and the second intermediate between meta- and submetacentric chromosome pairs have the same morphology as the two pairs of the freshwater diploid populations. This suggests that aneuploidy accompanied the ecological differentiation between marine and freshwater forms. The reproductive biology of the Roscoff population agrees with this hypothesis: when aneuploidy occurs it is the third subtelocentric chromosome pair that is eliminated. Besides, it seems that aneuploid meiosis in the male germ line produces aneuploid sperm and these must somehow be inhibited from fertilizing oocytes or their progeny must be inviable since there is no chromosomal polymorphism in the Roscoff population.

Modification of karyotype (ploidy level, aneuploidy, and change of chromosome structure) have evidently played an important role in the evolution of *G. hermaphroditus*. But this does not explain the reproductive isolation of the species. Indeed, Heitkamp (1978) claimed that no copulations occur in experiments attempting to cross specimens from different populations. According to him, reproductive isolation in *G. hermaphroditus* is caused by a sexual substance, different in each population, that inhibits copulation between animals of distinct species.

In the present state of our knowledge of *G. hermaphroditus*, this complex seems to be composed of at least seven sibling species – or even eight, if Heitkamp's (1978) 5 (Kehr) and Reuter's (1961) H (Hästhagen) populations, with apparently similar karyotypes, are also reproductively isolated. Because of the scanty data available on marine and estuarine populations, it seems premature to give a taxonomic name to each species of this complex.

Acknowledgments

I thank Seth Tyler and Reinhard Rieger for valuable corrections to and final review of the manuscript.

References

Ax, P., 1951. Die Turbellarien des Eulittorals der Kieler Bucht. Zool. Jb. Syst. Ökol. Geogr. Tiere 80: 227–378.

Ball, S. C., 1916. The development of Paravortex gemellipara (Graffila gemellipara Linton). J. Morph. 27: 453–558.

Bresslau, E., 1904. Beiträge zur Entwicklungsgechichte der Turbellarien. I. Die Entwicklung der Rhabdocoelen und Allocoelen. Z. wiss. Zool. 76: 213–332.

de Beauchamp, P., 1961. Classe des Turbellariés, Turbellaria (Ehrenberg, 1831). In P. P. Grassé (ed.), Traité de Zoologie, Masson, Paris, 4(1): 35–212.

Evdonin, L. A., 1971. The interstitial Kalyptorhynchia (Turbellaria Neorhabdocoela) from the bay of Great Peter of the Sea of Japan. Akad. Nauk S.S.S.R. Issled. Fauny Morei 8: 55–71 [in Russian].

Hallez, P., 1873. Observations sur le Prostomum lineare. Arch. Zool. exp. gén. 2: 559–586.

Hallez, P., 1879. Contribution à l'histoire naturelle des Turbellariés. Trav. Inst. zool. Lille Stat. marit. Wimereux 2: 1–213.

Hallez, P., 1909. Biologie, organisation, histologie et embryologie d'un Rhabdocoele parasite du Cardium edule L., Paravortex cardii n. sp. Arch. Zool. exp. gén. 9: 429–544.

Heitkamp, U., 1978. Speziationsprozesse bei Gyratrix hermaphroditus Ehrenberg, 1831 (Turbellaria, Kalyptorhynchia). Zoomorphologie 90: 221–251.

Hoxold, S., 1971. Eigebilde interstitieller Kalyptorhynchier (Turbellaria) von der deutschen Nordseeküste. Mikrofauna Meeresbodens 7: 1–143.

Hoxold, S., 1979. Populationsstruktur und Abundanzdynamik interstitieller Kalyptorhynchia. Mikrofauna Meeresbodens 41: 1–134.

Karling, T. G., 1963. Die Turbellarien Ostfennoscandiens. V. Neorhabdocoela 3. Kalyptorhynchia. Fauna fenn. 17: 1–59.

Kolasa, J., 1977. Remarks on the marine-originated Turbellaria in the freshwater fauna. Acta zool. fenn. 154: 81–87.

Kromhout, G. A., 1943. A comparison of the protonephridia of freshwater, brackish-water and marine specimens of Gyratrix hermaphroditus. J. Morph. 72: 167–181.

Levan, A., K. Fregda & A. A. Sandberg, 1964. Nomenclature for centromeric position on chromosomes. Hereditas 52: 201–220.

L'Hardy, J.-P., 1972. Recherches sur la reproduction et le développement des Turbellariés Calyptorhynques. Thèse Doct. Etat, Univ. Paris VI (A.O. CNRS 6546): 1–202.

L'Hardy, J.-P., 1979. Observations sur l'élaboration et la sclérotinisation de la capsule ovigère chez Gyratrix hermaphroditus Ehrenberg, 1831 (Turbellarié, Rhabdocoele). Vie Milieu 28–29: 237–251.

Marcus, E., 1946. Sobre Turbellaria Limnicos brasileiros. Bol. Fac. Fil. Cien. Letr. Univ. Sao Paulo, Zool. 11: 1–252.

Meixner, J., 1923. Über den Bau des Geschlechtsapparates bei Calyptorhynchiern und die Bildung des Eistieles bei diesen und einigen anderen rhabdocoelen Turbellarien. Zool. Anz. 57: 193–207.

Meixner, J., 1924. Über das Ovarium von Microstomum lineare und die Abscheidungsfolge des Schalen-und Dottermaterials bei rhabdocoelen Turbellarien. Zool. Anz. 58: 195–213.

Meixner, J., 1929. Morphologisch-ökologische Studien an neuen Turbellarien aus dem Meeressande der Kieler Bucht. Z. Morph. Ökol. Tiere 14: 765–791.

Meixner, J., 1943. Über die Umbildung einer Turbellarienart nach Einwanderung aus dem Meere ins Süsswasser. Int. Revue ges. Hydrobiol. Hydrogr. 43: 458–468.

Patterson, J. T., 1912. Early development of Graffila gemellipara. A supposed case of polyembryony. Biol. Bull. 22: 173–204.

Reuter, M., 1961. Untersuchungen über Rassenbildung bei Gyratrix hermaphroditus (Turbellaria Neorhadocoela). Acta zool. fenn. 100: 1–32.

Ruebush, T. K., 1938. A comparative study of the Turbellaria chromosomes. Zool. Anz. 122: 321–329.

Sharma, A. K. & A. Sharma, 1965. Chromosome techniques. Theory and Practice. Butterworths, Lond.: 1–575.

Chromosomal evolution in marine triclads and polyclads (Turbellaria)

Lodovico Galleni & Ileana Puccinelli
Istituto di Zoologia e Anatomia Comparata Università di Pisa, Italy

Keywords: Turbellaria, karyology, chromosome, evolution

Abstract

From an analysis of chromosomal mutations in seven species among five genera of marine triclads and polyclads, three different types of mechanisms of such mutations are identified: 1) rearrangement involving the centromere such that its position is changed, 2) rearrangement of whole chromosome arms, and 3) Robertsonian mechanisms. These mechanisms are the same as those reported for freshwater turbellarians, but aneuploidy and polyploidy, which are common in freshwater species, were not detected in these marine turbellarians.

Introduction

Since the publication of a revision of the karyology of Platyhelminthes by Benazzi & Benazzi Lentati (1976), several papers dedicated to karyology and cytotaxonomy of marine turbellarians have been published, mainly regarding benthic marine triclads and polyclads. Chromosome numbers of several species have been determined, but the most important gain in our knowledge is in karyometric analysis and idiograms, i.e. beta-level karyology, according to the classification of White (1978), as opposed to alpha-level karyology in which only numbers of chromosomes are reported.

This new material has allowed us to identify rearrangements of chromosomes that have occurred among related species in these marine groups and to propose a hypothesis on trends in the evolution of karyotypes in these species.

Material and methods

Comparisons have been made using karyometric analyses of species in the genus *Procerodes* (Tricladida) published by Galleni & Puccinelli (1979) and Galleni *et al.* (1984), of the polyclad species *Thysanozoon brocchii* and *Yungia aurantiaca* (see Galleni & Puccinelli, 1977, 1985), and of the polyclad genera *Notoplana* and *Leptoplana* (Galleni & Puccinelli 1975, 1981, 1984).

Relative length (r.l.) was calculated as mean total length of the two homologous chromosomes $\times 100$ divided by the total length of the haploid genome; centromeric index (c.i.) was calculated as length of the short arm $\times 100$ divided by total chromosome length.

Differences and similarities among the idiograms of these species have been evaluated with Student's test [see Benazzi & Puccinelli (1973) and Galleni & Puccinelli (1979) for application of the method to turbellarian karyology].

Results

For marine triclads, we examined the relationships among idiograms of *Procerodes lobata, P. littoralis,* and *P. dohrni* (Fig. 1). The numerical data were published by Galleni & Puccinelli (1979) and

Hydrobiologia 132, 239–242 (1986).

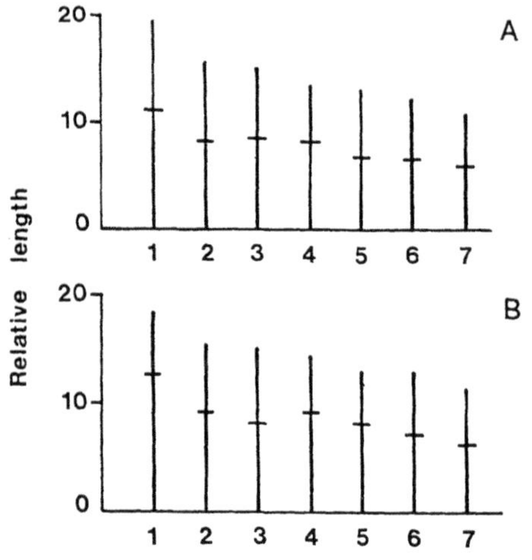

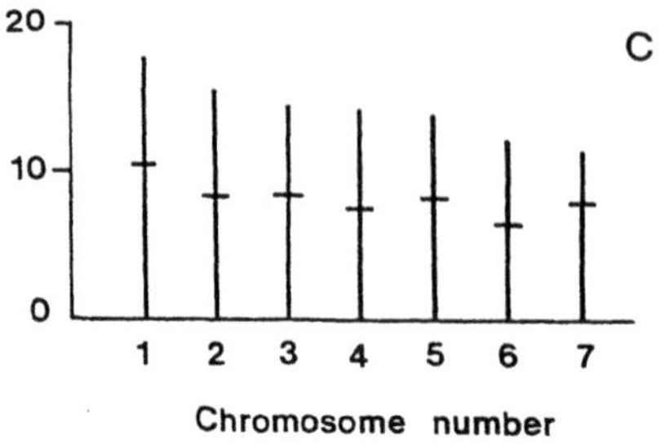

Fig. 1. Idiograms of (A) *Procerodes littoralis* (from Galleni & Puccinelli, 1979), (B) *P. lobata* (from Galleni & Puccinelli, 1979), and (C) *P. dohrni* (from Galleni *et al.* 1984).

Galleni *et al.* (1984). Numerical data reported by Ball (1976, 1979) on the same genus were not considered because he used different methods to order the chromosomes such that it was difficult to compare his data with ours.

The majority of the karyometric measurements regarding relative lengths of chromosomes in species of *Procerodes* are not significantly different, but measurements of centromeric indices are. This indicates that chromosomal rearrangements in this genus involve the centromere and thus changes in centromere position. Translocations, in contrast, seem to be rare (Table I, Fig. 1). The question of

Table 1. Comparisons of relative length (r.l.) and centromeric index (c.i.) in three species of *Procerodes*. Values are reported as means ± standard error of the mean. Numerical data from Galleni & Puccinelli (1979) and Galleni *et al.* (1984).

Chromosome number		*Procerodes littoralis* *Procerodes dohrni*	*Procerodes dohrni* *Procerodes lobata*	*Procerodes littoralis* *Procerodes lobata*
1	r.l.	1.85 ± 0.416[a]	0.68 ± 0.462[d]	1.17 ± 0.495[c]
	c.i.	6.43 ± 0.703[a]	7.61 ± 0.875[a]	14.04 ± 0.890[a]
2	r.l.	0.10 ± 0.349		
	c.i.	0.96 ± 0.875[d]		
2/3*	r.l.		0.67 ± 0.386[d]	0.77 ± 0.451[d]
	c.i.		1.24 ± 0.738[d]	0.28 ± 0.757
3/2*	r.l.		0.90 ± 0.437[d]	0.46 ± 0.385[d]
	c.i.		0.94 ± 1.178	3.19 ± 1.104[b]
3	r.l.	0.44 ± 0.451		
	c.i.	2.25 ± 0.839[b]		
4	r.l.	0.53 ± 0.236[c]	0.10 ± 0.231	0.43 ± 0.293[d]
	c.i.	8.01 ± 1.004[a]	8.28 ± 1.335[a]	0.23 ± 1.214
5	r.l.	0.86 ± 0.470[d]	1.05 ± 0.498[d]	0.19 ± 0.389
	c.i.	13.28 ± 0.789[a]	3.73 ± 1.282[b]	9.55 ± 1.045[a]
6	r.l.	0.02 ± 0.331	0.68 ± 0.239[b]	0.70 ± 0.282[c]
	c.i.	0.94 ± 0.737[d]	2.65 ± 0.967[b]	1.17 ± 0.968[d]
7	r.l.	0.76 ± 0.353[c]	0.22 ± 0.558	0.54 ± 0.530[d]
	c.i.	14.75 ± 1.460[a]	16.79 ± 1.238[a]	2.04 ± 1.622[d]

[a] = $p<0.001$; [b] = $0.01<p<0.02$; [c] = $0.05<p<0.02$; [d] = $p>0.05$
* Owing to small differences in r.l. in these cases, chromosomes with similar c.i. were compared.

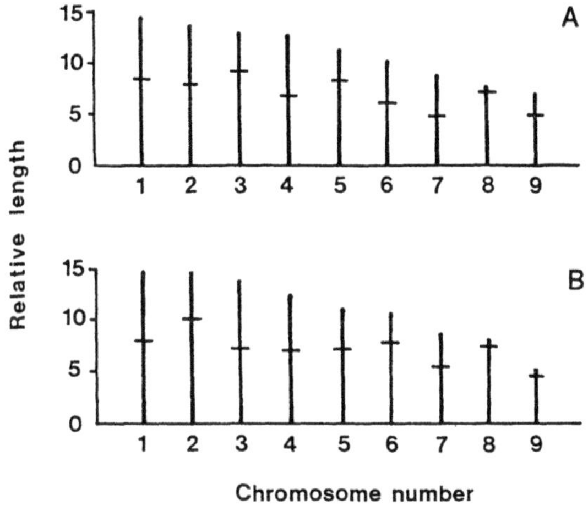

Fig. 2. Idiograms of (A) *Thysanozoon brocchii* (from Galleni & Puccinelli, 1979) and (B) *Yungia aurantiaca* (from Galleni & Puccinelli, 1985).

whether the rearrangements are due to pericentric inversions or to different levels of centromeric heterochromatin cannot be solved without the use of more detailed staining methods.

Karyometric comparisons between the polyclads *Thysanozoon brocchii* and *Yungia aurantiaca* reveal trends similar to those identified in *Procerodes* (see Galleni & Puccinelli, 1985). It is noteworthy that chromosome number 9 is submetacentric in *Thysanozoon brocchii* and acrocentric in *Yungia aurantiaca.* Since the long arm is of the same length in both species, the difference between the two must have arisen through rearrangement of the whole short arm (Fig. 2).

Comparisons of relative lengths and centromeric indices of *Notoplana igiliensis* and *Leptoplana tremellaris* (Table 2, Fig. 3) show that most measurements of relative lengths are not significantly different in the two species while those of centromeric index are.

Finally, the polyclad *Notoplana atomata* has a karyotype quite different from that of *Notoplana igiliensis* (see Galleni & Puccinelli, 1984). A chromosome number of 2n=18 in *Notoplana atomata* and of 2n=20 in *Notoplana igiliensis* and the lack of some pairs of heterobrachial chromosomes and the presence of a large metacentric chromosome in *Notoplana atomata* indicate that the difference in the karyotype of the two species could have arisen through a Robertsonian mechanism.

Table 2. Comparisons of relative length (r.l.) and centromeric index (c.i.) in *Notoplana igiliensis* and *Leptoplana tremellaris.* Values are reported as mean ± standard error of the mean. Numerical data from Galleni & Puccinelli (1981).

Chromosome number		
1	r.l.	0.48 ± 0.546
	c.i.	26.99 ± 2.678[a]
2	r.l.	0.02 ± 0.259
	c.i.	7.48 ± 1.875[b]
3	r.l.	0.11 ± 0.413
	c.i.	25.09 ± 1.044[a]
4	r.l.	0.53 ± 0.355[d]
	c.i.	9.37 ± 0.881[a]
5	r.l.	0.37 ± 0.310[d]
	c.i.	26.59 ± 1.346[a]
6	r.l.	0.64 ± 0.364[d]
	c.i.	1.91 ± 2.785
7/8*	r.l.	1.25 ± 0.383[c]
	c.i.	5.79 ± 2.913[d]
8/7*	r.l.	0.55 ± 0.486[d]
	c.i.	0.05 ± 1.631
9	r.l.	0.19 ± 0.540
	c.i.	0.27 ± 1.040
10	r.l.	0.07 ± 0.180
	c.i.	12.47 ± 2.540[c]

[a] = $p<0.001$; [b] = $0.001<p<0.01$; [c] = $0.01<p<0.02$; [d] = $p>0.05$
* = Owing to small differences in r.l., chromosomes with similar c.i. were compared.

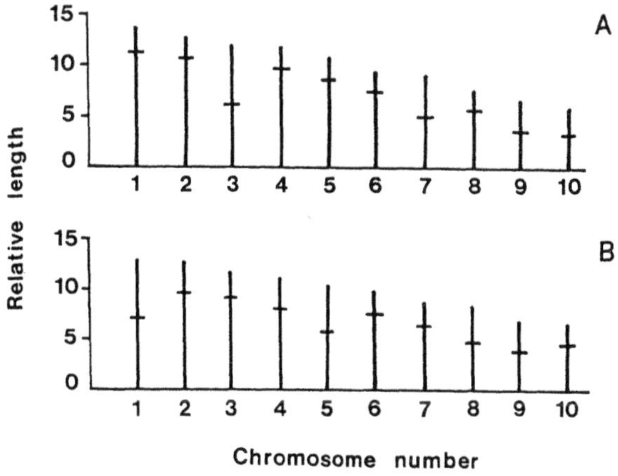

Fig. 3. Idiograms of (A) *Notoplana igiliensis* (from Galleni & Puccinelli, 1975) and (B) *Leptoplana tremellaris* (from Galleni & Puccinelli, 1981).

Discussion

Three different types of chromosomal rearrangements are evident among these five genera of marine turbellarians: 1) rearrangements involving the centromere which is thus shifted in position, 2) rearrangements of whole chromosome arms, and 3) Robertsonian rearrangements.

According to Benazzi (1982) 'only minor chromosome mutations such as small pericentric inversions have accompanied speciation' at least in some freshwater triclads such as *Dugesia gonocephala* and the *Dugesia* sugenus *Girardia*. The same seems to obtain to the *Procerodes* species and to *Notoplana igiliensis* and *Leptoplana tremellaris*.

Major rearrangements involving a whole chromosome arm in the *Thysanozoon-brocchii/Yungia-aurantiaca* example and a Robertsonian mechanism in *Notoplana atomata* are evident. A Robertsonian mechanism was also detected among biotypes E and F of the freshwater-triclad *Dugesia lugubris-polychroa* complex Benazzi & Puccinelli (1973). Thus chromosomal evolution among related species involves similar mechanisms whether they are marine or freshwater species. However, aneuploidy and polyploidy – which are commonly seen in freshwater turbellarians and are related to special forms of reproduction – both sexual (e.g. gynogenesis) and asexual – have not been detected in these marine polyclads and triclads.

It is not certain whether chromosomal rearrangements such as those we have found in these marine turbellarians are simply some of the many differences accumulated by differentiating populations or are the very cause of speciation itself as is hypothesized in the stasipatric model of speciation (White, 1978). Karyometric analysis by itself is not sufficient to determine how chromosomal differences among related species have arisen. Difficulties in comparing numerical data from different authors and in even conclusively identifying types of chromosomal rearrangements show the need for applying chromosome-banding techniques to turbellarians, techniques such as that Galleni *et al.* (1985) perfected in a study of *Dugesia polychroa.*

References

Ball, I.R., 1976. Les caryotypes de trois planaires marines nord-americaines: contribution à la phylogenie et la classification du groupe (Platyhelminthes, Turbellaria, Tricladida). Can. J. Zool. 54: 644–651.

Ball, I.R., 1979. The karyology of Procerodes lobata (Turbellaria, Tricladida, Maricola). Can. J. Zool. 57: 1820–1821.

Benazzi, M., 1982. Speciation events evidenced in Turbellaria. In C. Barigozzi (ed.), Mechanisms of speciation. Alan R. Liss, New York: 307–344.

Benazzi, M. & G. Benazzi Lentati, 1976. Plathyelminthes. In B. John (ed.), Animal Cytogenetics. Gebrüder Borntraeger, Berlin and Stuttgart, 182 pp.

Benazzi, M. & I. Puccinelli, 1973. A Robertsonian translocation in the freshwater triclad Dugesia lugubris: karyometric analysis and evolutionary inferences. Chromosoma 40: 193–198.

Galleni, L., R. Canovai, M. Gualandi, C. Pellicciari, S. Garagna & R. Stanyon, 1985. Fine characterization of turbellarian chromosomes I. Giemsa and Quinicrine banding in Dugesia polychroa (O. Schmidt). Genetica, in press.

Galleni, M., M. Nigro & I. Puccinelli, 1984. The chromosome complement of Procerodes dohrni (Tricladida: Maricola). J. mar. biol. Ass. U.K. 64: 227–231.

Galleni, L. & I. Puccinelli, 1975. Karyology of Notoplana igiliensis Galleni (Polycladida – Acotylea). Caryologia 28: 375–387.

Galleni, L. & I. Puccinelli, 1977. Karyometric analysis in Thysanozoon brocchii (Risso) (Turbellaria – Polycladida). Acc. Lincei Rend. Cl. Sci. fis., mat. nat. 63: 436–439.

Galleni, L. & I. Puccinelli, 1979. The karyology of the genus Procerodes (Tricladida: Maricola) in British waters. J. mar. biol. Ass. U.K. 59: 31–44.

Galleni, L. & I. Puccinelli, 1981. Karyological observation on polyclads. Hydrobiologia 84: 31–44.

Galleni, L. & I. Puccinelli, 1984. Karyology of five species of Turbellaria from the Øresund, Denmark. Ophelia 23: 141–148.

Galleni, L. & I. Puccinelli, 1985. Karyology of Yungia aurantiaca (Turbellaria: Polycladida). Trans. am. microsc. Soc. 104: 122–128.

White, M.J.D., 1978. Modes of speciation. W.H. Freeman & Company, San Francisco, 455 pp.

Variation in karyotypes of *Dugesia japonica japonica* (Turbellaria) from Ôsaka Prefecture, Central Japan

Sachiko Tamura
Ôsaka Prefectural Institute of Public Health, Nakamichi-1-chôme Higashinari-ku, Ôsaka 537, Japan

Keywords: Turbellaria, Tricladida, *Dugesia*, karyology, Japan

Abstract

An analysis of the karyotypes of *Dugesia japonica japonica* Ichikawa et Kawakatsu, 1964, from 30 localities of seven river systems in Ôsaka Prefecture, Central Japan, revealed a total of 26 karyotypes of which 12 are new-found varieties. More than two karyotypes were found in many localities. The mixoaneuploidic triploid karyotype showed the widest distribution (21 localities), orthoploidic diploid karyotype and orthoploidic mixoploid of diploid and triploid karyotype were next (13 localities, respectively), and other karyotypes (triploidic aneuploid, orthoploidic triploid and mixoploid of triploid & tetraploid) were also found. Some correlation of karyotype with elevation was also detected.

Introduction

In *Dugesia japonica* Ichikawa et Kawakatsu, 1964, the commonest freshwater planarian of the Far East including the Japanese Islands and the Ryûkyû Islands, two subspecies exist: *D. j. japonica* Ichikawa et Kawakatsu, 1964 (distributed in the Japanese Islands except the Amami and the Ryûkyû Islands, Taiwan, Hong Kong, China, and Korea; n=8, 2x=16, 3x=24) and *D. j. ryukyuensis* Kawakatsu, 1976 (distributed in the Ryûkyû Islands, the alpine region of Taiwan and the middle-eastern area of China; n=7, 2x=14, 3x=21) (Kawakatsu *et al.*, 1976, 1979).

The karyotypes of animals from 120 localities in Japan, Taiwan, and Korea have been examined by our team. A total of 29 karyotypes are now known in this species (Oki & Kawakatsu, 1983). We presume there is a relationship between the karyological variation of this species and geographical area though we have insufficient data to demonstrate this. To test this hypothesis in a restricted region, I have studied the variation of karyotypes of *D. j. japonica* from Ôsaka Prefecture in Central Japan.

Material and methods

Animals used for karyological analysis were collected from the following 30 localities classified into seven groups according to river system and further subdivided into 11 groups according to tributaries.

1. The Kanzaki-gawa River group is located in the northwestern part of Ôsaka Prefecture. 1) Mino'o stocks: No. 1 (elevation, 40 m); No. 2 (80 m); No. 3 (300 m); No. 4 (320 m); No. 5 (410 m). 2) Ai stock: No. 1 (140 m). 3) Shimo' otowa stock: No. 1 (140 m).
2. The Yodo-gawa River group is from the largest river system in Ôsaka Prefecture. 4) Akuta stocks: No. 1 (40 m); No. 2 (110 m); No. 3 (280 m). 5) Izuriha stock: No. 1 (340 m).
3. The Kyû-Yodo-gawa River group is located in the central part of Ôsaka Prefecture. 6) Gongen stocks: No. 1 (160 m); No. 2 (200 m).
4. The Yamato-gawa River group is located in the southeastern part of Ôsaka Prefecture. 7) Chihaya stocks: No. 1 (260 m); No. 2 (450 m); No. 3 (510 m); No. 4 (570 m); No. 5 (660 m).

Hydrobiologia 132, 243–249 (1986).

5. The Ôtsu-gawa River group is located in the southern part of Ôsaka Prefecture. 8) Makio stock: No. 1 (80 m). 9) Chichioni stocks: No. 1 (100 m); No. 2 (160 m); No. 3 (260 m); No. 4 (640 m). A single specimen of *Phagocata vivida* (Ijima et Kaburaki, 1916) occurred in the Station No. 4.

6. The Onosato-gawa River group is located in the southwestern part of Ôsaka Prefecture: 10) Kinŷuji stocks: No. 1 (60 m); No. 2 (100 m); No. 3 (110 m); No. 4 (120 m); No. 5 (125 m); No. 6 (140 m).

7. The Ôkawa River group is from a short stream located in the southernmost part of Ôsaka Prefec-

Table 1. Karyotypes of *Dugesia japonica japonica* inhabiting the seven river systems in Ôsaka Prefecture, Central Japan.

	Chromosomal numbers	Karyotypes	River systems*						
			1	2	3	4	5	6	7
Diploidy	16	2x	+	−	−	+	+	+	+
	16	2x & 2x + SB	−	−	−	−	+	−	−
Triploidy	24	3x	+	−	−	−	−	−	−
	25	3x + 1LB + SB	+	+	−	+	+	+	−
Mixo-ploidy	16 & 24	2x & 3x	+	−	+	+	+	+	−
	17 & 18 & 25 & 26**	2x + 1LB + SB & 2x + 2LB + SB & 3x + 1LB + SB & 3x + 2LB + SB	−	−	−	−	+	−	−
	25 & 33**	3x + 1LB + SB & 4x + 1LB + SB	−	+	−	−	−	−	−
Mixo-aneu-ploidy	25	3x + 1LB + SB & (3x − 1) + 2LB + SB	−	−	−	−	−	−	+
	25**	3x + 1LB + SB & (3x − 1 − 1) + 3LB + SB	−	−	−	−	−	+	−
	25	3x + 1LB + SB & (3x − 1) + 2LB + SB & (3x − 1 − 1) + 3LB + SB	−	−	−	−	+	+	+
	24 & 25**	(3x − 1) + 1LB & 3x + 1LB	−	+	−	−	−	−	−
	24 & 25**	(3x − 1 − 1) + 2LB & 3x + 1LB	+	−	−	−	−	−	−
	24 & 25	(3x − 1 − 1) + 2LB & (3x − 1) + 2LB	+	−	−	−	−	−	−
	24 & 25**	(3x − 1 − 1) + 2LB + SB & 3x + 1LB + SB & (3x − 1) + 2LB + SB	+	−	−	−	−	−	−
	24 & 25**	(3x − 1 − 1) + 2LB + SB & (3x − 1) + 2LB + SB & (3x − 1 − 1) + 3LB + SB	+	−	−	−	−	−	−
	25 & 26	3x + 1LB + SB & 3x + 2LB + SB	+	+	+	−	+	−	−
	25 & 26	3x + 1LB + SB & (3x − 1) + 3LB + SB	+	−	−	−	−	−	−
	25 & 26	3x + 1LB + SB & (3x − 1) + 2LB + SB & 3x + 2LB + SB	−	−	+	−	−	+	+
	25 & 26**	3x + 1LB + SB & (3x − 1 − 1) + 3LB + SB & 3x + 2LB	−	−	−	−	+	−	−
	25 & 26	3x + 1LB + SB & (3x − 1) + 2LB + SB & (3x − 1 − 1) + 3LB + SB & 3x + 2LB + SB	−	−	−	−	−	+	+
	25 & 26**	3x + 1LB + SB & (3x − 1) + 2LB + SB & 3x + 2LB + SB & (3x − 1) + 3LB + SB	−	−	−	−	−	+	−
	25 & 26	3x + 1LB + SB & (3x − 1) + 2LB + SB & (3x − 1 − 1) + 3LB + SB & 3x + 2LB + SB & (3x − 1) + 3LB + SB	−	−	−	−	−	+	+
	26 & 27	3x + 2LB + SB & 3x + 3LB	−	+	−	−	−	−	−
	24 & 25 & 26**	(3x − 1) + 1LB + SB & 3x + 1LB + SB & 3x + 2LB + SB	−	−	−	+	−	−	−
	24 & 25 & 27**	(3x − 1) + 1LB + SB & 3x + 1LB + SB & (3x − 1) + 4LB	+	−	−	−	−	−	−
	25 & 26 & 27**	3x + 1LB + SB & 3x + 2LB + SB & 3x + 3LB + SB	−	+	−	−	−	−	−

* River systems: 1, Kanzaki-gawa River; 2, Yodo-gawa River; 3, Kyû-Yodo-gawa River; 4, Yamato-gawa River; 5, Ôtsu-gawa River; 6, Onosato-gawa River; 7, Ôkawa River.
** New-found karyotypes of *Dugesia japonica japonica*.

ture. 11) Ôkawa stock: No. 1 (40 m).

The geographical location of these 30 localities is shown by Kawakatsu *et al.* (1984: 4, Fig. 2).

For chromosomal studies, a squash method was employed (Oki *et al.*, 1981).

Karyological results

A total of 26 karyotypes, classified according to differences in polyploidies, was found in 120 specimens (4 sexually mature and 116 immature) (Table 1). Karyograms of seven karyotypes found in seven animals from the Chichioni stock (four localities) of the Ôtsu-gawa River system are shown in Fig. 1 (A–G). Photomicrographs of the chromosomes of two specimens from the same stock (and of a single specimen from the Ôkawa stock) are shown by Kawakatsu *et al.* (1984: 9, Fig. 4).

The chromosome number of diploid specimens of *D. j. japonica* is 2x=16 (n=8). The karyotype consists of eight pairs of meta- or submetacentric chromosomes in descending order of size (see Fig. 1A). This diploid karyotype was found in animals from 13 localities of five river systems. Two animals having cells of the diploid karyotype (2x & 2x+SB; Fig. 1 B-1, B-2) were found in one locality.

The triploid karyotype (3x) was rather rare and was found only in two animals from one locality. Animals having cells of the triploid karyotype (3x+1LB+SB; Fig. 1D) were found from seven localities of five river systems.

The orthoploidic mixoploid karyotype (2x & 3x; Fig. 1 C-1, C-2) was rather common. It was found in 22 animals from 13 localities of five river systems. Animals having cells of the heteroploidic mixoploid karyotypes were very rare. Two kinds of karyotypes of this group were found in two animals from different localities; one animal had cells with B-chromosomes in addition to the diploid and the triploid chromosomes (see Fig. 1 E-1, E-2, E-3, E-4), and the other animal had cells with B-chromosomes in addition to the triploid and the tetraploid chromosomes.

Animals having cells of the mixoaneuploidy karyotypes include wide variations in their chromosome numbers (Table 1). Among these 19 karyotypes, five were found in animals from two to four river systems. Most common was the karyotype 3x+1LB & 3x+2LB, found in 10 animals from four river systems. The remaining 14 karyotypes were found only in a limited number of animals.

In brief, animals having karyotypes of the mixoaneuploidic triploidy were found in 21 localities in Ôsaka Prefecture (50 out of 120 specimens examined; approximately 42%).

Sperm and meiotic figures in the male line were observed in a single sexually mature specimen of the Ôkawa stock having the karyotype 3x+1LB+SB & (3x−1)+2LB+SB & (3x−1−1)+3LB+SB (Kawakatsu *et al.*, 1984: 9, Fig. 4 H, I). These figures may consist of univalent and polyvalent chromosomes. This is the first record of meiotic division in specimens of *D. j. japonica* having the mixoaneuploidic triploidy karyotype.

Discussion

The presence of a wide variation in chromosome numbers and karyotypes of *D. japonica* (*D. j. japonica* and *D. j. ryukyuensis*) has been reported in previous papers by our team (Oki *et al.*, 1981; and others). Most of the previous data are, however, based upon animals occasionally collected from various geographical areas of the Far East including the Japanese Islands. My finding two to six karyotypes in animals collected from a single locality (Table 1) shows the same variation can occur within one river system.

I have tested the correlation of variation of karyotypes with elevation of localities where animals were collected in the seven river systems (Fig. 2). The orthoploidic diploid karyotype (2x) is distributed widely in the elevation range of 40 to 660 meters. The mixoploid of orthoploidic diploid & diploidic aneuploidy karyotype (2x & 2x+SB) is, however, only found in one locality of the Ôtsu-gawa River system (elev. 260 m). The triploidic aneuploid karyotypes (3x+1LB+SB) and orthoploidic mixoploid of diploid & triploid karyotypes (2x & 3x) are distributed rather widely in five river systems (elev. 40–450 m). The mixoaneuploidic triploid karyotypes show a high degree of variation. Most of these 19 karyotypes are distributed in limited localities in each river system (elev. 40–410 m). The other three karyotypes are found only in one locality.

The geographical distribution of every known variation of the karyotypes of *D. japonica* from Ja-

Fig. 1. Karyograms of *Dugesia japonica japonica* from 4 localities of Ôtsu-gawa River system (Chichioni stock) in Ôsaka Prefecture, drawn from cells of 7 specimens (A–G). A: 2x. B: 2x & 2x+1SB. B-1, 2x; B-2, 2x+1SB. C: 2x & 3x. C-1, 2x; C-2, 3x. D: 3x+1LB. E: 2x+1LB+3SB & 2x+2LB+1SB & 3x+1LB & 3x+2LB+3SB. E-1, 2x+1LB+3SB; E-2, 2x+2LB+1SB; E-3, 3x+1LB; E-4, 3x+2LB+3SB. F: 3x+1LB+3SB & 3x+2LB+1SB. F-1, 3x+1LB+3SB; F-2, 3x+2LB+1SB. G: 3x+1LB & (3x−1−1)+3LB & 3x+2LB. G-1, 3x+1LB; G-2, (3x−1−1)+3LB; G-3, 3x+2LB.

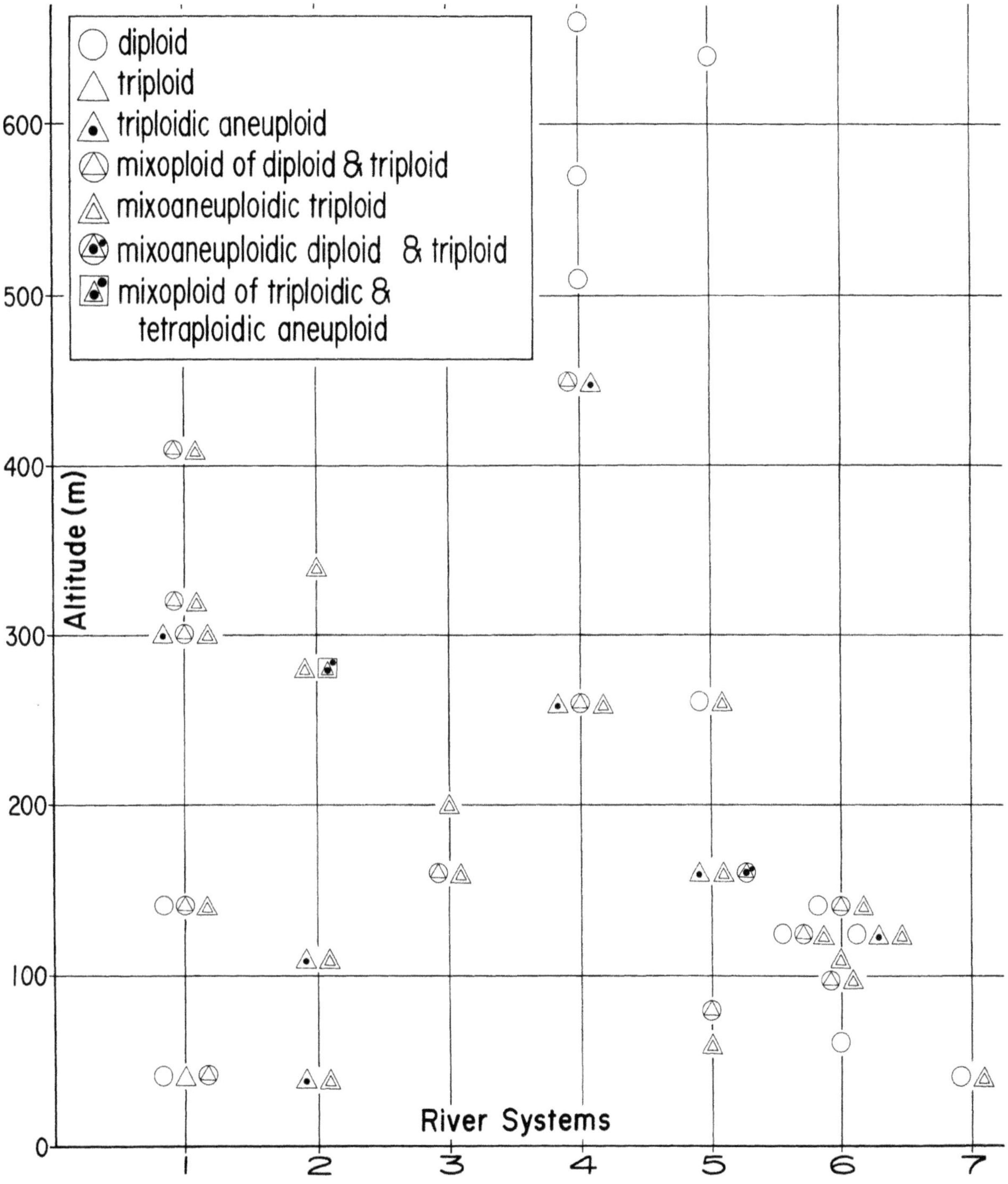

Fig. 2. Plot of karyotypes against elevation of localities of *Dugesia japonica japonica* in Ôsaka Prefecture. For names of river systems, see Table 1.

pan, Korea, and Taiwan (*D. j. japonica*) and from Okinawa or Ryûkyû in Southwest Japan (*D. j. ryukyuensis*) based upon the latest data of the chromosomal analysis of this species is shown in Table 2. In this table, various karotypes are simplified and reclassified into 11 fundamental forms of the two groups, i.e. the homokaryotype and the mixokaryotype. In interpreting this table, it must be noted that all the geographical areas in the Far East are not treated equally for chromosomal analysis. More data is from the central and southwestern areas of Honshû in Japan are larger than from other areas of Japan and the other countries.

Of all karyotypes, the orthoploidic diploid karyotype (2x) shows the widest distribution (ca 31% number of localities). The second most widespread is the orthoploidic mixoploid of diploid and triploid karyotype (2x & 3x; ca. 24%); third is the mixoaneuploidic triploidy karyotypes with different numbers of B-chromosomes [3x±(α&β)] (ca. 20%); and fourth is the triploidic aneuploidy karyotype with a few B-chromosomes (3x±α; ca. 14%). Although the orthoploidic triploid karyotype (3x) is found in very few localities, it reaches approximately 6% of the total. The numbers of localities of the other karyotypes are negligible (0.6–1%).

The mechanism of formation of various karyotypes and the origin of the B-chromosomes in this species have been discussed by Kawakatsu *et al.* (1976) and Oki *et al.* (1981).

It is noteworthy that animals having the diploid karyotype (2x) occur in the widest vertical distribution in Ôsaka Prefecture. In two river systems, the Yodo-gawa River and the Ôtsugawa River, the upper part of the streams (over 500 m in elevation) is occupied only by animals of the diploid form. In contrast, animals having the other karyotypes are distributed below an elevation of 450 meters (Fig. 2). Most of our previous data on the cytological analysis of this species are based upon the animals collected from the lowland areas. *D. japonica* is known as an indicator animal living in the oligosaprobic, β-mesosaprobic and α-mesosaprobic zones (Kawakatsu, 1977).

Analysis of the relationship between the variation of karyotypes of *D. japonica* and the physical and chemical properties of water where the planarians live is one of the future themes of studies in my laboratory.

Table 2. Geographical distribution of the eleven known forms of karyotypes of *Dugesia japonica*.

	Karyotypes	*D. j. japonica*						*D. j. ryukyuensis*	Total
		Japan				Korea	Taiwan	Japan	
		Hokkaidô	Honshû	Shikoku	Kyûshû			Okinawa	
Homokaryotypes	2x	–	39 (150)	3 (9)	4 (13)	1 (1)	3 (16)	2 (8)	52 (197)
	2x ± α	–	1 (2)	–	–	–	–	1 (1)	2 (3)
	3x	–	5 (16)	–	1 (1)	1 (1)	–	3 (8)	10 (26)
	3x ± α	–	12 (40)	–	5 (10)	–	3 (11)	3 (3)	23 (64)
Mixokaryotypes	2x & 2x ± α	–	2 (2)	–	–	–	–	–	2 (2)
	2x & 3x	3 (20)	33 (152)	–	1 (1)	2 (6)	–	1 (2)	40 (181)
	2x + (α & β) & 3x + (α & β)	–	1 (1)	–	–	–	–	–	1 (1)
	2x & 4x	–	2 (4)	–	–	–	–	–	2 (4)
	2x & 3x & 4x	–	1 (3)	–	–	–	–	–	1 (3)
	3x ± (α & β)	–	28 (54)	–	2 (5)	1 (2)	1 (1)	2 (5)	34 (67)
	3x + α & 4x + α	–	1 (1)	–	1 (1)	–	–	–	2 (2)

Values are numbers of localities; numbers of specimens examined are shown in parentheses. Large supernumerary chromosomes are shown by Greek letters (α or β), which indicate different numbers of supernumerary chromosomes found in cells of animals having the mixokaryotype.

Acknowledgments

I heartily thank Dr. Iwashiro Oki for his constant guidance in the course of the work. I am much indebted especially to Prof. Dr. Masaharu Kawakatsu for his invaluable advice and generous help with the preparation of this manuscript as a chief-leader of our team. Finally, I thank the staff of my home institution for their cooperation in collecting the animals in the field.

References

Kawakatsu, M., 1977. Ecology and distribution of freshwater planarians. The Heredity (Iden), Tôkyô 31 (10): 13–24. (In Japanese).

Kawakatsu, M., I. Oki, S. Tamura & H. Sugino, 1976. Studies on the morphology, karyology and taxonomy of the Japanese freshwater planarian Dugesia japonica Ichikawa et Kawakatsu, with a description of a new subspecies, Dugesia japonica ryukyuensis subspec. nov. Bull. Fuji Women's Coll. 14 (ser. II): 81–126.

Kawakatsu, M., I. Oki, S. Tamura, T. Yamayoshi, K.-Y. Lue & M. Hagiya, 1979. Additional report on freshwater planarians from Taiwan. Bull. Fuji Women's Coll. 17 (ser. II): 59–91.

Kawakatsu, M., S. Tamura & K.-Y. Lue, 1984. Preprint of papers given at the Fourth International Symposium on the Turbellaria. Occ. Publ., Biol. Lab. Fuji Women's Coll., Sapporo (Hokkaidô), Jap. 12: 1–20.

Oki, I. & M. Kawakatsu, 1983. Recent advancement of the turbellariology: Part 2. The Heredity (Iden), Tôkyô 37 (2): 40–49 (In Japanese).

Oki, I., S. Tamura, T. Yamayoshi & M. Kawakatsu, 1981. Karyological and taxonomic studies of Dugesia japonica Ichikawa et Kawakatsu in the Far East. Hydrobiologia 84: 53–68.

On the karyology of *Dugesia gonocephala* s.l. (Turbellaria, Tricladida) from Montpellier, France

Elizabeth J. de Vries
Institute of Taxonomic Zoology, University of Amsterdam, Plantage Middenlaan 53, Amsterdam, The Netherlands

Keywords: Turbellaria, *Dugesia gonocephala* s.l., karyology, aneuploidy, B-chromosomes, chromosomal heteromorphism

Abstract

Two species belonging to the *Dugesia gonocephala* group are found in the area of Montpellier, France. The karyology of these two species, *D. gonocephala* s. str. and *S. subtentaculata,* and of fissiparous *Dugesia* races has been studied.

Two populations belonging to *D. gonocephala* s. str. are diploids with a chromosome number of 16, whereas the specimens of a third population are sexual aneuploids; the majority of cells possess 24 chromosomes, but some cells contain 23 or 25. The specimens attributable to *D. subtentaculata* are triploids, the most notable karyological feature being the presence of a single unmatched acrocentric chromosome. The fissiparous *Dugesia* strains are all aneuploids, the most common chromosome number being 27 with up to three small B-chromosomes.

Introduction

Dugesia gonocephala (Dugès, 1830) is the type species of the genus *Dugesia* and hence is the type of the family Dugesiidae Ball, 1974. Thus it is problematic that the taxonomic status of this and of a closely related form, *D. subtentaculata* (Draparnaud, 1801), has never been properly resolved (De Vries & Ball, 1980).

In order to resolve this problem, I have carried out karyological and morphological studies on material from Montpellier, France. This is the area from where the original type material of *D. gonocephala* and *D. subtentaculata* was collected.

This paper focuses on the variety in the karyotypes found within *D. gonocephala* s.l. from Montpellier; the morphological aspects of the populations will be the subject of a separate paper.

Materials and methods

The specimens of *Dugesia gonocephala* studied were collected from the following localities and are now deposited in the Zoological Museum of Amsterdam.

ZMA V.Pl. 615 – River Vidourle between Sommières and Lecques, 15 July 1980.

ZMA V.Pl. 616 – River Hérault, ±4 km south of Ganges, 15 July 1980.

ZMA V.Pl. 617 – Tributary of the river Hérault, ±1 km north of St. Guilhem-le-désert, near the Grottes du Sergent, 16 July 1980.

ZMA V.Pl. 618 – In a fountain in the village of Montblanc, ±9 km south-west of Pézenas, 20 July 1980.

ZMA V.Pl. 619 – Tributary of the river Lerque, in the vicinity of Pégairolles-de-l'Escalette, ±8 km north of Lodève, 20 July 1980.

ZMA V.Pl. 620 – River Orb, near Sérieys, ±16 km north of Bédarieux, 16 July 1980 and 24 July 1981.

Hydrobiologia 132, 251–256 (1986).

ZMA V.Pl. 621 – River Berre, ±10 km west of Sigean, 30 July 1980.

ZMA V.Pl. 622 – River Orb, ±1 km south of site V.Pl. 620, 4 July 1981.

The karyological work was carried out on squashed regenerative blastemas as described by, for example, Ball & Gourbault (1975). The chromosomal nomenclature follows that proposed by Levan et al. (1964). The centromeric index (c.i.) was calculated as length of the short arm×100/total chromosome length. Relative length (r.l.) was calculated as the total chromosome length×100/total length of the haploid genome.

Results

Dugesia gonocephala *s. str.*

Sexually mature specimens and cocoons have been found in three of the sampled localities (V.Pl. 617, 619 & 620). Both the external and internal morphology of these sexual specimens are identical with those of *D. gonocephala* from western Europe as described by De Vries & Ball (1980).

Two of these populations belonging to *D. gonocephala* s. str. (V.Pl. 617 & 620) possessed a chromosome number of 16 (Fig. 1A). On one occasion meiosis was observed, showing eight fully paired bivalents. The chromosomes were all metacentric and they can be arranged according to length into a gradually graded series without clear-cut discontinuities in sequence, the greatest interval being between the first and second chromosome (Table 1). The longest chromosome was almost twice the length of the smallest.

The individuals of population V.Pl. 619 were aneuploids (Table 2A). The majority of cells studied contained 24 chromosomes (Fig. 1B), but some individuals were found to possess cells that com-

Table 1. Means and standard deviations of the relative lengths (r.l.) and centromeric indices (c.i.) of the eight chromosome pairs of *D. gonocephala* s. str., 2n = 16, from Montpellier, France.

Chromosome	A) V.Pl. 617 (10 cells/2 individuals)		B) V.Pl. 620 (23 cells/7 individuals)	
	r.l.	c.i.	r.l.	c.i.
1	17.04 ± 0.61	45.99 ± 2.05	17.31 ± 0.93	45.60 ± 2.33
2	14.67 ± 0.79	47.02 ± 2.18	14.89 ± 0.66	46.67 ± 2.95
3	13.84 ± 0.83	46.12 ± 2.77	13.01 ± 1.57	43.36 ± 4.06
4	12.43 ± 0.64	43.65 ± 1.98	12.60 ± 0.89	44.36 ± 3.97
5	11.68 ± 0.70	40.05 ± 2.61	11.47 ± 0.56	43.27 ± 3.18
6	11.25 ± 0.58	45.96 ± 2.70	10.77 ± 0.44	45.37 ± 3.66
7	10.38 ± 0.74	47.03 ± 1.99	10.38 ± 0.42	42.98 ± 4.53
8	9.91 ± 0.60	42.94 ± 3.67	9.52 ± 0.57	45.56 ± 4.68

Table 2. Means and standard deviations of the relative lengths (r.l.) and centromeric indices (c.i.) of the assumed eight chromosome sets of *D. gonocephala* s. str. and *Dugesia subtentaculata,* from Montpellier, France.

Chromosome	A) *D. gonocephala* s. str. V.Pl. 619 (7 cells/3 individuals)		B) *D. subtentaculata* V.Pl. 620 (5 cells/3 individuals)	
	r.l.	c.i.	r.l.	c.i.
1	16.95 ± 0.55	46.26 ± 2.70	17.37 ± 0.28	45.44 ± 2.99
2	15.21 ± 0.57	44.53 ± 3.77	15.18 ± 0.30	41.90 ± 4.06
3	13.35 ± 0.36	42.79 ± 8.63	13.95 ± 0.38	46.68 ± 2.43
4	12.21 ± 0.30	33.68 ± 7.97	12.42 ± 0.23	36.71 ± 11.27
5	11.52 ± 0.20	43.36 ± 3.38	11.43 ± 0.22	42.30 ± 5.38
6	10.92 ± 0.15	37.44 ± 6.98	10.89 ± 0.18	43.70 ± 3.62
7	10.23 ± 0.17	44.35 ± 7.23	10.17 ± 0.31	43.00 ± 6.99
8	9.69 ± 0.27	38.41 ± 6.79	9.06 ± 0.29	46.48 ± 3.20

prised chromosome numbers of 23 or 25. The accessory chromosomes were of the same length as the other chromosomes of the complement. Moreover, a variable number of five to seven heterobrachial chromosomes was observed in all the cells. Meiotic stages have not been found so the haploid chromosome number could not be established.

Dugesia subtentaculata

Locality V.Pl. 620 contained a mixed population of sexually mature diploids (cf. *D. gonocephala* s. str. above) and fissiparous triploids, whereas V.Pl. 622 contained only fissiparous triploids. The fissiparous specimens of the two populations all became sexually mature under laboratory conditions.

Population V.Pl. 620 was sampled twice in two successive years, and a number of specimens from both samples were cultured in the laboratory for karyological research. In both years, fissioning stopped spontaneously after several months in the laboratory, and large hyperplasic ovaries developed, which could be seen macroscopically as white areas occupying most of the posterior part of the animal. Over a period of about five months, the copulatory apparatus had developed and the animals remained sexually mature for approximately twelve months, during which time a large number of sterile cocoons were deposited. Subsequently, the copulatory apparatus regressed, and several specimens resumed fissioning. The specimens of population V.Pl. 622 showed an essentially similar development of sexual maturity.

The morphology of the copulatory apparatus of the ex-fissiparous specimens differed significantly from that of *D. gonocephala* s. str., so the ex-fissiparous specimens are given a separate status, viz., *D. subtentaculata*. The morphology of the copulatory apparatus will be described elsewhere.

All the investigated specimens of *D. subtentaculata* possessed an identical karyotype with 24 chromosomes (Table 2B). I found no meiotic stages, and so could not ascertain the haploid number. All the chromosomes could be arranged according to length into a gradually graded series. The majority of these chromosomes were metacentric, but some had a centromere that bordered on a submetacentric position. The most notable feature was the presence of a single subtelocentric chromosome (Fig. 1C).

Dugesia gonocephala *s.l.*

The animals from V.Pl. 615, 616, 618 & 621, differed in appearance from those of the other populations in that they were smaller and more ribbon-like, with a conspicuous mottled pigmentation and more pronounced auricles.

The studied animals possessed a chromosome number ranging from 26 to 29 (Table 3). In all but one cell a number of small B-chromosomes was present (Fig. 1D).

The majority of the chromosomes were metacentric with a variable number of submetacentric elements. Because the length of the chromosomes falls into a gradual series and the centromeric index varied considerably, no attempt was made to match the chromosomes into homologous sets.

Discussion

The animals studied can be grouped into the following four categories.

Table 3. Range and variation in numbers and chromosomes and B-chromosomes in four populations (V.Pl. 615, 616, 618 & 621) of *Dugesia gonocephala* s.l. from Montpellier, France (28 cells/10 individuals).

Population	Number of cells containing the following chromosome numbers:						
	26	26+2B	26+3B	27+1B	27+2B	27+3B	29+1B
V.Pl. 615		1			7		
V.Pl. 616			1	1	7		1
V.Pl. 618			1		3		
V.Pl. 621	1	2			2	1	
Total in all populations	1	3	2	1	19	1	1

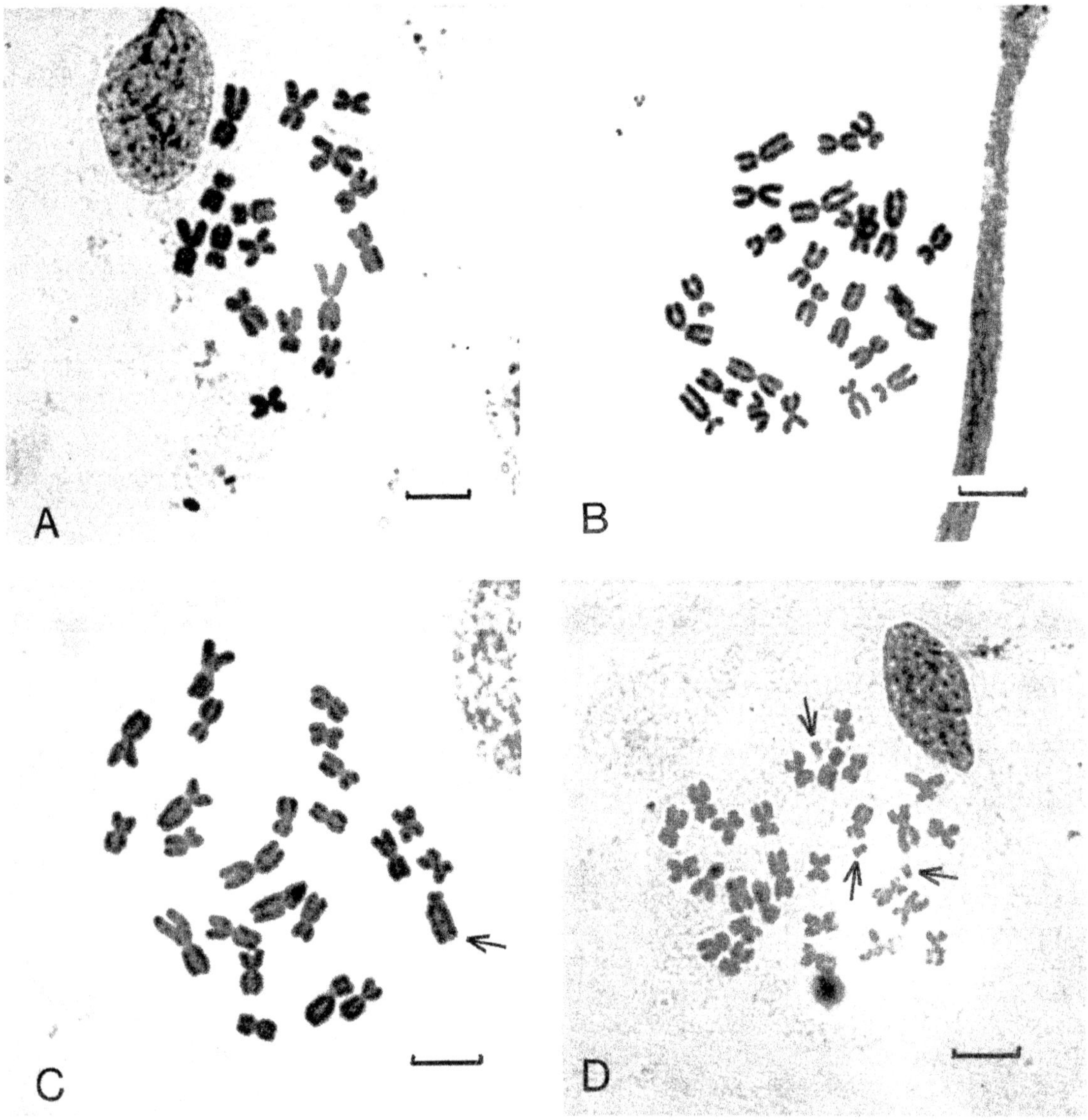

Fig. 1. Mitotic metaphase plates of *D. gonocephala* s.l. from Montpellier; A) V.Pl. 617, 2n = 16; B) V.Pl. 619, x = 24; C) V.Pl. 620, x = 24, arrow indicates aberrant chromosome; D) V.Pl. 621, x = 27 + 3B, arrows indicate B-chromosomes. Scale bars indicate 5 μm.

1) D. gonocephala *s. str., sexual diploids*

The karyological data obtained from the sexual diploid specimens are in accord with other published data on *D. gonocephala* from western Europe (Dahm, 1958, 1963; De Vries & Ball, 1980). The karyotypes represent the basic set of *D. gonocephala* from Europe (Benazzi & Benazzi-Lentati, 1976). Apart from the first and second chromosome, which could be classified on the basis of their different lengths, the lengths of all the chromosomes showed overlap to some extent, so that matching was difficult and in effect unreliable (Table 1).

2) D. gonocephala *s. str., sexual aneuploids*

Interpretation of the karyotype of V.Pl. 619 is

difficult due to the variation in chromosome number and centromeric index. My attempts to match the chromosomes into sets of three homologous elements were unreliable because of the overlap in length between the successive elements and the variable number of submetacentric chromosomes (Table 2A). These results were supported by Sluys & De Jong (1984), who, in working with the same population, proved that the chromosomes could not be matched into significant groups. Their findings are based on a scatter diagram of the complement of one single cell. This method is laborious but renders statistically more sound data because the values are calculated per chromosome and based on a relative length expressed as a percentage of the total complement length without assuming a haploid genome. However, because their data are derived from one single cell, they neglect the variation in chromosome number within this population.

A similar form of aneuploidy is found in *D. etrusca* (Benazzi-Lentati, 1957). During meiosis in the oocytes, eight bivalents and a variable number of univalents appear, whereas during spermatogenesis the accessory chromosomes were eliminated. Therefore, aneuploidy in *D. etrusca* is transmitted by the female line. The variable chromosome number in the somatic cells of *D. etrusca* arises from unequal distribution of the chromosomes over the daughter cells during mitotic division.

Future karyological studies on the meiosis of the aneuploid *D. gonocephala* s. str. should reveal whether the specimens function as sexual diploids with accessory chromosomes (as has been found in *D. etrusca*) or whether they reproduce by pseudogamy, by which the sperm functions only to stimulate cocoon production without fertilizing the oocytes [as has been found in *D. benazzii* (see Benazzi & Benazzi-Lentati, 1976)].

3) D. subtentaculata, *fissiparous triploids*

The high standard deviations of r.l. and c.i. (Table 2B) show that the elements cannot be reliably matched into triplets.

In each cell one single aberrant subtelocentric chromosome was found. In four out of five cells the aberrant chromosome could be grouped with the fourth chromosome on the basis of its relative length, whereas in the fifth cell it matched the seventh chromosome. The value for the centromeric index of the two chromosome triplets is influenced by that of the aberrant, subtelocentric, chromosome. The high values for the standard deviation in both occassions indicate that the chromosome is clearly aberrant and cannot be matched as has been done here.

This specific chromosomal aberration is most likely caused by a pericentric inversion, altering only the position of the centromere and not the relative length. More cytological research, including banding methods, will be necessary to elucidate the exact nature of the aberration. Presumably, this specific karyological aberration can only be maintained within an already asexually reproducing population, since problems arising during meiotic chromosome pairing would select against individuals possessing this mutation.

An identical chromosomal polymorphism was found in a population of *D. japonica japonica* from Taiwan (Oki *et al.*, 1981). According to these authors, the presence of this non-homologous chromosome is caused by the breaking away of one chromosome out of a set of chromosomes.

4) D. gonocephala *s.l., aneuploids with B-chromosomes*

The majority of cells studied possessed 27 chromosomes and 2 small B-chromosomes, although other numbers were also found (Table 3). The variation in chromosome number and B-chromosomes is caused by the unequal division of these elements during mitosis so that the B-chromosomes can therefore be classified as ones that are mitotically unstable (White, 1973).

In these populations the factors controlling fissioning seem to be more dominant than in the triploid populations of *D. subtentaculata* in that sexual maturation never occurred during the time they have been cultured in the laboratory. The influence of the B-chromosomes on this fissioning factor is not known (Bromley, 1977; Benazzi-Lentati & Deri, 1980). A variable number of B-chromosomes is known, furthermore, from *D. biblica* (Bromley, 1974) and *D. benazzii* (Benazzi-Lentati & Deri, 1977). A similar karyotype has also been found by Gourbault (1981) in some populations of *D. gonocephala* s.l. from Spain.

Conclusions

The occurrence of two species of the *gonocephala* group in one locality, as described here for *D. gonocephala* s. str. and *D. subtentaculata* in population V.Pl. 620, is the first record of the coexistence of two such closely related species. The possibility that the two species are in competition remains to be investigated.

Despite the different morphological features of *D. subtentaculata,* the normal members of the chromosome set are very similar to those found in *D. gonocephala* s. str. or other species of the *gonocephala* group (Bromley, 1974; Kawakatsu *et al.,* 1976; Gourbault, 1981). Morphological variation without apparent changes in the chromosomal morphology, and vice versa, seem to be general phenomena within the *gonocephala* group. Consequently, distinguishing asexual races on the basis of karyological data alone should be avoided.

Furthermore, matching of chromosomes into sets, as is usually done within the *gonocephala* group, is unreliable since several of the successive chromosomes show overlap in their relative length and centromeric index. This is demonstrated by the karyometric data presented here in Tables 1 & 2 but also in the karyometric data of, for example, *D. iberica* and *D. sicula* (Gourbault, 1981) and *D. gonocephala* s.l. from Corfu (Ball, 1979). This problem has been pointed out by Sluys & De Jong (1984), but their alternative, to present the data in the form of scatter diagrams, also has its disadvantages. It is obvious that other characters, such as banding patterns, are needed for a more reliable identification of the individual chromosomes.

Acknowledgments

I am grateful to Dr. Ronald A. Gase for his skilfull and enthusiastic help during the fieldwork. Prof. Dr. Ian R. Ball (University of Amsterdam) and Drs. Ronald Sluys (UvA) are thanked for their critical comments. Much practical support was received from L.R.T. It is a pleasure to thank Mr. and Dr. J. De Vries-Anderson (Montreal) for their hospitality during my stay in Canada. Grateful acknowledgement is made of a grant from the Catharine van Tussenbroek Foundation which made my attendance at the congress possible.

References

Ball, I. R., 1979. The karyotypes of two Dugesia species from Corfu, Greece (Platyhelminthes, Turbellaria). Bijdr. Dierk. 8: 187–190.

Ball, I. R. & N. Gourbault, 1975. The morphology, karyology and taxonomy of the genus Phagocata from California (Platyhelminthes: Turbellaria). Life Sci. Contr., R. Ont. Mus. 105: 1–19.

Benazzi, M. & G. Benazzi-Lentati, 1976. Animal Cytogenetics, I. Platyhelminthes. Gebr. Borntraeger, Berlin & Stuttgart: 1–182.

Benazzi-Lentati, G., 1957. Sul determinismo e sulla ereditarieta della aneuploidia in Dugesia etrusca Benazzi planaria a riproduzione anfigonica. Caryologia 10: 352–387.

Benazzi-Lentati, G. & P. Deri, 1977. Insorgenza di B-cromosomi in individui poliploidi di Dugesia benazzii (triclade paludicolo) allevati in laboratorio. Rc. Accad. naz. Lincei, ser. VIII, 62: 847–851.

Benazzi-Lentati, G. & P. Deri, 1980. On the origin of heterogeneous chromosome sets in some fissiparous planarians. Rc. Accad. naz. Lincei, Ser. VIII, 68: 318–326.

Bromley, H., 1974. Morpho-karyological types of Dugesia (Turbellaria, Tricladida) in Israel and their distribution patterns. Zool. Scr. 3: 239–242.

Bromley, H., 1977. Observations on fission, maturation and reproduction in Dugesia biblica (Turbellaria, Tricladida) from Israel. Zool. Scr. 6: 197–201.

Dahm, A., 1958. Taxonomy and ecology of five species groups in the family Planariidae. Nya Litografen, Malmö, 241 pp.

Dahm, A., 1963. The karyotypes of some freshwater triclads from Europe and Japan (Turbellaria, Tricladida, Paludicola). Ark. Zool., Ser. 2, 16: 41–67.

De Vries, E. J. & I. R. Ball, 1980. On Dugesia gonocephala from western Europe. Bijdr. Dierk. 50: 342–350.

Gourbault, N., 1981. The karyotypes of Dugesia species from Spain (Turbellaria, Tricladida). Hydrobiologia 84: 45–52.

Kawakatsu, M., I. Oki, S. Tamura & H. Sugino, 1976. Studies on the morphology, karyology and taxonomy of the Japanese freshwater planarian Dugesia japonica Ichikawa & Kawakatsu, with a description of a new subspecies, Dugesia japonica ryukyensis subspec. nov. Bull. Fuji Wom. Coll. 14(2): 81–126.

Levan, A., K. Fregda & A. A. Sandberg, 1964. Nomenclature for centromeric position on chromosomes. Hereditas 52: 201–220.

Oki, I., S. Tamura, T. Yamayoshi & M. Kawakatsu, 1981. Karyological and taxonomic studies of Dugesia japonica Ichikawa et Kawakatsu in the far east. Hydrobiologia 84: 53–68.

Sluys, R. & H. De Jong, 1984. Chromosome morphological studies of Dugesia gonocephala s.l. (Platyhelminthes, Tricladida). Caryologia, in press.

White, M. J. D., 1973. Animal cytology and evolution. 3rd ed. Cambridge University Press, Cambridge, 961 pp.

Taxonomic studies on marine triclads (Turbellaria, Tricladida, Maricola)

Ronald Sluys
Institute of Taxonomic Zoology, University of Amsterdam, P.O. Box 20125, 1000 HC Amsterdam, The Netherlands

Keywords: Turbellaria, Tricladida, *Procerodes*, taxonomy, biogeography

Abstract

The genus *Procerodes* forms a heterogeneous assemblage of taxa. Although it is premature to attempt phylogenetic weighting of the characters, the genus contains three well delimited subgroups, one with a subantarctic, panaustral distribution, one world-wide in distribution, and one occurring principally in the northern hemisphere. Outside of these subgroups, species in this genus are problematic since a number of their features also occurs in other procerodids. All recognized groups exhibit a wide distributional range. Many of the world-wide taxonomic relations lie on the species level and some at the level of genera. Poor dispersal capacity of extant species of *Procerodes* suggests that the biogeographic patterns may be very old.

Introduction

It has been noted on several occasions that many of the 'southern hemisphere' procerodids differ in a number of characters from their northern relatives (Böhmig, 1908; Meixner, 1928; Ball, 1977a) and also that the genus *Procerodes* constitutes a heterogeneous group (Ball, 1977a, b). The discovery of new procerodid species and detailed study of known species support these views.

In this paper, I examine the questions of which groups of species may be delimited within the currently accepted genus *Procerodes* and with which taxa within the Procerodidae they show taxonomic relations. I also discuss the biogeographic patterns resulting from the tentative groupings and from their relations with other members of the Procerodidae.

While I consider the distribution of taxonomic features, I do not assign to them any weight, i.e. relative apomorphy or plesiomorphy, as is customary in biogeographic studies based on phylogenetic principles. Since a monographic treatment of the Maricola, including a phylogenetic analysis, is in progress, an evaluation of biogeographic patterns based on phylogenetic analysis will be delayed until the entire group has been studied thoroughly.

Features of southern-hemisphere Procerodidae

Following and elaborating on Meixner's (1928) views, Ball (1977a) listed the characters indicating that the procerodids of the southern hemisphere may form a natural group, characters by which they differ from most of the northern triclads: 1) ventral testes, 2) shell glands opening in bursal canal ectally to the oviducts, and 3) lensed eyes. With respect to the testes, a distinction may be made between those forms in which the follicles are distributed throughout the body length and those species in which the testes are pre-pharyngeal, although 'intermediate' forms also occur.

In his analysis, Meixner (1928) paid much attention to the female copulatory apparatus. Concerning the Maricola, he distinguished three types of organization, which were considered to be part of

Hydrobiologia 132, 257–262 (1986).

a phylogenetic sequence. In type I, the oviducts open separately into the bursal canal, and the latter receives the openings of the shell glands. Type IIA is characterized by a common oviduct that opens directly into the bursal canal and by shell glands which discharge into the bursal stalk. The essential feature in type IIB is a diverticulum of the bursal canal, receiving the oviducts and the openings of shell glands. Meixner did not discuss in detail the geographic distribution of the various characters but only noted that type I and IIB seemed to have a world-wide distribution, whereas type IIA was confined to southern South America and Antarctica.

Exceptions to the rule

An interesting exception to the rule that northern hemisphere procerodids have dorsal testes and lack an eye lens, is the Uteriporidae. Ball (1975) enlarged the concept of the Uteriporidae and united within the family the type species *Uteriporus vulgaris* and *Foviella affinis*. Both species have ventral testes extending throughout the body length and lenses to the eyes, whereas the oviducts open separately into a so-called female genital duct. *Uteriporus* as well as *Foviella* occurs on both sides of the North Atlantic. Interestingly, *Micropharynx parasitica* also possesses ventral testes, although these are prepharyngeal. This species lacks eyes. *Micropharynx* is thought to be confined in range to northern Europe and the North Atlantic (*cf.* Ball & Reynoldson, 1981); however, if that range coincides with that of its host species, skates, then it may be expected to occur over a much vaster area (Ball & Sluys, in prep.).

Taxonomic groups within *Procerodes*

As has been mentioned above, Meixner (1928) used mainly characters of the female copulatory apparatus in his scheme. Detailed study of the species that he assembled under type IIA, viz. *Procerodes wandeli, P. hallezi,* and *P. ohlini*, revealed that they also agree in other unique features. The unity of type IIA has been strengthened by the finding of a new species on Auckland Island, New Zealand, showing all the essential features of that type. Therefore, it has been decided to remove *P. wandeli, P. hallezi,* and *P. ohlini* from the genus *Procerodes* and to place them with the new species in a newly erected genus (Sluys & Ball, in prep.). Position of the testes, presence of an eye lens, and the way in which the shell glands open into the bursal canal indicate that the new genus belongs to the typical southern-hemisphere procerodid fauna.

With respect to Meixner's type I and IIB the situation seems to be more complex in that they probably subsume various subgroups. But Meixner's point with respect to the openings of the oviducts is well taken: in the Procerodidae and even within the genus *Procerodes*, we may distinguish between 1) taxa in which the oviducts open separately into the bursal canal, 2) taxa with a diverticulum of the bursal canal into which opens a common oviduct, 3) taxa in which the oviducts open separately into the diverticulum, and 4) taxa in which the oviducts unite into a common oviduct, the latter opening directly into the bursal canal (Table 1). It is clear that, even after the removal of the afore-mentioned species from the genus, various morphological types are gathered under the generic name *Procerodes*.

An important difference concerns the presence or absence of a diverticulum of the bursal canal. I agree with Ball (1977a: 503) that the genus *Procerodes* should be restricted at least to those species in possession of such a diverticulum, since it occurs in the type species *P. littoralis*. Presence of a diverticulum runs parallel with other characters: 1) in all but one of these species, the testes are dorsal and distributed throughout the body length (*cf.* Meixner, 1928: 586; Ball, 1977a: 503), 2) all taxa but one lack eye lenses, and 3) all but two species show a constriction in the male atrium. It is clear from Table 1 that *P. kerguelenensis* does not conform to the groupings of other species having a diverticulum. In point of fact, it is doubtful whether the diverticulum and common oviduct in this species are homologous with those of the other species. Furthermore, the position of the testes in *P. kerguelenensis* indicates that the species belongs to the southern-hemisphere group of procerodids, whereas all the other species of the two subgroups have characteristics of the northern-hemisphere group. Therefore, *P. kerguelenensis* should not be listed with those species showing a diverticulum of the bursal canal, but its proper taxonomic position remains obscure.

Table 1. Four species groups within *Procerodes* and characters on which the division is based. The unicellular glands are different from shell glands.

Diverticulum + common oviduct			
Species	Testes	Constriction male atrium	Eye lens
littoralis	dorsal; throughout body length	+	–
plebeia	dorsal; throughout body length	+	–
dahli	dorsal; throughout body length	+	–
variabilis	dorsal; some ventral; throughout body length	–	–
kerguelenensis	ventral; prepharyngeal	–	–

Diverticulum + separate openings oviducts				
Species	Testes	Constriction male atrium	Eye lens	raised epithelium ejaculatory duct
lactea	dorsal; throughout body length	+	+	?
pacifica	ibid.	+	–	+
dohrni	ibid.	+	–	+
lobata	ibid.	+	–	±

Separate openings oviducts in bursal canal			
Species	Testes	Thick circular muscle layer around bursal canal	Eye lens
trigonocephala	ventral; half way pharyngeal pocket	+	+
harmsi	ventral; up to copulatory apparatus	+	+

Common oviduct opening in bursal canal				
Species	Testes	Thick circular muscle layer around bursal canal	Unicellular glands around bursal canal	Eye lens
macrostoma	ventral; prepharyngeal	+	+	+
japonica	ibid.	+	+	+
asahinai	ibid.	+	+	?

Those species that possess a diverticulum of the bursal canal may be divided into two subgroups on the basis of the way in which the oviducts open into the diverticulum (Table 1). One group is characterized by a long common oviduct, and the other group by oviducts opening separately into the diverticulum. Species of the latter group also have a raised epithelium of the ejaculatory duct (Table 1), i.e. the distal part of the duct shows an epithelium which is taller than in the rest of the duct and that is penetrated by the openings of penis glands. For *P. lactea*, no details are known about the histology of the ejaculatory duct; some specimens of *P. lobata* appear to lack this feature, although identity of these specimens as *P. lobata* is questionable.

The remaining *Procerodes* species, i.e. those without a diverticulum of the bursal canal, may be arranged into two groups on the basis of how the oviducts join the bursal canal (Table 1). These groups, however, are not so well delimited as the ones discussed so far. Members of these two groups show features some of which also occur in other procerodid taxa, outside *Procerodes*.

The group containing *P. macrostoma, P. japonica,* and *P. asahinai* seems to be rather well defined, but most of the features listed, including the common oviduct, also occur in *Nesion arcticum* (*Nesion* lacks the unicellular glands around the bursal canal). At the same time, *Nesion* is quite unlike these species in other features (see also Ball, 1973a: 403). In view of the restriction of the genus *Procerodes*, as discussed above, a new genus name may be necessary for the three last-mentioned species. But such a step is better delayed until a phylogenetic analysis has resolved the affinities between the taxa involved since the problem is even more complicated than discussed so far.

For example, a thick circular muscle layer around the bursal canal and unicellular glands discharging into the latter, also occur in such a morphologically distinct form as a new *Puiteca* species (*cf.* Sluys & Ball, in prep.). Moreover, *Procerodes macrostoma, P. asahinai, P. japonica* and *Nesion arcticum* also seem to be related to the genus *Ectoplana*. Representatives of this genus are, among others, characterized by a relatively thick circular muscle layer around the bursal canal and by pre-pharyngeal, ventral testes. *E. undata* may even possess unicellular glands around the bursal canal, but this could not be determined with certainty (Sluys, 1983). The two *Ectoplana* species lack eye lenses. Both species differ from *Nesion arcticum* and the above-mentioned three *Procerodes* species in that they lack a common oviduct; in *Ectoplana* the oviducts open separately into the bursal canal.

With respect to the openings of the oviducts, *Ectoplana* seems to be closer to the two remaining *Procerodes* species, viz. *P. harmsi* and *P. trigonocephala*. However, it must be realized that oviducts that open separately into the bursal canal can be found in various other genera such as *Tryssosoma* (*cf.* Ball, 1977a), *Sabussowia* (*cf.* Ball, 1973b), and *Stummeria* (*cf.* Böhmig, 1908). Both the state with separate oviducal openings and that with a common opening, even may be found within one genus. In *Vatapa gabriellae* the oviducts open separately into the bursal canal (Marcus, 1948), but in a recently discovered new *Vatapa* species they unite into a common oviduct which meets the bursal stalk (Sluys & Ball, in prep.). The same situation obtains in the genus *Dinizia* (*cf.* Ball, 1977a).

Several features suggest a close relationship between *P. trigonocephala* and *P. harmsi* (Table 1). These same features can be found in another Asian triclad, *Miroplana trifasciata* suggesting that it may be related to *P. trigonocephala* and *P. harmsi* and, more distantly, to *P. japonica, P. macrostoma,* and *P. asahinai.*

Biogeography

The new genus containing the former *Procerodes* species *wandeli, ohlini,* and *hallezi* and one new species, has a (sub)antarctic, panaustral distribution (Fig. 1). The species *wandeli* and *ohlini* are known from the entire (sub)antarctic region, whereas the new species is known only from Auckland Island and *P. hallezi* only from southern South America.

All the other, tentative, groups that have been distinguished within the genus *Procerodes* show a vast range (Fig. 1). For example, the group 'diverticulum + common oviduct' has members in the North Atlantic (*P. littoralis*), Mediterranean (*P. plebeia, P. dahli*), Azores (*P. dahli*), and the subantarctic region (*P. variabilis*).

The closely related group 'diverticulum + separate openings oviducts' also has a large area of distribution but with an eastern emphasis. *Procerodes lactea* is known only from Sakhalin, whereas *P. pacifica* occurs on both sides of the Pacific [*P. pacifica* was formerly only known from North America but has recently been found in New Zealand (Sluys & Ball, in prep.)]. Of this group *P. dohrni* is Mediterranean and *P. lobata* is known both from the Mediterranean and England (the correctness of the latter record should be checked on sectioned material since identification is based solely on living and squashed specimens).

The distribution of the species of the group 'common oviduct opening in bursal canal' is southern South America and Japan (Fig. 1), and the related *Nesion arcticum* is found in the Bering Sea and Gulf of Alaska. *P. harmsi* and *Miroplana* which as argued above may be related, are known from Indonesia and Japan, respectively. *P. trigonocephala* in the same group, was for a long time only known from Japan but has recently been found in Australia (Sluys & Ball, in prep.).

Thus, it appears that within the genus *Procerodes* there are biogeographic relations that cover a

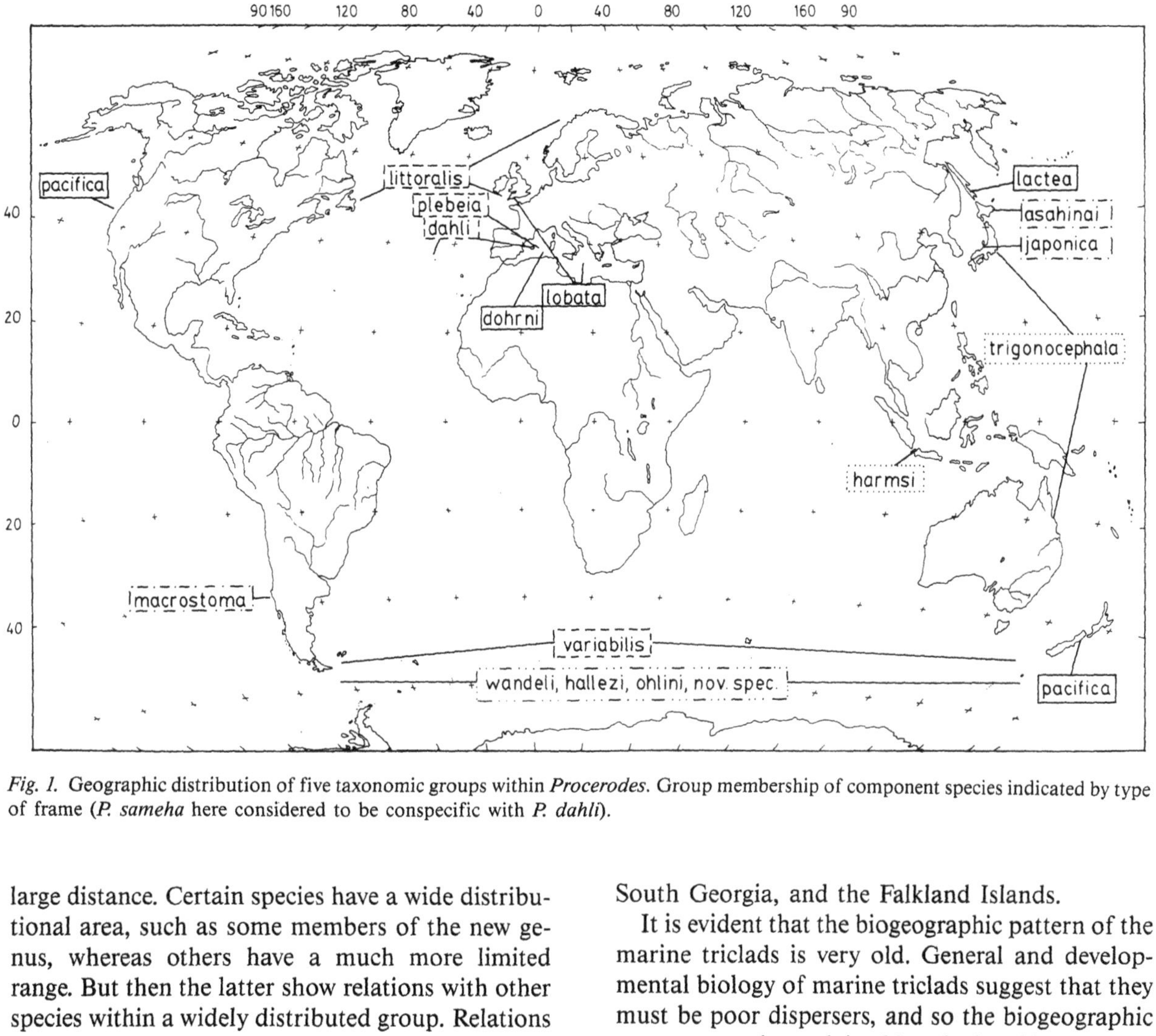

Fig. 1. Geographic distribution of five taxonomic groups within *Procerodes*. Group membership of component species indicated by type of frame (*P. sameha* here considered to be conspecific with *P. dahli*).

large distance. Certain species have a wide distributional area, such as some members of the new genus, whereas others have a much more limited range. But then the latter show relations with other species within a widely distributed group. Relations also appear to exist between each of the proposed groups and between such groups and other genera within the Procerodidae, resulting in a world-wide biogeographic pattern.

Conclusions

World wide connections involving lower taxonomic categories are by no means unique to the marine triclads (*cf.* Sterrer, 1973; Westheide, 1977). Karling (1977) discussed the biogeography of the genus *Austrorhynchus* (Kalyptorhynchia) and showed that one taxonomic group is formed by species from the Pacific (Hawaii and western North America), the Maldives, the Mediterranean Sea, South Georgia, and the Falkland Islands.

It is evident that the biogeographic pattern of the marine triclads is very old. General and developmental biology of marine triclads suggest that they must be poor dispersers, and so the biogeographic patterns may be explained best by invoking a factor that acted on a world-wide scale. The theory of plate tectonics provides such a factor.

Much of the foregoing is speculative. Only a detailed phylogenetic analysis of the entire Maricola will show whether the groups I propose here represent natural taxa and show what their relationships are with other groups within the Procerodidae. Nevertheless, the data already allow the conclusion that the marine triclads are a significant group for a world-wide phylogenetic and biogeographical analysis.

Acknowledgments

Material for this study was obtained through loans from the following museums: National Museum of Natural Sciences, Ottawa, Canada; National Museum of Victoria, Melbourne, Australia; Tasmanian Museum and Art Gallery, Hobart; The South Australian Museum, Adelaide; Australian Museum, Sydney. Professor P. Ax (Göttingen) is acknowledged for a loan of sectioned material. Thanks are due to Mr. L. Winsor and Dr. N. W. Riser for providing newly collected specimens. I am grateful to Prof. Dr. Ian R. Ball for arranging the loans of material and also for entrusting other specimens to me. He is also thanked for the facilities which he made available and, not in the last place, for introducing me into the field of planarian systematics and biogeography.

References

Ball, I. R., 1973a. On the anatomy and affinities of Nesion arcticum Hyman, a marine triclad from Alaska (Platyhelminthes, Turbellaria). Can. J. Zool. 51: 1299–1305.

Ball, I. R., 1973b. A new species of marine triclad Turbellarian of the genus Sabussowia from Prince Edward Island. J. Fish. Res. Bd Can. 30: 389–394.

Ball, I. R., 1975. Contributions to a revision of the marine triclads of North America: the monotypic genera Nexilis, Nesion, and Foviella (Turbellaria, Tricladida). Can. J.. Zool. 53: 395–407.

Ball, I. R., 1977a. La faune terrestre de l'Île de Sainte-Hélène VI-2. Turbellaria. Ann. Mus. r. Afr. Centr. in-8, Zool. 220: 492–511.

Ball, I. R., 1977b. On the phylogenetic classification of aquatic planarians. Acta Zool. fenn. 154: 21–35.

Ball, I. R. & T. B. Reynoldson, 1981. British Planarians. Cambridge University Press, London, 141 pp.

Ball, I. R. & R. Sluys, in prep. Marine triclads of Atlantic America.

Böhmig, L., 1908. Résultats de voyage de S.Y. Belgica. Turbellarien. Imprimerie J. E. Buschmann, Anvers: 32 pp., 2 pls.

Karling, T. G., 1977. Taxonomy, phylogeny and biogeography of the genus Austrorhynchus Karling (Turbellaria, Polycystididae). Mikrofauna Meeresboden 61: 153–165.

Marcus, E., 1948. Turbellaria do Brasil. Bol. Fac. Fil. Cien. Letr Univ. São Paulo, Zool. 13: 111–243.

Meixner, J., 1928. Der Genitalapparat der Tricladen und seine Beziehungen zu ihrer allgemeinen Morphologie, Phylogenie, Ökologie und Verbreitung. Z. Morph. Ökol. Tiere 11: 570–612.

Sluys, R., 1983. A new marine triclad ectoparasitic on Malaysian and Indonesian horseshoe crabs (Platyhelminthes, Turbellaria, Tricladida). Bijdr. Dierk. 52: 218–226.

Sluys, R. & I. R. Ball, in prep. A synopsis of the marine triclads of Australia and New Zealand.

Sterrer, W., 1973. Plate tectonics as a mechanism for dispersal and speciation in interstitial sand fauna. Neth. J. Sea Res. 7: 200–222.

Westheide, W., 1977. The geographical distribution of interstitial polychaetes. Mikrofauna Meeresboden 61: 287–302.

Ontogenetic patterns and phylogenetic trends in freshwater flatworms (Tricladida); constraint or selection?

P. Calow[1] & D. A. Read[2]
[1] *Department of Zoology, University of Sheffield S10 2TN, UK*
[2] *Department of Biology, University of Southampton S09 3TU, UK*

Abstract

Semelparity and iteroparity are unevenly distributed among the families of the Paludicola and this implies that there have been taxonomic restrictions on life-history evolution. Species differ in their investments in reproduction and high levels can be related, causally, to reduced life-spans; i.e. semelparous species invest more in reproduction than iteroparous species. However, there does not appear to be any fundamental reason why the extent and timing of these investments should not be open to modification by natural selection.

A major morphological difference between the predominantly semelparous Dendrocoelidae and the predominantly iteroparous Dugesiidae and Planariidae is the presence of an anterior adhesive organ; dendrocoelids have one, but members of the other families do not. A plausible scenario can be formulated relating this structure, causally, to enhanced juvenile survivorship which, in turn, favors the semelparous life history.

Introduction

Ontogeny is used here to mean all developmental events associated with a life history, including aging events; phylogeny is used to mean evolutionary history. Freshwater triclads (Paludicola) have a wide range of ontogenetic patterns, from those that do not involve gametes (fissiparous) to those that do; and in the latter category from semelparous (breed once and die) to iteroparous (breed repeatedly over several seasons) (Calow *et al.*, 1979). Which pattern evolves depends upon relative fitnesses (Stearns, 1976), yet ontogenetic mechanisms can themselves put constraints on evolutionary possibilities (Stearns, 1984).

Our aims here are:

1. To review the taxonomic distribution of these patterns, particularly semelparity and iteroparity since fissiparity has been treated elsewhere (Beveridge, 1982).

2. To review the ontogenetic correlates of semelparity and iteroparity and to investigate possible causal mechanisms behind them.

3. To use information derived from these reviews to evaluate the relative importance of selection and constraint in the phylogeny of life-history traits (as advocated by Ridley, 1983, and Stearns, 1984).

Original as well as previously published findings are included in what follows; and so, for the sake of continuity, methods, where appropriate, are described in appendices. Most work has been carried out on two species, the semelparous *Dendrocoelum lacteum* (Müller) and the iteroparous *Polycelis tenuis* (Ijima), but the findings from these appear to be applicable to other semelparous and iteroparous species (Woollhead, 1983, and Calow unpubl. obs.).

Hydrobiologia 132, 263–272 (1986).

Phylogenetic trends

Ball (1974, 1981) has proposed phylogenies for the Paludicola based on aspects of the morphology of the oviducts, pharynx, and eye structures and on the occurrence of an anterior adhesive organ (Fig. 1). There are two alternatives; one that assumes a monophyletic and the other a diphyletic origin of the group. Associated with both are interesting trends in reproductive patterns. Fissiparity occurs extensively in the Dugesiidae, less so in the Planariidae and not at all in the Dendrocoelidae (Beveridge, 1982). Not surprisingly, the capacity for regeneration after artificial transection is more limited in the Dendrocoelidae than in the other two families (Brøndsted, 1969).

Information on life-history patterns of gametic species is most complete for the European and particularly the British fauna (Reynoldson, 1983). Here, iteroparity is the dominant mode of reproduction in the Dugesiidae and Planariidae and semelparity is dominant in the Dendrocoelidae.

Phylogenies, such as the ones proposed in Fig. 1, group taxa, as far as possible, on the basis of shared, derived characters. A possible hypothesis is that these have evolved once under the influence of selection as adaptations, but then remain as features unresponsive to the influence of selection and possibly become a constraint on its action in any subsequent speciation within the taxonomic group sharing the trait (Ridley, 1983). Hence if the distribution of life-history patterns that is observed in the European triclad fauna is widespread, a possible interpretation, particularly on the basis of the monophyletic version of Fig. 1, is that iteroparity is primitive and that there has been some taxonomically correlated restriction on the evolution of semelparity. Once semelparity evolved in the dendrocoelids there was then a restriction on the evolution of iteroparity in this group.

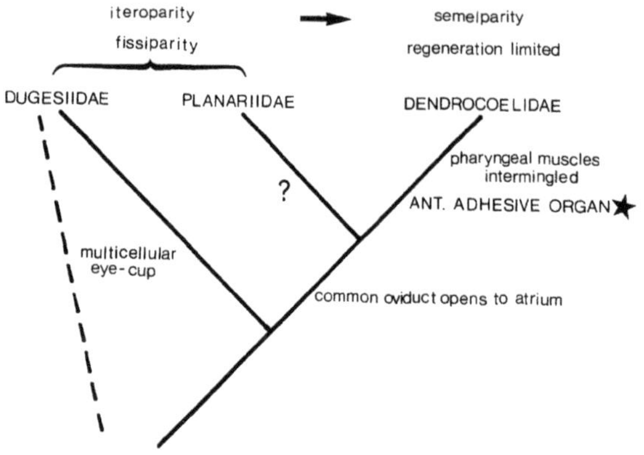

Fig. 1. Phylogenetic relationships of the Paludicola showing the defining autapomorphies (shared derived characters) for each phyletic line (modified from Ball, 1981). The question mark indicates that autapomorphies have yet to be defined. The star indicates an important character that is referred to again in the text. The broken line represents an alternative phylogeny for the Dugesiidae which implies a di- rather than mono-phyletic origin for the Paludicola (Ball, 1981). The arrow above the cladogram indicates a trend only.

Ontogeny

Investment in reproduction

Previous publications have established that adults of semelparous species invest a greater effort in reproduction per breeding season than those of iteroparous species (Woollhead & Calow, 1979; Woollhead, 1983). This has been shown by energy budget analysis and also by an analysis of the uptake and distribution of radiolabelled amino acids in the bodies of triclads (Read & Calow, 1984).

Further evidence has been obtained from a quantitative, histological analysis (Appendix 1) of how the body biomass was distributed between the somatic and reproductive tissues of *D. lacteum* (semelparous) and *P. tenuis* (iteroparous) at different points in the life histories of triclads taken from field populations. The somatic structures were gastrodermis and parenchyma (i.e., all somatic tissue between epidermis and gastrodermis) and the reproductive structures were testes and vitellaria (i.e., yolk-forming tissue). The techniques did not allow an accurate estimate of the biomass of the epidermis and the ovaries (see Appendix 1). Consequently, the epidermis, being a minor component of the total body biomass, was ignored; but an independent assessment was made of the developmental state of the ovaries. There were no major differences in the basic morphology of any of these tissues between species (a detailed description is in preparation).

The quantitative data are summarised in Fig. 2 together with the approximate extents of various

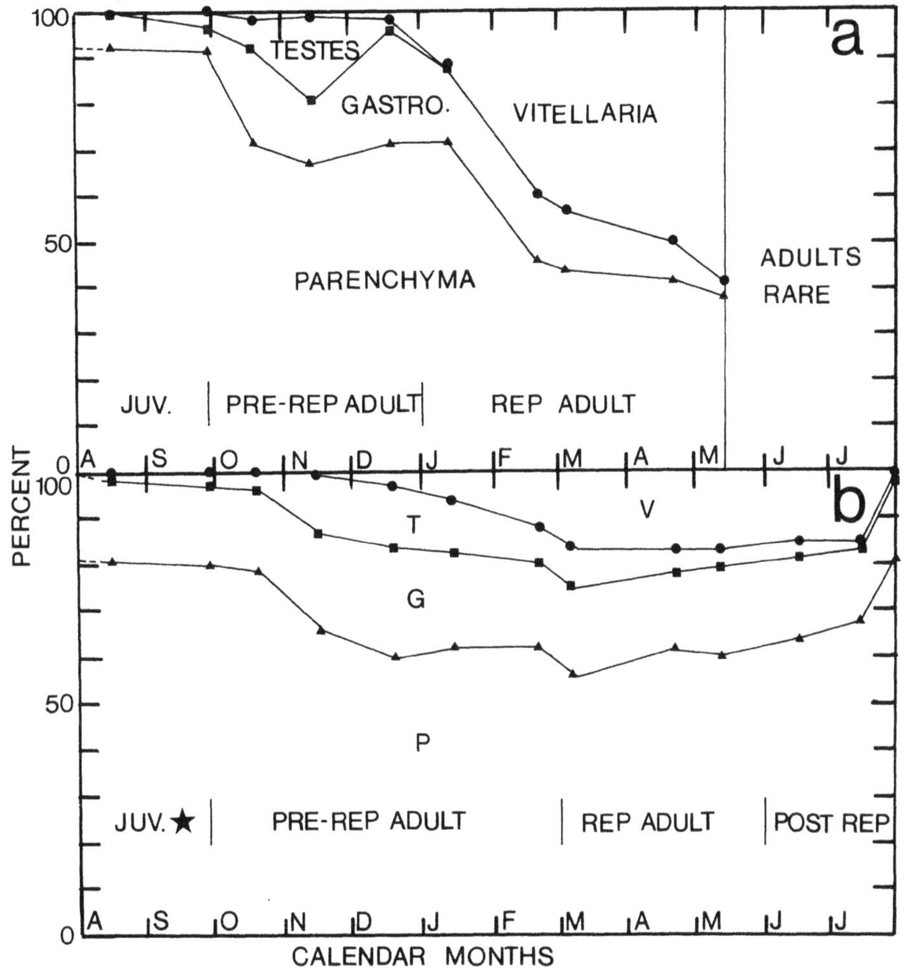

Fig. 2. Proportional distribution of body biomass between somatic and reproductive structures (V=Vitellaria, T=testes, G=Gastrodermis, P=Parenchyma) at different times of the year for *D. lacteum* (a) and *P. tenuis* (b). Reproduction (REP) indicates that cocoons were being released. Star indicates that juveniles could not be identified with complete certainty; see text. Divisions on the time scale are calendar months. Points represent actual sampling occasions.

life-history phases. That the material collected in late summer to early autumn was juvenile is certain for the annual *D. lacteum* (all adults die between May and July), but is less certain for the perennial *P. tenuis* since adults coexist with juveniles and can, by starvation, shrink to juvenile size (e.g., Calow, 1978). Nevertheless, none of the specimens used as juveniles in these and subsequent observations had a gonopore and this structure appears never to regress completely in adult *P. tenuis*, even after extensive shrinkage (Read, unpubl. obs.). The fact that cocoons started appearing earlier in the year for *D. lacteum* than for *P. tenuis* is typical (Reynoldson, 1979). The term 'post-reproductive adult' refers to specimens of *P. tenuis* that had gonads/vitellaria but which had ceased releasing cocoons. Some specimens of *D. lacteum* were in this state after May but, because of the high levels of adult mortality, were rare in the samples.

Testes formed before winter in both species but were present over a shorter interval of time in *D. lacteum* as compared with *P. tenuis*. Vitellaria occupied more of the body of *D. lacteum* than of *P. tenuis*. They remained extensive in *D. lacteum* until these animals died but disappeared from *P. tenuis* a short time after reproduction ceased and prior to the inter-breeding period. The ovaries developed at approximately the same rate in both species, but though they disappeared from *P. tenuis* at the end of the season they persisted in *D. lacteum* until death (Table 1). Differences between *D. lacteum* and *P. tenuis* in the effects of starvation on the tissue composition were dramatic (Fig. 3a, b). Under starvation, the testes and vitellaria of *P. tenuis* regressed, but the vitellaria of *D. lacteum* persisted (testes were not present at all – see above). These latter observations are consistent with the finding that cocoon production is more sensitive to feeding restrictions in *P. tenuis* than in *D. lacteum* (Calow & Woollhead, 1977).

Cellular and molecular consequences

It has been speculated that the increased investment in reproduction by semelparous species is directly related to their reduced post-reproductive survival, possibly by causing an enhanced deterioration of somatic tissues and thus leading to an accelerated senescence (Calow, 1979). This hypothesis derives from a model that relates rate of aging to the extent to which organisms invest resources in somatic repair processes and hence prevent the accumulation of damage within their somatic tissues (Kirkwood, 1981; Calow, 1978). The density of this damage is supposed to be a dynamic balance between processes of generation and repair:

DAMAGE DENSITY = GENERATION − REPAIR

It is presumed that organismic vitality is inversely proportional to the density of damage in the tissues, that this damage is continually generated at the suborganismic level by autoxidation, thermal

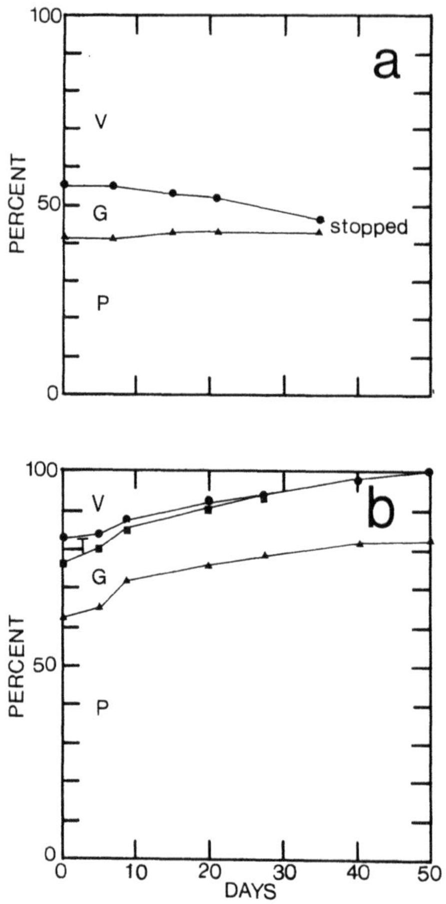

Fig. 3. Proportional distribution of body biomass between somatic and reproductive structures for reproductive adults suffering increasing starvation (days); a – for *D. lacteum* and b – for *P. tenuis*. Abbreviations as in Fig. 2.

Table 1. Summary of observations on the ontogeny of the ovaries of *D. lacteum* (*D.l.*) and *P. tenuis* (*P.t.*).

Collected in		August/ October	December	January/ February	March	April/ June	August
Ovaries	*D.l.*	UD	M−	(M)	M	M+	
	P.t.	UD	M−	M	(M)	M	A
Age		immature, possibly juveniles	pre-reproductive adults	cocoons start to appear			post-reproductive

A Organs were absent from all sample worms
UD Organs in an undifferentiated state, i.e. as clumps of neoblasts
M Mature organs present
− Few mature ova, many oogonia
\+ Few oogonia, predominantly mature ova

noise etc., and that the turnover of cells and molecules within organisms is, in part, a manifestation of the need to repair and replace damaged systems (Calow, 1978). The repair processes are metabolically expensive, and any other process that denies resources to these aspects of metabolism (e.g., because of the production of gametes) is likely to accelerate aging. This model predicts, therefore, that the cellular and molecular turnover of somatic tissues should reduce more in semelparous than in iteroparous triclads after the onset of the investment in reproduction.

Fig. 4 compares cell division rates, estimated by counting mitotic figures after colchicine treatment (Appendix 2), in the somatic cells of triclads sampled at different points in their life histories from the same field populations as those described above. Horizontal lines locate the means of more than 30 counts from usually three (in one case two) triclads, and vertical lines are 95% confidence limits. The number of counts varied between age classes and species. An analysis of deviance, using generalised linear models (McCullagh & Nelder, 1983), on the assumption of Poisson error distributions, nevertheless indicated significant differences between age classes ($P<0.001$) and between species ($P<0.05$), a significant interaction between species and age classes ($P<0.001$) but no significant differences between individuals of the same species within age classes ($P>0.05$). Making *a posteriori* comparisons between specific categories in these kinds of data is not straightforward (Sokal & Rohlf, 1969) but from Fig. 4 it is clear that the difference between species was largely due to the low rate of mitosis in reproductive *D. lacteum*. Indeed the difference between rates of mitosis in pre-reproductive and reproductive triclads is considerably more obvious for *D. lacteum* than for *P. tenuis*. This conclusion fits the expectation that somatic cell turnover should be more negatively influenced by reproduction in semelparous as compared with iteroparous animals.

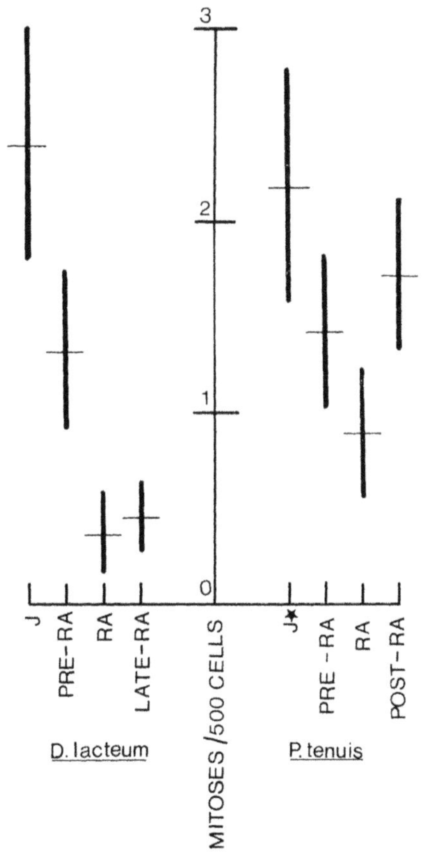

Fig. 4. Mean number of mitotic cells that had accumulated per 500 somatic cells over a 24 hour period of application of colchicine. Each is the mean of three triclads except for reproductive *P. tenuis* where n=2. Bars=95% confidence limits. Late reproductive adults were probably still releasing cocoons whereas post-reproductive adults were not. J* were *probably* juveniles (see text).

A further prediction from this model is that cellular and molecular damage should sharply increase in the tissues of *D. lacteum* with the onset of reproduction and the concomitant reduction in levels of repair. Histological changes in the somatic tissues of *D. lacteum* do indeed suggest a rapid increase in cellular disorganisation and damage as the breeding season proceeds (Table 2 and Read unpubl. results). Changes of this kind did not appear in *P. tenuis*.

To date, the molecular basis of these processes has not been researched very thoroughly. Nevertheless, preliminary observations on the uptake and loss of labelled amino acids (following the techniques of Calow & Read, 1981, and Read & Calow, 1984) do suggest a marked reduction in the protein turnover of adults as compared with juvenile *D. lacteum* but no difference between the same age categories of *P. tenuis* (Calow, unpubl.). Work on the accumulation of damaged molecules (e.g., enzymes with poor fidelity) has yet to be carried out.

Table 2. Changes in tissues of *D. lacteum* from pre- to post-reproductive condition.

		Pre-reproductive condition (December)	Reproductive condition (March)	Post-reproductive condition (May/June)
1.	Epidermis	Columnar cells, 25 – 30 μm thick.	Columnar cells, 25 μm thick	Cuboid/squamous cells < 8 μm thick, many containing no rhabdites.
2.	Sub-epidermal muscle fibers	25 – 35 μm thick	20 – 30 μm thick	Less than 5 μm thick.
3.	Dorso-ventral muscle fibers	Many and thick	Many and thick	Very sparse and thin
4.	Parenchyma	Densely cellular	Densely cellular	Low cell density. Large intercellular spaces. Large tracts of necrosis. Sperm often present (from ruptured spermiducal vesicles)
5.	Brain and nervous tissue	–	–	Nerve tracts (main branches) become less dense
6.	Anterior adhesive organ	Secretes heavily, much mucus	Secretes heavily, much mucus	Little secretion
7.	Vitellaria	Immature – cells not fully differentiated	Large vitelline cells with full vacuoles	Very large, but vitelline cells of poor quality, empty vacuoles and necrotic
8.	Ovaries	Few mature ova Many oogonia Many supporting cells	Many mature ova Some oogonia Many supporting cells	Few mature ova Very few oogonia or supporting cells
9.	Copulatory bursa	Cell lining thick. Bursa empty	Cell lining thick. Bursa empty	Cell lining very thick. Bursa lumen containing vitelline cells. Sperm often present
10.	Testes	Contain mature sperm + late stage spermatids, very few spermatogonia	Absent	Absent
11.	Spermiducal vesicles	Full of sperm	Full of sperm	Full of sperm. Often ruptured
12.	Gastrodermis	Thick vascuolated cells with filled food vacuoles	Thic vacuolated cells with filled food vacuoles	Very thin gastrodermis. Cells depleted of cytoplasm

Relationships between ontogeny and phylogeny for different life-history types

Two important assumptions of life-history theory are: (A) that there is a trade-off between somatic maintenance and reproduction (Calow & Sibly, 1983) and (B) that the relative emphasis on somatic maintenance and reproduction can be modified by natural selection (Sibly & Calow, 1983). It follows that if the costs of reproduction are paid *after* gamete release has been initiated, then (C) an improvement in juvenile survivorship should favor increased investment in gamete production with a concomitant reduction in parental life span, i.e., a shift towards semelparity, and *vice versa* with reduced juvenile survivorship (Sibly & Calow, 1983, 1984).

A mechanistic basis of assumption (A) was proposed above, and there is some evidence suggesting this is applicable to triclads. Two hypotheses (non-mutually exclusive) can be formulated about the distribution of life histories in Fig. 1 in terms of (B) and (C).

Hypothesis 1:
There are constraints that disallow (B); i.e. the dugesiid and planariid organisations are such as to prevent the evolution of semelparity and the dendrocoelid organisation is such as to prevent a return to iteroparity.

Hypothesis 2:
There are no constraints on (B) but the organisation and/or ecology of the dendrocoelids is such as

to allow improved juvenile survivorship compared with the dugesiids and planariids (C).

The first hypothesis sees evolutionary potential being restricted by certain organisations, whereas the second hypothesis sees evolutionary potential being opened up, possibly by the development of a novel trait. Plausible scenarios can be constructed around each hypothesis.

Hypothesis 1

This hypothesis implies that the form and function of the reproductive systems of the dugesiids and planariids are such that a high investment in reproduction is not possible. However, though differences certainly occur in the timing and extent of development of reproductive tissues in planariid and dendrocoelid triclads there are no obvious fundamental differences in the form of these systems (also Read, in prep.). It would be surprising if these ontogenetic differences were not open to modification by selection. Moreover, high levels of investment in reproduction relative to available resources are achieved in *Planaria torva* (Müller) (Woollhead, 1983), a planariid that is known to be semelparous in Britain (Reynoldson, 1983). Currently, we know little about the ontogeny of the dugesiid reproductive system.

Another possible hypothesis is that there are restrictions on the life-spans of dendrocoelids, probably through restrictions on repair processes, which make semelparous reproduction an inevitability. The model on p. 4 which forms a basis for the tradeoff between investment in reproduction and somatic repair and hence longevity (see assumption (A) above), allows, at least in principle, for modification under the influence of natural selection of lifespan (as required by (B) above) through adjustments in the allocation of resources between reproduction and repair. Restrictions have, nevertheless, been noted in the regenerative capacity of dendrocoelids and there is an associated lack of fissiparity. It has been suggested that these features might be associated with restrictions in the quantity and quality of regenerative cells – the so-called neoblasts (Brønsted, 1969); and the neoblast densities of *D. lacteum* and *P. torva* (both semelparous) are an order of magnitude less than those of *Dugesia lugubris* (Schmidt), an iteroparous triclad (Lange, 1983). But this only postpones the inevitable question of why neoblast densities are restricted and are not themselves open to the influence of natural selection. Moreover, there is evidence that some populations of *D. lacteum* in southern Europe are iteroparous (Herrmann, 1982) and it has been possible to substantially lengthen the adult life-span of *D. lacteum* beyond one year in the laboratory by providing a good, continuous food supply (Jennings & Calow, 1975), or by inhibiting cocoon production completely by periodic starvation or by culturing adults at low temperatures (Read, unpubl. obs.).

Hypothesis 2

The evolution of an anterior adhesive organ in the dendrocoelids has been implicated as a basis for improved juvenile survival and hence as a possible cause of the evolution of semelparity (Calow *et al.*, 1981). This structure enables dendrocoelids to exploit active prey and hence to adopt a sit-and-wait feeding behavior when there is no food. By contrast, dugesiids and planariids, which have no such organ, can only feed on dead and dying prey and in the absence of food have to adopt seek-out behavior. The sit-and-wait behavior is metabolically more economic than the seek-out behavior and in conditions of low food availability, as appears to be usual for freshwater triclads, particularly during the breeding season, can be related causally to reduced juvenile mortality (Calow & Woollhead, 1977; Calow *et al.*, 1981), and higher growth rates. The fact that reproduction is less responsive to feeding restrictions in *D. lacteum* than in *P. tenuis* is consistent with this hypothesis. Hence the evolution of an anterior adhesive organ has an effect on feeding behavior which, through its influence on physiology, has an effect on age-specific mortality (reducing juvenile mortality) appropriate to favor semelparity (Sibly & Calow, 1983) and cause divergence in life-history evolution between the dendrocoelids and the other families.

However, semelparity is not exclusively associated with the evolution of an anterior adhesive organ; for example, *P. torva* does not have one. Nevertheless, the semelparity of *P. torva* might be associated with an age-specific mortality that favors juveniles, but which is modulated by some factor other than food supply (Calow *et al.*, 1981).

Conclusions and prospects

At this stage the evidence seems to favor Hypothesis 2 more than 1. Moreover, this would imply a progressive shift in emphasis from somatic to reproductive processes and structures (and possibly a correlated reduction in neoblast density, as reported by Lange (1983)) in the ontogeny of the dendrocoelids. In turn, this might cause a progressive restriction of regenerative capacity and hence explain the more restricted ability of members of this family to regenerate and to be fissiparous. This could be analysed further by a consideration of age-specific regenerative capacity. Literature on this has already been reviewed by Lange (1968) and suggests a decrease in the regenerative abilities of large adult animals as compared with smaller juveniles. What is required now is a detailed interspecific analysis and comparison of more data of this kind.

Another question, that might benefit from further ontogenetic analysis, concerns the evolution of the anterior adhesive organ. This was an important development, on the basis of Hypothesis 2, that facilitated the evolution of semelparity in the dendrocoelids. However, it is not obvious why this structure is not more widespread. Are there ontogenetic constraints on its evolution? Ball (pers. comm.) has told us that anterior adhesive structures have been discovered in some Australian dugesiids. This implies that there are no taxonomically correlated constraints on the evolution of such a structure. It remains to be discovered whether these dugesiids are iteroparous or semelparous. More work, at both physiological and ecological levels, should also be carried out on the enigmatic *P. torva* and the possible iteroparous populations of *D. lacteum*.

Appendix 1: Histological analysis

All material was collected from a shallow Scottish loch (National grid ref: NS. 705 662) at intervals throughout the year, and was fixed and stored in Bouin's fluid. Each triclad was sectioned serially to 5 μm, stained with Mayer's haemalum/eosin, and observed using standard light microscopy. Either five longitudinal or ten transverse, evenly-spaced sections were examined per triclad, recorded by camera lucida and then the areas (μm^2) of each of the components – parenchyma, gastrodermis, vitellaria, and testes – were determined planimetrically. Hence the observations were two-dimensional but, since the third dimension was fixed, could be treated as indices of biomass. Approximately 1–2% of a triclad was analysed. The ovary, a restricted organ of relatively small size, was not well represented in the samples and so had to be excluded. Epidermis was also excluded.

Five triclads were considered per species on each sampling occasion. To obtain an estimate of the proportional composition of triclads, data from all replicates from each species on each sampling occasion were summed. The biomass of each tissue was expressed as a proportion (%) of the total biomass observed. This procedure was repeated for a group of reproductive adults (collected during April and May) that were subjected to complete starvation.

Appendix 2: Cell kinetics

We had no success with preliminary attempts to quantify cell kinetics using tritiated thymidine – the cells of triclads appear impervious to it (see also Martelly *et al.*, 1981). Cell division was therefore assessed by counting mitotic (metaphase) figures after the application of a stathmokinetic agent. We

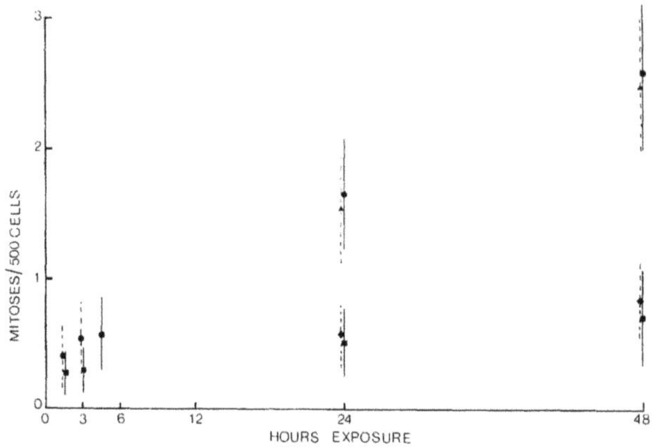

Fig. A1. Accumulation of mitotic cells with exposure time to colchicine (——) and vincristine sulphate (– – –) for adult *D. lacteum* (■ ◆) and *P. tenuis* (▲ ●).

routinely used colchicine (200 μg ml^{-1} copper-free tapwater) over 24 h at 10 °C but checked the linearity of the response with time (sampling at short intervals over 48 h) and compared the effects of colchicine with that of vincristine sulphate (1.0 μg ml^{-1} copper-free tapwater) since there has been some suggestion that colchicine applied over long periods of time is toxic and inhibits cell division (Mitchison, 1971). Fig. A1 shows that the response was linear and, judging by the overlap of confidence intervals, there was no significant difference between the effects of the two agents. Samples were taken from the field at appropriate times to include the juvenile/immatures (August to October) preproductive (late December, early January), reproductive (April), and late- or post-reproductive adults (May & June) (see Fig. 2). Three triclads were considered per sample and metaphase figures were counted in aceto-orcein squashes in 500 cells observed in each of at least 10 randomly chosen, non-overlapping fields per animal, giving a total of at least 30 counts. Mitotic counts were only recorded from somatic tissue (i.e., those in the gonads and vitelline glands were ignored) and were interpreted as mean division rates (per 24 h) per 500 total somatic cells (i.e. not just neoblasts (*cf.* Baguña, 1967a & b)).

Acknowledgments

This research was supported by a SERC Research Grant, GRB/12780. We are grateful to Professor Ian Ball for some useful advice and unpublished information that helped in the writing of this paper.

References

Baguña, J., 1976a. Mitosis in the intact and regenerating planarian Dugesia mediterranea n.sp. I. Mitotic studies during growth, feeding and starvation. J. exp. Zool. 195: 53–55.

Baguña, J., 1967b. Mitosis in the intact and regenerating planarian Dugesia mediterranea n.sp. II. Mitotic studies during regeneration and a possible mechanism of blastema formation. J. exp. Zool. 195: 65–80.

Ball, I. R., 1974. A contribution to the phylogeny and biogeography of the freshwater triclads (Platyhelminthes: Turbellaria). In N. W. Riser & M. P. Morse (eds), Biology of the Turbellaria. McGraw-Hill, New York: 339–401.

Ball, I. R., 1981. The phyletic status of the Paludicola. Hydrobiologia 84: 7–12.

Beveridge, M., 1982. Taxonomy, environment and reproduction in freshwater triclads (Turbellaria: Tricladida). Int. J. Invert. Repr. 5: 107–113.

Brøndsted, H. V., 1969. Planarian regeneration. Pergamon Press, Oxford, New York, 276 pp.

Calow, P., 1978. Life cycles. Chapman & Hall, London, 164 pp.

Calow, P., 1979. The cost of reproduction – a physiological approach. Biol. Rev. 54: 23–40.

Calow, P., M. Beveridge & R. Sibly, 1979. Heads and Tails: adaptational aspects of asexual reproduction in freshwater triclads. Am. Zool. 19: 715–727.

Calow, P., A. W. Davidson & A. S. Woollhead, 1981. Life-cycle and feeding strategies of freshwater triclads: a synthesis. J. Zool., Lond. 193: 215–237.

Calow, P. & D. A. Read, 1981. Transepidermal uptake of the amino acid leucine by freshwater triclads. Comp. Biochem. Physiol. 69A: 443–448.

Calow, P. & R. Silby, 1983. Physiological trade-offs and the evolution of life cycles. Sci. Prog., Oxf. 68: 177–188.

Calow, P. & A. S. Woollhead, 1977. The relationship between ration, reproductive effort and the evolution of life-history strategies – some observations on freshwater triclads. J. Anim. Ecol. 46: 765–781.

Herrmann, J., 1982. Food, reproduction and population ecology of Dendrocoelum lacteum (Turbellaria) in South Sweden. Ph.D. thesis, University of Lund.

Jennings, J. B. & P. Calow, 1975. The relationship between high fecundity and the evolution of entoparasitism. Oecologia 21: 109–115.

Kirkwood, T. B. L., 1981. Repair and its evolution: survival versus reproduction. In C. R. Townsend & P. Calow (eds), Physiological ecology. Blackwell Scientific Publs. Oxford: 165–189.

Lange, C. S., 1968. A possible explanation in cellular terms of the physiological ageing of a planarian. Exp. Gerontol. 3: 219–229.

Lange, C. S., 1983. Stem cells in planarians. In C. S. Potten (ed.), Stem Cells. Churchill Livingstone, London: 28–66.

Martelly, I., C. Rey & A. Le Moigne, 1981. Planarian regeneration: DNA metabolism in adults. Int. J. Invert. Reproduct. 4: 107–121.

McCullagh, P. & J. A. Nelder, 1983. Generalised linear models. Chapman and Hall, London, 261 pp.

Mitchison, J. M., 1971. The biology of the cell cycle. Cambridge University Press, Cambridge, 313 pp.

Read, D. A. & P. Calow, 1984. Uptake and distribution of the amino acid leucine by freshwater triclads in different reproductive states. Comp. Biochem. Physiol. 77A: 231–234.

Reynoldson, T. B., 1979. A key to the British species of freshwater triclads (Turbellaria, Paludicola). F.B.A. (U.K.) Sci. Publ. No. 23, 2nd (revised) edition, 31 pp.

Reynoldson, T. B., 1983. The population biology of the Turbellaria with special reference to the freshwater triclads of the British Isles. Adv. Ecol. Res. 13: 235–326.

Ridley, M., 1983. The explanation of organic diversity. Clarendon Press, Oxford. 272 pp.

Sibly, R. & P. Calow, 1983. An integrated approach to life-cycle evolution using selective landscapes. J. theor. Biol., 102: 527–547.

Sibly, R. & P. Calow, 1984. Direct and absorption costing in the evolution of life cycles. J. theor. Biol. 111: 463–473.

Sokal, R. R. & F. J. Rohlf, 1969. Biometry. W. H. Freeman & Co., San Francisco, 776 pp.
Stearns, S. C., 1976. Life-history tactics: a review of the ideas. Q. Rev. Biol. 51: 3–47.
Stearns, S. C., 1984. The tension between adaptation and constraint in the evolution of reproductive patterns. Adv. Invert. Reprod. 3: 387–398.
Woollhead, A. S., 1983. Energy partitioning in semelparous and iteroparous triclads. J. Anim. Ecol. 52: 603–620.
Woollhead, A. S. & P. Calow, 1979. Energy-partitioning strategies during egg production in semelparous and iteroparous triclads. J. Anim. Ecol. 48: 491–499.

Reproductive ecology of *Dendrocoelum lacteum* (Turbellaria) in a rapid stream in southern Sweden and comparisons with a lake population

Jan Herrmann
Department of Animal Ecology, University of Lund, Ecology Building, S-223 62 Lund, Sweden

Keywords: Turbellaria, stream, temperature, food, reproduction, phenology

Abstract

Rapid streams in southern Sweden are often inhabited by the triclad *Dendrocoelum lacteum*. Outside Fennoscandia, this species is mainly restricted to lentic habitats. The 'normal' food refuge for *D. lacteum*, the isopod *Asellus aquaticus*, is seldom found in the stream habitats, and probably the prey here is the amphipod *Gammarus pulex*. With respect to spatial and trophic niche components, it seems that *D. lacteum* has taken the place of *Crenobia alpina* in southern Sweden.

Two parameters of reproduction, namely production of cocoons and of hatchlings, both peaked approximately one month later in a stream than in an adjacent lake. This difference was attributed to a low temperature regime in the stream. Number of hatchlings per cocoon and cocoon sterility were higher in the stream than in the lake. Similar cocoon and hatchling output per adult was found in the two habitats, indicating a similar reproductive effort for the two triclad populations. I suggest that *D. lacteum*, by virtue of its choice of microhabitat, viz., under stones, is not so vulnerable to the more severe stream environment.

Introduction

Freshwater ecologists work mostly in either lentic or lotic habitats, less often in both. Few benthic invertebrate species are found in both stream riffles and weedy shores of lakes. Hence, comparisons of the reproductive ecology of species inhabiting these two habitats are very few (see, however, Calow (1981) on the snail *Lymnaea peregra* (Müller)).

Triclads as a group occur in both fast-flowing streams and all types of standing-waterbodies, but few species are found in both habitats. The lotic species reproduce mainly asexually, while most lentic species do not (Ball & Reynoldson, 1981; Beveridge, 1982). The wide-spread species *Dendrocoelum lacteum* Müller is predominantly lake-dwelling – totally so in Britain (Reynoldson, 1978). It is fairly common, however, in many rapid streams – both productive and unproductive ones – in southern Sweden.

The reproductive ecology of *D. lacteum* was thoroughly studied in Britain (see e.g., Young & Reynoldson, 1966; Macan & de Silva, 1979) and in Sweden (Herrmann, 1979, 1985a, 1985b), but all studies were confined to lake populations. This paper presents data for a population of *D. lacteum* in a fast-flowing stream and the findings are discussed in relation to the data for the nearby lake population (Herrmann, *op. cit.*).

Study area

The Skogsmölle stream (55° 37′ N, 13° 31′ E) is located in the very southernmost part of Sweden (Fig. 1). The bedrock is gneiss covered by moraine. The stream bottom is dominated by stones, pebbles, and gravel, but it also has some scattered sand banks. The stream is perennial, approximately 8 km long, 0.5–3 m wide and 0.1–0.6 m in depth.

Hydrobiologia 132, 273–277 (1986).

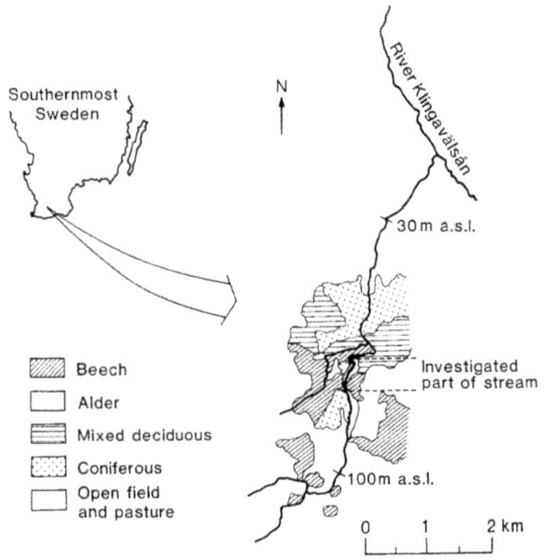

Fig. 1. The Skogsmölle stream, its nearby surroundings, and its location in Sweden.

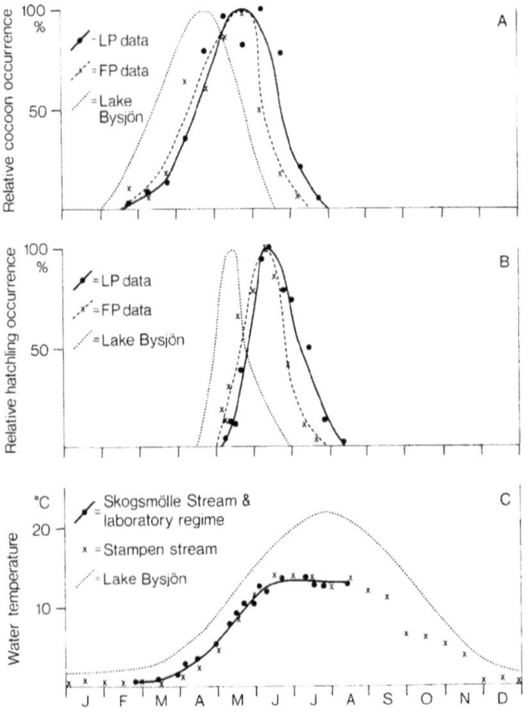

Fig. 2. Comparison between data from the Skogsmölle stream and Lake Bysjön. Lines are fitted by eye. (A) Cocoon occurrence, as percentage of maximum numbers. Stream data are LP- and FP-data from Table 2, lake data are from Herrmann (1985a). (B) Hatchling occurrence, as percentage of maximum numbers. Data origin as A. (C) Water temperature in the stream – also used in the laboratory rearings – in the nearby Stampen stream (from Otto, 1971) and in the lake (from Herrmann, 1985a).

Table 1. Relative abundance of macroinvertebrates in the Skogsmölle stream. x denotes sparse, xx fairly abundant and, xxx very abundant, respectively.

Turbellaria	
Dendrocoelum lacteum Müller	xx
Oligochaeta	
Eiseniella tetraedra (Sav.)	x
Gastropoda	
Ancylus fluviatilis Müller	xx
Crustacea	
Gammarus pulex L.	xxx
Ephemeroptera	
Baetis rhodani Pict.	xxx
Ephemera danica Müll.	x
Heptagenia sulphurea Müll.	xx
Plecoptera	
Nemoura sp.	x
Leuctra fusca L.	x
Coleoptera	
Gyrinus sp.	x
Limnius volckmari Panzer	x
Heteroptera	
Velia caprai Tam.	x
Trichoptera	
Agapetus sp.	x
Chaetopteryx villosa Fabr.	x
Hydropsyche angustipennis Curtis	xx
Polycentropus flavomaculatus Pictet	x
Plectrocnemia conspersa Curtis	xx
Potamophylax cingulatus Steph.	xx
Rhyacophila fasciata Hagen	x
Seriocostoma personatum K. & Sp.	x
Silo pallipes Fabr.	x
Diptera	
Tipula sp.	x
Eloeophila sp.	x
Dicranota sp.	x
Scleroprocta sp.	x
Simuliidae	x
Chironomidae	x

The surroundings are dominated by beech and to a lesser extent have also alder and mixed and spruce forests. Also some agricultural land drains into the stream. Aquatic vegetation is very sparse. No measurements of water chemistry were made, but the stream is moderately eutrophic and rather pristine. Its abiotic character very much resembles that of the nearby Stampen stream, described in detail by Hultin (1971) and Malmqvist *et al.* (1978).

Rather large fluctuations in water flow occur because of a considerable slope – on average 5% where the study was carried out – and the absence of swamps. Temperature was measured at each visit throughout February-August, and was found very similar to values recorded in the Stampen stream (Fig. 2C).

Material and methods

All cocoons and adults of *D. lacteum* were collected in 1980 in the Skogsmölle stream. The reproduction of the triclad population was studied in two ways.

A. Laboratory-produced cocoons (LP). 40 reproductive adults (size ≥14 mm) were collected at intervals of two weeks throughout the reproductive season from 15 February to 15 July. At the laboratory the triclads were kept in 2-L jars supplied with recirculating, thermostatically controlled lake water, each jar containing two specimens. The experimental design is described in Herrmann (1984). The temperature was gradually increased to follow the regime of the stream (Fig. 2C). One slightly crushed medium-sized *Gammarus pulex* L. per two *D. lacteum* specimens was given twice a week. Following the procedure given in Herrmann (1984), the number of cocoons deposited, the time required for these to hatch (in the same stream temperature regime as above), number of hatchlings per cocoon, and number of sterile cocoons were determined.

B. Field-produced cocoons (FP). 30 tiles (12×12 cm) were placed on the bottom along a 400 m stream section. The cocoons deposited under the tiles were counted and removed at intervals of two weeks. Some tiles were lost during spates, the rest remained in the stream until autumn. The cocoons were treated, after transference to the laboratory, in the same manner as the LP cocoons.

Handnet collections were taken in April and June, and the relative abundance of possible prey species was estimated.

Results

The benthic invertebrate fauna in the Skogmölle stream was dominated by the amphipod *G. pulex* (Table 1). Other potentially important food items for *D. lacteum* were mayfly (*Baetis rhodani* Pict.) nymphs and various caddisfly larvae. Significantly the isopod *Asellus aquaticus*, the normal food item of *D. lacteum* (see de Silva, 1976; Herrmann, 1984), was absent.

Cocoons, both those deposited in the laboratory (LP) and those taken from the stream (FP), appeared during the whole investigation period February-July (Table 2). A distinct peak was observed in late May (Fig. 2A). Also the patterns of hatchling occurrence were rather similar for the LP and the FP data sets (Table 2), both exhibiting a

Table 2. Total number of cocoons, total number of hatchlings from these cocoons, number of hatchlings per cocoon, and sterility percentage for half-month periods during the period of reproduction of *D. lacteum*. Data sets of two origins are presented; LP = laboratory-produced cocoons (per 20 pairs of *D. lacteum*), FP = field-produced cocoons (per an unknown number). Number of cocoons for which number of hatchlings per cocoon have been estimated equals the corresponding number of fertile cocoons. 95% confidence limits are given for the number of hatchlings per cocoon.

Period	Number of cocoons		Number of hatchlings from these cocoons		Hatchlings per cocoon (excl. steriles)		Sterility percentage	
	LP	FP	LP	FP	LP	FP	LP	FP
Febr. II	2	3	39	58	19.5	19.3 ± 16.7	0.0	0.0
March I	6	2	120	45	24.0 ± 14.6	22.5	16.7	0.0
March II	11	5	122	98	15.3 ± 4.0	24.5 ± 16.4	27.3	20.0
April I	27	18	383	207	15.3 ± 3.8	13.8 ± 3.1	7.4	16.7
April II	60	17	935	246	16.4 ± 2.3	14.5 ± 3.1	5.0	0.0
May I	70	24	981	313	15.3 ± 1.4	13.6 ± 2.7	8.6	4.2
May II	62	28	776	268	13.9 ± 1.2	10.7 ± 2.3	9.7	10.7
June I	76	14	730	133	10.9 ± 1.2	12.1 ± 6.6	11.8	21.4
June II	59	5	495	39	9.9 ± 1.1	9.8 ± 8.2	15.3	20.0
July I	16	2	136	9	9.7 ± 2.4	9.0	12.5	50.0
July II	4	-	20	-	5.0 ± 3.6	-	0.0	-
Total	393	118	4837	1416	$\bar{x} = 13.9$	$\bar{x} = 13.5$	$\bar{x} = 11.2$	$\bar{x} = 11.0$

peak just before mid June (Fig. 2B).

The number of hatchlings per cocoon (Table 2) showed no difference between LP and FP cocoons ($F=1.07$; $p>0.05$ for $df=1/337$), but both decreased significantly with time throughout spring ($F=10.24$; $p<0.001$ for $df=6/258$ and, $F=2.48$; $p<0.05$ for $df=6/79$, respectively).

Sterility figures were rather high for both LP and FP cocoons, although somewhat lower percentages were observed in April and May (Table 2).

As LP and FP data were very similar, they were treated as one set of 'stream data', unless otherwise stated.

Discussion

Continental and British streams with steep gradients are normally inhabited by triclad species other than *D. lacteum* (Pattee *et al.*, 1973; Reynoldson, 1978; Schwank, 1981). Of these species, only *Crenobia alpina* (Dana) occurs in Sweden at all (Dahm, 1956; Herrmann, unpubl.). In south Sweden, it can be found only in springs, whereas it exists in streams in northern Sweden (Dahm, 1958). Possibly most south Swedish streams are too warm for this species (Dahm, 1958; cf. also Pattee *et al.*, 1973; Reynoldson, 1983: 284).

Reynoldson (1983: 312) discussed why 'lake species are absent in lotic sites occupied by the stream species'. He suggested water current, competition, physiology or unsuitable life-cycle, and reproductive strategies as possible causes. As *D. lacteum* is found in many Swedish streams resembling the Skogsmölle stream, it seems it has taken at least the spatial niche of *C. alpina* in south Sweden. Furthermore, I suggest that the main food-niche component of *C. alpina*, the amphipod *G. pulex* (Reynoldson, 1983: 285), can form a major part of the diet of *D. lacteum* (cf. Table 1). Reynoldson (1983: 313) suggested food refuges to be less pronounced in streams than in lakes because the current restricts life-style diversity of the fauna. The bottom fauna in the studied stream (Table 1), however, shows a slightly more pronounced variation than that of the nearby Lake Bysjön (cf. Herrmann, 1979) if measured by proportions of functional groups (see Cummins, 1973).

It is possible that *C. alpina* is superior to *D. lacteum* in spatial and trophic niche components when temperature conditions permit this species to occur (cf. above), but this seems a less plausible explanation. The present data does not reveal anything about such an ecological release.

The temporal patterns for cocoon and hatchling occurrence were very clear, showing that reproduction in the stream evidently occurred and peaked approximately one month later than it did in Lake Bysjön (Fig. 2A–B). In this respect, LP data for the stream can be compared with the same type of LP data for the lake (from Herrmann, 1985a). Both cocoon- and hatchling-occurrence times were found to be significantly later in the stream ($F=9.83$; $p<0.01$ for $df=1/79$ and, $F=7.15$; $p<0.01$ for $df=1/79$, respectively).

I suggest that this difference depends on the considerably later increase in temperature as well as the overall lower temperature regime in the stream in comparison with the lake (Fig. 2C). These experiments confirm that temperature is the main factor ruling the timing of reproduction (Herrmann, 1985b). Deleterious high summer temperatures affecting the hatchlings in July in Lake Bysjön (Herrmann, 1985a, 1985b) are never met with in the stream. Consequently, no selective advantage to reproduce early exists in the stream.

Calow (1981) suggested that a higher probability of spates and hence mortality would genetically favour *Lymnaea peregra* to initiate reproduction earlier in the season in a stream population as compared with a canal population. Furthermore, it would put more effort into reproduction in the stream, this habitat being a relatively more dangerous place to live. Stream-dwelling snails also put fewer eggs in a greater number of small capsules than do the canal-dwelling snails, thus spreading the risk. Calow (1981) attributed all differences in parameters of reproduction to the more unpredictable situation in a stream, also with respect to food limitations here. The triclads differed from the snails in the following ways: (1) stream triclads reproduced later than lake triclads instead of the reverse; this I attribute to temperature, not discussed by Calow (1981); (2) hatchling output per adult was similar in lakes and streams, indicating a similar reproductive effort in the two habitats; (3) more hatchlings per cocoon are produced in the stream than in the lake.

Because of these differences it seems that triclads under stones are most probably less vulnerable to spates, droughts, and mechanical disturbance than are snails. Thus, there is perhaps no fundamental difference in selection pressures on stream and lake triclads in the way Calow (1981) suggested for *L. peregra*. Perhaps it is legitimate to suggest that the stream temperature regime is more optimal for *D. lacteum*, as the critically high summer values are never reached. Because of the difference in microhabitat choice between triclads and snails, it seems that Calow's model is less applicable to these triclads.

Thus, as shown for the lake-dwelling *D. lacteum* population (Herrmann, 1985b, c) the period of reproduction in the stream-living *D. lacteum* population also seems influenced chiefly by temperature. The stream triclads, however, do not seem to be heavily influenced by abiotic factors such as water current and velocity. The role of intraspecific competition for food, important in lake populations (Reynoldson, 1983; Herrmann, 1984), remains unknown.

Acknowledgments

For valuable criticism on the manuscript, I thank Drs L. M. Nilsson, C. Otto, and R. C. Petersen. Financial support was received from The Faculty of Mathematical and Natural Sciences of the University of Lund.

References

Ball, I. R. & T. B. Reynoldson, 1981. British Planarians. Synopsis of the British Fauna 19. Cambridge University Press, Cambridge, 141 pp.

Beveridge, M., 1982. Taxonomy, environment and reproduction in freshwater triclads (Turbellaria: Tricladida). Int. J. Invert. Reprod. 5: 107–113.

Calow, P., 1981. Adaptational aspects of growth and reproduction in Lymnaea peregra (Gastropoda: Pulmonata) from exposed and sheltered aquatic habitats. Malacologia 21: 5–13.

Cummins, K. W., 1973. Trophic relations of aquatic insects. Ann. Rev. Ent. 18: 183–206.

Dahm, A. G., 1956. Aneboda-områdets triclader. Södra Sveriges Fiskeriförening 1906–1955: 141–147. Lund.

Dahm, A. G., 1958. Taxonomy and Ecology of Five Species Groups in the Family Planariidae (Turbellaria Tricladida Paludicola). Malmö, 241 pp.

de Silva, P. K., 1976. The factors affecting the feeding of Dendrocoelum lacteum (Müller) (Turbellaria, Tricladida) on Asellus aquaticus (L.) (Crustacea, Isopoda). Arch. Hydrobiol. 77: 347–374.

Herrmann, J., 1979. Population dynamics of Dendrocoelum lacteum (O.F. Müller) (Turbellaria, Tricladida) in a South Swedish lake. Arch. Hydrobiol. 85: 482–510.

Herrmann, J., 1984. Prey choice and its impact on the reproductive output in Dendrocoelum lacteum (Turbellaria). Arch. Hydrobiol. 100: 355–370.

Herrmann, J., 1985a. Reproductive phenology in Dendrocoelum lacteum (Turbellaria) in South Sweden. Freshwat. Biol. 15: 127–132.

Herrmann, J., 1985b. Temperature-dependence of reproduction in Dendrocoelum lacteum (Turbellaria) – an experimental approach. Oikos 44: 268–272.

Herrmann, J., 1985c. Reproductive strategies in Dendrocoelum lacteum (Turbellaria) – comparisons between Swedish and British populations. Verh. int. Ver. Limnol. 22: 2974–2978.

Hultin, L., 1971. Upstream movement of Gammarus pulex pulex (Amphipoda) in a South Swedish stream. Oikos 22: 329–347.

Macan, T. T. & P. K. de Silva, 1979. On the occurrence of Dendrocoelum lacteum (Müller) and Asellus aquaticus (L.) as predator and prey in the stony substratum of Windermere. Arch. Hydrobiol. 86: 95–111.

Malmqvist, B., L. M. Nilsson & B. S. Svensson, 1978. Dynamics of detritus in a small stream in southern Sweden and its influence on the distribution of the bottom animal communities. Oikos 31: 3–16.

Otto, C., 1971. Growth and population movements of Potamophylax cingulatus (Trichoptera) larvae in a South Swedish stream. Oikos 22: 292–301.

Pattee, E., C. Lascombe & R. Delolme, 1973. Effects of temperature on the distribution of turbellarian triclads. In W. Wieser (ed.), Effects of Temperature on Ectothermic Organisms. Springer Verlag, Berlin: 201–207.

Reynoldson, T. B., 1978. A Key to the British species of freshwater triclads (Turbellaria, Paludicola). Freshwater Biological Association (U.K.) Sci. Publ. No. 23, 2nd (revised) edition, 31 pp.

Reynoldson, T. B., 1983. The population biology of Turbellaria with special reference to the freshwater triclads of the British Isles. Adv. ecol. Res. 13: 235–326.

Schwank, P., 1981. Turbellarien, Oligochaeten und Archianneliden des Breitenbachs und anderer oberheissischer Mittelgebirgsbäche. II. Die Systematik und Autökologie der einzelnen Arten. Arch. Hydrobiol. Suppl. 62: 86–147.

Young, J. O. & T. B. Reynoldson, 1966. A quantitative study of the population biology of Dendrocoelum lacteum (Müller) (Turbellaria, Tricladida). Oikos 15: 237–264.

Salinity-temperature tolerance of two closely related triclad species, *Dugesia lugubris* and *D. polychroa* (Turbellaria), in relation to their distribution in The Netherlands

G. Van der Velde, F. Hüsken & L. Van Welie
Laboratory of Aquatic Ecology, Catholic University, Toernooiveld, 6525 ED Nijmegen, The Netherlands

Keywords: Turbellaria, triclads, salinity, temperature, tolerance, distribution, habitat

Abstract

The tolerance of adult specimens of *Dugesia lugubris* and *D. polychroa* for 13 different chlorinities ranging from 15.0–3.8‰ and for two temperatures, viz. 4 and 23 °C, was tested.

At chlorinities of 7.5‰ and lower, the survival time of both species was considerably longer than at higher chlorinities (a few hours at 7.5‰, one to several days at 6.6‰ and lower concentrations). It is assumed that this is determined by the osmoregulatory capacity of the planarians.

It was found that at low chlorinities combined with a high temperature *D. polychroa* survived longer than *D. lugubris,* while at the same chlorinities the opposite was true for a low temperature. The effect of temperature on survival at low chlorinities was more drastic for *D. lugubris* than for *D. polychroa.*

The results correlate with data on the distribution of both species in The Netherlands. Outside areas with an average chlorinity below 2‰, the two species were rarely found.

Introduction

In 1973, Van der Velde started investigations on the distribution of triclad species in The Netherlands in order to answer two questions. First, what are the distribution patterns of the various species? And second, which factors determine these patterns?

For the first question, many waters in all parts of The Netherlands were investigated and their various substrata searched for triclads. With the data thus obtained, distribution maps were prepared, while reliable data from the Dutch literature and from unpublished hydrobiological reports were also added in order to get more detailed distribution maps. Subsequently, the data were arranged according to a rough water typology.

Answering the second question is more difficult since field observations as well as laboratory experiments are necessary to elucidate the complex of factors determining the distribution of a single species. Some of these factors are of direct influence on the occurrence of a species; others influence its distribution in an indirect way, e.g. by influencing the occurrence of its food and hence the presence or absence of the triclad species.

Salinity is a factor of the first type since the occurrence of the triclad species depends on their osmoregulatory capacities. We have investigated salinity-temperature tolerance in two closely related freshwater species that are widely distributed in The Netherlands and that are known to invade brackish water areas to a certain extent. Such experiments have never been carried out with freshwater species, while similar experimental studies on marine and brackish water species are scarce: data are available about *Procerodes littoralis* (Ström, 1768), in diluted salt water and fresh water from experiments by Pantin (1931a, b), Weil & Pantin (1931), Beadle (1934) and others. This species occurs in a special habitat called shock-habitat (Den Hartog, 1974). Schlieper *et al.* (1952) found

Hydrobiologia 132, 279–286 (1986).

that *Crenobia alpina* (Dana, 1766), a cold-water species, survived longer in tap water at a temperature of 25 °C when certain salts were added.

Investigations in the brackish water areas in the southwest of The Netherlands, the Delta area, were carried out by Krebs (1981, 1984a and b). It was found that only *Dugesia lugubris* (O. Schmidt, 1861) was present in the areas investigated and that most of the records were from isolated fresh waters within such areas. The chlorinity at localities where *D. lugubris* was present did not exceed 1‰ with one exception, a locality with a horizontal gradient situation where chlorinity varied from 0.84 to 2.76‰ (Krebs, 1984b). De Vos (1954) recorded *D. polychroa* (O. Schmidt, 1861) at a chlorinity of 2.35‰ in water near the IJssel lake (former Zuiderzee) in 1937; this water presumably had a stable chlorinity since in 1939 it was measured at 2.25‰ and in 1948 at 1.72‰. Cuppen & Dirkse (1978) and Cuppen (1980) found that in the eastern part of The Netherlands, *D. polychroa* was completely restricted to permanent waters while *D. lugubris* was highly characteristic of – and often even the only triclad species in – semi-permanent waters, such as shallow ponds and temporary brooks that fall dry during summer.

Materials and methods

For the laboratory tests, *Dugesia lugubris* and *D. polychroa* were collected near Nijmegen using liver-containing tubes placed at the bottom of waters in various localities. In the laboratory, the animals were removed from the tubes and stored at a temperature of 4 °C without food. For the tolerance tests 16 ml water was used starting with seawater of a chlorinity of 15‰. This seawater was diluted with tap water to obtain various lower chlorinities (Table 1) down to 3.8‰, the maximum chlorinity recorded for *D. 'lugubris'* (see Karling, 1974). At each chlorinity, tests were made at temperatures of 4 and 23 °C. The animals used for the tests at 23 °C were given the chance to acclimatize 24 hours at that temperature before the test started. For each test, 20 adult specimens of each species were used.

The animals were dropped in a petri dish with a specific chlorinity and were checked by means of a stereo microscope at regular intervals to see how many could be considered 'dead'. At high chlorinity levels, the animals were considered 'dead' when there were no visible movements even when the ani-

Table 1. Survival times at different chlorinities and temperatures of *Dugesia lugubris* and *D. polychroa*. (t100 means that 100% of the 20 adult specimens could be considered 'dead'.)

Chlorinity ‰	*D. lugubris*		*D. polychroa*	
	4 °C	23 °C	4 °C	23 °C
	t100 minutes	t100 minutes	t100 minutes	t100 minutes
15.0	60	23	74	29
14.1	63	48	69	33
13.2	95	51	91	49
12.2	117	71	116	55
11.3	146	77	286	60
10.3	227	83	223	67
9.4	249	111	250	93
8.5	306	264	351	136
7.5	390	327	408	187
	days	days	days	days
6.6	22.0	2.5	6.0	4.5
5.6	45.0	4.0	13.0	12.5
4.7	85.0	9.0	27.0	16.5
3.8	>200.0	18.0	87.0	20.5

mals were touched with tweezers or a needle. Certainly they are then not irreversibly damaged: when dropped into fresh water soon after the experiment, they resumed movement. During the experiments in the chlorinity range of 6.6–3.8‰, the animals started to disintegrate slowly and did not become immobilized. The caudal parts fragmented while the head part remained alive. Animals that started to fragment in such a way were considered 'dead' in these experiments and were removed. The time after which 10 specimens could be considered 'dead' was recorded as t50, and the time after which all specimens were 'dead' as t100.

Results

Reactions at different chlorinities and temperatures

1) 15–7.5‰.

a) 4 °C. The animals at first became highly mobile after being dropped into the saline water. They became shorter and thicker and sometimes coiled up. At the end of the experiments they were hard, as all the water was removed from the tissues by the high osmotic pressure of the surrounding medium.

b) 23 °C. The movements were more vigorous than at 4 °C. Most animals coiled up and then made no visible movements.

2) 6.6–3.8‰.

a) 4 °C. The behavior of the animals was different from that described above; they did not coil up and did not become hard through water extraction. During the test, however, they disintegrated to a slimy mass. In experiments that proceeded several days, the animals showed damage especially at the caudal part of the body and around the pharynx, both of which disintegrated. The head plus the remainder of the body remained alive. During the experiments the animals became smaller and lighter in color.

b) 23 °C. Animals disintegrated to a slimy mass sooner than at 4 °C. At this temperature animals with only a head and part of the body remaining alive were found less often.

Survival time

a) *The influence of chlorinity*

As could be expected for freshwater species, there was greater survival time for both species in lower chlorinity (Fig. 1, Table 1). Between chlorinities 7.5 and 6.6‰ the survival time of both species increased greatly, from several hours to days.

At chlorinities of more than 11.3‰, no clear difference between the two species studied could be discerned (Fig. 1, Table 2). At 11.3‰ and 4 °C there was a large difference between the survival times of the two species; *D. polychroa* lived longer under these conditions than *D. lugubris*. Between 11.3 and 7.5‰, at a temperature of 4 °C the differences in survival time were again not very clear, but at 23 °C *D. lugubris* survived distinctly longer than *D. polychroa* (Fig. 1, Table 2).

At lower chlorinity levels (between 6.6 and 3.8‰) the differences between the survival times of *D. lugubris* and *D. polychroa* were obvious. Under such circumstances and at a temperature of 4 °C, *D. lugubris* lived longer; and at a temperature of 23 °C *D. polychroa* lived longer (Fig. 1, Table 2).

It must be noted that in all tests all specimens could be considered 'dead', with the exception of *D. lugubris* at a chlorinity of 3.8‰ and a temperature of 4 °C. Under these conditions specimens of *D. lugubris* were still alive after 200 days, while all specimens of *D. polychroa* were 'dead' after 87 days. No observations were made after 200 days.

b) *The influence of temperature*

Survival times of each species at similar chlorinities were longer at 4 °C than at 23 °C (Fig. 1, Table 3). Differences in survival time at the two temperatures were better correlated with chlorinity for *D. polychroa* than for *D. lugubris*. Most interesting are the differences caused by the two temperatures at lower chlorinity levels. The differences in survival time between 4 °C and 23 °C were larger for *D. lugubris* than for *D. polychroa*.

Distribution in The Netherlands

Both *D. lugubris* and *D. polychroa* are widely distributed in The Netherlands (Fig. 2); they are frequently found in lentic waters. Both species have only been found in waters with snails and with a pH above 6.0.

The distributional data for the two species show

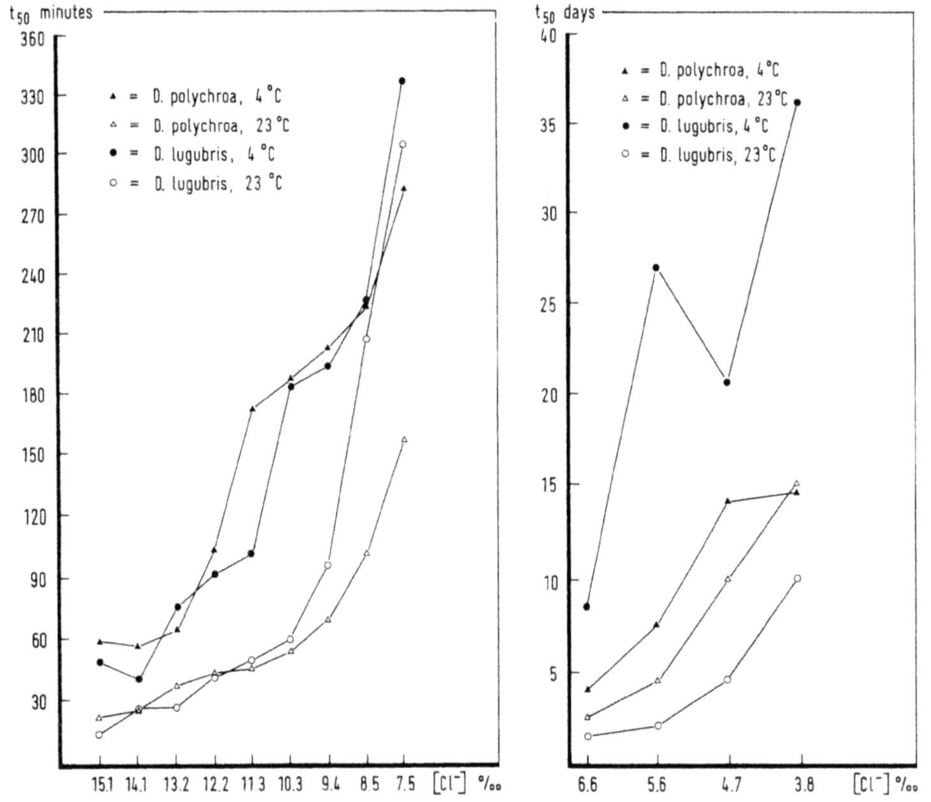

Fig. 1. Survival times at different chlorinities and temperatures of *Dugesia lugubris* and *Dugesia polychroa* (t50 means that 50% of the 20 adult specimens could be considered 'dead').

Table 2. Differences in survival times at different chlorinities between *Dugesia lugubris* (l.) and *D. polychroa* (p.) at two temperatures. (For the meaning of t50 and t100 see caption Fig. 1 and Table 1.)

Chlorinity ‰	4 °C		23 °C	
	t50 l.-t50 p. minutes	t100 l.-t100 p. minutes	t50 l.-t50 p. minutes	t100 l.-t100 p. minutes
15.0	−19	−14	−8	−6
14.1	−17	−6	0	+15
13.2	−12	+4	−10	+2
12.2	−12	+1	−2	+16
11.3	−68	−140	+4	+17
10.3	−6	+4	+6	+16
9.4	−9	−1	+28	+18
8.5	+4	−45	+104	+128
7.5	+54	−18	+149	+140
	days	days	days	days
6.6	+4.5	+16.0	−1.0	−2.0
5.7	+19.5	+32.0	−2.5	−8.5
4.7	+7.0	+58.0	−5.5	−7.5
3.8	+21.5	> +113.0	−5.0	−2.5

Table 3. Differences in survival times for *Dugesia lugubris* and *D. polychroa* caused by different temperatures at different chlorinity levels. (For the meaning of t50 and t100 see caption Fig. 1 and Table 1.)

Chlorinity ‰	*D. lugubris*		*D. polychroa*	
	t50 (4 °C)-t50 (23 °C) minutes	t100 (4 °C)-t100 (23 °C) minutes	t50 (4 °C)-t50 (23 °C) minutes	t100 (4 °C)-t100 (23 °C) minutes
15.0	26	37	37	45
14.1	15	15	32	36
13.2	50	44	28	42
12.2	51	46	61	61
11.3	53	69	125	226
10.3	122	144	134	156
9.4	97	138	134	157
8.5	21	42	121	215
7.5	32	63	127	221
	days	days	days	days
6.6	7.0	19.5	1.5	1.5
5.6	25.0	41.0	3.0	0.5
4.7	16.5	76.0	4.0	10.5
3.8	26.0	>182.0	0.5	66.5

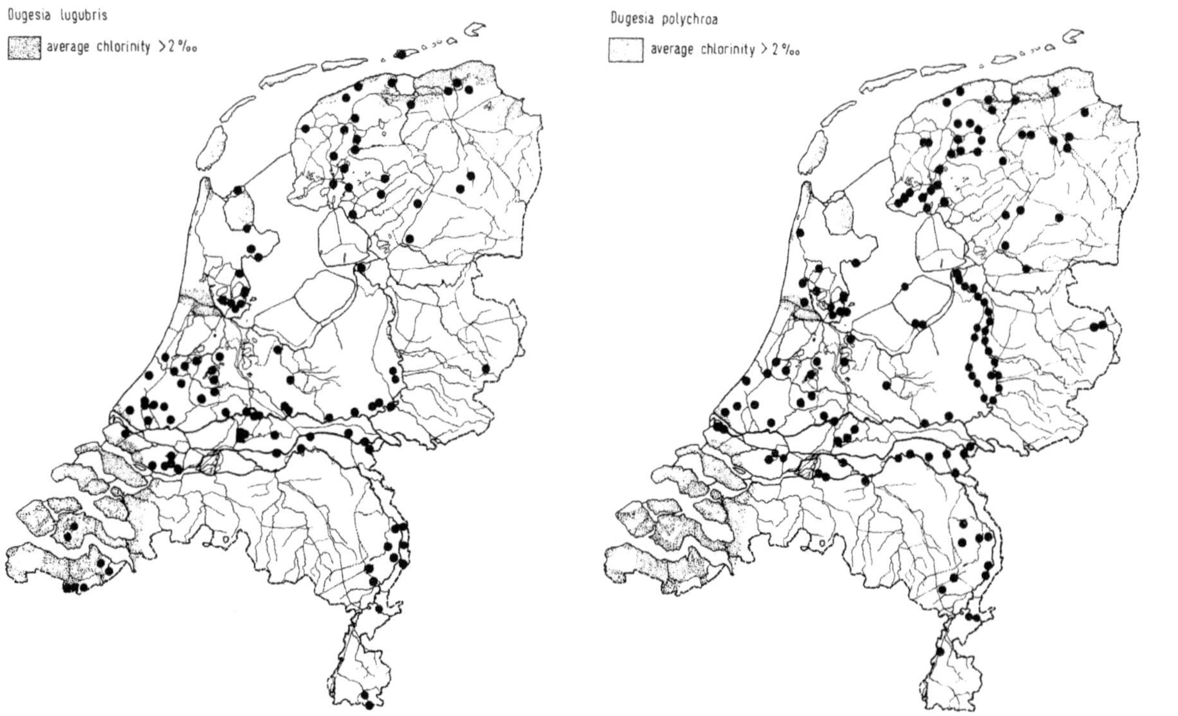

Fig. 2. Distribution of *Dugesia lugubris* and *D. polychroa* in The Netherlands.

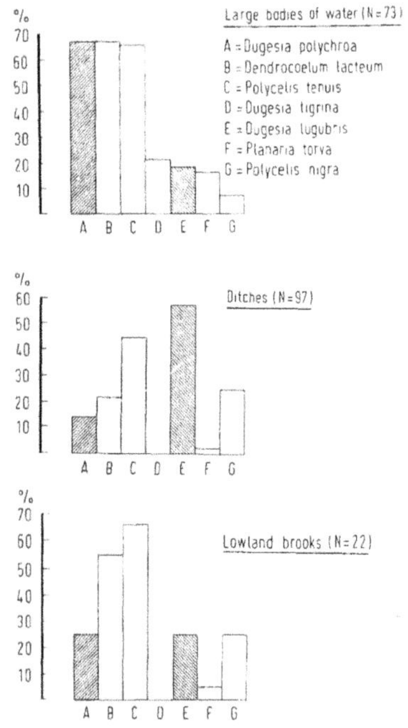

Fig. 3. The frequency of occurrence of triclad species in three types of water in The Netherlands. N is the number of water bodies investigated.

that both were seldom found in areas where average chlorinity is above 2‰ (Fig. 2).

The two species were found to occupy different habitats. *D. polychroa* was found most frequently in larger bodies of water such as lakes and canals, while *D. lugubris* was found most frequently in smaller bodies of water such as ditches (Fig. 3), as already indicated by Komárek (1926) and Den Hartog & Van der Velde (1973). In large water bodies, *D. polychroa* was sometimes associated with *D. lugubris*. *D. lugubris* in such waters was more frequently associated with *D. polychroa* (Fig. 4). In the ditches, *D. lugubris* appeared to be sometimes associated with *D. polychroa*, but *D. polychroa* was here relatively often associated with *D. lugubris*. Since lakes, canals, and ditches in The Netherlands are usually interconnected, both species can coexist in intermediate situations or migrate to each other's habitats. A remarkable phenomenon in the ditches was the relatively high frequency of situations in which *D. lugubris* was found as the only triclad species. *D. polychroa*, on the other hand, was mostly associated with other triclad species in the larger waters (Fig. 4).

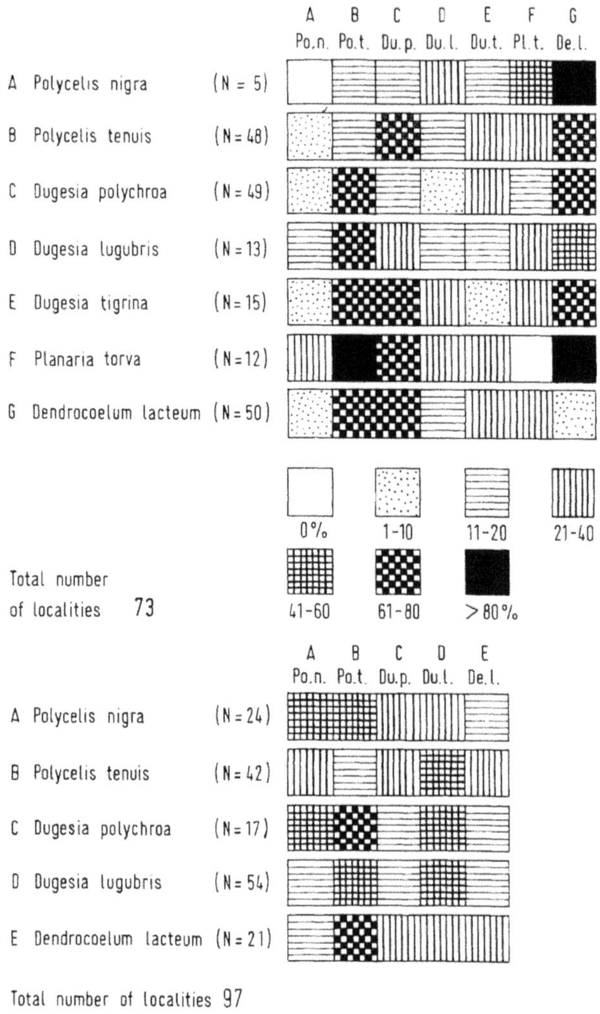

Fig. 4. Coexistence of triclad species in 73 larger bodies of water (above) and in 97 ditches (below). The squares give the frequency of species combinations in percentages of N, which is the number of water bodies in which each species was found (cf. Fig. 3). For instance *Dugesia lugubris* was found in 13 larger bodies of water: in 11–20% of these localities the species was found with *Polycelis nigra*, in 61–80% with *Polycelis tenuis*, in 21–40% with *D. polychroa*, in 11–20% as the sole species without accompanying triclad species, etc.

Discussion and conclusions

From our results the following conclusions can be drawn:

a. Continuous survival of *D. lugubris* and *D. polychroa* is only possible at a chlorinity level below 3.8‰; higher constant chlorinities are only tolerated for a limited time. Above a chlorinity level of 6.6‰ the changes of survival become very slim, probably because above this concentration the osmoregulatory capacities of both species are insufficient.

b. At a low temperature some specimens of *D. lugubris* can survive for extended periods at a chlorinity of 3.8‰.

c. Field data suggest that *D. lugubris* as well as *D. polychroa* can survive between 1–3‰ Cl^-, while distribution data suggest that the occurrence of both species is limited by a maximum average chlorinity level of 2‰, so that both species can occur only in isolated localities in the brackish-water areas. Karling (1974) mentions that *D. 'lugubris'* can be found in water with a salinity of 6‰, i.e. a chlorinity of about 3.5‰. It is, however, not clear for which species he gives this upper limit, since in the past there was a great deal of confusion concerning the separation and identification of *D. lugubris* and *D. polychroa* (Reynoldson & Bellamy, 1970).

d. Survival chances of both species below a chlorinity of 6.6‰ were much higher at low temperature than at high temperature. This could be a result of the better oxygen conditions in cooler water. Osmoregulation takes a lot of energy, so it would be interesting to measure respiration during salinity-temperature tests, especially in the critical chlorinity range of 1–3.8‰.

e. The difference in habitat between *D. lugubris* and *D. polychroa* perhaps results in an avoidance of possible competition as both species use the same kind of snails for their main food (food refuge) according to Reynoldson & Piearce (1979). In view of the separate habitats occupied by *D. lugubris* and *D. polychroa,* it is interesting to note that at low chlorinities combined with a higher temperature the former species survived shorter than the latter, while the opposite was true at the lower temperature. Since both species live at low chlorinities the wide variation in survival time between both temperatures under such circumstances for *D. lugubris* as compared to *D. polychroa* can perhaps explain why *D. lugubris* can occur in small and semi-permanent waters with low temperatures in winter time and high temperatures and/or drying during summer and why *D. polychroa* is restricted to larger permanent waters where temperature fluctuations are not so extreme. It seems that *D. polychroa* is a relatively thermophilous species which can complete its life cycle only in a temperature range of 10–23 °C (Ball & Reynoldson, 1981). Hadderingh (1977) found that *D. polychroa* occurred in larger numbers at the outlets of power plants, where heated cooling water is discharged, than at the inlets. The same is true for *D. tigrina* (Girard, 1850) but not for other triclad species. Unfortunately, such data are not known for *D. lugubris.* The results from our experiments indicate that *D. lugubris* is a 'cold-water' species in comparison with *D. polychroa.* Temperature is one of the key factors that play a role in the distribution of triclad species (Pattee *et al.,* 1973).

Acknowledgments

The authors are much indebted to Prof. Dr. C. den Hartog for critical remarks and to the Beijerinck-Popping Fund for financial support.

References

Ball, I. R. & T. B. Reynoldson, 1981. British planarians. Platyhelminthes: Tricladida. Keys and notes for the identification of the species. Synopsis of the British Fauna (New Series) No. 19: 1–141.

Beadle, L. C., 1934. Osmotic regulation in Gunda ulvae. J. exp. Biol. 11: 382–397.

Cuppen, H. P. J. J., 1980. De macrofauna in een aantal droogvallende- en permanente stilstaande wateren in het ruilverkavelingsgebied Brummen-Voorst. Een analyse van de effecten op het natuurlijk milieu van de in voorbereiding zijnde ruilverkaveling Brummen-Voorst met behulp van een hydrobiologisch onderzoek naar de macrofauna. Report Regionale Milieuraad Oost-Veluwe. 112 pp.

Cuppen, H. P. J. J. & G. M. Dirkse, 1978. Biologisch onderzoek van de watergangen in het Woold (Gem. Winterswijk). Report Rijks Instituut voor Natuurbeheer, Arnhem en Leersum. 68 pp., 3 appendices.

Den Hartog, C., 1974. Brackish water classification, its development and its problems. Hydrobiol. Bull. (Amsterdam) 8: 15–29.

Den Hartog, C. & G. van der Velde, 1973. Systematische notities over de Nederlandse platwormen (Tricladida). Levende natuur 76: 41–45.

De Vos, A. P. C., 1954. Over de oever- en bodemfauna der binnendijkse kolken langs het IJsselmeer. Chapter 20. In L. F. de Beaufort (ed.) Veranderingen in de flora en fauna van de Zuiderzee (thans IJsselmeer) na de afsluiting in 1932. Verslag van de onderzoekingen door de Zuiderzee-commissie der Nederlandse Dierkundige Vereniging. De Boer Jr., Den Helder: 277–282.

Hadderingh, R. H., 1977. Resultaten van vier jaar onderzoek (1972–1975) naar de bodemfauna van het inlaat- en uitlaatgebied van de Flevocentrale. Report KEMA, Arnhem: 1–22, 8 tables, 12 Figs.

Karling, T. G., 1974. Turbellarian fauna of the Baltic proper. Identification, ecology and biogeography. Fauna fennica 27: 1–101.

Komárek, J., 1926. Was ist Planaria polychroa und Planaria lugubris? Zool. Anz. 65: 29–33.

Krebs, B. P. M., 1981. Aquatische macrofauna van binnendijkse wateren in het Deltagebied deel 1: Zuid-Beveland. Delta Instituut voor Hydrobiologisch Onderzoek. Rapporten en Verslagen nr. 1981-8: 1–158.

Krebs, B. P. M., 1984a. Waterkwaliteitsbeoordeling van enkele Zeeuwse watergangen op grond van hun macrofaunasamenstelling. Delta Instituut voor Hydrobiologisch Onderzoek. Rapporten en Verslagen nr. 1984-1: 1–59.

Krebs, B. P. M., 1984b. Aquatische macrofauna van binnendijkse wateren in het Deltagebied deel 2: Zeeuws-Vlaanderen, oostelijke deel. Delta Instituut voor Hydrobiologisch Onderzoek. Rapporten en Verslagen nr. 1984-2: 1–124.

Pantin, C. F. A., 1931a. The adaptability of Gunda ulvae to salinity. I. The environment. J. exp. Biol. 8: 63–72.

Pantin, C. F. A., 1931b. The adaptability of Gunda ulvae to salinity. III. The electrolyte exchange. J. exp. Biol. 8: 82–94.

Pattee, E., C. Lascombe & R. Delolme, 1973. Effects of temperature on the distribution of turbellarian triclads. In W. Wieser (ed.). Effects of temperature on ectothermic organisms. Springer-Verlag Berlin, Heidelberg, New York: 201–207.

Reynoldson, T. B. & L. S. Bellamy, 1970. The status of Dugesia lugubris and D. polychroa (Turbellaria, Tricladida) in Britain. J. Zool., Lond. 162: 157–177.

Reynoldson, T. B. & B. Piearce, 1979. Predation on snails by three species of triclad and its bearing on the distribution of Planaria torva in Britain. J. Zool., Lond. 189: 459–484.

Schlieper, C., J. Bläsing & E. Halsband, 1952. Experimentelle Veränderungen der Temperaturtoleranz bei stenothermen und eurythermen Wassertieren. Zool. Anz. 149: 163–169.

Weil, E. & C. F. A. Pantin, 1931. The adaptability of Gunda ulvae to salinity. II. The water exchange. J. exp. Biol. 8: 73–81.

Distribution, abundance, and size of rhabdoids in *Dugesia polychroa* (Turbellaria: Tricladida)

Frederick J. Wrona
Aquatic Ecology Group, Department of Biology, University of Calgary, Calgary, Alberta, Canada T2N 1N4

Keywords: Turbellaria, Tricladida, *Dugesia,* rhabdoids

Abstract

The ontogenetic changes in the distribution, abundance, and size of rhabdoids were examined in an undisturbed laboratory population of *Dugesia polychroa.* Irrespective of triclad age, rhabdoids in the epidermis and parenchyma were more abundant on the dorsal than the ventral side of the body. No significant differences were found in the abundance of epidermal or parenchymal rhabdoids among the anterior, medial, and posterior regions of the body. Rhabdoid number and size changed significantly with triclad age, with a marked depression coinciding with the onset of cocoon production. Rhabdoid discharge was correlated with physical and/or physiological disturbance and occurred in the absence of any overt environmental disturbance. Simple allometric relationships were observed between rhabdoid size and number on one hand and body plan area on the other. Different allometric trends were observed from field-collected individuals compared to the undisturbed laboratory population. The potential function of rhabdoids in the Tricladida is discussed in light of these findings.

Introduction

Numerous small rod-shaped secretions known as rhabdoids are characteristically found in the epidermis and parenchyma of most species of the Tricladida (Hyman, 1951). A recent comparative study by Smith *et al.* (1982) on the ultrastructure and histochemistry of rhabdoids in the Turbellaria suggests that two distinct types of rhabdoids occur in the Tricladida: 1) epidermal rhabdoids, secretions originating in epidermal cells or epidermal-replacement cells and 2) true rhabdites, secretions with a characteristic ultrastructure produced by gland cells located within the parenchyma and discharged to the exterior of the animal through the gland necks penetrating in the epidermis. Both epidermal rhabdoids and true rhabdites have been referred to as rhabdites in earlier literature, however because they are not easily distinguishable by light microscopy; the more general term rhabdoids should be used in light-microscopical investigations (Smith *et al.,* 1982).

To date, the functional role and adaptive significance of rhabdoids remain uncertain although many theories have been forwarded. Prenant (1919) summarized earlier theories which ranged in terms of rhabdoids functioning as poisons, as structures involved in copulation, as tactile organs or levers whose movements influence nerve ends, as organs of support for the epidermis, or as being involved in the production of mucus for prey capture or protection. Prenant (1919), and later Skaer (1961), postulated that rhabdoids contained a metabolic waste product. A frequently proposed ecological function for rhabdoids is that they serve to deter predators and/or protect the animal from adverse environmental conditions (Hyman, 1951; Jennings, 1957; Young & Reynoldson, 1965; Coward & Vitale-Calpe, 1971; Coward & Piedilato, 1972; Ball & Reynoldson, 1981). This hypothesis is based on the fact

Hydrobiologia 132, 287–293 (1986).

that rhabdoids discharge upon exposure to noxious chemical and/or physical stimuli (Hyman, 1951; Alexander, 1979). Upon discharge the rhabdoids swell and dissolve, producing mucus which is toxic and consequently invokes a bad-taste response in predators (Hyman, 1951; Jennings, 1957; Young & Reynoldson, 1965; Coward & Vitale-Calpe, 1971; Coward & Piedilato, 1972; Ball & Reynoldson, 1981). The rhabdoid-mucus matrix has also been proposed to be an anti-fungal agent serving as protection against infection (Ball & Reynoldson, 1981).

In this study, I examine the ontogenetic changes in the distribution, abundance, and size of rhabdoids in an undisturbed laboratory population of the freshwater triclad *Dugesia polychroa* (O. Schmidt, 1861) and compare these data with similar measures of a field population which has presumably been exposed to natural biotic and/or abiotic environmental disturbances. The potential functional role(s) of rhabdoid bodies are discussed in light of the findings.

Materials and methods

A population of 150 *Dugesia polychroa* of approximately the same size were collected from a small lake near Bangor, North Wales, and maintained in the laboratory. Cocoons laid were removed and separately reared until hatching, thus producing cohorts of known age. Cultures were maintained at 20 °C and left undisturbed other than being fed weekly on chopped earthworm.

Eighteen triclads were selected at random from a cohort culture at each sampling occasion, and the plan area of each individual measured using a calibrated grid. Following the protocols outlined below, nine individuals were used to obtain age-specific (body-size) data on rhabdoid size while rhabdoid distribution and abundance patterns were determined on the nine remaining individuals. Two cohort populations of triclads were followed, each for a period of 16 weeks.

(a) Rhabdoid size

Individual triclads were placed on microscope slides, and excess water was removed. One drop of a 25% sodium-chloride solution was placed on the triclad for two minutes, inducing the rhabdoids to be discharged without appreciable distortion. The discharged rhabdoids were stained using Alcian blue (in 3% aqueous acetic acid), covered with a cover slip, and the lengths and breadths of the first 25 rhabdoids found for each triclad measured under oil immersion (×1 000).

(b) Rhabdoid distribution and abundance

To determine the age-specific number and distribution of rhabdoids in the various regions of the body of *D. polychroa,* individual specimens were placed on a rectangular block of 0.6% agar and excess water removed with filter paper. With the triclad on its dorsal surface and extended, the agar block was quickly immersed in liquid air before being placed on the chuck of a cryostat, reimmersed briefly in liquid air, and then left in the cryostat at −20 °C for two hours before sectioning. Transverse sections, 8 μm thick, were cut from the anterior, medial, and posterior thirds of the body and the rhabdoids in 100 μm linear length of epidermis and associated underlying parenchyma counted in each body half (dorsal versus ventral) under oil immersion (×1 000) (Fig. 1). A minimum of five sections

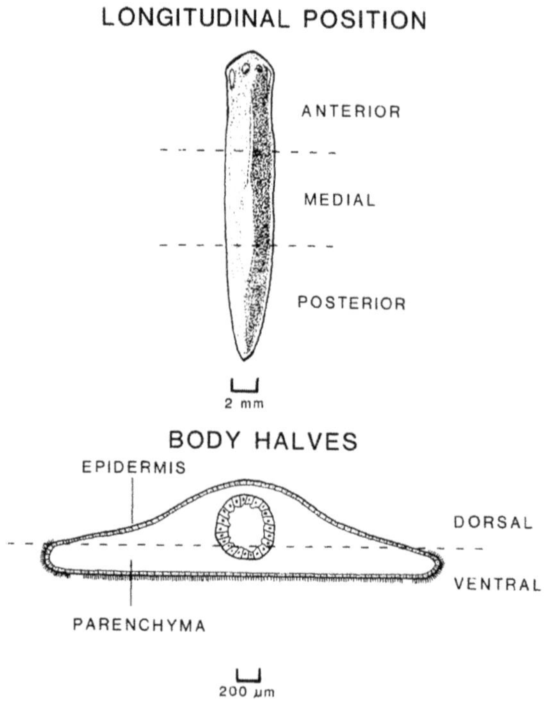

Fig. 1. Schematic representation of the body regions examined for rhabdoid distribution and abundance.

per body region were examined for each specimen.

Using similar protocols, twelve field-collected individuals were also analyzed for body-size specific data on rhabdoid size, distribution, and abundance patterns. Allometric models between the rhabdoid variable and triclad body size (plan area) were derived for both the laboratory and field-collected individuals using linear regression procedures (Peters, 1983).

Statistical analyses were conducted using the SPSS (Nie *et al.*, 1975) statistical package implementing standard Model I ANOVA and linear least-squares regression procedures. The Student-Newman-Keuls multiple range comparison was used to determine group differences when performing a one-way ANOVA. All hypothesis testing was performed at the P=0.05 level of probability.

Results

Irrespective of triclad age, no significant differences were found in the number of epidermal rhabdoids in the anterior, medial, or posterior regions of the body (Table 1). Similarly no significant two or three-way interactions were found among the independent variables tested with respect to epidermal rhabdoid number (Table 1). However, with respect to age, rhabdoids were consistently more abundant in the dorsal than in the ventral epidermis where their numbers increased significantly with increasing age (Table 1, Fig. 2a). The dorsal epidermal rhabdoids increased in number from age 0 to age 8 weeks and then decreased significantly at 12 weeks when cocoon production commenced. Rhabdoid numbers in the dorsal epidermis regained their pre-reproductive levels at 16 weeks (Fig. 2a). A similar but less pronounced trend was observed for the numbers of rhabdoids in the ventral epidermis.

Rhabdoid abundance was consistently significantly higher in the dorsal than in the ventral parenchyma (Table 1, Fig. 2b). The number of rhabdoids in the dorsal parenchyma showed significant differences with age, with a marked depression in number occurring at 12 weeks (Fig. 2b). In contrast, rhabdoids in the ventral parenchyma re-

Table 1. Three-factor (age, longitudinal position (LP), dorsal/ventral (DV)) Model I ANOVA, where the dependent variable is the density ($\log_{10}$ number/100 μm) of rhabdoids in the epidermis or parenchyma. (DF = degrees of freedom, NS = not significant at P = 0.05, ** = significant at P = 0.01).

Source of variation	DF	Epidermis F	Parenchyma F
Main effects:			
Age	3	25.1 **	16.7 **
Longitudinal position	2	0.1 NS	0.6 NS
Dorsal/ventral	1	741.0 **	167.8 **
2-Way interactions:			
Age × LP	6	0.3 NS	0.3 NS
Age × DV	3	0.1 NS	1.2 NS
LP × DV	2	0.1 NS	1.3 NS
3-Way interactions:			
Age × LP × DV	6	0.5 NS	0.5 NS
Error	480		

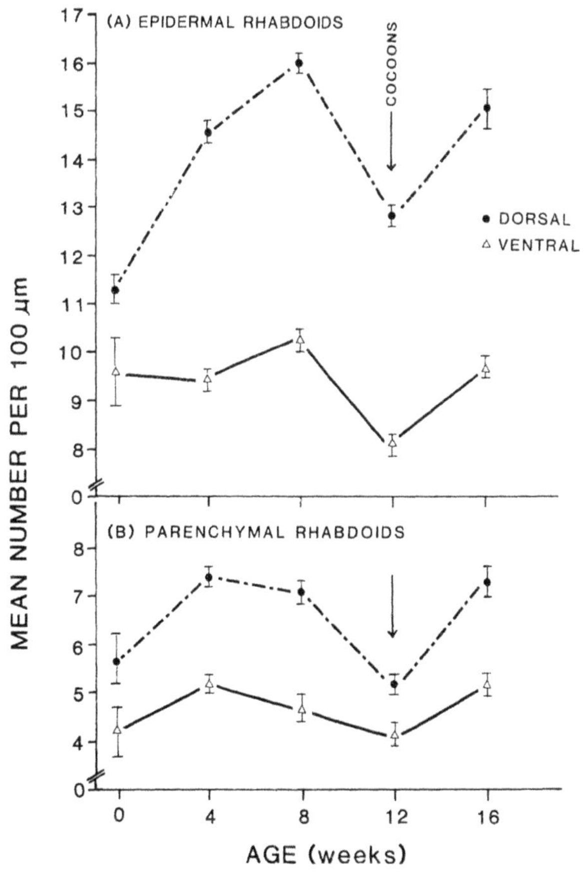

Fig. 2. (a) Mean number (±SE) of rhabdoids found in the dorsal and ventral epidermis as a function of triclad age. (b) Mean number (±SE) of rhabdoids found in the dorsal and ventral parenchyma as a function of triclad age.

mained relatively constant with age (Fig. 2b). Parenchymal rhabdoid numbers were not significantly different among the three longitudinal body regions and no significant two or three-way interactions among the experimental variables were observed (Table 1).

Mean rhabdoid size was found to significantly differ with triclad age (Table 2, Fig. 3). Multiple-range comparisons delineated five cohort age groups with similar trends in rhabdoid size (Table 2). In general, older age classes had larger rhabdoids, although a high degree of overlap occurred among the age groups (Table 2, Fig. 3). Additionally, rhabdoid size in the sexually reproducing (12-week) cohort was significantly smaller than in all other age groups (Table 2).

Mean triclad body size (plan area) increased over time but decreased at age 12 weeks with cocoon production (Fig. 4).

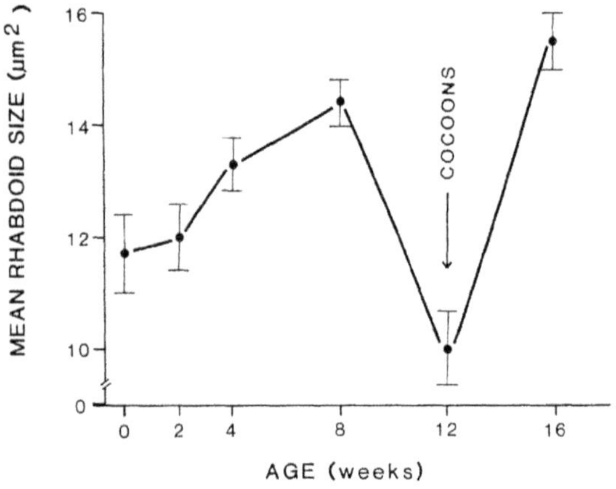

Fig. 3. Mean rhabdoid size (±SE) versus triclad cohort age.

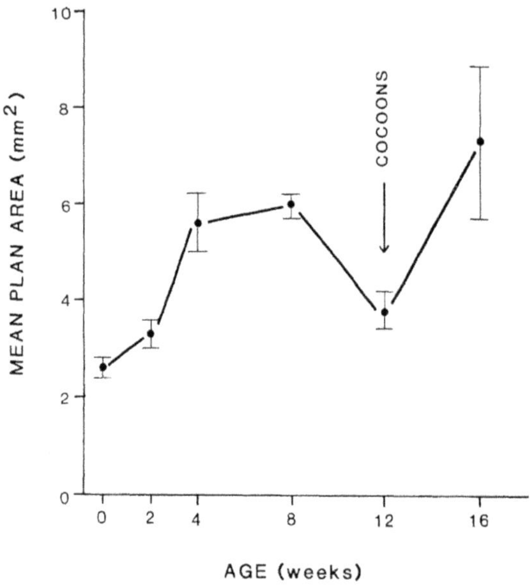

Fig. 4. Mean body plan area (±SE) versus cohort age.

The plan-area size range of the undisturbed laboratory population only slightly overlapped that of field-collected individuals which were generally larger (Table 3). Additionally, mean rhabdoid size and number from the field individuals were higher than from the laboratory cohort population (Table 3).

Significant positive allometric relationships were found between rhabdoid size and rhabdoid number on one hand and body plan area with respect to body position on the other hand in the undisturbed laboratory population (Table 3). To determine whether the allometric models derived from the un-

Table 2. One-factor (Age) Model I ANOVA, where the dependent variable is rhabdoid size ($\log_{10}$ μm), with the corresponding Student-Newman-Keuls (SNK) multiple range comparisons of the various triclad age groups. Means with a common suprascore are not significantly different. (DF = degrees of freedom, ** = significant at P = 0.01).

Source of variation	DF				Significance F	
Age	5				15.7 **	
Error	150					
	SNK comparisons					
Age groups ranked by means	Age 12	Age 0	Age 2	Age 4	Age 8	Age 16
Mean rhabdoid size	0.98	1.05	1.08	1.12	1.15	1.19

Table 3. Triclad body plan area ($\log_{10}$ mm^2) size range and mean ± SE rhabdoid size ($\log_{10}$ μm^2) and body-position-specific rhabdoid number (number/100 μm) for the laboratory and field populations. The corresponding allometric regression models for the laboratory (LAB) and laboratory + field (LAB + FIELD) populations are given. All regression models are significant at P = 0.05. Equality of the slopes between the regression models tested with a Student's t-test. (a = intercept, b = slope, epi = epidermis, paren = parenchyma, DF = degrees of freedom, NS = not significant at P = 0.05, ** = significant at P = 0.01).

Triclad plan area range	Population		Regression models				DF	t
	LAB 0.27 – 1.21	FIELD 1.19 – 1.42	LAB		LAB + FIELD			
			a	b	a	b		
Rhabdoid size:	1.11 ± 0.01	1.44 ± 0.02	1.01	0.16	0.91	0.33	326	3.27 **
Number:								
Dorsal-epi	1.16 ± 0.01	1.37 ± 0.01	0.96	0.27	0.94	0.30	560	1.57 NS
Ventral-epi	0.97 ± 0.01	1.14 ± 0.01	0.85	0.16	0.81	0.22	560	1.94 NS
Dorsal-paren	0.82 ± 0.01	0.96 ± 0.03	0.57	0.35	0.61	0.28	560	1.36 NS
Ventral-paren	0.66 ± 0.01	0.70 ± 0.03	0.39	0.37	0.50	0.21	560	3.31 **

disturbed laboratory population (LAB) were applicable to field-collected individuals, the regressions were recalculated including the field population data (i.e. LAB+FIELD) (Table 3).

As the slope of the allometric model describing the functional relationship between rhabdoid size and body plan area significantly increased upon inclusion of field individuals (Table 3), rhabdoid size in field individuals was higher than would have been predicted by the allometric relationship derived from the undisturbed laboratory population. Although not significant, similar trends were observed for rhabdoid number in the dorsal and ventral epidermis, indicating a tendency for field individuals to have larger numbers of epidermal rhabdoids on a plan-area basis than would have been predicted from laboratory cultures. In contrast, the regression slopes decreased in value in the LAB+FIELD models describing rhabdoid number in the dorsal and ventral parenchyma, with a significant change only occurring in the ventral-parenchyma model (Table 3). Thus, on a plan-area basis, field-collected individuals had a lower number of parenchymal rhabdoids than would have been predicted using the relationships derived from the undisturbed laboratory population.

Discussion

This study has shown that significant changes occur during ontogeny in the distribution, abundance, and size of rhabdoids in an undisturbed laboratory population of *Dugesia polychroa.* The occurrence of a greater number of rhabdoids in the dorsal half of the animal is consistent with previous studies on rhabdoid distribution in other turbellarians (Hyman, 1951; Coward & Vitale-Calpe, 1971). Additionally, the number of epidermal and parenchymal rhabdoids were found to be equally numerous in the anterior, medial, and posterior thirds of the body (Table 1), indicating no concentration in any of these body regions.

Rhabdoid number and size changed significantly with triclad age (Figs. 2a, 2b, 3). Both increased from age 0 to age 8 weeks, then decreased at age 12 weeks, which coincided with cocoon production, and then increased again upon completion of cocoon production at age 16 weeks.

Changes in mean body plan area over time show that the laboratory cohorts increased in size from hatching to 12 weeks when cocoon production and degrowth occurred (Fig. 4). After the reproductive period, the mean individual size in the laboratory cohorts regained their pre-reproductive level (Fig. 4). Similar results were obtained by Boddington & Mettrick (1977) who recorded a loss of mean individual weight (size) during cocoon production in a laboratory population of *D. polychroa* and by Reynoldson (1965) who observed the same trends in a sexually reproducing field population.

Thus, both body size and reproductive state influence the abundance and size patterns of rhabdoids in *D. polychroa.* The shrinkage of adult body size accompanying sexual reproduction may be partly responsible for the observed decrease in rhabdoid abundance. However, since physiological changes must also be occurring prior and/or during the reproductive period, it is hypothesized that these changes in some way are also involved in either decreasing rhabdoid production or increasing rhabdoid discharge. As both size and numbers of rhabdoids were found to significantly decrease during cocoon production, the latter explanation seems more likely.

Significant positive allometric relationships were found between body plan area and rhabdoid size and body-position-specific number in the undisturbed laboratory population. Recalculation of the allometric models including the data obtained on field individuals resulted in changes, although not all significant, in each of the model's parameter values (Table 3). On a plan-area basis, field individuals had significantly larger and tended to have higher numbers of dorsal and ventral epidermal rhabdoids than would have been predicted for similar sized individuals reared under undisturbed laboratory conditions. In contrast, rhabdoid number in both the dorsal and ventral parenchyma of field individuals was generally less than would be predicted under laboratory conditions. Therefore, it is postulated that the differences in rhabdoid size and abundance patterns between laboratory-reared and field individuals is primarily due to the latter being exposed to higher frequencies or intensities of environmental disturbances. The fewer dorsal and ventral parenchymal rhabdoids found in field specimens may thus be a consequence of higher levels of rhabdoid migration into the epidermis as a result of individuals being exposed to environmental disturbances that induce higher rates of rhabdoid discharge from the epidermis. Therefore, the observed rhabdoid size and abundance patterns displayed by field individuals may be considered to be supportive of the hypothesis that rhabdoids are involved in protecting the triclad from adverse environmental conditions.

Although this study found rhabdoid discharge to occur in the presence of physical and/or physiological disturbance, discharge also occurred in the absence of any external environmental disturbance. Additionally, rhabdoid size and abundance were found to be allometrically related to body plan area, and the size and abundance of rhabdoids in field specimens were generally higher than would be predicted from undisturbed laboratory reared individuals. However, these results do not allow for any definitive statement to be made regarding the actual functional role(s) of rhabdoids in triclads.

It is therefore suggested that previously proposed theories on rhabdoid function, in particular the commonly forwarded hypothesis that rhabdoids serve an anti-predator role, be more rigorously tested in laboratory and field experiments. Since previous laboratory and field studies have shown the incidence of predation on triclads to be highly variable and dependent on the type of predator (Jennings, 1957; Brown & Jenkins, 1962; Young & Reynoldson, 1965; Davies, 1969; Davies & Reynoldson, 1971; Coward & Piedilato, 1972; Wright, 1975; Beveridge, 1982), the proposition that rhabdoids generally act as anti-predator devices by being distasteful must be seriously questioned.

Biochemical studies on turbellarian rhabdoid bodies have shown that many consist primarily of porphyrin compounds which are believed to be metabolic pigments and end-products (McRae, 1959, 1961, 1963). A detailed review by Kennedy (1975) reveals that porphyrin-type pigments occur widely among invertebrates, and they may serve a number of functions, such as being involved in the fixation of energy, respiration, nutrition, excretion, or protection against external abiotic factors, one of which being light. This, this unique biochemical structure may impart properties that would allow rhabdoids to serve a number of concurrent physiological and ecological functions.

Acknowledgments

It is a pleasure to acknowledge the useful discussions with D. Baird, J. M. Culp, R. W. Davies, L. Linton, and J. Smith. Comments on earlier drafts of the manuscript from S. Tyler and two anonymous reviewers and the technical assistance of J. Dixon are also greatly appreciated. This research was supported by an N.S.E.R.C. operating grant.

References

Alexander, R. M., 1979. The invertebrates. Cambridge University Press, Cambridge, 562 pp.

Ball, I. R. & T. B. Reynoldson, 1981. British Planarians. Platyhelminthes: Tricladida. Cambridge University Press, Cambridge, 141 pp.

Beveridge, M. C. M., 1981. Adaptive aspects of bioenergetics on sexual and asexual species of freshwater triclads. Unpubl. Ph.D. thesis, University of Glasgow, 276 pp.

Boddington, M. J. & D. F. Mettrick, 1977. A laboratory study of the population dynamics and productivity of *Dugesia polychroa* (Turbellaria: Tricladida). Ecology 58: 109–118.

Brown, H. P. & M. M. Jenkins, 1962. A protozoan (*Dileptus*–Ciliata) predatory upon Metazoa. Science 136: 710.

Coward, S. J. & R. O. Vitale-Calpe, 1971. Behaviorally significant surface specialization of the planarian *Dugesia dorotocephala.* J. Biol. Physiol. 13: 3–10.

Coward, S. J. & J. W. Piedilato, 1972. The behavioural significance of rhabdite cells in planarians. J. Biol. Psychol. 14: 5–7.

Davies, R. W., 1969. Predation as a factor in the ecology of triclads in a small weedy pond. J. Anim. Ecol. 38: 577–584.

Davies, R. W. & T. B. Reynoldson, 1971. The incidence and intensity of predation on lake-dwelling triclads in the field. J. Anim. Ecol. 40: 191–214.

Hyman, L. H., 1951. The invertebrates. Platyhelminthes and Rhynchocoela, the Acoelomate Bilateria. Vol. II. McGraw-Hill, N.Y., 550 pp.

Jennings, J. B., 1957. Studies on feeding, digestion, and food storage in free-living flatworms (Platyhelminthes: Turbellaria). Biol. Bull. 112: 63–80.

Kennedy, G. Y., 1975. Porphyrins in invertebrates. Ann. N.Y. Acad. Sci. 244: 662–673.

McRae, E. K., 1959. Porphyrin synthesis in the planarian. I. Conversion of delta-aminolevulinic acid to porphobilinogen. J. exp. Zool. 142: 667–674.

McRae, E. K., 1961. Localization of porphyrin fluorescence in planarians. Science 134: 331–332.

McRae, E. K., 1963. Chromatographic identification of porphyrins in several species of Turbellaria. Experientia 19: 1–4.

Nie, N. H., C. H. Hull, J. G. Jenkins, K. Steinbrenner & D. H. Brent, 1975. SPSS Statistical Package for the Social Sciences, 2nd ed., McGraw-Hill, N.Y., 675 pp.

Peters, R. H., 1983. The ecological implications of body size. Cambridge University Press, Cambridge, 329 pp.

Prenant, M., 1919. Rhabdites des turbellaries. Arch. Zool. exp. gén. 58: 219–245.

Reynoldson, T. B., 1964. Evidence for intra-specific competition in field populations of triclads. J. Anim. Ecol. 33: 187–201.

Skaer, R. J., 1961. Some aspects of the cytology of *Polycelis nigra.* Quart. J. Microsc. Sci. 102: 295–317.

Smith, J. III, S. Tyler, M. B. Thomas & R. M. Rieger, 1982. The morphology of turbellarian rhabdites: Phylogenetic implications. Trans. am. microsc. Soc. 101: 209–228.

Wright, J. F., 1975. Observations on some predators of stream-dwelling triclads. Freshwat. Biol. 5: 41–50.

Young, J. O. & T. B. Reynoldson, 1965. A laboratory study of predation on lake-dwelling triclads. Hydrobiologia 26: 307–313.

The importance of turbellarians in the marine meiobenthos: a review

Paul M. Martens & Ernest R. Schockaert
Department SBM, Limburgs Universitair Centrum, B-3610 Diepenbeek, Belgium

Keywords: Turbellaria, Meiofauna, Ecology

Abstract

Recent data and opinions on meiofaunal ecology are briefly reviewed; and from scattered data, the place of turbellarians in the meiobenthic community is discussed. Turbellarian diversity, density, and biomass are higher in sandy habitats than in muddy bottoms. In sand, turbellarian diversity is of the same magnitude as that of other important meiofaunal taxa, while densities range between 7–25% of the total meiofauna. Mean individual turbellarian dry weight seems to be four times that of nematodes and in sandy habitats turbellarian biomass may be equal to or excede that of nematodes. Most turbellarian species may be considered as predators and in this respect may take the place occupied by macrofaunal species in muddy sediments. mens.

Introduction

Ecological studies on the meiofauna have expanded considerably during the last 20–25 years. Much attention has been focused on nematodes and harpacticoids which are often considered the dominating meiofaunal representatives. The soft-bodied fauna is proportionally disregarded because most soft-bodied animals must be studied alive and/or with light histological techniques in order to be identified. Turbellarian ecology has been dealt with in some autoecological studies (Ax, 1951, 1956, 1959; Ax & Ax, 1970; Dörjes, 1968; Ehlers, 1973; Faubel, 1976; Schmidt, 1972a, b; Sopott, 1973) and less in a more general perspective (Bilio, 1964, 1967; Ax, 1969, 1977; Pawlak, 1969; Gray & Rieger, 1971; Boaden, 1981; Purschke, 1981; Riese, 1984; Xylander & Reise, 1984).

In the present paper we attempt to determine a possible role of the Turbellaria within the meiobenthic community in terms of species-diversity, density, biomass, place in the trophic chain. Our speculations are based on the scattered (qualitative and quantitative) data that could be found in the literature on meiofauna and in some of our own preliminary observations.

Turbellarians as members of the meiobenthos

The term 'meiobenthos' (meiofauna) was introduced by Mare (1942) to indicate those benthic metazoans smaller than the 'macrobenthos' but larger than the 'microbenthos'. In practice the meiofauna consists of animals which pass through a 0.5–2 mm sieve but are retained on a sieve of 0.1–0.04 mm mesh width (McIntyre, 1969; Coull & Bell, 1979; Platt, 1981). However, many turbellarian species are larger than 2 mm (even up to more than 1 cm) and are nevertheless considered meiofaunal elements.

In a recent study, Warwick (1984) compared life strategies of several taxa in different marine benthic communities and concluded that animals larger than ±45 μg dry weight (macrofauna) have a planktonic development, disperse in the larval

Hydrobiologia 132, 295–303 (1986).

stage, have a continuous growth throughout life with a generation time of more than one year, and feed in indiscriminate fashion on particles (but often selected on the basis of particle size). Species smaller than ±45 μg dry weight (meiofauna) have no planktonic development, disperse in the adult stage, have an asymptotic growth to a maximum body size with a generation time of less than one year, and feed by selection on size but also on shape or quality of the particles.

From these observations, meio- and macrofauna can be redefined on the basis of biological criteria *and* of body size. Except for some triclads and polyclads all marine benthic turbellarians fit very well into this new definition, and some of them must be considered 'large meiobenthos'.

Two factors – among others – characterize the benthic habitat: grain size of the sediment and water movement (dynamism); and these are related to each other to some extent.

Grain size and packing of the grains determine the space available between grains for the so called *interstitial meiofauna* (Coull & Bell, 1979). The coarser the sand, the larger the interstices and vice versa. In a very fine sediment, such as mud, animals move by displacement of the sediment and so constitute *burrowing meiofauna*. Intertidal animals are small and/or elongate and exhibit a number of adaptive features (Remane, 1933; Wilson, 1935; Swedmark, 1964); there seems to be a relation between body shape and grain size (Wieser, 1959), but there is, of course, a limit: in mud-living meiofauna body size and shape are not related to grain size (Coull & Bell, 1979).

Sandy bottoms, especially on beaches exposed to wave action, are mostly dynamic environments. Muddy bottoms, found in estuaries, lagoons, saltmarshes and sublittoral places, are much less dynamic. Turbellarians are not very abundant in muddy bottoms: we find a lower diversity, as in other meiofauna taxa, and a lower density, contrary to many other meiofaunal taxa (see below).

In a sandy habitat, both diversity and density of turbellarian species are high (see also below) and all orders are represented. They often share a number of characteristics: they are haptic and may be quite contractile or move fast by muscular action, some of them have a tail appendage, many have long sensory bristles, a chordoid structure can be present, and some are extremely elongate. All these characteristics are considered adaptations to the interstitial way of life or to the dynamism of the environment or to both (Remane, 1933; Hyman, 1951; Ax, 1963, 1966; Boaden, 1968; Bush, 1968). The functional significance of these features has been discussed by Ax (1963, 1966) and Bush (1968). Some of these characteristics, such as adhesive papillae, sensory bristles, and body musculature, are well developed in Otoplanidae which is the dominating metazoan group in the swash zone, the most dynamic zone of the beach and called the Otoplana zone by Remane (1933).

As with other interstitial meiofaunal groups, there seems to be a relation between grain size of the sand and body length of the turbellarians. On the beaches of the island of Sylt (Germany) with coarse sand (400–500 μm; Schmidt, 1968) Proseriata is the dominating group, with many representatives of the Monocelididae, Coelogynoporidae, and Nematoplanidae reaching up to more than 1 cm in length (Sopott, 1973; Noldt & Wehrenberg, 1984). On the other hand at Robin Hood's Bay in fine sand (Gray & Rieger, 1971) and at the Belgian coast also in fine sand (Martens, 1984) Neorhabdocoela up to ±2 mm (Dalyellioida and Kalyptorhynchia, respectively) dominate the turbellarian fauna (together with the Acoela). According to Douglas (1984), there might be a relation between body size and grain size in the acoel *Convoluta roscoffensis*.

Diversity

The diversity in the marine meiofauna is generally high, and the discovery of new species (and even higher taxa) goes on even in intensively investigated areas. Sandy bottoms (especially fine and medium sand) contain more meiofaunal species than does mud (especially in brackish water) (Platt & Warwick, 1980; Remane & Schlieper, 1971).

Ecological studies on meiofauna, have paid most attention to the so-called 'hard-bodied' meiofauna such as nematodes (see Gerlach, 1980) and harpacticoids; the 'soft-bodied' meiofauna, to which the turbellarians belong, is mostly neglected. The reason is obvious: turbellarians must be extracted from the sediment and studied alive as soon as possible after sampling. This is also the main reason why sublittoral turbellarians are relatively poorly known. Moreover, taxonomists working with marine turbellarians are few. They are often compelled to focus their attention on some orders and to disregard others. Most areas have been investigated only over a short period of time. Finally, it is known that the methods used to extract turbellarians are far from being 100% efficient (Martens, 1984) and are almost of no use at all in a muddy sediment. For all or some of these reasons, the numbers of turbellarian species occurring in a given area are undoubtedly underestimated, and lists of species are necessarily incomplete, especially in muddy sediments. In mud, generally the most fragile turbellarians such as acoels and some smaller species of Macrostomida, Dalyellioida, and Monocelididae form the majority (own obs., see also Karling, 1974; Riedl, 1956; Steinböck & Reisinger, 1930): they are often damaged during extraction or simply lost or overlooked.

Nevertheless, when the data on turbellarian diversity (Table 1) are compared with the diversity of the much better investigated nematodes (Platt & Warwick, 1980: 735, Table III), it is obvious that in many instances numbers of turbellarian species are of about the same magnitude as those for nematodes. For the whole North Sea, 735 species of nematodes and 515 of harpacticoids are known (Heip *et al.*, 1983). A survey of the literature shows that over 400 turbellarian species are known from the same area but almost exclusively from littoral stations. Considering the data on the German Wadden Sea, where nematodes and harpacticoids as well as turbellarians have been investigated intensively, the following figures emerge: there are about 300 turbellarian, 300 nematode, and 150 harpacticoid species reported here (Wolff & Dankers, 1983). And for the island of Sylt: 230 turbellarian (Wehrenberg, 1983) and 178 nematode (Blome, 1982, 1983) species are reported.

To conclude we can safely state that species diversity of turbellarians in the meiofauna is at least similar to that of nematodes and may perhaps even be higher than that of harpacticoids, at least in some marine sandy habitats. Species diversity of turbellarians in mud is considerably lower than in sand, but as stated above, this is also valid for many other meiofaunal taxa.

Densities

The densities of the total meiofauna vary according to kind of sediment, latitude, depth, salinity, wave action, etc. On the average, one can expect to find 1 000–2 000 individuals 10 cm^{-2} (Coull & Bell, 1979). Densities tend to increase in detritus-rich muddy sediments with the highest values in intertidal mudflats of estuaries, lagoons, or salt-marshes. Densities are intermediate in muddy or silty sands and lowest in clean sand, especially on very exposed sites. There is also a decrease in meiofaunal density with depth into the ocean (Coull & Bell, 1979).

Table 1. Numbers of species of Turbellaria known from various localities.

Number of species	Location	Authors
111	Baltic Sea, Sandy habitats	Karling, 1974
56	Baltic Sea, Muddy habitats	Karling, 1974
302	German Wadden Sea*	Wolff & Dankers, 1983
230	Island of Sylt, Germany	Wehrenberg, 1983
60	Sylt, Sublittoral	Wehrenberg, 1983
39	Netherlands Delta area, Sublittoral	Martens & Schockaert, 1981
91	Netherlands Delta area, Salt-Marsh	Den Hartog, 1977
110	Belgian coast, Littoral	Own observations
200	Bay of Calvi, Corsica, Sandy habitats	Own observations
56	Bay of Calvi, Corsica, Periphyton	Own observations
70	Romanian Littoral waters	Mack-Fira, 1974
78	Kieler Bucht, Germany	Ax, 1951
45	Coastal lakes of French Mediterranean*	Ax, 1956
72	Ponto-Caspian brackish water*	Ax, 1959

* Some species of the periphyton included.

Table 2. Some maximum meiofaunal densities for various detritus-rich and muddy sediments.

Habitat	Locality	Max. Density	Authors
Estuary	Lynher Estuary, England	22×10^6 ind.m^{-2}	Warwick & Price, 1979
Estuary	Westerschelde Estuary, Belgium	17×10^6 ind.m^{-2}	Van Damme *et al.*, 1980
Lagoon	Bay of Arcachon, France	12×10^6 ind.m^{-2}	Lasserre *et al.*, 1975
Salt-marsh	Georgia, USA	16×10^6 ind.m^{-2}	Teal & Wieser, 1966
Salt-marsh	Rhode Island, USA	10×10^6 ind.m^{-2}	Nixon & Oviatt, 1973

In mudflats, maximum total meiofauna densities may reach ten times the average (Table 2) and nematodes may account for more than 90% of the total meiofaunal density (Bilio, 1966; Coull & Bell, 1979; Elmgren, 1976; Heip *et al.*, 1982; Platt & Warwick, 1980; Skoolmun & Gerlach, 1971; Warwick, 1971).

Turbellarians may occur in densities of 100–500 individuals 10 cm^{-2} and may represent between 7–25% of the meiofauna in sandy bottoms and even more in some particular places (Table 3). In muddy systems, however, turbellarians constitute only a few percent of the total meiofauna (McIntyre, 1969; Straarup, 1970; Watzin, 1983). From our own experience with muddy sediments, in both littoral and sublittoral stations in the Mediterranean (Calvi, Corsica) and in the North Sea, we can confirm that they are poor in turbellarians (but see above).

On the other hand, some data are known from sites showing extremely high (absolute or relative) densities of turbellarians (Table 4; all the species here are rather large, 1.35–7 mm long). The well-known Otoplana zone is another example.

Concerning densities of turbellarians in the meiofauna, we can conclude that in sandy habitats they may constitute from 7 to 25% of the total meiofauna but are much less abundant in mud. In some restricted areas they may become the dominant metazoan group.

Table 3. Comparison of meiofaunal and turbellarian densities in various localities.

Individuals 10 cm^{-2}		% of the total meiofauna	Habitat	Locality	Authors
Meiofauna	Turbellarians				
38– 655	27–591	28 –95	Exposed beaches	Spitsbergen	Radziejewska & Stankowska-Radziun, 1979
67–4092	3–101	1 –10	Sheltered, detritus-rich beaches	Spitsbergen	Radziejewska & Stankowska-Radziun, 1979
523–2828	96–273	10 –22	Steep Northern beach	Firemore, Scotland	McIntyre & Murison, 1973
833–4284	124–306	5 –21	Less steep Southern beach	Firemore, Scotland	McIntyre & Murison, 1973
2466–4415	60–389	2 –13	Sublittoral	Firemore, Scotland	McIntyre & Murison, 1973
51– 230	4– 37	4 –31	Exposed beach	Robin Hood's Bay, England	Gray & Rieger, 1971
420–3815	10– 48	0.3– 8	Estuary	Porto-Novo, India	McIntyre, 1968
603	245	41	Sandy beach	Porto-Novo, India	McIntyre, 1968
1125	37	3	Muddy beach	Porto-Novo, India	McIntyre, 1968
741–4904	24–408	2 – 9	Muddy-sand beach	Dalnezelenetsky Bay, Barents Sea	Galtsova & Platonova, 1980
80– 375	–	5 –15	Sheltered coarse-sand beaches	Sylt, Germany	Schmidt, 1968
3152	236	7.5	Sand flat	Sylt, Germany	Reise, 1983c
454–4682	54–407	7 –24	Meidum-exposed sandy beaches	Belgian coast	Own observations

Table 4. Turbellarian species showing high (absolute or relative) densities.

Species	Locality	Max. Density	Authors
Paromalostomum coronum	Reid State Park, USA High-energy beach	dominating metazoan species	Spurr, 1983
Paromalostomum fusculum	Island of Sylt, Germany	115 500 ind. m^{-2}	Pawlak, 1969
Paromalostomum fusculum	Belgian Coast	61 000 ind. m^{-2}	Martens, 1984
Pseudomonocelis ophiocephala	Black Sea, Russia	115 800 ind. m^{-2}	Murina, 1981
Pseudomonocelis ophiocephala	Protected sand beaches, Corsica and Italy	dominating metazoan species	Own obs.
Triporoplana synsiphonioides	Protected gravel beaches, Corsica	dominating metazoan species	Own obs.

Biomass

On the average, the biomass of meiobenthos is between 1 and 2 g dwt (dry weight) m^{-2} (Coull & Bell, 1979). In areas with high densities, it may rise to 4.9 g dwt m^{-2} (Van Damme *et al.*, 1980).

Mean individual dry weight of nematodes in the meiofauna averages between 0.15 and 0.45 μg (Wieser, 1960; Juario 1975; McLachlan, 1977; McIntyre, 1978; Heip *et al.*, 1982) and the individual dry weight of turbellarians appears to be about four times that of nematodes (Table 5). We have seen (Table 3) that in sandy habitats turbellarians may represent 7–25% of total meiofauna density, and can conclude in light of this 4:1 ratio in dry weight that turbellarian biomass may thus be equal to or exceed that of nematodes in some instances. Nematodes often constitute much less than half of the total meiofauna biomass in sandy habitats (Guille & Soyer, 1969; McIntyre & Eleftheriou, 1968; McIntyre & Murison, 1973; McLachlan, 1977; Platt & Warwick, 1980), and so it is clear that in those systems turbellarians may be relatively more important in terms of biomass.

Table 5. Average individual dry-weight biomass for turbellarians.

Author	Size classes in mm	μg dwt/ind.
Ankar & Elmgren, 1976	/	1.4
Coull, 1970	/	2
Van Damme & Heip, 1977	/	2.4
Faubel, 1982	0.5 – 1.0	0.63
Faubel, 1982	1.5 – 2.5	1.73
Faubel, 1982	3.0 – 4.0	8.81

Place in the food web

Turbellarians with a pharynx simplex or a pharynx doliiformis may swallow relatively large food particles. Those with a pharynx rosulatus or a pharynx plicatus can also swallow particles or suck out a prey even larger than themselves (see Straarup, 1970 with references). Thus, on the whole, turbellarians feed on a variety of food items.

Food requirements of turbellarians may be known by direct observations, feeding-experiments (e.g. in culture), or by the analysis of gut contents. According to Straarup (1970) and Reise (1983a, b) the diet of many acoels and dalyellioids consists mainly of diatoms, but the majority of turbellarians appear to be predators. Turbellarians are known to eat ciliates, hydroids, nematodes, turbellarians, copepods, amphipods, annelids, and also dead animals (Bilio, 1967; Elmgren, 1976; Hyman, 1951; Jennings, 1957, 1974, 1977; Karling, 1974; Murina, 1981; Pawlak, 1969; Reise, 1983a, b; Remane, 1933; Spurr, 1983; Watzin, 1983; Westblad, 1923; own observations). In cultures they can be kept alive with tubificids, *Artemia* nauplii, copepods, or meat (Curini-Galletti, Riser, pers. commun.; Spurr, 1983).

Prey consumption may be considerable. A population of *Pseudomonocelis ophiocephala* with a mean annual density of 14 640 individuals m^{-2} may consume about 589 000 amphipods (*Gammarus insensibilis*) a year (Murina, 1981). Watzin (1983) tested the predation of turbellarians on settling macrofauna larvae and juveniles. She found that increased densities of turbellarians significantly reduce densities of juvenile spionids and some other deposit feeders.

For the time being there is no strong evidence

that turbellarians feed on bacteria. Reise (1983a, b) considered some species to do so, but his conclusion was based on the absence of any recognisable material in gut contents (pers. comm.). Straarup (1970) considered the bacteria found in some species to be engulfed along with the prey or even as food of the prey animals themselves. Contrary to some statements (Jennings, 1977), we have found no evidence that turbellarians feed on detritus.

From the data now available we conclude that bacteria and detritus play a minor role (if any at all) in turbellarian diet, that some turbellarians prefer diatoms, and that the majority of them are mainly predators or scavengers. This implies that they belong to a high trophic level (McIntyre & Murison, 1973; Straarup, 1970) and exert predation pressure on the rest of the meiofauna and on small animals of the macrofauna.

Is there predation on turbellarians? Turbellarians are known to be eaten by other turbellarians (see above); some polychaetes, like *Nereis diversicolor* and some macrofauna deposit feeders seem to eat turbellarians (Bilio, 1967; Straarup, 1970; Watzin, 1983), but it is generally accepted that there is no heavy predation on turbellarians because of their defensive use of rhabdites and kleptocnids (Hyman, 1951; Karling, 1966). It is, however, evident that turbellarians would not be easy to identify in the gut contents of possible predators.

Possible role of turbellarians in the meiofauna

Much attention has been paid to the role of the meiofauna in an ecosystem, and much research has been carried out to establish the trophic relations between meio- and macrofauna. From an extensive review of the literature on this subject, Coull & Bell (1979 with many references) suggested that 'the mud/detrital meiofauna [may]... serve as a significant source of food for higher trophic levels... [meaning macrofaunal predators including fishes and natant crustaceans], whereas in sandy substrates the members of the meiofauna serve primarily as relatively rapid metazoan nutrient regenerators.' In other words, in a muddy bottom a significant portion of the biomass production realised in the meiofauna [with a production rate at least five times that of the macrofauna (Gerlach, 1971)] is transferred to the macrofauna, whereas biomass transfer is much less significant in a sandy substrate.

In a sandy habitat, especially clean sand of an exposed beach, macrofauna density is relatively low because the life strategies of the animals do not fit very well into such a disturbed environment (Gerlach, 1971; McIntyre, 1968, 1978; Renaud-Debyser & Salvat, 1963; Warwick, 1981, 1984; Whitlatch, 1980). The 'high trophic levels' (predators, secondary predators, etc.) are to be found within the meiofauna itself.

We have concluded in previous sections that diversity, density, and biomass of turbellarians may become rather important in sandy habitats and that many turbellarians (especially the larger ones) are most probably voracious predators. We may assume that in sandy habitats these turbellarians belong to the 'higher trophic levels,' the end of the trophic chain. Also, among nematodes e.g. more predators are found in sandy than in muddy bottoms (Platt & Warwick, 1980).

As Coull & Bell (1979) state, '... it is becoming increasingly apparent that [apart from physical properties of the habitat] biological interactions ... [such as] predatory controls play a ... role in structuring meiofaunal assemblages...' Perhaps we have here an example of how a low predatory pressure from the macrofauna induces an increase in meiofaunal predators. However, one can wonder whether those meiofaunal predators themselves control the macrofaunal assemblages, as a result of either food competition or predation from the meiofauna on young members of the macrofauna (Watzin, 1983).

Conclusions

Our ideas about the importance of turbellarians in the meiobenthic community can be summarized as follows:

1. Concerning their life strategies and body size (<45 μg dwt), turbellarians certainly belong to the meiofauna. They are well adapted to the interstitial way of life in sandy (dynamic) habitats.

2. Diversity, density, and biomass of turbellarians are higher in such habitats than in muddy bottoms. In sand, the diversity of turbellarians can be of the same magnitude as that of nematodes, and turbellarian biomass can be even higher.

3. The majority of turbellarian species must be considered to be predators and thus to represent the higher trophic levels of the trophic chain, at least in sandy bottoms where they may take the place of the macrofaunal predators of a muddy substratum.

Our main aim with this paper is to attract the attention of meiofaunal ecologists to a component too much neglected in the study of meiofaunal assemblages. Much remains to be investigated in turbellarian ecology, in the field as well as in laboratory experiments: diversity and density in habitats with different dynamism, respiration and reproduction rates, feeding and food requirements, competition, etc. But it is clear that turbellarians must be taken into consideration when assessing any of the parameters of the ecosystem.

Acknowledgments

We gratefully acknowledge Prof. Dr. Van Poucke and Dr. Snoeckx for critical reading of the manuscript. We also thank Hilde Zurings for her careful typing of the manuscript.

References

References marked with an asterisk (*) appeared shortly before this review was sent to press, and their data have not been considered here.

Ankar, S. & R. Elmgren, 1976. The benthic macro- and meiofauna of the Askö-Landsort area (northern Baltic proper). A stratified random sampling survey. Contr. Askö Lab. Univ. Stockholm 11: 1–115.

Ax, P., 1951. Die Turbellarien des Eulitorals der Kieler Bucht. Zool. Jb. Syst. 80: 276–378.

Ax, P., 1956. Les Turbellariés des étangs côtiers du littoral méditerranéen de la France méridionale. Vie Milieu, Suppl. 5: 1–215.

Ax, P., 1959. Zur Systematik, Ökologie und Tiergeographie der Turbellarienfauna in den ponto-kaspischen Brackwassermeeren. Zool. Jb. Syst. 87: 43–184.

Ax, P., 1963. Die Ausbildung eines Schwanzenfadens in der interstitiellen Sandfauna und die Verwertbarkeit von Lebensformcharakteren für die Verwantschaftsforschung. Zool. Anz. 174: 51–76.

Ax, P., 1966. Die Bedeutung der interstitiellen Saundfauna für allgemeine Probleme der Systematik, Ökologie und Biologie. Veröff. Insts. Meeresforsch. Bremerh. 2: 15–66.

Ax, P., 1969. Populationsdynamik, Lebenszyklen und Fortpflanzungsbiologie der Mikrofauna des Meeressandes. Verh. dt zool. Ges. Innsbruck 1968: 66–113.

Ax, P., 1977. Life cycles of interstitial Turbellaria from the eulittoral of the North Sea. Acta zool. fenn. 154: 11–20.

Ax, P. & R. Ax, 1970. Das Verteilungsprinzip des subterranen Psammon am Übergang Meer-Süsswasser. Mikrofauna Meeresboden 1: 1–51.

Bilio, M., 1964. Die agnatische Bodenfauna von Salzwiesen der Nord- und Ostsee. I Biotop und ökologische Faunenanalyse: Turbellaria. Int. rev. ges. Hydrobiol. 49: 509–562.

Bilio, M., 1966. Die aquatischen Bodenfauna von Salzwiesen der Nord- und Ostsee II. Ökologische Faunenanalyse: Hydrozoa, Nematoda, Rotatoria, Gastrotricha, Nemertini, Polychaeta, Oligochaeta, Halacaridae, Ostracoda, Copepoda. Int. Revue ges. Hydrobiol. Hydrogr. 51: 147–195.

Bilio, M., 1967. Nahrungsbeziehungen der Turbellarien in Küstensalzwiesen. Helgoländer wiss. Meeresunters. 15: 602–621.

Blome, D., 1982. Nematode-Systematics of a Sandy Beach on the North Sea Island of Sylt. Mikrofauna Meeresboden 86: 1–194.

Blome, D., 1983. Ökologie der Nematoda eines Sandstrandes der Nordseeinsel Sylt. Mikrofauna Meeresboden 88: 1–76.

Boaden, P. J. S., 1968. Water movement – a dominant factor in interstitial ecology. Sarsia 34: 125–136.

Boaden, P. J. S., 1981. Oxygen availability, redox and the distribution of some Turbellaria Schizorhynchidae and other forms. Hydrobiologia 84: 103–112.

Bush, L., 1968. Characteristics of interstitial sand Turbellaria: the significance of body elongation, muscular development and adhesive organs. Trans. am. microsc. Soc. 87: 244–251.

Coull, B. C., 1970. Shallow-water meiobenthos of the Bermuda platform. Oecologia 4: 325–357.

Coull, B. C. & S. S. Bell, 1979. Perspectives of marine meiofaunal ecology. In R. J. Livingston (ed.), Ecological processes in coastal an marine systems. Plenum Publishing Corporation, London: 189–216.

Den Hartog, C., 1977. Turbellaria from intertidal flat and saltmarshes in the estuaries of the south-western part of the Netherlands. Hydrobiologia 52: 29–32.

Dörjes, J., 1968. Zur Ökologie der Acoela (Turbellaria) in der Deutschen Bucht. Helgoländer wiss. Meeresunters. 18: 78–115.

Douglas, A. E., 1984. Relationship between sediment characteristics and size of the acoel turbellarian Convoluta roscoffensis Graff. Hydrobiologia 109: 207–210.

Ehlers, U., 1973. Zur Populationsstruktur interstitieller Typhloplanoida und Dalyellioida (Turbellaria, Neorhabdocoela). Mikrofauna Meeresboden 19: 1–105.

Elmgren, R., 1976. Baltic benthos communities and the role of the meiofauna. Contr. Askö Lab. Univ. Stockholm 14: 1–31.

Faubel, A., 1976. Populationsdynamik und Lebenszyklen interstitieller Acoela und Macrostomida (Turbellaria). Untersuchungen an einem mittel-lotischen Sandstrand der Nordseeinsel Sylt. Mikrofauna Meeresboden 56: 1–107.

Faubel, A., 1982. Determination of individual meiofauna dry weight values in relation to definite size classes. Cah. Biol. mar. 23: 339–345.

Galtsova, V. V. & T. A. Platonova, 1980. Distribution of meiofauna from muddy-sand beach of Dalnezelenetsky Bay, Barents Sea. Russian SFSR Biol. Morya 2: 15–17.

Gerlach, S. A., 1971. On the importance of marine meiofauna for benthos communities. Oecologia 6: 176–190.

Gerlach, S. A., 1980. Development of marine nematode taxonomy up to 1979. Veröff. Inst. Meeresforsch. Bremerh. 18: 249–255.
Gray, J. S. & R. M. Rieger, 1971. A quantitative study of the meiofauna of an exposed sandy beach at Robin Hood's Bay, Yorkshire. J. mar. biol. Ass. U.K. 51: 1–19.
Guille, A. & J. Soyer, 1969. La faune benthique des substrats meubles de Banyuls-sur-Mer. Premières données qualitatives et quantitatives. Vie milieu 19: 323–360.
Heip, C., R. Herman & M. Vincx, 1983. Subtidal meiofauna of the North Sea: a review. Biol. Jb. Dodonaea 51: 116–170.
Heip, C., M. Vincx, N. Smol & G. Vranken, 1982. The systematics and ecology of free-living marine nematodes. Helminth. Abstr. Series B, 51: 1–31.
Hyman, L. H., 1951. The invertebrates, vol. 2. Platyhelminthes and Rhynchocoela. The acoelomate Bilateria. McGraw-Hill, New York, 550 pp.
Jennings, J. B., 1957. Studies on feeding, digestion and food storage in free-living flatworms (Platyhelminthes: Turbellaria). Biol. Bull. 112: 63–80.
Jennings, J. B., 1974. Digestive physiology of the Turbellaria. In N. W. Riser & P. M. Morse (eds), Biology of the Turbellaria. McGraw-Hill, New-York: 173–197.
Jennings, J. B., 1977. Patterns of nutritional physiology in free-living and symbiotic Turbellaria and their implications for the evolution of entoparasitism in the phylum Platyhelminthes. In T. G. Karling & M. Meinander (eds), The Alex. Luther centennial symposium on Turbellaria. Acta zool. fenn. 154: 63–80.
Juario, J. V., 1975. Nematode species composition and seasonal fluctuation of a sublittoral meiofauna community in the German Bight. Veröff. Inst. Meeresforsch. Bremerh. 15: 283–337.
Karling, T. G., 1966. On nematocysts and similar structures in turbellarians. Acta zool. fenn. 116: 1–28.
Karling, T. G., 1974. Turbellarian fauna of the baltic proper. Identification, ecology and biogeography. Fauna fenn. 27: 1–101.
Lasserre, P., J. Renaud-Mornant & J. Castel, 1975. Metabolic activities of meiofaunal communities in a semi-enclosed lagoon. Possibilities of trophic competition between meiofauna and mugilid fish. 10th European Symp. Mar. Biol. Ostend 2: 393–414.
Mack-Fira, V., 1974. The turbellarian fauna of the Romanian littoral waters of the Black Sea and its annexes. In N. W. Riser & M. P. Morse (eds), Biology of the Turbellaria. McGraw-Hill, New York: 248–290.
Mare, M. F., 1942. A study of the marine benthic community with special reference to the micro-organisms. J. mar. biol. Ass. U.K. 25: 517–554.
Martens, P. M., 1984. Comparison of three different extraction methods for Turbellaria. Mar. Ecol. Prog. Ser. 14: 229–234.
Martens, P. M. & E. R. Schockaert, 1981. Sand dwelling Turbellaria from the Netherlands Delta area. Hydrobiologia 84: 113–127.
McIntyre, A. D., 1968. The meiofauna and macrofauna of some tropical beaches. J. Zool., Lond. 156: 377–392.
McIntyre, A. D., 1969. Ecology of marine meiobenthos. Biol. Rev. 44: 245–290.
McIntyre, A. D., 1978. The benthos of the Western North Sea. Rapp. P.-v. Réun. Cons. int. Explor. Mer, 1972: 405–417.
McIntyre, A. D. & A. Eleftheriou, 1968. The bottom fauna of a flatfish nursery ground. J. mar. biol. Ass. U.K. 48: 113–142.
McIntyre, A. D. & D. J. Murison, 1973. The meiofauna of a flatfish nursery ground. J. mar. biol. Ass. U.K. 53: 93–118.
McLachlan, A., 1977. Studies on the psammolittoral meiofauna of Algoa Bay, South Africa. II. The distribution, composition and biomass of the meiofauna and macrofauna. Zool. Afric. 12: 33–60.
Murina, G. V., 1981. Notes on the biology of some psammophile Turbellaria of the Black Sea. Hydrobiologia 84: 129–130.
Nixon, S. W. & C. A. Oviatt, 1973. Ecology of a New England salt marsh. Ecol. Monogr. 43: 463–498.
Noldt, U. & C. Wehrenberg, 1984. Quantitative extraction of living Plathelminthes from marine sands. Mar. Ecol. Prog. Ser. (in press).
Pawlak, R., 1969. Zur Systematik und Ökologie (Lebenszyklen, Populationsdynamik) der Turbellarien-Gattung Paromalostomum. Helgoländer wiss. Meeresunters. 19: 417–454.
Platt, H. M., 1981. Meiofaunal dynamics and the origin of the Metazoa. In P. L. Forey (ed.), The Evolving Biosphere. Cambridge University Press, Cambridge: 207–216.
Platt, H. M. & R. M. Warwick, 1980. The significance of free-living nematodes to the littoral ecosystem. In J. H. Price, D. E. G. Irvine & W. F. Farnham (eds), The shore environment. 2. Ecosystems. Academic Press, London: 729–759.
Purschke, G., 1981. Tolerance to freezing and supercooling of interstitial Turbellaria and Polychaeta from a sandy tidal beach of the island of Sylt (North Sea). Mar. Biol. 63: 257–267.
Radziejewska, T. & M. Stankowska-Radziun, 1979. Intertidal meiofauna of Recherchefjorden and Mablukta, Vest. Spitsbergen. Sarsia 64: 253–258.
Reise, K., 1983a. Biotic enrichment of intertidal sediments by experimental aggregates of the deposit-feeding bivalve Macoma balthica. Mar. Ecol. Prog. Ser. 12: 229–236.
Reise, K., 1983b. Sewage, green algal mats anchored by lugworms, and the effects on Turbellaria and small Polychaeta. Helgoländer wiss. Meeresunters. 36: 151–162.
Reise, K., 1983c. Experimental removal of lugworms from marine sand affects small zoobenthos. Mar. Biol. 74: 327–332.
Reise, K., 1984*. Free-living Platyhelminthes (Turbellaria) of a marine sand flat: an ecological study. Microfauna Marina 1: 1–62.
Remane, A., 1933. Verteilung und Organisation der benthonischen Mikrofauna der Kieler Bucht. Wiss. Meeresunters. (Abt. Kiel) 21: 161–221.
Remane, A. & C. Schlieper, 1971. Biology of Brackish Water. J. Wiley & Sons, N.Y., 372 pp.
Renaud-Debyser, J. & B. Salvat, 1963. Eléments de prospérité des biotopes des sédiments meubles intertidaux et écologie de leurs populations en microfaune et macrofaune. Vie Milieu 14: 463–550.
Riedl, R., 1956. Zur Kenntnis der Turbellarien adriatischer Schlammböden sowie ihrer geographischen und faunistischen Beziehungen. Thalassia jugoslav. 1: 69–184.
Schmidt, P., 1968. Die quantitative Verteilung und Populationsdynamik des Mesopsammons am Gezeiten-Sandstrand der Nordseeinsel Sylt. I. Faktorengefüge und biologische Gliederung des Lebensraumes. Int. Revue ges. Hydrobiol. Hydrogr. 53: 723–779.

Schmidt, P., 1972a. Zonierung und jahreszeitliche Fluktuationen des Mesopsammons im Sandstrand von Schilksee (Kieler Bucht). Mikrofauna Meeresboden 10: 1–60.

Schmidt, P., 1972b. Zonierung und jahreszeitliche Fluktuationen der interstitiellen Fauna in Sandstränden des Gebiets von Tromso (Norwegen). Mikrofauna Meeresboden 12: 1–86.

Skoolmun, P. & S. A. Gerlach, 1971. Jahreszeitliche Fluktuationen der Nematodenfauna im Gezeitenbereich des Weser-Ästuars (Deutsche Bucht). Veröff. Inst. Meeresforsch. Bremerh. 13: 119–138.

Sopott, B., 1973. Jahreszeitliche Verteilung und Lebenszyklen der Proseriata (Turbellaria) eines Sandstrandes der Nordseeinsel Sylt. Mikrofauna Meeresboden 15: 1–106.

Spurr, S. J., 1983. A new species of the genus Paromalostomum (Turbellaria: Macrostomida) from New England. Trans. am. microsc. Soc. 102: 333–348.

Steinböck, O. & E. Reisinger, 1930. Ergebnisse einer von E. Reisinger & O. Steinböck mit Hilfe des Rask-Orsted Fonds durchgeführten zoologischen Reise in Grönland 1926. 1. Reisebericht. Vidensk. Meddr dansk naturh. Foren. 90: 13–43.

Straarup, B. J., 1970. On the ecology of turbellarians in a sheltered brackish shallow water bay. Ophelia 7: 185–216.

Swedmark, B., 1964. The interstitial fauna of marine sand. Biol. Rev. 39: 1–42.

Teal, J. M. & W. Wieser, 1966. The distribution and ecology of nematodes in a Georgia salt marsh. Limnol. Oceanog. 11: 217–222.

Van Damme, D. & C. Heip, 1977. Het meiobenthos in de zuidelijke Noordzee. In C. F. Nihoul & L. A. P. De Coninck (eds), Nationaal Onderzoeks- en ontwikkelingsprogramma – Project Zee 7: 1–113.

Van Damme, D., R. Herman, Y. Sharma, M. Holvoet & P. Martens, 1980. Benthic studies of the Southern Bight of the North Sea and its adjacent continental estuaries. Progress Report II. Fluctuations of the meiobenthic communities in the Westerschelde estuary. ICES. C.M./L. 23: 131–170.

Warwick, R. M., 1971. Nematode associations in the Exe estuary. J. mar. biol. Ass. U.K. 51: 439–454.

Warwick, R. M., 1981. Survival strategies of meiofauna. In N. V. Jones & W. J. Wolff (eds), Feeding and survival strategies of estuarine organisms. Plenum, New-York: 39–52.

Warwick, R. M., 1984. Species size distribution in marine benthic communities. Oecologia 61: 32–41.

Warwick, R. M. & R. Price, 1979. Ecological and metabolic studies on free-living nematodes from an estuarine mud-flat. Estuar. coast. mar. Sci. 9: 257–271.

Watzin, M. C., 1983. The effects of meiofauna on settling macrofauna: meiofauna may structure macrofaunal communities. Oecologia 59: 163–166.

Wehrenberg, C., 1983. Besiedlungsstruktur der Turbellaria in sublitoralen Sänden der Nordsee bei Sylt. Thesis, University of Göttingen, 93 pp.

Westblad, E., 1923. Zur Physiologie der Turbellarien. I. Die Verdauung. II. Die Exkretion. Acta Univ. lund., N.F., 2 Avd. 33: 1–212.

Whitlach, R. B., 1980. Patterns of resource utilization and coexistence in marine intertidal deposit-feeding communities. J. mar. Res. 38: 743–765.

Wieser, W., 1959. The effect of grain size on the distribution of small invertebrates inhabiting the beaches of Puget Sound. Limnol. Oceanog. 4: 181–194.

Wieser, W., 1960. Benthic studies in Buzzards Bay II. The meiofauna. Limnol. Oceanog. 5: 121–137.

Wilson, G. B., 1935. A new and important copepod habitat. Smithsonian Misc. Coll. 94: 1–13.

Wolff, W. J. & N. Dankers, 1983. Preliminary checklist of the zoobenthos and nekton species of the Wadden Sea. In N. Dankers, H. Kühl & W. J. Wolff (eds), Invertebrates of the Wadden Sea. A. A. Balkema, Rotterdam: 24–60.

Xylander, W. E. .R. & K. Reise, 1984*. Free-living Platyhelminthes (Turbellaria) of a rippled sand bar and a sheltered beach: a quantitative comparison at the island of Sylt (North Sea). Microfauna Marina 1: 207–277.

An overview of surface specializations in the digenetic trematodes

Burton J. Bogitsh
Department of Biology, Vanderbilt University, Nashville, TN 37235, USA

Keywords: Digenea, *Schistosoma mansoni*, tegument, miracidium, sporocyst, cercaria

Abstract

The outer covering of digenetic trematodes undergoes a number of changes in the course of the life cycle in adaptation to parasitism and to the host-seeking requirements of the various stages. Using *Schistosoma mansoni* as a model, I review morphological, physiological, and cytochemical features of the coverings of miracidia, sporocysts, cercariae, and adults. Such features should be used with caution in evaluating phylogenetic relationships of any flatworms since the environment is itself a powerful determinant.

Introduction

The life cycles of digenetic trematodes are complex, involving at least two hosts and exposure to a variety of environmental conditions. The only interface between the organism and the environment is its outer covering. In the discussion that follows, I direct attention to this covering in four stages common to the life cycles of all digeneans, i.e. the miracidium, sporocyst, cercaria, and adult. These stages are chosen since each exemplifies specializations essential to the survival and success of the organism. A model life cycle for this purpose is that of the human blood fluke belonging to the genus *Schistosoma*. Several factors make this genus a good choice, two of which are of particular significance. First, its medical importance has prompted extensive investigations of the various stages of the life cycle, and second, the involvement of only two hosts provides a convenient fit for the confines of the current discussion. One major drawback of this choice is that, in a number of respects, these organisms are more highly specialized than the typical digenetic trematode. To compensate, attention will be focussed upon particular specializations that illustrate aspects of a larger general pattern of specialization(s) essential to successful adaptation of the organisms to changing environmental conditions.

Miracidium

Morphologically, this form represents perhaps the closest link to what many consider the ancestral form of the Digenea (e.g., a rhabdocoel turbellarian). The miracidium is first observed in the comfortable confines of an eggshell. The eggshell of the schistosomes is a porous structure and probably provides a means for the exchange of nutrients and/or metabolic wastes between the environment and the larval outer covering since these forms do not possess a functional digestive tract. To test this capacity, Kawanaka & Carter (1984) have exposed the eggshell-enclosed miracidium of *S. mansoni* to [^{14}C]glucose *in vitro*. Their analysis of the larvae and the surrounding media reveals that the miracidium can incorporate the label into proteins and glycoproteins indicating that the surface of the organism has at least a limited capacity for absorption and/or transport. Upon reaching a suitable aquatic medium, the schistosome egg hatches, free-

Hydrobiologia 132, 305–310 (1986).

ing the miracidium. The hatching phenomenon apparently depends upon the favorable interactions of a number of factors such as osmolarity, temperature, etc. (Smyth, 1976). These optimum requirements vary according to the species studied; in fact, eggs of some groups will hatch only after ingestion by a suitable snail intermediate host.

The miracidium of *S. mansoni* has been described in detail by Pan (1980). Its outer covering consists of tiers of 21 ciliated epidermal plates which are separated from each other by ridges of cytoplasm whose cyton areas are sunk into the parenchyma (Wilson, 1969; Meuleman, 1972; Wikel & Bogitsh, 1974). This orientation of cyton area to plate may vary, however, from that arrangement to one in which the nucleus is more superficially situated (Smyth & Halton, 1983). The anteriormost portion of the miracidium is free of cilia, its most prominent feature being the apical papilla at which site gland cells empty their contents to the exterior. The apical papilla and the cytoplasmic ridges are equipped with sensory organelles of which there are four major types (Brooker, 1972; Pan, 1980). Each ciliated epidermal plate possesses scattered surface microvilli, a nucleus, mitochondria, and scattered glycogen. The mitochondria appear to be metabolically active, evidenced by cytochemically demonstrable cytochrome oxidase and peroxidase activities (Bogitsh, 1975). Becker (1977), however, has reported that a complete Krebs cycle is most likely absent in hatched miracidia and that most energy is derived anaerobically. The cytoplasmic ridges likewise contain numerous metabolically active mitochondria, some surface microvilli and numerous membrane-bound organelles. Once the miracidium of *S. mansoni* hatches, its survival depends upon its ability to (1) locate a suitable snail intermediate host, (2) attach to and penetrate the intermediate host's tissues, (3) evolve an outer covering from one primarily specialized for motility to one specialized for absorption. The outer covering of the miracidium displays remarkably specialized features which allow the organism to accomplish all of these vital functions. For instance, (1) the ciliated epidermal plates provide the necessary motility to locate the suitable intermediate host, (2) the apical papilla contains sensory organelles which aid in location of the snail intermediate host as well as providing the mechanism for attachment and penetration (Brooker, 1972; Wikel & Bogitsh, 1974), and (3) upon penetration, or shortly thereafter, the miracidial epidermal plates are shed, and the cytoplasmic ridges spread to form the covering, or tegument, of the transformed miracidium, now called the sporocyst (see Bogitsh & Carter, 1982, for a review). The ciliated epidermal plates of the miracidium also apparently afford the hatched organism osmotic protection since, following their loss, the resulting sporocyst will swell when placed in freshwater.

Sporocyst

The sporocyst tegument, derived from the cytoplasmic ridges, is a syncytium with the proximal cyton areas sunken beneath the surface and connected to the distal portion by cytoplasmic connectives. This cytoarchitectural pattern persists through the remaining larval and adult stages.

The function of the sporocyst tegument is twofold. First, it is protective in that it shields the organism against the defense mechanisms of the snail (e.g. attack by amebocytes). The nature of this protective function has been investigated by Smyth (1973), Basch (1976) and others who report that, during the miracidium-sporocyst transformation process, a concomitant transformation occurs in the carbohydrate composition of the sporocyst tegumental glycocalyx. Heyneman *et al.* (1971), Basch (1976) and Yoshino *et al.* (1977) hypothesize that this change in the glycocalyx simulates snail tissue, preventing the snail from recognizing the sporocyst as a foreign entity. Second, the sporocyst tegument is highly absorptive. While direct evidence of its absorptive nature is scant, there is ample indirect evidence based on morphological and histochemical studies to justify this conclusion. Carbohydrate metabolism appears to be a major source of energy for this stage. Carter & Bogitsh (1975) and Becker (1980) report that schistosome sporocysts can divert glucose from the hemocoel of the snail through the tegument. Hoskin & Cheng (1974), McManus & James (1975a, 1975b), and Smyth & Halton (1983) report that sporocysts of other genera also can actively transport monosaccharides from the tissues of their snail intermediate hosts through their teguments. In addition, a number of investigators have documented the presence of nonspecific phosphatase activity associated with the extensive microvillar surface of a number of sporocysts (Dusanic, 1959; Dike & Read, 1971; Koie, 1971; Krupa & Bogitsh, 1972). Membrane-associated nonspecific phosphatase activity has been observed in tissues active in carbohydrate transport. This activity is associated with the outer surface of microvilli and is usually complemented by the action of specific ATPases located in the membrane (Crane, 1968; Bogitsh, 1975; Bogitsh & Carter, 1982). It is hypothesized that the nonspecific phosphatase hydrolyzes phosphate esters of the monosaccharide, while the ATPase is responsible for the transport of the carbohydrate into the cytoplasm. In support of this concept, Krupa *et al.* (1975), using nitrophenyl phosphate as a substrate, have demonstrated the presence of a ouabain-sensitive, potassium activated ATPase system in *S. mansoni* sporocysts.

Cercaria

In most digenetic trematode life cycles, the cercaria develops asexually from germinal cells within either a second generation of sporocysts or from within another intramolluscan stage, the redia. In either instance, the tailed cercaria emerges from the enclosing tissues to seek an aquatic environment. As with the hatched miracidium, the cercaria must find a suitable host quickly or perish. Once the cercaria reaches the next host in its life cycle, it sheds its tail and develops into the next stage. For the schistosomes, man serves as the next host. The cercaria penetrates the human skin and migrates via the blood stream to its definitive site, mesenteric venules surrounding various visceral organs, the particular organ depending on the species of schistosome. While the cercaria is in the water, it is covered by an extensive surface coat which apparently forms during its intrasporocyst development and may be described chemically as a mucosubstance (Kemp, 1970), either a glycoprotein or a neutral or slightly acidic mucopolysaccharide (Hockley, 1973). One of the most distinctive properties of this surface coat is its impermeability, which probably protects the cercaria from variations in osmotic pressure as it moves from snail tissue to vertebrate tissue *via* freshwater (Morris, 1971). In addressing the impermeability of the cercarial tegument, Carter & Bogitsh (1975) observe that developing cercariae within sporocysts, when exposed to radiolabeled glucose, display a heterogeneous pattern of incorporation of the label. Mature cercariae with their surface coats formed display little if any incorporation, while somewhat immature cercariae within the same sporocyst display a uniform distribution of label. The question then arises whether the surface coat is responsible for the quiescent nature of the mature cercaria tegument or whether the tegument itself is metabolically inactive. Cytochemical investigation of the cercarial mitochondria favors the latter premise. While mitochondria of the cercarial parenchyma are reactive for cytochrome oxidase and ATPase activities, tegumental mitochondria are not. A similar phenomenon is observed in the cercaria following its penetration of human skin, leading to speculation that the outer covering may be limited in its absorptive capability. Once the cercaria penetrates the skin of its human host, the protective surface coat is lost, leaving only a thin glycocalyx. The resulting juvenile, or schistosomule, after a brief period in the skin, migrates and develops into an adult.

Adult

Threadgold (1963) wrote an early treatise on the ultrastructure of the adult tegument of digenetic trematodes. During the last twenty years, numerous studies on the teguments of a wide variety of adult digeneans have been undertaken. These studies pertain to the significance of the structure, encompassing morphology, cytochemistry, biochemistry, and immunology. While there is a multiplicity of functions attributed to the tegument such as absorption, protection, sensory perception, and secretion, I will address only the first two.

As for the sporocyst stage, there is ample indirect evidence that the tegument of adult digeneans functions as an absorptive surface. Further, the ease in isolating the adult worm from its environment makes possible *in vitro* biochemical studies, not feasible with the sporocyst stage, and these substantiate the indirect evidence. Also, unlike the sporocyst stage where the only means of nutrition is the transport of material over the surface membrane, all adult Digenea possess a functional digestive tract which probably serves as the primary nutritional mechanism for the organism, with the tegument serving in an ancillary capacity. However, certain materials essential to the well-being of the organism apparently are available only by membrane transport. For example, the schistosomes depend on glycolysis for most of their metabolic energy, and they obtain glucose by tegumental transport. Rogers & Bueding (1975), using an ingenious chamber enabling them to observe the mouth region of the organism apart from the remaining portion of the body, demonstrate that the tegument is the main pathway for the uptake of radiolabeled glucose *in vitro*. Other studies show the digestive tract to be the primary site for the digestion of macromolecules, such as hemoglobin, with little or no involvement in the transport of glucose or amino acids (Uglem & Read, 1975; Bogitsh & Carter, 1977). Bogitsh & Krupa (1971) note the distribution of a number of alkaline nucleoside di- and triphosphatases in the schistosome tegument. According to their findings, enzyme activity is limited to the deeper regions of the extensive invaginations of the tegument, and they conclude that this localization indicates sites of absorption. In a study to correlate the extent of distribution of alkaline phosphatase reaction product with the ability to transport glucose, the schistosome tegument has been compared to the teguments of two other digeneans, *Haematoloechus medioplexus* and *Megalodiscus temperatus*. *H. medioplexus* resides in the lungs of frogs, an environment relatively poor in free glucose. Even so, it can transport glucose across its tegument, although to a far lesser degree than the schistosomes (Parkening & Johnson, 1969). *M. temperatus* resides in the rectum of frogs, a hostile region devoid of glucose (Halton, 1967). It is not surprising that *M. temperatus* is incapable of such transport (Shannon & Bogitsh, 1971). *H. medioplexus* displays a uniform distribution of reaction product as compared to the schistosomes, indicating involvement of the entire surface, while *M. temperatus* manifests no alkaline phosphatase activity. Other indications gleaned from this study are (a) the extent of invaginations appears directly proportional to either transport ability and/or availability of glucose, and (b) the presence of mitochondria in tegumental matrix correlates with the absorptive capability of the tegumental surface (e.g. *M. temperatus* possesses no tegumental mitochondria).

In all adult digeneans studied, it is well established that synthetic activities occur in the cyton area. A number of studies demonstrate that the material involved in the maintenance of the glycocalyx or surface is formed *via* the Golgi complex in the cyton areas and is translocated to the surface (Bogitsh, 1968; Threadgold, 1976; Shannon & Bogitsh, 1971; Bogitsh & Carter, 1980). In the schistosomes, this translocation is microtubule-dependent (Bogitsh & Carter, 1980). There is some debate as to whether a glycocalyx is present on the tegumental surface of adult schistosomes. Smyth & Halton (1983) hypothesize that, since there is no morphological evidence of a schistosome glycocalyx, its function is assumed by a multi-layered surface plasma membrane (Hockley & McLaren 1973; McLaren & Hockley, 1977). In most adult digeneans, the glycocalyx is considered the primary protection of the organism against hostile elements of its environment, such as the digestive enzymes of the host's intestinal tract and/or the immunological defenses of the host. Obviously, the survival of the blood flukes depends, in part, on their ability to evade the host's antibodies. The traditional concept that schistosomes acquire a coating of host material which masks the antigenic determinants on the worm's surface protecting the worm from recognition by parasite-specific host responses (Smithers *et al.*, 1969) has remained basically intact with minor modifications. For instance, newly penetrated larvae (schistosomules) are labile to the host's immune attack during the first four to five days following cercarial penetration, after which time, they do not bind host antibody (Goldring *et al.*, 1976). The time frame for this protection coincides precisely with the time it takes for the worms to begin feeding (Bogitsh & Carter, 1977). *In vitro* studies of the effects of feeding upon the ability of schistosomules to express host erythrocyte antigens on their outer surfaces show non-feeding larvae to be incapable of such expression while fed larvae acquire the antigenicity of the host blood. Butterworth *et al.* (1982) report that, in addition to this incorporation of blood-group glycolipids and glycoproteins into the outer tegumental membrane, the developing larvae may also release a variety of materials that either cleave various globulins or interfere with the functions of a number of formed elements such as lymphocytes and mast cells.

In conclusion, digenetic trematodes have a number of specializations in their outer covering in order to fulfill their 'destinies', i.e. the parasitization of another organism. Read (1970) has proposed the plausible hypothesis that relatively generalized rather than specialized members of various phyla assume a parasitic existence and that specializations evolve only subsequently. If this is true, then while a study of the nature of the outer coverings of parasitic helminths may provide insight into their kinship to free-living species (Threadgold, 1979), specific conclusions drawn from such studies are fraught with pitfalls. Such studies hold real promise for providing clues to phylogenetic relationships as mentioned above, but it should be stressed that the host environment is in itself a powerful determinant for which specializations of the outer coverings will persist. The ability of the adult schistosome to evade the immunological defenses of its host illustrates this point. Specialization of the apical plasma membrane of the tegument to this end appears to be peculiar to this group of trematodes. One must be cautious, therefore, in attributing specific characteristics of the helminth's outer covering to phylogenetic adaptation rather than being merely a response to a very specific environmental condition.

References

Basch, P. F., 1976. Intermediate host specificity in Schistosoma mansoni. Exp. Parasitol. 39: 150–169.

Becker, W., 1977. Zur Stoffwechselphysiologie der miracidien von Schistosoma mansoni während ihrer Aktivierung innerhalb der Eischale. Z. Parasitenkd 52: 69–79.

Becker, W., 1980. Metabolic interrelationship of parasite trematodes and molluscs, especially Schistosoma mansoni in Biomphalaria glabrata. Z. Parasitenkd 63: 101–111.

Bogitsh, B. J., 1968. Cytochemical and ultrastructural observations on the tegument of Megalodiscus temporatus. Trans. am. microsc. Soc. 87: 477–486.

Bogitsh, B. J., 1975. Cytochemical localization of peroxidase activity in the miracidium of Schistosoma mansoni. J. Parasitol. 61: 621–626.

Bogitsh, B. J. & O. S. Carter, 1977. Developmental studies in the digestive tract of schistosomules (Schistosoma mansoni) grown in vitro. I. Ultrastructure. Trans. am. microsc. Soc. 96: 219–227.

Bogitsh, B. J. & O. S. Carter, 1980. Schistosoma mansoni: radioautography of colchicine's effect on [^{3}H]proline incorporation into adults *in vitro*. Exp. Parasitol. 49: 319–327.

Bogitsh, B. J. & O. S. Carter, 1982. Developmental biology of Schistosoma mansoni, with emphasis on the ultrastructure and enzyme histochemistry of intramolluscan stages. In F. W. Harrison & R. R. Cowden (eds), Developmental biology of freshwater invertebrates. Alan R. Liss, New York: 221–248.

Bogitsh, B. J. & P. L. Krupa, 1971. Schistosoma mansoni and Haematoloechus medioplexus: nucleosidediphosphatase localization in tegument. Exp. Parasitol. 30: 418–425.

Brooker, B. E., 1972. The sense organs of trematode miracidia. In E. U. Canning & C. A. Wright (eds), Behavioural aspects of Parasitic Transmission. Academic Press, London: 171–180.

Butterworth, A. E., D. W. Taylor, M. C. Veuth, M. A. Vadas, A. Dessein, R. F. Sturrock, & E. Wells, 1982. Studies on the mechanisms of immunity in human schistosomiasis. Immunol. Rev. 61: 5–39.

Carter, O. S. & B. J. Bogitsh, 1975. Histologic and cytochemical observations of the effects of Schistosoma mansoni on Biophalaria glabrata. Ann. N.Y. Acad. Sci. 266: 380–393.

Crane, R. K., 1968. A concept of the digestive-absorptive physiology. Am. phys. Soc. 5: 2535–2542.

Dike, S. C. & C. P. Read, 1971. Relation of tegumentary phosphohydrolase and sugar transport in Hymenolepis diminuta. J. Parasitol. 57: 1251–1255.

Dusanic, D. G., 1959. Histochemical observations of alkaline phosphatase in Schistosoma mansoni. J. inf. Dis. 105: 1–8.

Goldring, O. L., J. A. Clegg, S. R. Smithers, & R. J. Terry, 1976. Acquisition of human blood group antigens by Schistosoma mansoni. Clin. exp. Immunol.

Halton, D. W., 1967. Observations on the nutrition of digenetic trematodes. Parasitology 57: 639–660.

Heyneman, D., W. P. Faulk & H. H. Fudenberg, 1971. Echinostoma lindowense: larval antigens from the snail intermediate host, Biomphalaria glabrata. Exp. Parasitol. 29: 480–492.

Hockley, D. J., 1973. Ultrastructure of the tegument of Schistosoma. In B. Dawes (ed.), Advances in Parasitology. Vol. II. Academic Press, New York (N.Y.): 233–305.

Hockley, D. J. & D. J. McLaren, 1973. Schistosoma mansoni: changes in the outer membrane of the tegument during development from cercaria to adult worm. Int. J. Parasitol. 3: 13–25.

Hoskin, G. P. & T. C. Cheng, 1974. Himasthla quissetensis: uptake and utilization of glucose by rediae as determined by autoradiography and respirometry. Exp. Parasitol. 35: 61–67.

Kawanaka, M. & C. E. Carter, 1984. Schistosoma japonicum egg metabolism: utilization of glucose and amino acids during growth and development in vitro. J. Parasitol. (Ms. submitted).

Kemp, W. M., 1970. Ultrastructure of the cercarienhullen reaktion of Schistosoma mansoni. J. Parasitol. 56: 713–723.

Koie, M., 1971. On the histochemistry and ultrastructure of the tegument and associated structures of the cercaria of Zoogonoides viviparus in the first intermediate host. Ophelia 9: 165–206.

Krupa, P. L. & B. J. Bogitsh, 1972. Ultrastructural phosphohydrolase activities in Schistosoma mansoni sporocysts and cercariae. J. Parasitol. 58: 495–514.

Krupa, P. L., L. M. Lewis, & P. DelVecchio, 1975. Electron microscopy of ouabain-inhibited potassium-dependent transport adenosine triphosphatase activity in schistosome sporocysts. Ann. N.Y. Acad. Sci. 266: 465–479.

McLaren, D. J. & D. J. Hockley, 1977. Blood flukes have a double outer membrane. Nature 269: 147–149.

McManus, D. P. & B. L. James, 1975a. Tricarboxylic acid cycle enzymes in the digestive gland of Littorina saxatalis rudis (Maton) and in the daughter sporocysts of Microphallis similis (Jag.) (Digenea: Microphallidae). Comp. Biochem. Physiol. 50B: 491–495.

McManus, D. P. & B. L. James, 1975b. Anaerobic glucose metabolism in the digestive gland of Littorina saxatalis rudis (Maton) and in the daughter sporocysts of Microphallis similis (Jag.). Comp. Biochem. Physiol. 51B: 293–297.

Meuleman, E. A., 1972. Host-parasite inter-relationships between the freshwater pulmonate Biomphalaria pfeifferi and the trematode Schistosoma mansoni. Neth. J. Zool. 22: 355–427.

Morris, G. P., 1971. The fine structure of the tegument and associated structures of the cercaria of Schistosoma mansoni. Z. Parasitenkd 36: 15–31.

Pan, S. C-T., 1980. The fine structure of the miracidium of Schistosoma mansoni. J. invert. Path. 36: 307–372.

Parkening, T. A. & A. D. Johnson, 1969. Glucose uptake in Haematoloechus medioplexus and Gorgoderina trematodes. Exp. Parasitol. 25: 358–367.

Read, C.P., 1970. Parasitism and symbiology. The Ronald Press, N.Y., 316 pp.

Rogers, S. H. & E. Bueding, 1975. Anatomical localization of glucose uptake by Schistosoma mansoni adults. Int. J. Parasitol. 5: 369–371.

Shannon, W. A. & B. J. Bogitsh, 1971. Megalodiscus temperatus: comparative auto-radiography of glucose-^{3}H and galactose-^{3}H incorporation. Exp. Parasitol. 29: 309–319.

Smithers, S. R., R. J. Terry, & D. J. Hockley, 1969. Host antigens in schistosomiasis. Proc. R. Soc. (B.) 171: 483–494.

Smyth, J. D., 1973. Some interface phenomena in parasitic protozoa and platyhelminthes. Can. J. Zool. 51: 367–377.

Smyth, J. D. 1976. An introduction to animal parasitology, 2nd ed. John Wiley & Sons, N.Y., 466 pp.

Smyth, J. D. & D. W. Halton, 1983. The physiology of trematodes, 2nd ed. Cambridge Univ. Press, N.Y., 446 pp.

Threadgold, L. T., 1963. The ultrastructure of the 'cuticle' of Fasciola hepatica. Exp. Cell Res. 30: 240–242.

Threadgold, L. T., 1976. Fasciola hepatica: ultrastructure and histochemistry of the glycocalyx of the tegument. Exp. Parasitol. 39: 119–134.

Threadgold, L. T., 1979. A bit of a fluke. New Lecture Series No. 116, Mayne, Boyd & Son, Belfast: 3–34.

Uglem, G. & C. P. Read, 1975. Sugar transport and metabolism in Schistosoma mansoni. J. Parasitol. 61: 390–397.

Wikel, S. K. & B. J. Bogitsh, 1974. Schistosoma mansoni: penetration apparatus and epidermis of the miracidium. Exp. Parasitol. 36: 342–354.

Wilson, R. A., 1969. Fine structure of the tegument of the miracidium of Fasciola hepatica L. J. Parasitol. 55: 124–133.

Yoshino, T. P., T. C. Cheng & L. R. Renwrantz, 1977. Lectin and human blood group determinants of Schistosoma mansoni: alteration following in vitro transformation of miracidium to mother sporocyst. J. Parasitol. 63: 818–824.

Occurrence of symbiotic turbellarians in the oyster, *Crassostrea virginica*

Lesley C. Fleming
Department of Biology, University of New Brunswick, Fredericton, N.B., E3B 6El, Canada

Keywords: *Urastoma*, Turbellaria, Mollusca, commensalism, *Paravortex*

Abstract

Crassostrea virginica was collected from several locations where it is cultured, both along the Northumberland Strait of New Brunswick and Malpeque Bay on the coast of Prince Edward Island. The oysters were found with two turbellarians on their gills. *Urastoma cyprinae* (Graff) was found in the oysters mostly during the warmer months of the year in numbers averaging as high as 50 worms per host (N = 50) and with as much as 78% of the host population infected (N = 100). *Paravortex gemellipara* (Linton) was also found during warmer months, but much less frequently or abundantly.

Both male and female oysters were found to have *U. cyprinae*. No eggs or recent young of *U. cyprinae* were found in hosts; female-mature individuals of *P. gemellipara* with young were found from June through August.

Introduction

Urastoma cyprinae (Prolecithophora, Hypotrichinidae) and occasionally *Paravortex gemellipara* (Rhabdocoela, Graffillidae) are found as commensals on the gills of the oyster *Crassostrea virginica* (Gmelin). This mollusc is cultured in Shediac, on the Northumberland Strait off the north coast of New Brunswick and in the Malpeque Bay on the north coast of Prince Edward Island.
land.

U. cyprinae may not live year round in the oysters, as no eggs or egg cases have ever been found in them, and as *Urastoma* spp. have been reported previously as free-living (Marcus, 1951; Westblad, 1955). *Paravortex*, however, may complete its life cycle in the oyster host. The female worm retains the embryos within the gut and these are released as free-swimming embryos.

In the present paper, I report some preliminary data on the occurrence of *U. cyprinae* and *P. gemellipara* in oysters of Shediac and Malpeque Bay.

Material and methods

Samples were collected from Shediac, N.B. as follows: 20 oysters each month during July to October, 1978, and May, 1979; 50 oysters each month during August to November, 1983, and 100 oysters each month during June to August and November, 1982.

Samples were collected from Malpeque Bay, P.E.I. as follows: 25 oysters in July, 1983; 50 oysters in September, October, and December, 1982, and in January and May, 1983; 100 oysters each month during June to August and November, 1982.

No collections were possible during months when the ice was present in the collecting sites.

Oysters were examined as soon as possible after they were brought into the laboratory and were kept refrigerated with their shells closed until examination.

The shells, bodies, and gills of the oysters were examined using a Wild M5 dissecting microscope.

Turbellarians removed from oysters for histologi-

Hydrobiologia 132, 311–315 (1986).

cal study were fixed in 5% formalin in seawater.

Whole mounts were stained with acetic acid alum carmine or Gower's borax carmine. Specimens were embedded in wax, sectioned at 8–10 μm, and stained with haematoxylin and eosin, or Mallory's trichrome strain.

Sections were examined with a Wild M20 microscope and drawings made with the aid of a Wild drawing tube.

Results

Urastoma cyprinae

U. cyprinae (Fig. 1) is found in mucus on the gill surfaces and amongst the gill filaments of *Crassostrea virginica* (Figs 2 & 3).

This worm is a protandrous hermaphrodite. Juveniles found from June to January lacked either male or female reproductive organs. Some adult specimens found from September to January had developing testes, but none had developing eggs, and those collected during May and June had well-developed eggs and vitellaria, abundant sperm stored in the penis, but no visible testes. Sperm could be observed fertilizing the eggs at this time

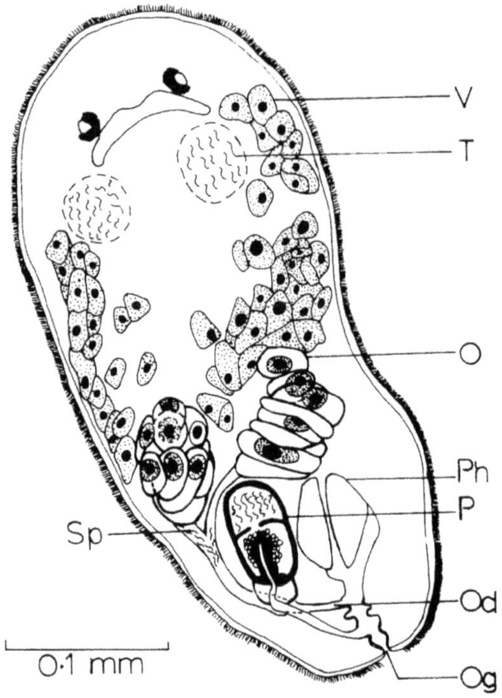

Fig. 1. Urastoma cyprinae. Composite diagram. (From Fleming *et al.*, 1981.) Abbreviations: A, Atrium; An, Antrum; G, Gill; GP, Gonopore; I, Intestine; MS, Mucus Strings; O, Ovary; OG, Ora-genital pore; Ph, Pharynx; Sp, Sperm; SV, Seminal Vesicle; T, Testes; V, Vitellaria; Va, Vagina.

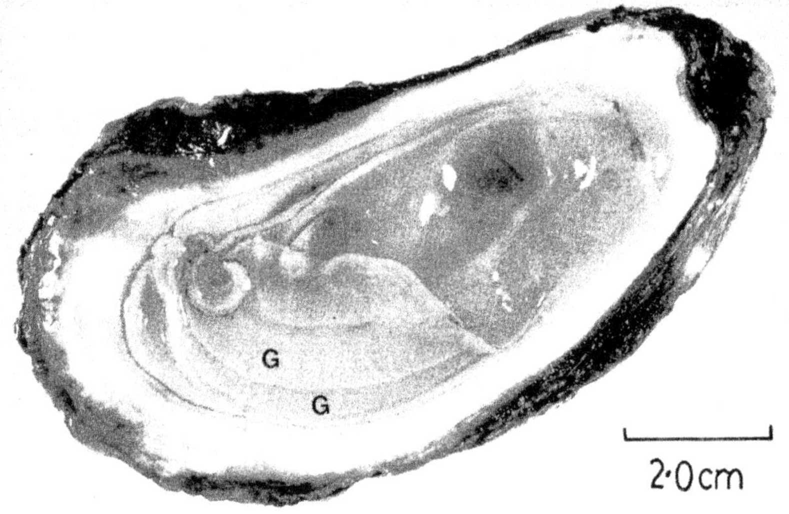

Fig. 2. Crassostrea virginica. One shell removed to expose gills. Abbreviations see Fig. 1.

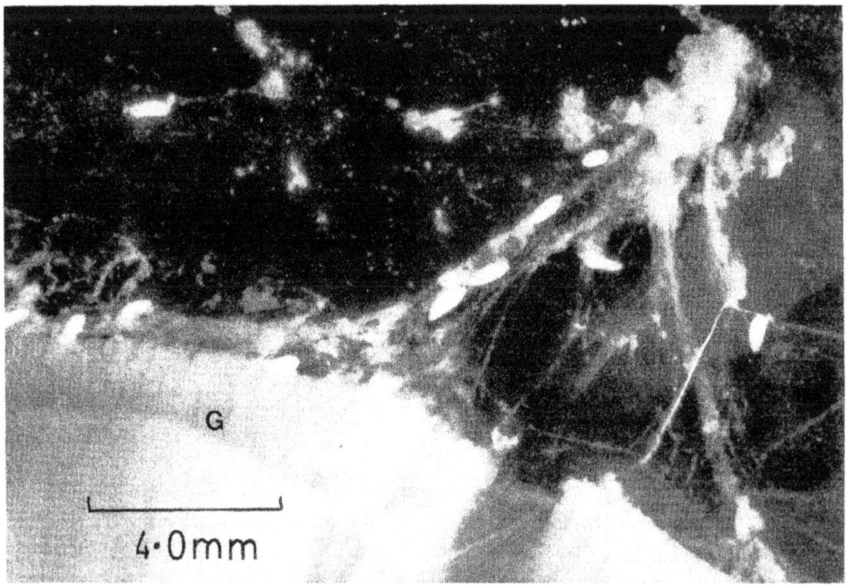

Fig. 3. Crassostrea virginica. Gills to show mucus and *Urastoma.* Abbreviations see Fig. 1.

(Fig. 1), but no eggs or egg cases were found in the host. The number of specimens of *U. cyprinae* decreased sharply from 1 416 in 61 of 100 oysters collected in August, 1982, to 11 worms in 6 of 50 oysters collected in January, 1983 (See Table I for prevalence and intensity). Lower prevalences and intensities of infection were noted immediately after the oysters had spawned, when their gills were fibrous and the shells porous, indicating a poor general condition. Also, fewer worms were obtained in those samples collected during the winter months.

Table 1. Occurrence, Prevalence, and Intensity of *Urastoma cyprinae* in *Crassostrea virginica.*

Date (& Number) Collected	Location*	Prevalence (%)	Intensity (Range)
June, 1978 (20)	S	50	2.3 (1 – 5)
July, 1978 (20)	S	60	4.0 (1 – 10)
August, 1978 (20)	S	10	1.0 (1 – 1)
September, 1978 (20)	S	-	-
October, 1978 (20)	S	-	-
February, 1979 (20)	S	-	-
July, 1979 (20)	S	10	1.5 (1 – 2)
June, 1982 (100)	M	18	5.3 (1 – 25)
July, 1982 (100)	M	2	2.5 (1.4)
August, 1982 (100)	M	61	23.2 (1 – 125)
September, 1982 (50)	M	78	20.5 (3 – 110)
October, 1982 (50)	M	38	34.9 (1 – 100)
November, 1982 (100)	M	52	23.8 (1 – 78)
December, 1982 (50)	M	12	4.6 (1 – 20)
January, 1982 (50)	M	12	1.8 (1 – 4)
May, 1983 (50)	M	-	-
July, 1983 (25)	M	-	-

*Location: S = Shediac, N.B.; M = Malpeque, P.E.I.

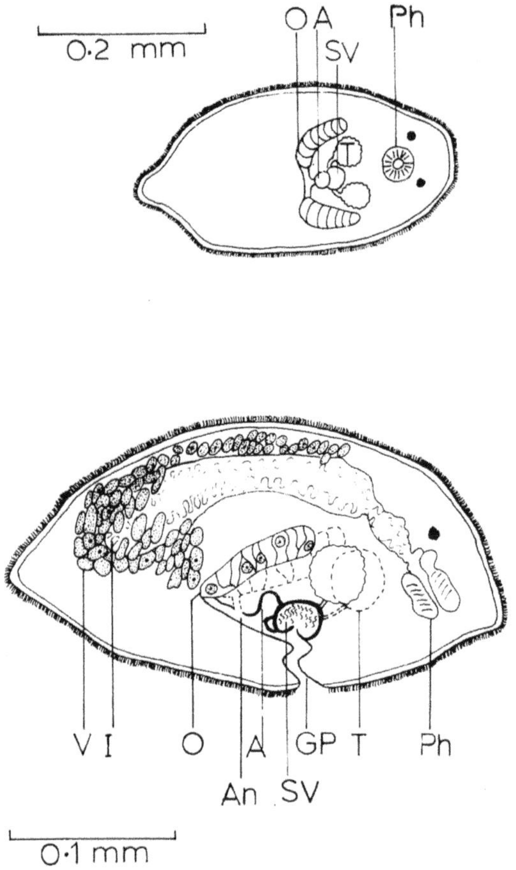

Fig. 4. Paravortex gemellipara. Composite diagram. Abbreviations see Fig. 1.

Paravortex gemellipara

P. gemellipara (Fig. 4), like *U. cyrinae* is found in mucus on the gill surfaces and amongst the gill filaments of *C. virginica.*

P. gemellipara was found only occasionally commensal with *C. virginica.* At least 1 000 oysters were collected from various locations in 1978–1979 and 1982–1983, including over 800 used in this study, from Shediac, N.B., and Malpeque, P.E.I. In the study locations, only four oysters collected in June, one oyster collected in July, and one collected in August were found to have *P. gemellipara.* Few specimens were collected from these oysters. Female worms, with encapsulated young, were found in June, July, and August. During July and August, in these same few oysters, both juveniles and worms with developing testes were also found.

Discussion

C. virginica is found in the Canadian maritime region (N.B., N.S., P.E.I.) and in the waters of the Atlantic Ocean and the Gulf of Mexico off the coast of the USA. These oysters do not spawn until the temperature reaches at least 18 °C and is rising. The spawning season usually lasts four to six weeks in any locality (Medcof, 1961).

Oysters may change sex, releasing eggs one year and sperm the next (Yonge, 1966). Contrary to an earlier report (Fleming *et al.*, 1980) that only female oysters, or those of indeterminate sex (prespawning), were found to have *U. cyprinae*, I have found that oysters of either sex carry the worm.

Although *C. virginica* will not feed when the water temperature is below 5 °C, and cannot feed fast enough for growth at 5–15 °C, they feed almost continuously at 15–18 °C (Medcof, 1961). In spring, spawn develops under these 15–18 °C conditions, and in autumn food is stored in tissues under similar conditions. In this Atlantic area, feeding is only possible for half the year [May (10 °C) until October (10 °C)]. During the rest of the year, oysters cease activity, their shells are closed, and they do not filter water (Medcof, 1961).

This cycle correlates with the occurrence of *U. cyprinae*: I found larger numbers during the warmer months (up to 100 *Urastoma* per oyster in 61 of the 100 oysters collected in August, 1982) and smaller numbers during the colder months (11 *Urastoma* in 6 oysters collected in January, 1983). Few worms were found in the other winter samples. However, ice conditions precluded a sufficient number of samples during the winter months to provide statistically significant figures.

Most individuals of both *U. cyprinae* and *P. gemellipara* apparently leave the oysters during colder months. Marcus (1951) and Westblad (1955) reported *Urastoma* spp. on red algae and in mud.

It was often noted during this study that oysters exhibiting clean, mucus-free gills were never found to contain *U. cyprinae* or *P. gemellipara.* It seems probable that these worms feed on the mucus produced by the oyster as part of its feeding activity (Fig. 4), and leave the oyster when feeding is no longer possible.

Acknowledgments

Throughout this study assistance has been received from the following people: Dr. G. B. Bacon and C. P. Bagnell from Research and Productivity Council in Frederiction for assistance in obtaining *Crassostrea virginica* from New Brunswick sites, the Burleigh Brothers for oysters collected from Malpeque Bay, and Roger Smith for photographic assistance. Financial assistance was provided, during the earlier part of this study (1978–79), by an NSERC Operating Grant and a Fisheries Research Board Grant, both to Dr. M. D. B. Burt. During 1982–83, financial assistance was provided by a U.N.B. Research Grant to the author.

References

Fleming, L. C., M. D. B. Burt & G. B. Bacon, 1981. On some commensal Turbellaria of the Canadian East Coast. Hydrobiologia 84: 131–137.

Marcus, E., 1951. Turbellaria Brazileiros (9). Bolm. Fac. Filos. Cienc. Univ. São Paulo. (Zool.) 16: 5–215.

Medcof, J. C., 1961. Oyster Farming in the Maritimes. Bulletin. No. 131 of the Fisheries Research Board of Canada, 258 pp.

Westblad, E., 1955. Marine 'Alloecocoels' (Turbellaria) from North Atlantic and Mediterranean coasts. I. Ark. Zool. 7(24): 491–526.

Yonge, C. M., 1966. Oysters. Collins. London, 209 pp.

History of the study of Turbellaria in China. Part 1: Ages of Materia Medica and of early expeditions by westerners

Kuang Yang Lue[1] & Masaharu Kawakatsu[2]
[1] *Department of Biology, National Taiwan Normal University, 88 Section 5, Roosevelt Road, Taipei, Taiwan 117, Republic of China*
[2] *Biological Laboratory, Fuji Women's College, Kita-16, Nishi-2, Kita-ku, Sapporo (Hokkaidô) 001, Japan*

Keywords: Turbellaria, planaria, history, China, Taiwan, Japan

Abstract

The history of the study of turbellarians in China and Japan through the early twentieth Century is reviewed. Up to the middle of the 19th Century, knowledge of the natural history of these countries, including several records of planarians from each, is to be found only in manuscripts and books of the so-called 'Materia Medica.' During the latter half of the 19th Century and into the early part of the 20th Century, several Western zoologists laid the foundation for the Western approach to the study of turbellarians in China and Japan.

Introduction

In 1969, Kawakatsu published an article on records of planarians found in old Chinese and Japanese books of natural history (see also Kawakatsu, 1973; 1983). Since this publication, Kawakatsu has obtained additional references, the discovery of these newer references has been aided especially by the availability in recent years of annotated republications cf old Chinese and Japanese books in Japanese libraries. We present here a historical review of the discovery of turbellarians in prescientific China and Japan, together with some corrections to the 1969 article of Kawakatsu. In this account, the names of Chinese persons, books, and places are romanized according to Wade's method. Their sounds in Japanese are also given in parentheses following each romanized Chinese name according to Hepburn's method.

Records of planarians found in Chinese and Japanese Materia Medica

There are many kinds of manuscripts and wood-block books of natural history, collectively called Materia Medica, prepared and published in China. The oldest of such manuscripts is known as "Shên-Nung P'ên-Ts'ao-Ching" (Jin'no-Honzô-kyô in Japanese phonetics), which was compiled in the 5th Century by Hong-Ying Tao (Kôkei Tô). It has been republished both in China (1891) and Taiwan (1969). A more comprehensive manuscript entitled "Hsin-Hsiu P'ên-Ts'ao" (Shinshû-Honzô) was written in the 6th Century by Chin Su (Kei So) and his collaborators. Several handwritten copies of these manuscripts in Chinese were introduced into Japan by envoys to China in the Sui and T'ang Dynasties (ca. 581–907; according to Japanese records, envoys to China were dispatched between 607 and 894). The contents of these old Chinese manuscripts can also be found in several editions of wood-block books published in the 18th Century in Japan based upon the manuscript written by the Japanese herb doctor Sukehito Fukae (or Fukane) in 918. These contain short notes on two species of water-leech and a grass-leech.

In the middle of the 9th Century, Ch'êng-Shih T'uan (Seishiki Dan; ?–863?), the literary man of the T'ang Dynasty, wrote "Yu-Yang Tsa-Tsu" (Yûyô

Hydrobiologia 132, 317–322 (1986).

Zasso), a collection of tales of a wizard including essays. The final manuscript was probably completed in 860. The following description is found in volume 17 of this serial manuscript (literal translation from the Japanese edition of T'uan's book published in 1981):

"T'u-K'u" or "To-K'u" (Doko). "It is like a belt in shape. Its coloration is similar to that of an earthworm, and it is approximately 60 cm in length. The head is of a flat chisel-shape, and the body has black and yellow folds (or patterns?) on the dorsal surface. If we touch it, the animal will easily separate into several pieces. The animal usually chases earthworms. When the earthworm is captured, the animal covers it. The captured earthworm melts within a few hours; its body becomes a syrupy slaver. The animal is poisonous. Chickens that peck the animal die immediately. This animal is called T'u-K'u."

The Chinese animal mentioned above appears to be a large-sized species of land planarian to the genus *Bipalium*. As far as we know, this is the oldest record of planarians in the world, approximately 1 000 years prior to their recording by any Western people. Land planarians are frequently found in the warm and humid areas of China, and also in Central and Southern Japan. The author Ch'êng-Shih T'uan lived in the Prefecture of Shan-Tung Hsing (Santô-shô) in mid-eastern China (ca. lat. 34° to 38° N), while the "T'ien-Shê" of Shin-Chên Li is from the more southern Prefecture of Shih-Chiang Hsing (Sekikô-shô; ca. lat. 27° to 31° N), so Ch'êng-Shih T'uan may not have had direct personal knowledge of this animal.

In the Sung Dynasty (ca. 960–1279), several woodblock books of Materia Medica were also completed (for example, Tsung-Shieh K'ou's "Pên-Ts'ao Yen-I" written in 1185; Sôseki Kô's Honzô-Engi). The contents of these are almost identical with those of the previous manuscripts written in the Sui and T'ang Dynasties.

In 1590 (or 1593), during the Ming Dynasty, a monumental woodblock book of Materia Medica was published by Shin-Chên Li (Jichin Li; 1518–1593) and his son. This bulky book (52 volumes) in Chinese entitled "P'ên-Ts'ao Kang-Mu" (Honzô Kômoku) includes 1892 entries describing various plants, animals, minerals, and natural phenomena. Volume 40 contains descriptions of several species of water-leech, mud-leech, grass-leech, stone-leech, and mountain-leech. A description of the "T'ien-Shê" (Tenja), presumably a land planarian (see below), is also included in volume 43. This volume consists of two sections, 'Dragons' (9 species) and 'Snakes' (17 species). Although some of these are imaginary Chinese animals, most are apparently real species.

Read (1934: 348) presented an English translation of the "T'ien-Shê" article in "P'ên-Ts'ao Kang-Mu." He translated its name as the "Chekiang Snake" (i.e. a snake of Shin-Chiang Hsing (Sekikô-shô), or a Prefecture of mid-eastern China). He identified the "T'ien-Shê" as a land planarian of the genus *Bipalium* (Read, 1934: 300, 348, 361). Tu (1940: 202), who cited Read's (*op. cit.*) translation of this "T'ien-Shê" article, also considered the animal to be a species of *Bipalium* (see also Kawakatsu, 1969: 31–32). Because Read's English version is a free translation, we present here a literal English translation of the original: "The T'ien-Shê lives in shaded and moist places. It appears when rains have ceased. It is greatly feared by the Chekiang people. The animal is like a chopstick in shape, flat, and measures up to 90–120 cm in length. The coloration is yellowish red. If vinegar is poured on it, the animal dissolves very soon. It also will die if lime is sprinkled on the body." Following this portion of the manuscript, there are uncertain and puzzling descriptions concerning injury to man caused by the "T'ien-Shê." We fully agree with Tu's (1940) opinion that this is a large *Bipalium*.

A set of original copies of Li's book was introduced into Japan in 1607, and woodblock reprints were published in Japan in the 17th and 18th Centuries (1637, 1653, 1669, 1672, 1714, etc.). In addition, various new editions of 'Japanized' Materia Medica were published in Japan during the 17th through the 19th Centuries: "Tashiki-Hen" in 1612 by Razan Hayashi (1583–1657), "Kinmô-Zui" in 1666 by Tekisai Nakamura (1629–1702), "Kôsei-Honzô-Kômoku" in 1672 and "Yamato-Honzô" by Ekiken Kaibara (1630–1714), "Wakan-Sansai-Zue" in 1713 by Ryôan Terajima (herb doctor of the 18th Century), "Zôho-Kashiragaki Kinmô-Zui-Taisei" in 1789 by Kanbei Murakami and others, "Honzô-Kômoku-Keimô" in 1803 by Ranzan Ono (1729–1810), etc. These old Japanese books were written in Chinese and/or in Japanese with Chinese characters. Chinese characters, it should be noted, were reserved for the ruling ('samurai') class in the

Edo era when Japan was under a policy of national isolationism; cursive letters were used by educated women of that class and by the Court, and square syllabaries were used by the trading class.

The contents of these old Japanese books (and manuscripts) are almost identical with those of the Chinese books cited above. However, in some of these Japanese books there appear several new observations based on Japanese leeches and land planarians. The first Japanese to distinguish land planarians from leeches was Nakamura (1666). His book was the first illustrated encyclopedia of natural history published in Japan. Figures of leeches are found in volume 11 (p. 26) as well as figures of *Bipalium* (p. 27).

In 1713, Terajima employed Nakamura's classification system in his book, which is a more comprehensively illustrated encyclopedia than Nakamura's. A literal translation of Terajima's discussion of land planarians is as follows:

"Doko" or "Toku". "The animal has the shape of a Japanese belt in general appearance and is without legs. It measures up to 12 to 15 cm in length; a large specimen attains about 30 cm. The body is flattish in shape as a leaf of leek. There are yellow and black folds on the dorsal surface. The animal has a head shaped like a Japanese forceps. The creeping trace of a living animal appears as a white stripe. It lives in moist places. If the animal is touched, fission may occur. The animal usually chases an earthworm, and the captured worm becomes water. The animal is poisonous; chickens that peck it will die.... The author [*i.e.* Terajima] considered the shape of the head as similar to a Japanese ornamental bodkin for women's hair (*i.e.* 'kôgai'). The Japanese name for the animal is 'kôgai-biru.' The animal may belong to the group of leeches." If Terajima had not, himself, known of the Japanese *Bipalium* species, he might not have written this last statement.

Ono (1803) also classified *Bipalium* with the leeches (vol. 36, pp. 17–18), and he added several local names used for this animal in central and shoutwestern Japan. He cited the Chinese "T'ien-Shê" as an 'uncertain species' (vol. 39, p. 18). Some Japanese naturalists in the early 19th Century recognized two or three species of *Bipalium*. Several manuscript sketches of these animals in color remain (Kawakatsu, 1983: 52, Fig. 2).

It is possible that the discussions of water-leeches found in old Chinese and Japanese books may include some records of freshwater planarians. One of the European planariid species, *Crenobia alpina* (Dana, 1766), was described originally under the name "*Hirudo alpina.*" It is interesting that animal taxonomy of the 17th and 18th Centuries had turbellarians classified as members of the Hirudinea in both Eastern and Western societies (Kawakatsu, 1969, 1973, 1983).

Age of expeditions into the Far East by Westerners

Near the end of the 17th Century, countries of the Far East were visited by a considerable number of Western people such as diplomats, missionaries, medical doctors, botanists, zoologists, and explorers. The earliest systematic study of oriental flora and fauna was carried out by naturalists among them.

Cantor, in 1842, reported on the fauna and flora of Chusan (Choushan), the Choushan Islands located on the Bay of K'angchou (Kôshû), near Shanghai, in mid-eastern China. Cantor's species of "*Hirudo?*" are now known to be bipaliids, or land planarians (see below).

A much greater contribution to knowledge of the invertebrate fauna of China and Japan was given by William Stimpson who visited these countries during 1853 and 1854 as a zoologist accompanying the U.S. Surveying Expedition to the North Pacific, the Seas of Japan, etc., under Commander John Rodgers. In 1855, Stimpson published his first paper on new marine invertebrates from the Chinese and Japanese Seas. He described 17 species of turbellarians, of which 8 were from China (including Hong-Kong), and the others from Loo Choo (*i.e.* Ryûkyû or Okinawa) and Kikaishima Island in southwestern Japan. His Chinese species, all polyclads, are as follows: *Stylochus corniculatus, Leptoplana acuta, L. obscura* (non *S. obscurus* Stimpson 1857, known as *Pseudostylochus obscurus* (Stimpson, 1857), from North Japan), and *L. trullaeformis.*

In Stimpson's second paper on turbellarians, published in 1857, 42 species of polyclad turbellarians were described, of which most were from Hong-Kong, China, and the Seas of Japan. He listed 9 species from Hong-Kong: *Cryptocoelum opacum, Elasmodes actus (olim L. acuta), L. fusca, L.*

trullaeformis Stimpson, 1855, *L. delicatula, Prosthiostomum obscurum (olim L. obscura), P. affine, P. tenebrosum,* and *S. corniculatum* Stimpson, 1855.

Reidentification of Stimpson's (1855, 1857) species of Chinese polyclads is impossible because his descriptions are very meager and are without figures. Although Yeri & Kaburaki (1918) reidentified five Japanese species described by Stimpson, no Chinese species is included among them. Tu (1940) listed Stimpson's 10 species of Chinese polyclads in the section of "Zweifelhafte Arten," except for three in the foregoing list: *P. obscurum, P. tenebrosum,* and *L. obscura* Stimpson, 1855, the last being omitted from his paper. There is not now sufficient knowledge available to provide a more detailed discussion of the current taxonomic status of these Chinese polyclad species.

Stimpson (1857) also described 10 species of triclad turbellarians (freshwater, marine, and terrestrial) from China, Japan and Kamtschatca, including two from Hong-Kong: *Planaria Sinensis* and *Fo via graciliceps. Planaria Sinensis* and the other two species Stimpson reported from "Ousima" (i.e. Amami-Ôshima Island) in Southwest Japan, *P. cinerea* and *P. badia*, seem to be identical with *Dugesia japonica* Ichikawa et Kawakatsu, 1964, a polymorphic form and the most common planarian found in the northeastern portion of the Far East, including the Japanese Islands. Our opinion is that it would be better to supress the names of Stimpson's species as *nomina oblita* than to rename the now well-established *D. japonica* (see Kawakatsu & Wong, 1975; Kawakatsu *et al.*, 1976). *'Fovia graciliceps'* is now classified as *Procerodes graciliceps* (Stimpson, 1857), a free-living marine triclad (Wilhelmi, 1909; Tu, 1940).

We return now to the taxonomic history of the Chinese and Japanese land planarians. Stimpson (1857) described five species of land planarians from Japan: *Geoplana lapidicola* and *Bipalium virgatum* from "Loo Choo" (i.e. Ryûkyû or Okinawa in Southwest Japan), *B. maculatum* from "Ousima" (i.e. Amami-Ôshima Island), *B. fuscatum* from 'Simoda' (i.e. Shimoda City, Izu Peninsula, in Shizuoka Prefecture, Central Japan), and *B. trilineatum* from "Jesso" (i.e. Hakodate City in Hokkaidô in North Japan). All of these descriptions were based upon external features.

Cantor's (1842) two species of Annelides, or "*Hirudo?*" (see above) were described by Wright (1860) as "*Dunlopea Grayia*" (=*Bipalium grayi* (Wright, 1860)). The former was recorded from Chou-Shan in China (Cantor, 1841, 1842; see also Wright, 1860; Diesing, 1862; Moseley, 1875, 1877), from Assam (Naga Hills), Bengal in India, and Ceylon (Cantor, *op. cit.*; Wright, *op. cit.*; Diesing, *op. cit.*; Stimpson, 1861; Moseley, *op. cit.*, 1878; Loman, 1888; Von Graff, 1899). The latter was discovered first by Mr. Fortune in the 1830's from the Island of Nin-P'o, Chou-Shan, in China (Cantor, *op. cit.*; Wright, *op. cit.*; Diesing, *op. cit.*; Stimpson, *op. cit.*; Humbert & Claparède, 1862; Moseley, *op. cit.*; Loman, *op. cit.*; Von Graff, *op. cit.*; Tu, 1940).

Stimpson (1861) noted a species of *Bipalium* that he discovered in 1854 in the garden of Sir J. Browring in Hong-Kong. He presented a generic definition of his genus *Bipalium* Stimpson, 1857, and he gave a description of the external appearance of this species (Browring's 'ground-leech'), but he did note name it. Stimpson (1861: 135) wrote, "The original species observed in China was not found named in the Synopsis, as the specimens of that species were unfortunately lost." This species was named "*Bipalium Stimpsoni*" by Diesing (1862) (see also Moseley, 1875, 1877; Loman, 1888; Von Graff, 1899; Tu, 1940).

Some of the collecting sites of Chinese and Japanese turbellarians listed by Stimpson (1855, 1857) are described by Heine (1858–1859).

Reidentification of these Chinese and Japanese land planarians is most difficult because their genital anatomy is not known. Moreover, there are several bipaliid species having one to five longitudinal stripes on the dorsal and/or ventral surface(s) now known from various areas of Japan, including the Southwest Islands (Islands of Amami, Okinawa and Yaeyama), and from Taiwan (Kawakatsu, 1981, 1983). A more detailed discussion of the Chinese and Japanese (as well as Korean) bipaliid species can be found in Kawakatsu *et al.* (1982).

In the latter part of the 19th Century and the early part of the 20th Century, several papers on Chinese and Japanese turbellarians were published by westerners who came to these countries and/or studied materials from them based on the collections of European naturalists.

As for polyclads, Plehn (1896) reported on the collection made by the expedition of Captain Chierchia. He described *Plagiotata promisca*

Plehn, 1896, from the coast of Hong-Kong. Later, two Chinese polyclads were described by the Swedish zoologist, Bock: *Stylochus pusilla* Bock, 1913, from Hong-Kong, and *S. orientalis* Bock, 1913, from off Fou-Chou (Fukushû) of the Tai-Wan Hai-Hsia, or the Formosa Strait (Bock, 1913, see also Katô, 1944). Bock, who visited Japan during 1914, also described six Japanese species of polyclads in 1922, 1923 (a, b), and 1924 (see Katô, 1944; Kawakatsu, 1983).

As for land planarians, Von Graff (1899) reported the occurrence of *Bipalium kewense* Moseley, 1878, in Hong-Kong under the name of *Placocephalus kewensis. Bipalium hilgendorfi* (Von Graff, 1899) (*olim P. hilgendorfii*), a Japanese land planarian from the vicinity of Tôkyô, was also described in his monograph based upon the collection of Hilgendorf, who was in Japan from 1873 to 1876 as a visiting professor of the Tôkyô Medical College (now the Faculty of Tôkyô University).

As for freshwater turbellarians, several specimens were taken during the expedition to collect aquatic animals of Tibet (Hsi-Tang) made by Captain F. H. Stewart (Stewart, 1911: 68). Meixner & Muth (1911), who studied Stewart's collection, reported two species: *Mesostoma craci* Schmidt, 1858, from Mang-Tsa, and *Sorocelis* sp. from Tsuang, Tering Gompa, and High Hill Gompa. The latter was again mentioned by Muth (1912). This form was classified as a synonym of *Polycelis tibetica* Hyman, 1934, a freshwater triclad from Indian Tibet.

Arndt (1918) reported the occurrence of '*Planaria alpina* Dana' in the vicinity of Manchouli (Manshûri), on the western side of the Tai-Hsing-An-Ling, or the Great Kingan Range, in the northeast of China. This record was listed in Tu's (1940) paper as *Crenobia alpina* (Dana, 1766), a species distributed only in Europe and Asia Minor. Arndt's form from the Far East is a *species inquirenda*.

Acknowledgments

The authors wish to express their gratitude to Dr. Robert W. Mitchell of Texas Tech University, Lubbock, Texas, U.S.A., for his careful reading of the manuscript of this paper and his suggestions on English usage.

References

References of Chinese and Japanese Materia Medica and their republications are not listed here (see Kawakatsu, 1969, 1973, 1983; Kawakatsu & Lue, 1984).

Arndt, W., 1918. Zur Kenntnis Verbreitung von Planaria alpina Dana. Zool. Anz. 50: 100–105.

Bock, S., 1913. Studien über Polycladen. Zool. Bidr. Upps. 2: 31–344.

Bock, S., 1922. Two new cotylean genera of polyclads from Japan and remarks on some other cotyleans. Ark. Zool. 14: 1–31.

Bock, S., 1923a. Two new acotylean polyclads from Japan. Ark. Zool. 15: 1–39.

Bock, S., 1923b. Boninia, a new polyclad genus from the Pacific. Nova Acta Reg. Soc. Sci. Upps. (4) 6: 1–32.

Bock, S., 1924. Eine neue Stylochoplana aus Japan. Ark. Zool. 16: 1–24.

Cantor, T., 1841. Conspectus of collections made by Dr. Cantor, assistant surgeon, during his employment with H.M. 26th regiment of the expedition to China, 1840. The Chinese Repository 10: 434–438. See also Calcutta J. nat. Hist. (5) for 1840.

Cantor, T., 1842. General features of Chusan, with remarks on the flora and fauna of that Island. Ann. Mag. nat. Hist. 9: 265–278.

Dana, J. P. M., 1766. De Hirudinis nova specie, noxa, remedi-isque adhibendis. Mélanges Philos. Mathém. Soc. r. Turin (Misc. Taurin.) 3: 199–205, pl. 6.

Diesing, K. M., 1862. Revision der Turbellarien. Abt. Dendrocoelen. Sitz. mat.-nat. Cl. Akad. Wiss., Wien 44: 485–578.

Heine, W., 1858–1859. Die Expedition in die Seen von China, Japan und Ochotsk, etc. Herman Gostenoble, Leipzig; 1: 20, 326 pp.; 2: 8, 365 pp.; 3: 7, 376 pp.

Humbert, M. A. & M. E. Claparède, 1862. Description de quelques espèces nouvelles de planaires terrestres de Ceylan par M. Aloïs Humbert suivie d'observations anatomiques sur le genre Bipalium par M. Edouard Claparède. Mém. Soc. Phys. d'Hist. nat. Genève 16: 293–311, pl. 1.

Hyman, L. H., 1934. Report on triclad Turbellaria from Indian Tibet. Mem. Connect. Acad. Arts Sci. 10: 5–12, pls. 1–2.

Katô, K., 1944. Polycladida of Japan. J. Sigenkagaku Kenkyûsyo, Tôkyô 1: 257–318, pls. 26–29. See also Kawakatsu, 1982.

Kawakatsu, M., 1969. A list of publications on Japanese Turbellarians (1968) – Including titles of publications on foreign Turbellarians written by the Japanese authors. Part 2: On some old records of Turbellarians found in the Japanese books printed in wood-block. Bull. Fuji Women's Coll. (7) (2): 30–43 (pls. 1–7).

Kawakatsu, M., 1973. Taxonomy and ecology of planarians, 1–4. The Nature & Animals (Dôbutsu-to-Shizen), Tôkyô 3 (5): 9–16, pl. 1; (6): 11–16; (7): 8–11; (8): 7–13. (In Japanese).

Kawakatsu, M., 1981. A history of the taxonomic study on Japanese land planarians and future problems. Zool. Mag., Tôkyô 90: 684 (in Japanese).

Kawakatsu, M., 1982. A commentary on 'Polycladida of Japan', a paper by the late Dr. Kôjirô Katô, published in 1944, with its reprint edition. Special Publ., Biol. Lab. Fuji Women's Coll., Sapporo (Hokkaidô), Japan. Toppan Printing Co., Sapporo, 88 pp.

Kawakatsu, M., 1983. Notes on land planarians. The Heredity (Iden), Tôkyô 37 (9): 52–61 (in Japanese).

Kawakatsu, M. & K.-Y. Lue, 1984. History of the study of Turbellaria in China. Part 2: Age of studies by Japanese and Chinese turbellariologists. Bull. Fuji Women's Coll. (22) (2): 105–117.

Kawakatsu, M., N. Makino & Y. Shirasawa, 1982. Bipalium nobile sp. nov. (Turbellaria, Tricladida, Terricola), a new land planarian from Tokyo. Annot. zool. Japan. 55: 236–262.

Kawakatsu, M., I. Oki, S. Tamura & H. Sugino, 1976. Studies on the morphology, karyology and taxonomy of the Japanese freshwater planarian Dugesia japonica Ichikawa et Kawakatsu, with a description of a new subspecies, Dugesia japonica ryukyuensis subspec. nov. Bull. Fuji Women's Coll. (14) (2): 81–126.

Kawakatsu, M. & M.-H. Wong, 1975. The freshwater planaria from Hong Kong. Annot. zool. Japon. 48: 262–273.

Loman, J. C. C., 1888. Über den Bau von Bipalium Stimpson, nebst Beschreibung neuer Arten aus dem indischen Archipel. Bijdr. Dierk. 14: 61–88.

Meixner, A. & A. Muth, 1911. Report on a collection of aquatic animals made in Tibet by Captain F. H. Stewart, I.M.S., during the year 1907. Part 3. Turbellaria and summary. Rec. ind. Mus. 6: 57–65, pl. 4.

Moseley, H. N., 1875. On the anatomy and histology of the landplanarians of Ceylon, with some account of their habits, and a description of two new species, and with notes on the anatomy of some European aquatic species. Phil. Trans. r. Soc., Lond. for 1875: 105–171, pls. 10–15.

Moseley, H. N., 1877. Notes on the structure of several forms of land planarians, with a description of two new genera and several new species, and a list of all species at present kwown. Quart. J. Micros. Sci. N.S. 17: 273–292, pl. 20.

Moseley, H. N., 1878. Description of a new species of landplanarian from the hothouses at Kew Gardens. Ann. Mag. nat. Hist., 1: 237–239.

Muth, A., 1912. Beiträge zur Kenntnis der Gattung Sorocelis Grube. Mitt. naturw. Ver. Steiermark 8: 381–410.

Plehn, M., 1896. Neue Polycladen, gesammelt von Hernn Kapitän Chierchia bei der Erdumschiffung de Korvette Vettor Pisani, von Herrn Prof. Dr. Kükenthal im nördlichen Eismeer und von Herrn Prof. Dr. Semon in Java. Z. Naturwiss. 30: 137–176.

Read, B. E., 1934. Chinese Materia Medica, 7. Dragons and snakes. Peking nat. Hist. Bull. 8: 297–362.

Stewart, F. H., 1911. List of the aquatic animals hitherto recorded from the provinces of Tsang and Ü in Central Tibet, with a table showing their geographical distribution. Rec. ind. Mus. 6: 67–72.

Stimpson, W., 1855. Descriptions of some of the new marine invertebrata from the Chinese and Japanese Seas. Proc. Acad. nat. Sci. Philad. 7: 375–384.

Stimpson, W., 1857. Prodromus descriptionis animalium evertebratorum quae in expeditione ad Oceanum, Pacificum Septentrionalem a Republica Federata missa, Johanne Rodgers Duce, observavit et descripsit. Pars. 1. Turbellaria Dendrocoela. Proc. Acad. nat. Sci. Philad. 9: 19–31.

Stimpson, W., 1861. On the genus Bipalium. Am. J. Sci. Arts. 2 Ser. 31: 134–135.

Tu, T.-J., 1940. Geschichtlicher Überblick über das Studium der Turbellarien in Ostasien und Stand unserer Kenntnisse von Diesen. Zool. Jb. Syst. Okol. Geogr. Tiere 73: 201–260.

Von Graff, L., 1899. Monographie der Turbellarien. 2. Tricladida Terricola (Landplanarien); Atlas von achtundfünfzig Tafeln zur Monographie der Turbellarien, etc. Verlag von Wilhelm Engelmann, Leipzig, 14, 574 pp.; pls. 1–58.

Wilhelmi, J., 1909. Tricladen. Fauna und Flora des Golfes von Neapel und der Angrenzenden Meeres-Abschnitte, Monogr. (32). Verlag von R. Friedländer & Sohn, Berlin, 12, 405 pp., pls. 1–16.

Wright, E. P., 1860. Notes on Dunlopea. Ann. Mag. nat. Hist. 6 (3): 54–56.

Yeri, M. & T. Kaburaki, 1918. Descriptions of some Japanese polyclad Turbellaria. J. Coll. Sci. imp. Univ. Tokyo, 28 (9): 1–54, pls. 1–2.

Electron microscopy of turbellarian platyhelminths: a bibliography

Compiled by Seth Tyler[1], Julian P. S. Smith III[1], Reinhard M. Rieger[2], Ulrich Ehlers[3] & Vittorio Gremigni[4]
[1] *Department of Zoology, University of Maine, Orono, ME 04469, U.S.A.*
[2] *Department of Biology, University of North Carolina, Chapel Hill, NC 27514, U.S.A. [present address: Zoologisches Institut, Universität Innsbruck, A-6020 Innsbruck, Austria]*
[3] *II. Zoologisches Institut der Universität, Berliner Strasse 28, D-3400 Göttingen, Federal Republic of Germany*
[4] *Istituto di Zoologia e Anatomia Comparata, Università di Pisa, Via A. Volta 4, 56100 Pisa, Italy*

Keywords: Turbellaria, ultrastructure, electron microscopy

Achtelik, W., 1972. Electron microscopic studies on the genesis of cortical granules in the oocytes of Dendrocoelum lacteum (Müll.). Zool. pol. 21: 59–66.

Afzelius, B., 1966. Anatomy of the Cell. University of Chicago Press, Chicago & London. 127 pp.

Ax, P., 1984. Das phylogenetische System. Systematisierung der lebenden Natur auf Grund ihrer Phylogenese. G. Fischer, Stuttgart, New York. 349 pp.

Baccetti, B. & B. A. Afzelius, 1976. The biology of the sperm cell. Monographs in Developmental Biology 10: 1–254.

Banchetti, R. & V. Gremigni, 1973. Indirect evidence for neoblast migration and for gametogonia dedifferentiation in ex-fissiparous specimens of Dugesia gonocephala s.l. Accad. naz. Lincei 55: 107–115.

Bautz, A., 1975. Dégénérescences cellulaires naturelles ou provoquées en cas d'inhibition de la régénération chez Dendrocoelum lacteum. These de doctorat en specialité. Université Nancy I.

Bautz, A., 1977. Structure fine de l'epiderme chez des planaires triclades terrestries et paludicoles. Arch. Zool. exp. gén. 118: 155–172.

Bautz, A. & F. Stephan-Dubois, 1974. Aspects ultrastructuraux des teguments chez les planaires. Bull. Soc. zool. Fr. 100: 248.

Bedini, C., E. Ferrero & A. Lanfranchi, 1973a. Fine structure of the eyes in two species of Dalyelliidae (Turbellaria: Rhabdocoela). Monit. zool. ital. 7: 51–70.

Bedini, C., E. Ferrero & A. Lanfranchi, 1973b. The ultrastructure of the ciliary sensory cells in two Turbellaria Acoela. Tissue Cell 5: 359–372.

Bedini, C., E. Ferrero & A. Lanfranchi, 1975. Fine structural observations on the ciliary receptors in the epidermis of three otoplanid species (Turbellaria, Proseriata). Tissue Cell 7: 253–266.

Bedini, C., E. Ferrero & A. Lanfranchi, 1977. Fine-structural changes induced by circadian light-dark cycles in photoreceptors of Dalyelliidae (Turbellaria: Rhabdocoela). J. ultrastruct. Res. 58: 66–77.

Bedini, C. & A. Lanfranchi, 1974. The fine structure of photoreceptors in two otoplanid species (Turbellaria, Proseriata). Z. Morph. Tiere 77: 175–186.

Bedini, C. & A. Lanfranchi, 1981. Ultrastruttura degli occhi di Mesostoma ehrenbergii (Turbellaria, Typhloplanidae). Boll. zool. ital. 48: 20.

Bedini, C. & F. Papi, 1970. Peculiar patterns of microtubular organisation in spermatozoa of lower Turbellaria. In B. Baccetti (ed.), Comparative Spermatology. Academic Press, New York: 363–368. [Accad. naz. Lincei 137: 363–368.]

Bedini, C. & F. Papi, 1974. Fine structure of the turbellarian epidermis. In N. W. Riser & M. P. Morse (eds), Biology of the Turbellaria. McGraw-Hill, New York: 108–147.

Benazzi, M. & V. Gremigni, 1982. Developmental biology of triclad turbellarians (Planaria). In F. W. Harrison & R. R. Cowden (eds), Developmental biology of freshwater invertebrates. Alan R. Liss, New York: 151–211.

Hydrobiologia 132, 323–343 (1986).

Bergstrom, B. H., 1973. An ultrastructural study of spermiogenesis in Mesostoma georgianum. Ph.D. thesis, University of North Carolina, Chapel Hill.

Bergstrom, B. H., C. Henley & D. P. Costello, 1973. Particulate flagellar and ciliary necklaces revealed by the use of freeze-etch. Cytobios 7: 51–60.

Best, J. B., M. Morita & J. Noel, 1968. Fine structure and function of planarian goblet cells. J. ultrastruct. res. 24: 385–397.

Best, J. B. & J. Noel, 1969. Complex synaptic configurations in planarian brain. Science 164: 1070–1071.

Boguta, K. K. & V. K. Lebskii, 1975. Study of invertebrate soft tissues using the smear technique in the light and electron microscope. Tsitologiya 17: 735–736.

Boguta, K. K. & Yu. V. Mamkaev, 1972. Kvoprosu o strukture parenkhimy beskishechnykh turbellyarii. [Structure of the parenchyma of acoelous turbellarians (In Russian). English abstract in Biol. Abstr. 55: 4486.] Vestn Leningr. Univ. Ser. Biol. 27: 15–29.

Bowen, I. D., 1980. Phagocytosis in Polycelis tenuis. In D. C. Smith & Y. Tiffon (eds), Nutrition in the lower metazoa. Pergamon Press, Oxford. 180 pp.

Bowen, I. D., J. E. de Hollander & G. H. J. Lewis, 1982. Cell death and acid phosphatase activity in the regenerating planarian Polycelis tenuis Iijima. Differentiation 21: 160–167.

Bowen, I. D. & T. A. Ryder, 1973. The fine structure of the planarian Polycelis tenuis Iijima. I. The pharynx. Protoplasma 78: 223–241.

Bowen, I. D. & T. A. Ryder, 1974a. Cell autolysis and deletion in the planarian Polycelis tenuis Iijima. Cell Tissue Res. 154: 265–274.

Bowen, I. D. & T. A. Ryder, 1974b. The fine structure of the planarian Polycelis tenuis Iijima. III. The epidermis and external features. Protoplasma 80: 381–392.

Bowen, I. D. & T. A. Ryder, 1976. Use of the p-nitrophenyl method for the demonstration of acid phosphatase during starvation and cell autolysis in the planarian Polycelis tenuis (Iijima). Histochem. J. 8: 319–323.

Bowen, I. D., T. A. Ryder & C. Dark, 1976. The effects of starvation on the planarian worm Polycelis tenuis (Iijima). Cell Tissue Res. 169: 193–210.

Bowen, I. D., T. A. Ryder & J. A. Thompson, 1974. The fine structure of the planarian Polycelis tenuis Iijima. II. The intestine and gastrodermal phagocytosis. Protoplasma 79: 1–17.

Bowen, I. D., T. A. Ryder & C. Winters, 1975. The distribution of oxidizable mucosubstances and polysaccharides in the planarian Polycelis tenuis (Iijima). Cell Tissue Res. 161: 263–275.

Boyer, B. C., 1972. Ultrastructural studies of differentiation in the oocyte of the polyclad turbellarian, Prostheceraeus floridanus. J. Morph. 136: 273–296.

Boyer, B. C. & G. W. Smith, 1982. Sperm morphology and development in two acoel turbellarians from the Phillipines. Pac. Sci. 36: 365–380.

Brandenburg, J., 1966. Die Reusenformen der Cyrtocyten: Eine Beschreibung von fünf weiteren Reusengeisselzellen und eine vergleichende Betrachtung. Zool. Beiträge (NF) 12: 345–417.

Brandenburg, J., 1975. The morphology of the protonephridia. Fortschr. Zool. 23: 1–17.

Brandenburg, J. & G. Kümmel, 1961. Die Feinstruktur der Solenocyten. J. ultrastruct. Res. 5: 437–452.

Bresciani, J. & M. Koie, 1970. On the ultrastructure of the epidermis of the adult female of Kronborgia amphipodicola Christensen & Kanneworff, 1964 (Turbellaria, Neorhabdocoela). Ophelia 8: 209–230.

Brüggemann, J., 1984a. Ultrastruktur und Differenzierung den prostatoiden Organe von Polystyliphora filum (Platyhelminthes, Proseriata). Zoomorphology 104: 86–95.

Brüggemann, J., 1984b. Ultrastruktur und Bildungsweise genitaler Hartstrukturen bei freilebenden Plathelminthen. Ph.D. thesis, Universität Göttingen.

Brüggemann, J., 1985. Ultrastructural investigations on the differentiation of genital hard structures in free-living platyhelminths and their phylogenetic significance. This volume.

Bunke, D., 1981a. Ultrastruktur der Schalendrusen von Microdalyellia fairchildi (Turbellaria, Neorhabdocoela). Zoomorphology 97: 101–119.

Bunke, D., 1981b. Ultrastruktur-Untersuchungen an Vitellocyten von Microdalyellia fairchildi (Turbellaria, Neorhabdocoela). Zoomorphology 99: 71–86.

Bunke, D., 1982. Ultrastruktur-Untersuchungen zur Eischalenbildung bei Microdalyellia fairchildi (Turbellaria). Zoomorphology 101: 61–70.

Burt, M. D. B. & G. H. Bance, 1981. Ultrastructure of the eye of Urastoma cyprinae (Turbellaria, Alloeocoela). Hydrobiologia 84: 276.

Burt, M. D. B. & L. C. Fleming, 1978. Phylogenetic clues in the ultrastructure of Turbellaria and Cestoda tegument. Parasitology 77: xxx–xxxi.

Burt, M. D. B. & W. L. Phillips, 1969. The fine structure of Urastoma cyprinae, a microturbellarian found in oysters. Am. Zool. 9: 622.

Burton, P. R. & M. Silveira, 1971. Electron microscopical and optical diffraction studies of negatively stained axial units of certain platyhelminth sperm. J. ultrastruct. Res. 36: 757–767.

Carpenter, K. S., M. Morita & J. B. Best, 1974a. Ultrastructure of the photoreceptor of the planarian Dugesia dorotocephala. I. Normal eye. Cell Tissue Res. 148: 143–158.

Carpenter, K. S., M. Morita & J. B. Best, 1974b. Ultrastructure of the photoreceptor of the planarian Dugesia dorotocephala. II. Changes induced by darkness and light. Cytobiologie 8: 320–338.

Chandebois, R., 1965. Cell transformation systems in planarians. In V. Kiortsis & H. A. L. Trampusch (eds), Regeneration in animals and related problems. North-Holland Publ., Amsterdam: 131–142.

Chandebois, R., 1973. The syncytial and intercellular nature of dedifferentiated material in the parenchyma of regenerated and starved planarians. Oncology 27: 356–384.

Chandebois, R., 1976. Histogenesis and morphogenesis in planarian regeneration. Monographs in Developmental Biology 11: 1–182.

Chien, P. & H. Koopowitz, 1972. The ultrastructure of neuromuscular systems in Notoplana acticola, a free-living polyclad flatworm. Z. Zellforsch. mikrosk. Anat. 133: 277–288.

Chien, P. K. & H. Koopowitz, 1977. Ultrastructure of nerve plexus in flatworms. III. The infraepithelial nervous system. Cell Tissue Res. 176: 335–347.

Christensen, A. K., 1961. Fine structure of an unusual spermatozoan in the flatworm Plagiostomum. Biol. Bull. 121: 416.

Clark Nichols, K., 1968. Comparative studies with Temnocephala and Udonella. M.A. Thesis. University of California, Berkeley, 105 pp.

Clark Nichols, K., 1975a. Observations on lesser-known flatworms: Temnocephala. Int. J. Parasitol. 5: 245–252.

Clark Nichols, K., 1975b. Observations on lesser-known flatworms: Udonella. Int. J. Parasitol. 5: 475–482.

Coomans, A., 1981. Phylogenetic implications of the photoreceptor structure. Atti Conv. Lincei 49: 23–68.

Costello, D. P., 1973a. A new theory on the mechanics of ciliary and flagellar motility. I. Supporting observations. Biol. Bull. 145: 279–291.

Costello, D. P., 1973b. A new theory on the mechanics of ciliary and flagellar motility. II. Theoretical considerations. Biol. Bull. 145: 292–309.

Costello, D. P., C. Henley & C. Ault, 1969. Microtubules in spermatozoa of Childia (Turbellaria, Acoela) revealed by negative staining. Science 163: 678–679.

Coward, S. J., 1973. Precocious myogenesis during regeneration in the planarian Dugesia dorotocephala. Am. Zool. 13: 1353–1354.

Coward, S. J., 1974. Chromatoid bodies in somatic cells of the planarian: observations on their behaviour during mitosis. Anat. Rec. 180: 533–546.

Coward, S. J., 1979. On the occurrence and significance of annulate lamellae in gastrodermal cells of regenerating planarians. Cell Biol. Int. Rep. 3: 101–106.

Coward, S. J., C. E. Bennett & B. L. Hazlehurst, 1974. Lysosomes and lysosomal activity in the regenerating planarian; evidence in support of dedifferentiation. J. exp. Zool. 189: 133–146.

Coward, S. J. & E. D. Hay, 1972. Fine structure of cells involved in physiological and reconstitutive regeneration in planarians. Anat. Rec. 172: 296.

Coward, S. J. & J. W. Piedilato, 1972. The behavioural significance of the rhabdite cells in planarians. J. biol. Psychol. 14: 5–7.

Coward, S. J. & R. O. Vitale-Calpe, 1971. Behaviorally significant surface specializations of the planarian, Dugesia dorotocephala. J. biol. Psychol. 13: 3–10.

Crezee, M., 1975. Monograph of the Solenofilomorphidae (Turbellaria: Acoela). Int. Revue ges. Hydrobiol. 60: 769–845.

Crezee, M. & S. Tyler, 1976. Hesiolicium inops n.g. n. sp. (Turbellaria, Acoela) with observations on its ultrastructure. Zool. Scr. 5: 207–216.

Curtis, S. K., R. R. Cowden, J. D. Moore & J. L.

Robertson, 1983. Histochemical and ultrastructural features of the epidermis of the land planarian Bipalium adventitium. J. Morph. 175: 171–194.

Czubaj, A., 1979. Ultrastructural distribution of AchE in Catenula leptocephala (Nuttycombe, 1956). Histochemistry 61: 189–198.

DeEguilior, M. & R. Valvassori, 1975. Fine structure of flatworm muscles: Dugesia tigrina (Girard) and Dugesia lugubris (Schmidt). Monit. zool. ital. 9: 37–50.

Doe, D., 1976. The proboscis hooks in Karkinorhynchidae and Gnathorhynchidae (Turbellaria, Kalyptorhynchia) as basement membrane or intracellular specializations. Zool. Scr. 5: 105–115.

Doe, D., 1977. Fine structure of 'cuticular' structures in Platyhelminthes. Am Zool. 17: 970.

Doe, D., 1978. Comparative ultrastructure of the pharynx simplex in Turbellaria. Ph.D. Dissertation, University of North Carolina, Chapel Hill.

Doe, D. A., 1981. Comparative ultrastructure of the pharynx simplex in Turbellaria. Zoomorphology 97: 133–193.

Doe, D. A., 1982. Ultrastructure of copulatory organs in Turbellaria. I. Macrostomum sp. and Microstomum sp. (Macrostomida). Zoomorphology 101: 39–59.

Doe, D. A., 1985. Ultrastructure of the copulatory stylet and accessory spines in Haplopharynx quadristimulus (Turbellaria). This volume.

Doe, D. A. & R. M. Rieger, 1977. A new species of the genus Retronectes (Turbellaria, Catenulida) from the coast of North Carolina, U.S.A. Mikrofauna Meeresbodens 66: 1–10.

Domenici, L., L. Galleni & V. Gremigni, 1975. Electron microscopical and cytochemical study of egg-shell globules in Notoplana alcinoi (Turbellaria, Polycladida). J. submicr. Cytol. 7: 239–247.

Domenici, L. & V. Gremigni, 1974. Electron microscopical and cytochemical study of vitelline cells in the fresh-water triclad Dugesia lugubris. II. Origin and distribution of reserve materials. Cell Tissue Res. 152: 219–228.

Domenici, L. & V. Gremigni, 1977. Fine structure and functional role of the coverings of the eggs in Mesostoma ehrenbergii (Focke) (Turbellaria, Neorhabdocoela). Zoomorphologie 88: 247–257.

Dorey, A. E., 1963. The cytology and microanatomy of acoelous turbellarians. Ph.D. Thesis, Cambridge University.

Dorey, A. E., 1965. The organization and replacement of the epidermis of acoelous turbellarians. Quart. J. microsc. Sci. 106: 147–172.

Douglas, A. E., 1983. Establishment of the symbiosis in Convoluta roscoffensis. J. mar. biol. Ass. U.K. 63: 419–434.

Duffy, J. E. & S. Tyler, 1983. Differences in mitochondrial ultrastructure between a thiobiotic and an oxybiotic turbellarian. Am. Zool. 23: 1026.

Duffy, J. E. & S. Tyler, 1984. Quantitative differences in mitochondrial ultrastructure of a thiobiotic and an oxybiotic turbellarian. Mar. Biol. 83: 95–102.

Duma, A., 1980. Activity in cAMP phosphodiesterase in the early regeneration stage of planarian Dugesia lugubris (O. Schmidt). Ultracytochemical studies. Acta Med. pol. 21: 317–318.

Duma, A. & J. Moraczewski, 1980. Ultracytochemistry of cyclic 3′,5′-nucleotide phosphodiesterase activity in the planarian Dugesia lugubris (O. Schmidt). Histochemistry 66: 211–220.

Durand, J. & N. Gourbault, 1975. Etude cytologique des organes photorécepteurs de Dendrocoelopsis (Amyadenium) chattoni (Triclade Paludicole Hypoge). Ann. Speleol. 30: 129–135.

Durand, J. P. & N. Gourbault, 1977. Etude cytologique des organes photorécepteurs de la planaire australienne, Cura pinguis. Can. J. Zool. 55: 381–390.

Eakin, R. M., 1963. Lines of evolution of photoreceptors. In D. Mazia & A. Tyler (eds), General Physiology of Cell Specialization. McGraw-Hill, N.Y.: 393–425.

Eakin, R., 1965. Evolution of photoreceptors. Cold Spring Harbor Symp. Quant. Biol. 30: 363–370.

Eakin, R. M., 1972. Structure of invertebrate photoreceptors. In H. J. A. Dartnell (ed.), Handbook of sensory Physiology, Vol. VII/1, Springer, Berlin, N.Y.: 625–684.

Eakin, R. M. & J. L. Brandenburger, 1981a. Fine structure of the eyes of Pseudoceros canadensis (Turbellaria, Polycladida). Zoomorphology 98: 1–16.

Eakin, R. M. & J. L. Brandenburger, 1981b. Unique

eye of probable evolutionary significance. Science 211: 1189–1190.

Ehlers, B., 1977. 'Trematoden-artige' Epidermisstrukturen bei einem freilebenden proseriaten Strüdelwurm. Acta zool. fenn. 154: 129–136.

Ehlers, B. & U. Ehlers, 1977a. Ultrastruktur pericerebraler Cilienaggregate bei Dicoelandropora atriopapillata Ax und Notocaryoplanella glandulosa Ax (Turbellaria, Proseriata). Zoomorphologie 88: 163–174.

Ehlers, B. & U. Ehlers, 1977b. Die Feinstruktur eines ciliaren lamellarkörpers bei Parotoplanina geminoducta Ax (Turbellaria: Proseriata). Zoomorphologie 87: 65–72.

Ehlers, B. & U. Ehlers, 1980. Struktur und Differenzierung penialer Hartgebilde von Carenscoilia bidentata Sopott (Turbellaria, Proseriata). Zoomorphologie 95: 159–167.

Ehlers, U., 1977. Vergleichenden Untersuchungen über Collar-Rezeptoren bei Turbellarien. Acta zool. fenn. 154: 137–148.

Ehlers, U., 1981. Fine structure of the giant aflagellate spermatozoon in Pseudostomum quadrioculatum (Leuckart) (Platyhelminthes, Prolecithophora). Hydrobiologia 84: 287–300.

Ehlers, U., 1984. Das phylogenetische System der Plathelminthes. Habilitationsschrift, Universität Göttingen.

Ehlers, U., 1985a. Phylogenetic relationships within the Platyhelminthes. In S. Conway Morris, J. D. George, R. Gibson & H. M. Platt (eds), Origins and relationships of lower invertebrates. University Press, Oxford.

Ehlers, U., 1985b. Comments on a phylogenetic system of the Platyhelminthes. This volume.

Ehlers, U. & B. Ehlers, 1977. Monociliary receptors in interstitial Proseriata and Neorhabdocoela (Turbellaria, Neoophora). Zoomorphologie 86: 197–222.

Ehlers, U. & B. Ehlers, 1978. Paddle cilia and discocilia – genuine structures? Cell Tissue Res. 192: 489–501.

Farnesi, R. M., 1975. Alcune osservazioni sul faringe di esemplari di Dugesia lugubris s.l. trattati con colcemid. [Some ultrastructural observations on the pharynx of Dugesia lugubris treated with colcemid.] Boll. Soc. ital. Biol. Sper. 51: 1177–1183.

Farnesi, R. M., M. Marinelli, S. Tei & D. Vagnetti, 1977. Ultrastructural research on spermatogenesis in Dugesia lugubris s.l. Riv. Biol. 70: 113–136.

Farnesi, R. M., M. Marinelli, S. Tei & D. Vagnetti, 1979. Localization de la phosphatase acide dans les cellules parietales de la bourse copulatrice chez Dugesia lugubris s.l. Etude ultrastructurale. Acta. embryol. exp. 1: 69–78.

Farnesi, R. M., M. Marinelli & D. Vagnetti, 1975. Osservazioni al ME sui rapporti tra neurosecrezione e ghiandole del cemento in Dugesia lugubris s.l. Boll. Soc. ital. Biol. Sper. 51: 329–334.

Farnesi, R. M. & R. Pascolini, 1973. Investigations on the ultrastructure of the pharyngeal glands in Dugesia lugubris s.l. Boll. zool. ital. 40: 137–147.

Farnesi, R. M. & S. Tei, 1980. Dugesia lugubris s.l. auricles: research into the ultrastructure and on the functional efficiency. Riv. Biol. 73: 65–77.

Ferrero, E., 1973. A fine structure analysis of the statocyst in Turbellaria Acoela. Zool. Scr. 2: 5–16.

Ferrero, E., C. Bedini & A. Lanfranchi, 1982. Ultrastructure of the statocyst in otoplanid turbellarians. Boll. zool. ital. 49: 76.

Ferrero, E., A. Lanfranchi & C. Bedini, 1983. Ultrastructural features of the nervous system in otoplanid turbellarians. Nova Thalassia 6: 43.

Franquinet, R., 1976. Etude comparative de l'évolution des cellules de planaires d'eau douce Polycelis tenuis (Iijima) dans des fragments dissociés en culture in vitro aspects ultrastructuraux incorporations de leucine et d'uridine tritiée. J. Embryol. exp. Morph. 36: 41–54.

Franquinet, R. & T. Lender, 1972. Quelques aspects ultrastructuraux de la spermiogenèse chez Polycelis tenuis et Polycelis nigra (Planaires). Z. mikrosk. anat. Forsch. 86: 481–495.

Franquinet, R. & T. Lender, 1973. Etude ultrastructurale des testicules de Polycelis tenuis et Polycelis nigra (planaires). Evolution des cellules germinales males avant la spermiogenèse. Z. mikrosk. anat. Forsch. 87: 4–22.

Franzén, A., 1967. Remarks on spermiogenesis and morphology of the spermatozoan among lower Metazoa. Ark. Zool. 19: 342–355.

Franzén, A., 1970. Phylogenetic aspects of the morphology of spermatozoa and spermiogenesis. In B. Baccetti (ed.), Comparative spermatology. Academic Press, N.Y.: 29–46. [Accad. naz. Lincei 137: 29–46.]

Franzén, A., 1977. Sperm structure with regard to fertilization biology and phylogenetics. Verh. dt zool. Ges. 1977: 123–138.

Gabriel, A. & A. Le Moigne, 1971. Action de l'actinomycine D sur la différenciation cellulaire au cours de la régénération de planaires qui viennent d'éclore. I. Etudes morphologiques, histologiques et ultrastructurales du pouvoir de régénération en presence de l'antibiotique. Z. Zellforsch. mikrosk. Anat. 115: 426–441.

Galleni, L. & V. Gremigni, 1984. Platyhelminthes-Turbellaria. In K. G. Adiyodi & R. G. Adiyodi (eds), Reproductive biology of invertebrates. Vol. IV. Fertilization, development and parental care. John Wiley & Sons, Ltd, Chichester, (in press).

Gargiulo, A. M., R. Pascolini & F. Panara, 1976a. Acid phosphatase in the triclad Dugesia lugubris. I. Enzyme localization during the first stages of intracellular digestion. IRCS Med. Sci. 4: 361.

Gargiulo, A. M., R. Pascolini & F. Panara, 1976b. Acid phosphatase in the triclad Dugesia lugubris. II. Enzyme localization in the food vacuoles. IRCS Med. Sci. 4: 362.

Grasso, M., 1974. Some aspects of sexuality and agamy in planarians. Boll. zool. ital. 41: 379–393.

Grasso, M., 1975. Sexuality and neurosecretion in freshwater planarians. In R. Reinboth (ed.), Intersexuality in the animal kingdom. Springer-Verlag, N.Y.: 20–29.

Grasso, M., L. Montanaro & A. Quaglia, 1975. Studies on the role of neurosecretion in the induction of sexuality in a planarian organic strain. J. ultrastruct. Res. 52: 404–408.

Grasso, M. & A. Quaglia, 1970a. Studies on neurosecretion in planarians. I. Neurosecretory fibers near the testes of Dugesia lugubris. J. submicrosc. Cytol. 2: 119–125.

Grasso, M. & A. Quaglia, 1970b. Studies on neurosecretion in planarians. II. Observations on the ovaries of Dugesia lugubris. J. submicrosc. Cytol. 2: 127–132.

Grasso, M. & A. Quaglia, 1971. Studies on neurosecretion in planarians. III. Neurosecretory fibres near the testes and ovaries of Polycelis nigra. J. submicrosc. Cytol. 3: 171–180.

Gremigni, V., 1969a. Recherche istochimiche e ultrastrutturali sull'ovogenesi dei Tricladi. I. Inclusi deutoplasmatici in Dugesia lugubris e Dugesia benazzii. Accad. naz. Lincei 47: 101–108.

Gremigni, V., 1969b. Ricerche istochimiche e ultrastrutturali sull'ovogenesi dei tricladi. II. Inclusi citplamatici in Planaria torva, Dendrocoelum lacteum e Polycelis nigra. Accad. naz. Lincei 47: 397–404.

Gremigni, V., 1969c. Origine del vitello negli ovociti Dugesia lugubris e D. benazzii (Tricladida Paludicola). Atti VII Congr. ital. Microsc. Elettr.: 68–70.

Gremigni, V., 1974. The origin and cytodifferentiation of germ cells in the planarians. Boll. zool. ital. 41: 359–377.

Gremigni, V., 1976. Genesis and structure of the so-called 'Balbiani body' or 'yolk nucleus' in the oocyte of Dugesia dorotocephala (Turbellaria, Tricladida). J. Morph. 149: 265–278.

Gremigni, V., 1979. An ultrastructural approach to planarian taxonomy. Syst. Zool. 28: 345–355.

Gremigni, V., 1981. The problem of cell totipotency, dedifferentiation and transdifferentiation in Turbellaria. Hydrobiologia 84: 171–179.

Gremigni, V., 1983. Platyhelminthes – Turbellaria. In K. G. Adiyodi & R. G. Adiyodi (eds), Reproductive biology of invertebrates, Vol. 1: Oogenesis, oviposition and oosorption. John Wiley & Sons, Ltd, Chichester: 67–102.

Gremigni, V. & R. Banchetti, 1972a. Submicroscopic morphology of hyperplasic ovaries of ex-fissiparous individuals in Dugesia gonocephala s.l. Accad. naz. Lincei 52: 539–543.

Gremigni, V. & R. Banchetti, 1972b. The origin of hyperplasia in the ovaries of ex-fissiparous specimens of Dugesia gonocephala s.l. Accad. naz. Lincei 53: 477–485.

Gremigni, V. & M. Benazzi, 1969a. Zoologia-ricerche istochimiche e ultrastrutturali sull' ovogenesi dei tricladi. I. Inclusi deutoplasmatici in Dugesia lugubris and Dugesia benazzii. Accad. naz. Lincei Rc Sci. VII 47: 102–108.

Gremigni, V. & M. Benazzi, 1969b. Zoologia-ricerche istochimiche e ultrastruttural sull'ovogenesi dei Triclad. II. Inclusi citoplasmatici in Planaria torva, Dendrocoelum lacteum, e Polycelis nigra. Accad. naz. Lincei Rc Sci. VIII 47: 397–401.

Gremigni, V. & L. Domenici, 1974. Electron microscopical and cytochemical study of vitelline cells in the fresh-water triclad Dugesia lugubris s.l. I.

Origin and morphogenesis of cocoon shell globules. Cell Tissue Res. 150: 261–270.
Gremigni, V. & L. Domenici, 1975. Genesis, composition, and the fate of cortical granules in the eggs of Polycelis nigra (Turbellaria, Tricladida). J. ultrastruct. Res. 50: 277–283.
Gremigni, V. & L. Domenici, 1976. Uptake of maternal nutrients during embryonic development in the subitaneous eggs of the turbellarian Mesostoma ehrenbergii (Focke). Monit. zool. ital. 10: 229–238.
Gremigni, V. & L. Domenici, 1977. On the role of specialized, peripheral cells during embryonic development of subitaneous eggs in the turbellarian Mesostoma ehrenbergii (Focke): an ultrastructural and autoradiographic investigation. Acta Embryol. exper. 2: 251–265.
Gremigni, V. & M. Nigro, 1983. An ultrastructural study of oogenesis in a marine triclad. Tissue Cell 15: 405–416.
Gremigni, V. & M. Nigro, 1984. Ultrastructural study of oogenesis in Monocelis lineata (Turbellaria, Proseriata). Int. J . invert. Reprod. 7: 105–118.
Gremigni, V., M. Nigro & M. Settembrini, 1985. Ultrastructural features of oogenesis in some marine neoophoran turbellarians. This volume.
Gremigni, V. & I. Puccinelli, 1977. A contribution to the problem of the origin of the blastema cells in planarians: a karyological and ultrastructural investigation. J. exp. Zool. 199: 57–72.
Hay, E. D., 1968a. Fine structure and origin of regenerating cells in planarians. Anat. Rec. 160: 363.
Hay, E. D., 1968b. Dedifferentiation and metaplasia in vertebrate and invertebrate regeneration. In H. Ursprung (ed.), The stability of the differentiated state. Springer-Verlag, Berlin, New York: 85–108.
Hay, E. D., 1974. Cellular basis of regeneration. In J. W. Lash & J. R. Whittaker (eds), Concepts in developmental biology. Sinauer, Stamford, Connecticut: 404–428.
Hay, E. D. & S. J. Coward, 1975. Fine structure studies on the planarian Dugesia. I. Nature of the 'neoblast' and other cell types in noninjured worms. J. ultrastruct. Res. 50: 1–21.
Hendelberg, J., 1965. On different types of spermatozoa in Polycladida, Turbellaria. Ark. Zool. 18: 267–304.
Hendelberg, J., 1969a. On the development of different types of spermatozoa from spermatids with two flagella in the Turbellaria with remarks on the ultrastructure of the flagella. Zool. Bidr. Upps. 38: 1–50.
Hendelberg, J., 1969b. Flatworm spermiogenesis and spermatozoa studied by light and electron microscope methods. (In: Abstracts of Uppsala Dissertations in Science.) Acta Univ. Upsaliensis 140: 1–12.
Hendelberg, J., 1970. On the number and ultrastructure of the flagella of flatworm spermatozoa. In B. Baccetti (ed.), Comparative spermatology. Academic Press, New York: 367–374.
Hendelberg, J., 1974. Spermiogenesis, sperm morphology, and biology of fertilization in the Turbellaria. In N. W. Riser & M. P. Morse (eds), Biology of the Turbellaria. McGraw-Hill, New York: 148–164.
Hendelberg, J., 1975. Functional aspects of flatworm sperm morphology. In B. A. Afzelius (ed.), The functional anatomy of the spermatozoan. Pergamon Press, Oxford, New York: 299–309.
Hendelberg, J., 1977a. Comparative morphology of turbellarian spermatozoa studied by electron microscopy. Acta zool. fenn. 154: 149–162.
Hendelberg, J., 1977b. Ultrastructure of the cytoplasmic region in spermatozoa of Cryptocelides (Polycladida, Turbellaria). Zoon 5: 107–114.
Hendelberg, J., 1981. The system of epidermal ciliary rootlets in Turbellaria. Hydrobiologia 84: 240.
Hendelberg, J., 1983a. Platyhelminthes – Turbellaria. In K. G. Adiyodi & R. G. Adiyodi (eds), Reproductive biology of invertebrates, Vol. II: Spermatogenesis and sperm function. J. Wiley & sons, Ltd, Chichester: 75–104.
Hendelberg, J., 1983b. Trends in the evolution of flatworm spermatozoa. In J. Andre (ed.), The sperm cell. M. Nijhoff Publ., The Hague: 450–453.
Hendelberg, J., 1985. The phylogenetic significance of sperm morphology in the Platyhelminthes. This volume.
Hendelberg, J. & K.-O. Hedlund, 1974. On the morphology of the epidermal ciliary rootlet system of the acoelous turbellarian Childia groenlandica. Zoon 2: 13–24.

Henley, C., 1968. Refractile bodies in the developing and mature spermatozoa of Childia groenlandica (Turbellaria: Acoela) and their possible significance. Biol. Bull. 134: 382–397.

Henley, C., 1970. Changes in microtubules of cilia and flagella following negative staining with phosphotungstic acid. Biol. Bull. 139: 265–276.

Henley, C., 1974. Platyhelminthes (Turbellaria). In A. C. Giese & J. S. Pearse (eds), Reproduction of marine invertebrates. Vol. I. Acoelomate and pseudocoelomate metazoans. Academic Press, New York: 267–343.

Henley, C. & D. P. Costello, 1969. Microtubules in spermatozoa of some turbellarian flatworms. Biol. Bull. 137: 403.

Henley, C., D. P. Costello & C. R. Ault, 1968. Microtubules in the axial filament complexes of acoel turbellarian spermatozoa as revealed by negative staining. Biol. Bull. 135: 422–423.

Henley, C., D. P. Costello, M. B. Thomas & W. D. Newton, 1969. The 9+1 pattern of microtubules in the spermatozoa of Mesostoma (Platyhelminthes Turbellaria). Proc. natn Acad. Sci. U.S.A. 64: 849–856.

Hermans, C. O., 1983. The duo-gland adhesive system. Oceanogr. mar. Biol. ann. Rev. 21: 283–339.

Holt, P. A. & D. F. Mettrick, 1975. Ultrastructural studies of the epidermis and gastrodermis of Syndesmis franciscana (Turbellaria: Rhabdocoela). Can. J. Zool. 53: 536–549.

Hori, I., 1975. Electron microscopic studies on the subepidermal basement membrane in the planaria, Dugesia japonica. [In Japanese, English abstract and captions] Zool. Mag. 84: 123–131.

Hori, I., 1978. Possible role of rhabdite-forming cells in cellular succession of the planarian epidermis. J. electr. Microsc. 27: 89–102.

Hori, I., 1979a. Regeneration of the epidermis and basement membrane of the planarian Dugesia japonica after total-body X-irradiation. Radiat. Res. 77: 521–533.

Hori, I., 1979b. Structure and regeneration of the planarian basal lamina: An ultrastructural study. Tissue Cell 11: 611–621.

Hori, 1980. Localization of newly synthesized precursors of basal lamina in the regenerating planarian as revealed by autoradiography. Tissue Cell 12: 513–521.

Hori, I., 1982. An ultrastructural study of the chromatoid body in planarian regenerative cells. J. electr. Microsc. 31: 63–72.

Hori, I., 1983a. Differentiation of myoblasts in the regenerating planarian Dugesia japonica. Cell Different. 12: 155–163.

Hori, I., 1983b. Ultrastructural variation of rough endoplasmic reticulum – Relation to the types of planarian regenerative cells. J. submicrosc. Cytol. 15: 975–989.

Hori, I., 1983c. Cytological studies on rhabdite formation in the planarian differentiating cells. J. submicrosc. Cytol. 15: 483–494.

Hori, I., 1985. Ultrastructural observations on gastrodermal regeneration in the planarian Dugesia japonica (Turbellaria). This volume.

Ishida, S., T. Gotoh & W. Teshirogi, 1981. Oogenesis and egg-shell formation in polyclads. [In Japanese, English summary] Rep. Fukaura mar. biol. Lab. 9: 32–48.

Ishida, S. & W. Teshirogi, 1985. Eggshell formation in polyclads (Turbellaria). This volume.

Ishii, S., 1962/1963. Electron microscopic observations on the planarian tissues. I. A survey of the pharynx. Fukushima J. med. Sci. 9–10: 51–73.

Ishii, S., 1964. The ultrastructure of the outer epithelium of the planarian pharynx. Fukushima J. med. Sci. 11: 109–125.

Ishii, S., 1965. Electron microscopic observations on the planarian tissues. II. The intestine. Fukushima J. med. Sci. 12: 67–87.

Ishii, S., 1966. The ultrastructure of the insunk epithelium lining the planarian pharyngeal cavity. J. ultrastruct. Res. 14: 345–355.

Ishii, S., 1972. Electron microscopic observations on intracellular digestion in the gastrodermis of the lower invertebrate, Bdellocephala brunnea. Fukushima J. med. Sci. 18: 91–139.

Ishii, S., 1980a. The ultrastructure of the protonephridial tubules of the freshwater planarian Bdellocephala brunnea. Cell Tissue Res. 206: 451–458.

Ishii, S., 1980b. The ultrastructure of the protonephridial flame cell of the freshwater planarian Bdellocephala brunnea. Cell Tissue Res. 206: 441–450.

Ivanov, A. V., 1971. On the primitivism of Acoela. [In Russian; English abstract]. Zool. Zh. 50: 621–632.

Ivanov, A. V. & Yu. V. Mamkaev, 1977. Über die Struktur des Digestionsparenchyme bei Turbellaria Acoela. Acta zool. fenn. 154: 59–61.

Ivanov, A. V., Yu. V. Mamkaev & R. A. Pevzner, 1972. Electron microscopic study on the statocyst in the acoelan turbellarian Convoluta convoluta. [In Russian] Zh. Evol. Biokhim. Fiziol. 8: 189–193.

Jespersen, A. & J. Luetzen, 1972. Triloborhynchus psilastericola n. sp., a parasitic turbellarian (Fam. Pterastericolidae) from the starfish Psilaster andromeda (Mueller and Troschel). Z. Morph. Tiere 71: 290–298.

Kelbetz, S., 1862. Zum Feinbau der Rhabditen. Naturwissenschaften 49: 548.

Kishida, Y., 1965. The ultrastructure of the eye in Dugesia japonica. I. The distal portion of the visual cell. [In Japanese, with English summary] Zool. Mag. 74: 149–155.

Kishida, Y., 1966. The ultrastructure of the eye in Dugesia japonica. II. The pigment cell layer. [In Japanese, with English summary] Zool. Mag. 75: 40–46.

Kishida, Y., 1967a. Electron microscopic studies on the planarian eye. I. Fine structures of the normal eye. Sci. Rep. Kanazawa Univ. 12: 75–110.

Kishida, Y., 1967b. Electron microscopic studies on the planarian eye. II. Fine structures of the regenerating eye. Sci. Rep. Kanazawa Univ. 12: 111–142.

Kishida, Y. & E. Asai, 1977. Les cristeaux intranucléaires des cellules pharyngiennes chez la Planaire, Dugesia japonica Ichikawa et Kawakatsu. J. electr. Microsc. 26: 145–147.

Klauser, M. D., J. P. S. Smith & S. Tyler, 1985. Ultrastructure of the frontal organ in Convoluta 'pulchra' and Macrostomum spp.: significance for models of the turbellarian archetype. This volume.

Klima, J., 1959. Das Bild des endoplasmatischen Reticulums vom Planaria alpina in Abhängigkeit vom pH-Wert des Fixierungsmittels (I). Protoplasma 51: 415–435.

Klima, J., 1961. Elektronenmikroskopische Studien über die Feinstruktur der Tricladen (Turbellaria). Protoplasma 54: 101–162.

Klima, J., 1967. Zur Feinstruktur des acoelen Süsswasser-Turbellars Oligochoerus limnophilus Ax & Dörjes. Ber. Nat. Med. Ver. Innsbruck 55: 107–124.

Klug, H., 1960. Über die Funktionelle Bedeutung der Feinstrukturen der Exokrinen Drüsenzellen (Untersuchungen an Euplanaria). Z. Zellforsch. mikrosk. Anat. 51: 617–632.

Koie, M. & J. Bresciani, 1973. On the ultrastructure of the larva of Kronborgia amphipodicola Christensen and Kanneworff, 1964 (Turbellaria, Neorhabdocoela). Ophelia 12: 171–203.

Komatsu, H., 1971. Electron microscopic observation of the cilia in a freshwater planarian, Dugesia japonica. II. Fine structure of the cilia of pharynx epidermis. In: Comn. Compil. Sci. Papers Publ. Occ. Retir. Prof. H. Sugino. Ser. Turbellarians, pp. 157–162.

Koopowitz, H., 1974. Some aspects of the physiology and organization of the nerve plexus in polyclad flatworms. In N. W. Riser & M. P. Morse (eds), Biology of the Turbellaria. McGraw-Hill, New York: 198–212.

Koopowitz, H. & P. Chien, 1974. Ultrastructure of the nerve plexus in flatworms. I. Peripheral organization. Cell Tissue Res. 155: 337–351.

Koopowitz, H. & P. Chien, 1975. Ultrastructure of nerve plexus in flatworms. II. Sites of synaptic interactions. Cell Tissue Res. 157: 207–216.

Kozloff, E., 1972. Selection of food, feeding, and physical aspects of digestion in the acoel turbellarian Otocelis luteola. Trans. am. microsc. Soc. 91: 556–565.

Kubo-Irie, M. & M. Ishikawa, 1983. Spermiogenesis in the flatworm, Notoplana japonica, with special attention to the organization of an acrosome and flagella. Develop. Growth Differentiation 25: 143–152.

Kümmel, G., 1962. Zwei neue Formen von Cyrtocyten. Vergleich über bisher bekannten Cyrtocyten und Erörterung des Begriffes 'Zelltyp'. Z. Zellforsch. mikrosk. Anat. 57: 172–201.

Kümmel, G. & J. Brandenburg, 1961. Die Reusengeisselzellen (Cyrtocyten). Z. Naturforsch. 166: 692–697.

Lacalli, T. C., 1982. The nervous system and ciliary band of Muller's larva. Proc. r. Soc. Lond. B 217: 37–58.

Lacalli, T. C., 1983. The brain and central nervous system of Muller's larva. Can. J. Zool. 61: 39–51.

Lanfranchi, A., 1978. Morphology and taxonomy of two new otoplanids (Turbellaria, Proseriata) from the Ligurian Sea. Zool. Scr. 7: 249–254.

Lanfranchi, A. & C. Bedini, 1982. The ultrastructure of the sense organs of some Turbellaria

Rhabdocoela. I. The eyes of Polycystis naegelii Kölliker (Eukalyptorhynchia Polycystididae). Zoomorphology 101: 95–102.

Lanfranchi, A. & C. Bedini, 1985. Electron microscopic study of larval eye development in Turbellaria Polycladida. This volume.

Lanfranchi, A., C. Bedini & E. Ferrero, 1981. The ultrastructure of the eyes in larval and adult polyclads. Hydrobiologia 84: 267–275.

Lanzavecchia, G., 1972. The 'smooth' muscle of invertebrates. Boll. zool. ital. 39: 159–172.

Lanzavecchia, G., 1977. Morphological modulations in helical muscles (Aschelminthes and Annelida). Int. Rev. Cytol. 51: 133–186.

Lanzavecchia, G., 1981. Evolution of muscle fibers in lower Metazoa. Atti Conv. Lincei 49: 11–22.

Le Moigne, A., 1966. Etude au microscope electronique de cellules d'embryons de Polycelis (Turbellarie, Triclade), au debut de leur developpement. C. r. Acad. Sci., Paris 263: 550–553.

Le Moigne, A., 1967a. Etude au microscope electronique de la differenciation des principaux types cellulaires chez l'embryon de la planaire Polycelis nigra. Bull. Soc. zool. Fr. 92: 617–628.

Le Moigne, A., 1967b. Présence d'émissions nucleaires fréquemment associées a des mitochondries dans les cellules embryonnaires de Planaires. C. r. Soc. Biol. Fr. 161: 508–511.

Le Moigne, A., 1967c. Mise en évidence au microscope electronique de la persistance de cellules indifferenciées au cours du developpement embryonnaire de la planaire Polycelis nigra. C. r. Acad. Sci., Paris 265: 242–244.

Le Moigne, A., 1968. Etude au microscope electronique de l'evolution des structures embryonnaires de planaires apres irradiation aux rayons X. J. Embryol. exp. Morph. 19: 181–192.

Le Moigne, A., 1969. Etude du developpement et de la régénération embryonnaires de Polycelis nigra (Ehr.) et Polycelis tenuis (Iijima); Turbellaries, Triclades. Ann. Embryol. Morph. 2: 51–69.

Le Moigne, A. & A. Gabriel, 1969. Mise en evidence du charactére 'active' des cellules indifferenciées du parenchyme chez des jeunes Planaires en croissance. C. r. soc. Biol. Fr. 163: 1070–1073.

Le Moigne, A. & A. Gabriel, 1971a. Action de l'actinomycine D sur la differenciation cellulaire au cours de la régénération de Planaires qui viennent d'éclore. II. Etudes autoradiographiques histologiques et ultrastructurales de l'action de l'antibiotiques sur les syntheses d'ARN. Z. Zellforsch. mikrosk. Anat. 115: 442–460.

Le Moigne, A. & A. Gabriel, 1971b. Action de l'actinomycine D sur la differenciation cellulaire au cours de la régénération de Planaires qui viennent d'eclore. III. Etude ultrastructurale de la pénétration d'actinomycine-3H dans les cellules de régénération. Z. Zellforsch. mikrosk. Anat. 122: 26–35.

Le Moigne, A., A. Gabriel & M. J. Sauzin, 1972. Aspects ultrastructuraux et biochimiques de la régénération chez les planaires: comparaison entre adultes et jeunes. Bull. Soc. Zool. Fr. 97: 283–293.

Le Moigne, A. & M. J. Sauzin-Monnot, 1971. Etude au microscope électronique d'inclusions nucleaires chez les planaires (Turbellarie, Triclades). J. Microsc. (Paris) 10: 107–112.

Le Moigne, A., M. J. Sauzin & T. Lender, 1966. Comparaison de l'ultrastructure du néoblaste et de la cellule embryonnaire des Planaires d'eau douce. C. r. Acad. Sci., Paris 263: 627–629.

Le Moigne, A., M. J. Sauzin, T. Lender & C. Delavault, 1965. Quelques aspects des ultrastructures du blastème de régénération et des tissus voisins chez Dugesia gonocephala (Turbellarie, Triclade). C. r. Soc. Biol. Fr. 159: 530–538.

Lee, D. L., 1972. The structure of the helminth cuticle. In B. Davis (ed.), Advances in parasitology. Academic Press, London, New York: 347–379.

Lender, T., 1974. The role of neurosecretion in freshwater planarians. In N. W. Riser & M. P. Morse (eds), Biology of the Turbellaria. McGraw-Hill, New York: 460–475.

Lender, T., 1980. Endocrinologie des planaires. Bull. Soc. Zool. Fr. 105: 173–191.

Lentz, T. L., 1967a. Rhabdite formation in planaria: the role of microtubules. J. ultrastruct. Res. 17: 114–126.

Lentz, T. L., 1967b. Fine structure of nerve cells in a planarian. J. Morph. 121: 323–338.

Lentz, T. L., 1968. Primitive nervous systems. Yale University Press, New Haven, 148 pp.

Lopez Camps, J. & R. Bargallo, 1976. El axonema del gameto masculino de Typhloplana viridata (Turbellaria, Typhloplanidae). VII. Reunion. Soc. esp. Microsc. electr. Cordoba: 32.

Lord, B. A.P. & D. R. DiBona, 1976. Role of the

septate junction in the regulation of paracellular transepithelial flow. J. Cell Biol. 71: 967–972.
Lyons, K., 1973. The epidermis and sense organs of the Monogenea and some related groups. Adv. Parasitol. 11: 147–232.
McKanna, J. A., 1968a. Fine structure of the protonephridial system in planaria. I. Flame cells. Z. Zellforsch. mikrosk. Anat. 92: 509–523.
McKanna, J. A., 1968b. Fine structure of the protonephridial system in planaria. II. Ductules, collecting ducts, and osmoregulatory cells. Z. Zellforsch. mikrosk. Anat. 92: 524–535.
MacKinnon, B. M., M. D. B. Burt & A. W. Pike, 1981. Ultrastructure of the epidermis of adult and embryonic Paravortex species (Turbellaria, Eulecithophora). Hydrobiologia 84: 241–252.
MacRae, E. K., 1961. The fine structure of planarian rhabdites. Am. Zool. 1: 460.
MacRae, E. K., 1963. Observations on the fine structure of the pharyngeal muscle in the planarian Dugesia tigrina. J. Cell Biol. 18: 651–662.
MacRae, E. K., 1964a. Observations on the fine structure of photoreceptor cells in the planarian Dugesia tigrina. J. ultrastruct. Res. 10: 334–349.
MacRae, E. K., 1964b. Organized structures in the nucleolus. J. Cell Biol. 23: 195–200.
MacRae, E. K., 1965a. Fine structure of the basement lamella in a marine turbellarian. Am. Zool. 5: 247.
MacRae, E. K., 1965b. The fine structure of muscle in a marine turbellarian. Z. Zellforsch. mikrosk. Anat. 68: 348–362.
MacRae, E. K., 1966. The fine structure of photoreceptors in a marine flatworm. Z. Zellforsch. mikrosk. Anat. 75: 469–484.
MacRae, E. K., 1967. The fine structure of sensory receptor processes in the auricular epithelium of the planarian Dugesia tigrina. Z. Zellforsch. mikrosk. Anat. 82: 479–494.
Mainitz, M., 1977. The fine structure of the stylet apparatus in Gnathostomulida Scleroperalia and its relation to the turbellarian stylets. Acta zool. fenn. 154: 163–174.
Malczewska, M., A. Czubaj, E. Morawska & J. Moraczewski, 1980a. Changes in the acetylcholinesterase activity during the regeneration of planarian Dugesia lugubris (O. Schmidt). Ultrastructural studies. Acta med. pol. 21: 381–382.
Malczewska, M., A. Czubaj, E. Morawska & J. Moraczewski, 1980b. Changes in AChE activity during regeneration of planarian Dugesia lugubris (O. Schmidt). Ultracytochemical study. Acta med. pol. 21: 318–320.
Mamkaev, Yu. V., 1979. On the histological organization of the turbellarian digestive system. [In Russian – translation available] Proc. zool. Inst. Acad. Sci. USSR 84: 13–24.
Mamkaev, Yu. & V. P. Ivanov, 1970. Elektronmikroskopitjesko issledovanie spermatozoidov beskisjetjnoj turbellarii Convoluta convoluta. in: Ottjetnaja nautnaja sessija poitoyam rabot 1969 goda, 23–25 Marta. 1970 G., tezisy dokladoc, A. N. SSSR, Zool. Inst., Leningrad: 12–13.
Mamkaev, Yu. V. & T. G. Markosova, 1979. Electron microscopic studies of the parenchyma in some representatives of the Acoela. [In Russian – translation available] Proc. zool. Inst. Acad. Sci. USSR 84: 7–12.
Marinelli, M. & R. M. Farnesi, 1974. The cement glands in Dugesia lugubris: histochemical and ultrastructural investigations. Riv. Biol. 67: 301–317.
Marinelli, M. & D. Vagnetti, 1973a. Electron microscopic investigation on the yolk cells in the cocoon of Dugesia lugubris s.l. Boll. zool. ital. 40: 367–369.
Marinelli, M. & D. Vagnetti, 1973b. Investigation into the nature of yolk cell lipids in cocoons of Dugesia lugubris s.l. and individuals at hatching. Boll. zool. ital. 42: 91–93.
Marinelli, M. & D. Vagnetti, 1977a. Ultrastructural research on Dugesia lugubris s.l. male copulatory complex. Riv. Biol. 70: 207–223.
Marinelli, M. & D. Vagnetti, 1977b. Osservazioni ultrastrutturali sulla formazione del guscio del bozzolo in Dugesia lugubris s.l. Accad. naz. Lincei 63: 431–345.
Markosova, T. G., 1985. An ultrastructural and cytochemical study of the digestive system in Oxyposthia praedator (Turbellaria: Acoela). This volume.
Martens, E. E., 1983. Differentiation in the copulatory organ of Turbellaria (Platyhelminthes). In W. Engels (ed.), Advances in invertebrate reproduction 3. Proc. Third Int. Symp. Int. Soc. Invert. Reprod., Tübingen: 611.
Martens, E. E., 1984a. Ultrastructure of the spines in the copulatory organ of some Monocelididae

(Turbellaria, Proseriata). Zoomorphology 104: 261–265.
Martens, E. E., 1984b. Cirrifera genitoducto sp. n. (Proseriata, Coelogynoporidae) from the Belgian coast, with observations on its ultrastructure. Zool. Scr. (in press).
Martens, E., 1985. Comparative ultrastructure of copulatory organs having a stylet in the Proseriata (Turbellaria). This volume.
Martens, E. E. & E. Schockaert, 1981. Observations on the ultrastructure of the copulatory organ of Archilopsis unipunctata (Fabricius, 1826) (Proseriata, Monocelididae). Hydrobiologia 84: 277–285.
Martens, E. E. & E. R. Schockaert, 1984. Studies on the ultrastructure of the genital organs in Proseriata (Turbellaria). I. Cirrifera aculeata (Ax, 1951) (Coelogynoporidae). Zool. Scr. (in press).
Martin, G. G., 1978a. The duo-gland adhesive system of the archiannelids Protodrilus and Saccocirrus and the turbellarian Monocelis. Zoomorphologie 91: 63–75.
Martin, G. G., 1978b. A new function of rhabdites: Mucus production for ciliary gliding. Zoomorphologie 91: 235–248.
Martin, G. G., 1978c. Ciliary gliding in lower invertebrates. Zoomorphologie 91: 249–261.
Matricon-Gondran, M., 1980. Gap junctions and particle aggregates in the tegumentary syncytium of a trematode. Tissue Cell 12: 385–394.
Moraczewski, J., 1977a. Asexual reproduction and regeneration of Catenula (Turbellaria, Archoophora). Zoomorphologie 88: 65–80.
Moraczewski, J., 1977b. Submicroscopic differentiation in lower Turbellaria during paratomy and regeneration. Proc. XV. Czechoslovak. Conf. E. M., Prague: 207–208.
Moraczewski, J., 1978. Cell differentiation in turbellarians during regeneration. [In Polish] Postepy Biol. Kormoki 5: 93–107.
Moraczewski, J., 1981a. Cell activation during regeneration of planarians. Hydrobiologia 84: 203–207.
Moraczewski, J., 1981b. The fine structure of some Catenulida (Turbellaria Archoophora). Zool. pol. 28: 367–415.
Moraczewski, J., J. Bakowska, J. Kaczmarczyk & A. Czubaj, 1975. Ultrastructure of brain and neurosecretory cells in Catenulida (Turbellaria). Ann. med. Sect. pol. Acad. Sci. 20: 101–103.
Moraczewski, J. & A. Czubaj, 1974. Muscle ultrastructure in Catenula lemnae (Dug.) (Turbellaria, Archaeophora). J. submicrosc. Cytol. 6: 29–38.
Moraczewski, J., A. Czubaj & J. Bakowska, 1977. Organization and ultrastructure of the nervous system in Catenulida (Turbellaria). Zoomorphologie 87: 87–95.
Moraczewski, J., A. Czubaj & J. Kwiatkowska, 1977. Localization of neurosecretion in the nervous system of Catenulida (Turbellaria). Zoomorphologie 87: 97–102.
Moraczewski, J. & A. Duma, 1978. Ultrastructural localization of adenyl cyclase activity in neoblasts of turbellarian Dugesia lugubris (O. Schmidt). J. exp. Zool. 203: 491–496.
Moraczewski, J. & A. Duma, 1981. Localization of adenylate cyclase activity in the tissues of an intact planarian Dugesia lugubris (O. Schmidt). Histochemistry 71: 301–311.
Morawska, E., J. Moraczewski, M. Malczewska & A. Duma, 1981. Adenylate cyclase in regenerating tissues of the planarian Dugesia lugubris (O. Schmidt). Hydrobiologia 84: 209–212.
Morita, M., 1964. Fine structure and enzyme activity in planarian nervous tissue. Anat. Rec. 148: 418–419.
Morita, M., 1965. Electron microscopic studies on planaria I. Fine structure of muscle fiber in head of the planarian Dugesia dorotocephala. J. ultrastruct. Res. 13: 383–395.
Morita, M., 1967. Observations on the fine structure of the neoblast and its cell division in the regenerating planaria. Sci. Rep. Tohoku Univ. Biol. 33: 399–406.
Morita, M., 1968. Electron microscopic studies on planaria IV. Fine strucure of some secretory gland in the planarian Dugesia dorotocephala. Fukushima J. med. Sci. 15: 13–33.
Morita, M. & J. B. Best, 1965. Electron microscopic studies on planaria II. Fine structure of the neurosecretory system in the planarian Dugesia dorotocephala. J. ultrastruct. Res. 13: 396–408.
Morita, M. & J. B. Best, 1966. Electron microscopic studies on planaria III. Some observations on the fine structure of planarian nervous tissue. J. exp. Zool. 161: 391–412.
Morita, M. & J. B. Best, 1974. Electron microscopic studies of planarian regeneration. II. Changes

in epidermis during regeneration. J. exp. Zool. 187: 345–373.

Morita, M. & J. B. Best, 1984a. Electron microscopic studies of planarian regeneration. III. Degeneration and differentiation of muscles. J. exp. Zool. 229: 413–424.

Morita, M. & J. B. Best, 1984b. Electron microscopic studies of planarian regeneration. IV. Cell division of neoblasts in Dugesia dorotocephala. J. exp. Zool. 229: 425–436.

Morita, M., J. B. Best & J. Noel, 1969. Electron microscopic studies of planarian regeneration. I. Fine structure of neoblasts in Dugesia dorotocephala. J. ultrastruct. Res. 27: 7–23.

Newton, W. D., 1974. The accessory cell and yolk halo of the oocyte of the freshwater turbellarian Hydrolimax grisea (Platyhelminthes; Plagiostomidae). Biol. Bull. 147: 618–629.

Newton, W. D., 1975. The ultrastructure of the aflagellate spermatozoon of the freshwater turbellarian Hydrolimax grisea (Platyhelminthes: Plagiostomidae). Biol. Bull. 149: 551–567.

Newton, W. D., 1979. The motile apparatus of the flagellate spermatozoon of Macrostomum tubum. Am. Zool. 19: 998.

Newton, W. D., 1980. Ultrastructural analysis of the motile apparatus of the aflagellate spermatozoon of Macrostomum tubum. J. ultrastruct. Res. 73: 318–330.

Oakley, H. A., 1982. Meiosis in Mesostoma ehrenbergii ehrenbergii (Turbellaria, Rhabdocoela). II. Synaptonemal complexes, chromosome pairing and disjunction in aschiasmate oogenesis. Chromosoma 87: 133–147.

Oakley, H. A. & G. H. Jones, 1982. Meiosis in Mesostoma ehrenbergii (Turbellaria Rhabdocoela). I. Chromosome pairing, synaptonemal complexes and chiasma localisation in spermatogenesis. Chromosoma 85: 311–322.

Olsen, G. E. & J. B. Rattner, 1975. Observations on the substructure of ciliary rootlets. J. ultrastruct. Res. 51: 409–417.

Oosaki, T. & S. Ishii, 1965. Observations on the ultrastructure of nerve cells in the brain of the planarian, Dugesia gonocephala. Z. Zellforsch. mikrosk. Anat. 66: 782–793.

Oschman, J. L., 1966. Development of the symbiosis of Convoluta roscoffensis and its endosymbiotic algae. J. Phycol. 2: 105–111.

Oschman, J., 1967. Microtubules in the subepidermal glands of Convoluta roscoffensis (Acoela, Turbellaria). Trans. am. microsc. Soc. 86: 159–162.

Oschman, J. L., 1969. Endonuclear viruslike bodies in Convoluta roscoffensis (Turbellaria, Acoela). J. invertebr. Pathol. 13: 147–148.

Oschman, J. L. & P. Gray, 1965. A study of the fine structure of Convoluta roscoffensis and its endosymbiotic algae. Trans. am. microsc. Soc. 84: 368–375.

Ott, J., G. Rieger, R. Rieger & F. Enders, 1982. New mouthless interstitial worms from the sulfide system: symbiosis with prokaryotes. Mar. Ecology 3: 313–333.

Palladini, G., V. Margotta, A. Carolei, S. Chiarini, M. Delpiano, G. Lauro, L. Medolago-Albani & G. Venturini, 1983. Cerebrum of Dugesia gonocephala s.l. Platyhelminthes, Turbellaria, Tricladida. Morphological and functional observations. J. Irnsforsch. 23: 165–172.

Palladini, G., V. Margotta, L. Medolago-Albani, M. C. Hernandez, A. Carolei & T. De Michele, 1981. Photosensitization and the nervous system in the planarian Dugesia gonocephala. A histochemical, ultrastructural and behavioral investigation. Cell Tissue Res. 215: 271–279.

Palladini, G., L. Medolago-Albani, V. Margotta, A. Conforti & A. Carolei, 1979a. The pigmentary system of planaria. I. Morphology. Cell Tissue Res. 199: 197–202.

Palladini, G., L. Medolago-Albani, V. Margotta, A. Conforti & A. Carolei, 1979b. The pigmentary system of planaria. II. Physiology and functional morphology. Cell Tissue Res. 199: 203–211.

Palmberg, I., 1985. Cell migration and differentiation during wound healing and regeneration in Microstomum lineare (Turbellaria). This volume.

Palmberg, I. & M. Reuter, 1983. Asexual reproduction in Microstomum lineare (Turbellaria). I. An autoradiographic and ultrastructural study. Int. J. Invert. Reprod. 6: 197–206.

Palmberg, I., M. Reuter & M. Wikgren, 1980. Ultrastructure of epidermal eyespots of Microstomum lineare (Turbellaria, Macrostomida). Cell Tissue Res. 210: 21–32.

Parke, M. & I. Manton, 1967. The specific identity of the algal symbiont in Convoluta roscoffensis. J. mar. biol. Ass. U.K. 47: 445–464.

Pascolini, R., L. Facchini & R. M. Farnesi, 1971a.

Secrezioni ghiandolari del faringe di Dugesia lugubris: prime osservazioni al M. E. Boll. Soc. ital. Biol. Sperim. 47: 378–379.

Pascolini, R., L. Facchini & R. M. Farnesi, 1971b. Alcune osservazioni ultrastrutturali sulle fibre nervose del faringe in Dugesia lugubris. Boll. Soc. ital. Biol. Sperim. 47: 380–381.

Pascolini, R. & A. M. Gargiulo, 1974. Relationships between neurosecretion and glandular secretion in the pharynx of Dugesia lugubris s.l. Boll. zool. ital. 41: 43–46.

Pascolini, R. & A. M. Gargiulo, 1975. Ultrastructural modification of cells in the intestine of Dugesia lugubris. Boll. zool. ital. 41: 123–131.

Pascolini, R., A. M. Gargiulo, A. Petris & F. Panara, 1977. Acid phosphatase in the triclad Dugesia lugubris. III. Localization in the pharynx. IRCS med. Sci. 5: 22.

Pascolini, R., A. M. Gargiulo, A. Spreca & P. Ceccarelli, 1982. Cytochemical localization of adenylate cyclase activity in planarian tissues. J. exp. Zool. 223: 193–196.

Pascolini, R., A. M. Gargiulo, A. Spreca & A. Orlacchio, 1979. Studies on the cytochemical localization of adenylate-cyclase activity in Dugesia lugubris s.l. Experientia 35: 1315–1316.

Pascolini, R., A. Orlacchio, R. Floridi, J. A. Reinosa & E. A. Munn, 1979. The localization of N-acetyl-ß-D-glucosaminidase in planarian Dugesia. In R. Gilles (ed.), Animals and environmental fitness. Vol. 2. Pergamon Press, Oxford: 145–146.

Pascolini, R., A. Orlacchio & A. M. Gargiulo, 1981. N-acetyl-ß-D-hexosaminidase activity levels and cytochemical localization in intact and regenerating planarian Dugesia lugubris. Comp. Biochem. Physiol. 69B: 869–872.

Pascolini, R., S. Tei, D. Vagnetti & C. Bondi, 1984. Epidermal cell migration during wound healing in Dugesia lugubris. Observations based on scanning electron microscopy and treatment with cytochalasin. Cell Tissue Res. 236: 345–349.

Patzner, A. M. & H. Adam, 1979. Der Einfluss eines nichtionischen Tensids, Lutensol AO 10, auf Dugesia gonocephala (Plathelminthes, Tricladida) – Eine licht und elektronen mikroskopische Untersuchung. Zool. Anz. 203: 220–229.

Pedersen, K. J., 1959a. Some features of the fine structure and histochemistry of planarian subepidermal gland cells. Z. Zellforsch. mikrosk. Anat. 50: 121–142.

Pedersen K. J., 1959b. Cytological studies on the planarian neoblast. Z. Zellforsch. mikrosk. Anat. 50: 799–817.

Pedersen, K. J., 1961a. Studier over ferskvandstricladernes cytologi. [In Danish with English summary]. Munksgaard, Copenhagen.

Pedersen, K. J., 1961b. Studies on the nature of planarian connective tissue. Z. Zellforsch. mikrosk. Anat. 53: 569–608.

Pedersen, K. J., 1961c. Some observations on the fine structure of planarian protonephridia and gastrodermal phagocytes. Z. Zellforsch. mikrosk. Anat. 53: 609–628.

Pedersen, K. J., 1963. Slime-secreting cells of planarians. Ann. N.Y. Acad. Sci. 106: 424–442.

Pedersen, K. J., 1964. The cellular organization of Convoluta convoluta, an acoel turbellarian: a cytological, histochemical and fine structural study. Z. Zellforsch. mikrosk. Anat. 64: 655–687.

Pedersen, K. J., 1965. Cytological and cytochemical observations on the mucous gland cells of an acoel turbellarian, Convoluta convulta [sic]. Ann. N.Y. Acad. Sci. 118: 930–965.

Pedersen, K. J., 1966. The organization of the connective tissue of Discocoelides langi (Turbellaria, Polycladida). Z. Zellforsch. mikrosk. Anat. 71: 94–117.

Pedersen, K. J., 1972. Studies on regeneration blastemas of the planarian Dugesia tigrina with special reference to differentiation of the muscle-connective tissue filament system. Wilhelm Roux Arch. EntwMech 169: 134–169.

Pedersen, K. J., 1976. Scanning electron microscopical observations on epidermal wound healing in the planarian Dugesia tigrina. Wilhelm Roux Arch. EntwMech 179: 251–273.

Pedersen, K. J., 1983. Fine structural observations on the turbellarians Stenostomum sp. and Microstomum lineare with special reference to extracellular matrix and connective tissue systems. Acta zool. 64: 177–190.

Peters, A., S. L. Palay, H. de F. Webster, 1976. The fine structure of the nervous system: the neurons and supporting cells. W. B. Saunders Co., Philadelphia.

Phillips, J. J., 1978. The occurrence and distribution of haemoglobin in the entosymbiotic rhabdocoel Paravortex scrobiculariae (Graff) (Platyhelminthes: Turbellaria). Comp. Biochem. Physiol. 61A: 679–683.

Pigou, A., M. Morita & J. B. Best, 1974. Cephalic mechanism for social control of fissioning in planarians. II. Localization and identification of the receptors by electron micrographic and ablation studies. J. Neurobiol. 5: 443–462.

Platzer, I., 1965. Zur ultrastruktur von Planarienrestitutions-Körpern. Protoplasma 60: 34–60.

Press, N., 1959. Electron microscopic study of the distal portion of a planarian retinular cell. Biol. Bull. 117: 511–517.

Provasoli, L., T. Yamasu & I. Manton, 1968. Experiments on the resynthesis of symbiosis in Convoluta roscoffensis with different flagellate cultures. J. mar. biol. Ass. U.K. 48: 465–479.

Quick, D. C. & R. G. Johnson, 1977. Gap junctions and rhombic particle arrays in planaria. J. ultrastruct. Res. 60: 348–361.

Reisinger, E., 1964. Das Integument der Coelenteraten, acoelomaten und pseudocoelomaten Würmer. Stud. gen. 17: 125–142.

Reisinger, E., 1968. Xenoprorhynchus ein Modellfall für progressiven Funktionswechsel. Z. zool. Syst. Evolutionsforsch. 6: 1–55.

Reisinger, E., 1969. Ultrastrukturforschung und Evolution. Ber. Phys-Med Ges. Wurzburg 77: 1–43.

Reisinger, E., 1970. Zur Problematik der Evolution der Coelomaten. Z. zool. Syst. Evolutionsforsch. 8: 81–109.

Reisinger, E., 1972. Die Evolution des Orthogons der Spiralier und das Archicoelomatenproblem. Z. zool. Syst. Evolutionsforsch. 10: 1–43.

Reisinger, E. & S. Kelbetz, 1964. Feinbau und Entladungsmechanismus der Rhabditen. Z. wiss. Mikrosk. 65: 472–508.

Reuter, M., 1975a. Viruslike particles in Gyratrix hermaphroditus (Turbellaria: Rhabdocoela). J. invert. Pathol. 25: 79–95.

Reuter, M., 1975b. Ultrastructure of the epithelium and the sensory receptors in the body wall, the proboscis, and the pharynx of Gyratrix hermaphroditus (Turbellaria, Rhabdocoela). Zool. Scr. 4: 191–204.

Reuter, M., 1977. Ultrastructure of the stylet protractor muscle in Gyratrix hermaphroditus (Turbellaria, Rhabdocoela). Acta zool. 58: 179–184.

Reuter, M., 1978. Scanning and transmission electron microscopic observations on surface structures of three turbellarian species. Zool. Scr. 7: 5–11.

Reuter, M., 1981. The nervous system of Microstomum lineare (Turbellaria, Macrostomida). 2. Ultrastructure of synapses and neurosecretory release sites. Cell Tissue Res. 218: 375–387.

Reuter, M., 1983. Developmental changes and lytic activity in aminergic neurons of Microstomum lineare (Turbellaria). Acta zool. 64: 85–95.

Reuter, M. & P. Lindroos, 1979a. The ultrastructure of the nervous system of Gyratrix hermaphroditus (Turbellaria, Rhabdocoela) I. The brain. Acta zool. 60: 139–152.

Reuter, M. & P. Lindroos, 1979b. The ultrastructure of the nervous system of Gyratrix hermaphroditus (Turbellaria, Rhabdocoela) II. The peripheral nervous system and the synapses. Acta zool. 60: 153–161.

Reuter, M. & I. Palmberg, 1983. Asexual reproduction in Microstomum lineare (Turbellaria). II. The nervous system in the division zone. Int. J. invert. Reprod. 6: 207–217.

Reuter, M., M. Wikgren & I. Palmberg, 1980. The nervous system of Microstomum lineare. 1. A fluorescence and electron-microscopic study. Cell Tissue Res. 211: 31–40.

Rieger, R. M., 1976. Monociliated epidermal cells in Gastrotricha: Significance for concepts of early metazoan evolution. Z. zool. Syst. Evolutionsforsch. 14: 198–226.

Rieger, R. M., 1978. Multiple ciliary structures in developing spermatozoa of marine Catenulida (Turbellaria). Zoomorphologie 89: 229–236.

Rieger, R. M., 1980. A new group of interstitial annelids, Lobatocerebridae nov. fam. and its significance for metazoan phylogeny. Zoomorphologie 94: 41–84.

Rieger, R. M., 1981a. Morphology of the Turbellaria at the ultrastructural level. Hydrobiologia 84: 213–229.

Rieger, R. M., 1981b. Fine structure of the body wall, nervous system and digestive tract in the Lobatocerebridae Rieger and the organization of the gliointerstitial system in Annelida. J. Morph. 167: 139–165.

Rieger, R. M., 1984. Evolution of the cuticle in the lower Eumetazoa. In J. Bereiter-Hahn, A. G. Matoltsy & K. S. Richards (eds), Biology of the integument, Vol. 1, Invertebrates. Springer-

Verlag, Berlin, Heidelberg, New York, Tokyo: 389–399.

Rieger, R. M. & D. Doe, 1975. The proboscis armature of Turbellaria-Kalyptorhynchia, a derivative of the basement lamina? Zool. Scr. 4: 25–32.

Rieger, R. M. & G. Rieger, 1976. Fine structure of the archiannelid cuticle and remarks on the evolution of the cuticle within the Spiralia. Acta zool. 57: 53–68.

Rieger, R. M. & W. Sterrer, 1975. New spicular skeletons in Turbellaria, and the occurrence of spicules in marine meiofauna. Z. zool. Syst. Evolutionsforsch. 13: 207–248.

Rieger, R. M. & S. Tyler, 1974. A new glandular sensory organ in interstitial Macrostomida. I. Ultrastructure. Mikrofauna Meeresboden 42: 1–41.

Rieger, R. M. & S. Tyler, 1979. The homology theorem in ultrastructural research. Am. Zool. 19: 654–666.

Rieger, R. M. & S. Tyler, 1985. Das Homologietheorem in der Ultrastrukturforschung. In J. A. Ott, G. P. Wagner & F. M. Wudetits (eds), Evolution, Ordnung und Erkenntnis: Rupert Riedl zum 60. Geburtstag am 22. Februar 1985. Verlag Paul Parey, Berlin, Hamburg: 21–36.

Röhlich, P., 1966. Sensitivity of regenerating and degenerating planarian photoreceptors to osmium fixation. Z. Zellforsch. mikrosk. Anat. 73: 165–173.

Röhlich, P. & E. Tar, 1968. The effect of prolonged light-deprivation on the fine structure of planarian photoreceptors. Z. Zellforsch. mikrosk. Anat. 90: 507–518.

Röhlich, P. & L. J. Török, 1961. Elektronenmikroskopische Untersuchungen des Auges von Planarien. Z. Zellforsch. mikrosk. Anat. 54: 362–381.

Röhlich, P. & L. J. Török, 1962. The effect of light and darkness on the fine structure of the retinal club in Dendrocoelum lacteum. Ouart. J. Microsc. Sci. 103: 543–548.

Ruppert, E. E., 1978. A review of metamorphosis of turbellarian larvae. In F. S. Chia & M. Rice (eds), Settlement and metamorphosis of marine invertebrate larvae. Elsevier North-Holland, New York: 65–81.

Ruppert, E. E. & S. P. Schreiner, 1980. Ultrastructure and potential significance of cerebral light-refracting bodies of Stenostomum virginianum (Turbellaria, Catenulida). Zoomorphology 96: 21–31.

Ryder, T. A. & I. D. Bowen, 1974. The fine structural localization of alkaline phosphatase in Polycelis tenuis (Iijima). Protoplasma 79: 19–29.

Ryder, T. A. & I. D. Bowen, 1975. The fine structural localization of acid phosphatase activity in Polycelis tenuis (Iijima). Protoplasma 83: 79–90.

Sano, S., 1971. Electron microscopic observation of the cilia in a freshwater planarian Dugesia japonica. I. Fine structure of the cilia of epidermis. Comm. Comp. Sci. Papers Publ. Occ. Retir. Prof. H. Sugino. Series Turbellarians, Osaka: 149–156.

Sarfatti, G. & C. Bedini, 1965. The symbiont alga of the flatworm Convoluta psammophila Bekl. observed at the electron microscope. Caryologia 18: 207–223.

Sauzin, M. J., 1966. Etude au microscope électronique du néoblaste de la planaire Dugesia gonocephala (Turbellarie, Triclade) et de ses changements ultrastructuraux au cours des premieres stades de la régénération. C. r. Acad. Sci., Paris 263: 605–608.

Sauzin, M. J., 1967a. Etude ultrastructurale de la différenciaton du néoblaste au cours de la régénération de la planaire Dugesia gonocephala. I. Différenciation en cellule nerveuse. Bull. Soc. zool. Fr. 92: 313–318.

Sauzin, M. J., 1967b. Etude ultrastructurale de la différenciation au cours de la régénération de la planaire Dugesia gonocephala. II. Différenciation musculaire. Bull. Soc. zool. Fr. 92: 613–616.

Sauzin, M. J., 1968. Présence d'émissions nucléaires dans les cellules différenciées et en différenciation de la planaire adulte Dugesia gonocephala. C. r. Acad. Sci., Paris 267: 1146–1148.

Sauzin-Monnot, M. J., 1972. Etude ultrastructurale du tissue nerveux et des produits de sécretion nerveuse, au course des premieres heures de régénération de la planaire Polycelis nigra (Turbellarie-Triclade) au niveau de la blessure. Ann. Embryol. Morph. 5: 257–265.

Sauzin-Monnot, M. J., 1973a. Etude ultrastructurale des néoblasts de Dendrocoelum lacteum au cours de la régénération. J. ultrastruct. Res.

45: 206–222.
Sauzin-Monnot, M. J., 1973b. Les néoblastes de la planaire Dendrocoelum lacteum: ultrastructure, évolution morphologique apres section et étude des synthèse d'ARN dans le cas de régénération et non régénération céphalique. C. r. Acad. Sci., Paris 276: 1337–1340.
Sauzin-Monnot, M. J., 1974. Etude autoradiographique en microscopie électronique de la synthèse d'acides ribonucléiques dans les néoblastes de la planaire Dendrocoelum lacteum après section a différent niveaux. Ann. Embryol. Morph. 7: 51–59.
Sauzin-Monnot, M. J., 1975a. La régénération des planaires d'eau douce. Etude ultrastructurale des premiers étapes influence du système nerveux. Thèse d'Etat AO-CNRS 10977, 164 pp.
Sauzin-Monnot, M. J., 1975b. Etude ultrastructurale de la régénération chez Dugesia gonocephala et Polycelis nigra, planaires a pouvoir de régénération totale et chez Dendrocoelum lacteum, planaire a pouvoir de régénération céphalique limitée (Turbellaries, Triclades). Bull. Soc. zool. Fr. 100: 293–304.
Sauzin-Monnot, M. J., T. Lender & A. Gabriel, 1970. Etude autoradiographique et ultrastructurale des néoblastes dans la blastème de régénération de 24 heures et dans les tissus en arriére du blastème, chez la planaire Dugesia gonocephala (Turbellarie, Triclade). C. r. Acad. Sci., Paris 271: 1892–1895.
Sayre, R. M. & W. P. Wergin, 1979. The use of SEM to classify and evaluate the parasites and predators of pest nematodes. Scanning electr. Microsc. 1979/III: 89–96.
Schilt, J., 1976. Aspect ultrastructural de l'ovovitelloducte de Dugesia lugubris O. Schmidt (Turbellarie, Triclade). Bull. Soc. zool. Fr. 101: 527–533.
Schilt, J., 1978. Structure de l'extrémité céphalique de l'ovovitelloducte de Dugesia lugubris O. Schmidt. (Turbellarie, Triclade). Bull. Soc. zool. Fr. 103: 107–111.
Schilt, J. & F. Stephan, 1978. L'ovovitelloducte de Dugesia lugubris (Turbellarie, Triclade): étude ultrastructurale. 103 Congr. nat. Soc. Sav. Nancy (3): 99–110.
Schockaert, E. R. & C. Bedini, 1977. Observations on the ultrastructure of the proboscis epithelia of Polycystis naegelii Kölliker (Turbellaria Eukalyptorhynchia) and some associated structures. Acta zool. fenn. 154: 175–191.
Shirasawa, Y. & N. Makino, 1980. Light and electron microscopic studies on early regeneration of land planarian Bipalium sp. type III. [In Japanese with English abstract] Bull. Tokyo. med. Coll. 6: 73–86.
Shirasawa, Y. & N. Makino, 1981. Light and electron microscopic studies on the normal and the regenerating photoreceptor of a land planarian (Bipalium fuscatum). Bull. Tokyo. med. Coll. 7: 35–50.
Shirasawa, Y. & N. Makino, 1982. The effect of light and darkness on the regenerating eyes of a land planarian (Bipalium sp. type III). Bull. Tokyo. med. Coll. 8: 75–87.
Shirasawa, Y. & N. Makino, 1985. Histological study on the intestinal tissue and the parenchyma in the early regeneration of the land planarian, Bipalium nobile Kawakatsu et Makino. [In Japanese; English abstract and captions] Bull. Tokyo med. Coll. 11: 87–99.
Silveira, M., 1962. Contribution a l'étude de l'ultrastructure des muscles chez Dugesia tigrina Girard (Turbellaria). C. r. Acad. Sci., Paris 255: 168–170.
Silveira, M., 1967. Formation of structured secretory granules within the Golgi complex in an acoel turbellarian. J. Microsc. (Paris) 6: 95–100.
Silveira, M., 1968. Action de la pepsine sur un flagelle du type 9+1. Experientia 24: 1243–1245.
Silveira, M., 1969. Ultrastructural studies on a 'nine plus one' flagellum. J. ultrastruct. Res. 26: 274–288.
Silveira, M., 1970. Characterization of an unusual nucleus by electron microscopy. J. submicr. Cytol. 2: 13–24.
Silveira, M., 1972. Association of polysaccharide material with certain flagellar and ciliary structures in turbellarian flatworms. Rev. Microsc. electr. 1: 96–97.
Silveira, M., 1973. Intra-axonemal glycogen in '9+1' flagella of flatworms. J. ultrastruct. Res. 44: 253–264.
Silveira, M., 1975. The fine structure of 9+1 flagella in turbellarian flatworms. In B. A. Afzelius (ed.), The functional anatomy of the spermatozoon. Pergamon Press, Oxford, New York: 289–298.
Silveira, M. & A. Corinna, 1976. Fine structural ob-

servations on the protonephridium of the terrestrial triclad Geoplana pasipha. Cell Tissue Res. 168: 455–463.

Silveira, M. & K. R. Porter, 1964. The spermatozoids of flatworms and their microtubular systems. Protoplasma 59: 240–265.

Skaer, R. J., 1961. Some aspects of the cytology of Polycelis nigra. Quart. J. microsc. Sci. 102: 295–318.

Skaer, R. J., 1965. The origin and continuous replacement of epidermal cells in the planarian Polycelis tenuis (Iijima). J. Embryol. exp. Morph. 13: 129–139.

Sleigh, M. A., 1973. The biology of Protozoa. Edward Arnold Ltd, London, 315 pp.

Smales, L. R. & H. D. Blankespoor, 1978. The epidermis and sensory organs of Dugesia tigrina (Turbellaria: Tricladida) a scanning electron microscope study. Cell Tissue Res. 193: 35–40.

Smith, D. C., 1979. From extracellular to intracellular: the establishment of a symbiosis. Proc. r. Soc. Lond. 204: 115–130.

Smith, J. & M. B. Thomas, 1976. The structure and histochemistry of the epidermis and subepidermis of Stylochus zebra (Turbellaria, Polycladida). J. Elisha Mitchell Sci. Soc. 92: 64.

Smith, J. P. S., 1981a. Fine-structural observations on the central parenchyma in Convoluta sp. Hydrobiologia 84: 259–265.

Smith, J. P. S., 1981b. Fine-structural anatomy of the parenchyma in the Acoela and Nemertodermatida (Turbellaria). Ph.D. Dissertation, University of North Carolina, Chapell Hill.

Smith, J. P. S. & S. Tyler, 1984. Ultrastructure of the frontral organ in Diopisthoporus (Turbellaria Acoela). Am. Zool. 23: 1023.

Smith, J. P. S. & S. Tyler, 1985a. The acoel turbellarians: kingpins of metazoan evolution or a specialized offshoot? In S. Conway-Morris, D. G. George, R. Gibson & J. M. Platt (eds), Origins and relationships of lower invertebrates. Oxford University Press, Oxford. (in press).

Smith, J. P. S. & S. Tyler, 1985b. Fine-structure and evolutionary implications of the frontral organ in Turbellaria Acoela. I. Diopisthoporus gymnopharyngus sp. n. Zool. Scr. 14: 91–102.

Smith, J. P. S. & S. Tyler, 1985c. Frontal organs in the Acoelomorpha (Turbellaria): ultrastructure and phylogenetic significance. This volume.

Smith, J. P. S., S. Tyler & R. M. Rieger, 1985. Is the Turbellaria polyphyletic? This volume.

Smith, J. S., S. Tyler, M. B. Thomas & R. Rieger, 1982. The nature of turbellarian rhabdites: phylogenetic implications. Trans. am. microsc. Soc. 101: 209–228.

Soltynska, M., B. Mroczka & J. Moraczewski, 1975. Ultrastructure of epithelium in some Catenulida. Ann. med. Sec. pol. Acad. Sci. 20: 121–122.

Soltynska, M. S., B. Mroczka & J. Moraczewski, 1976. Ultrastructure of epidermis in Turbellaria from the family Catenulida (Archoophora). J. submicr. Cytol. 8: 293–301.

Sopott-Ehlers, B., 1979. Ultrastruktur der Haftapparate von Nematoplana coelogynoporoides (Turbellaria, Proseriata). Helgoländer wiss. Meeresunters. 32: 365–373.

Sopott-Ehlers, B., 1981. Ultrastructural observations on paracnids. I: Coelogynopora axi Sopott (Turbellaria, Proseriata). Hydrobiologia 84: 253–257.

Sopott-Ehlers, B., 1982. Ultrastruktur potentiell photoreceptorischer Zellen unterschiedlicher Organisation bei einen Proseriat (Platyhelminthes). Zoomorphology 101: 165–176.

Sopott-Ehlers, B., 1984a. Feinstruktur pigmentierter und unpigmentierter Photoreceptoren bei Proseriata (Plathelminthes). Zool. Scr. 13: 9–17.

Sopott-Ehlers, B., 1984b. Epidermale Collar-Receptoren der Nematoplanidae und Polystyliphoridae (Unguiphora, Plathelminthes). Zoomorphology 104: 226–230.

Sopott-Ehlers, B., 1985. Fine-structural characters in female and male germ cells of Proseriata Otoplanidae (Platyhelminthes). This volume.

Spiegelman, M. & P. L. Dudley, 1973. Morphological stages of regeneration in the planarian Dugesia tigrina: A light and electron microscopic study. J. Morph. 139: 155–184.

Stephan-Dubois, F. & F. Bautz, 1975. Etude histologique et cytologique des néoblastes et des cellules fixes du parenchyme chez des planaires Dendrocoelum lacteum privées de région anterieurs. Arch. anat. microsc. Morphol. exp. 64: 75–89.

Stephan-Dubois, F. & M. Gusse, 1974. Origine et differenciation des cellules vitellines lors de la régénération sai sonniere des vitellogénes, chez la planaire Dendrocoelum lacteum. Wilhelm Roux Arch. devl Biol. 174: 181–194.

Stephan-Dubois, F., R. Lanot & A. M. Bautz, 1974. Aspects structuraux des dégénérescences cellulaires au cours des processus morphogénetiques. Ann. Biol. 13: 27–34.

Sterrer, W. & R. Rieger, 1974. Retronectidae – a new cosmopolitan marine family of Catenulida (Turbellaria). In N. W. Riser & M. P. Morse (eds), Biology of the Turbellaria. McGraw-Hill, New York: 63–92.

Stewart, A. M., 1966. Ultrastructure of the eyes in three freshwater triclad flatworms. Am. Zool. 6: 615–616.

Storch, V., 1973. Vergleichende elektronenmikroskopische Untersuchungen über Receptoren von Wirbellosen (Nemertinen, Turbellarien, Molluscen, Anneliden, Aschelminthen). Verh. dt zool. Ges. 66: 61–65.

Storch, V. & R. Abraham, 1972. Elektronenmikroskopische Untersuchungen über die Sinneskante des terricolen Turbellars Bipalium kewense Moseley (Tricladida). Z. Zellforsch. mikrosk. Anat. 133: 267–275.

Storch, V. & U. Welsch, 1977. Septate junctions in the cephalic epidermis of turbellarians (Bipalium). Cell Tissue Res. 184: 423–425.

Sun, C. N., S. B. Cheng Chew & H. J. White, 1979. The fine structure of smooth muscle in Bipalium kewense Moseley: its possible relation to evolution of skeletal muscle. Cytologia 44: 181–190.

Taylor, D. L., 1971. On the symbiosis between Amphidinium klebsii (Dinophyceae) and Amphiscolops langerhansi (Turbellaria, Acoela). J. mar. biol. Ass. U.K. 51: 301–313.

Thomas, M. B., 1970. Transitions between helical and protofibrillar configurations in doublet and singlet microtubules in spermatozoa of Stylochus zebra (Turbellaria – Polycladida). Biol. Bull. 138: 219–234.

Thomas, M. B., 1975. The structure of the 9+1 axonemal core as revealed by treatment with trypsin. J. ultrastruct. Res. 52: 409–422.

Thomas, M. B., 1985. Embryology of the Turbellaria and its phylogenetic significance. This volume.

Thomas, M. B. & C. Henley, 1971. Substructure of the cortical singlet microtubules in spermatozoa of Macrostomum (Platyhelminthes Turbellaria) as revealed by negative staining. Biol. Bull. 141: 592–601

Török, L. J. & P. Röhlich, 1959. Contributions to the fine structure of the epidermis of Dugesia lugubris O. Schm. Acta Biol. Acad. Sci. Hung. 10: 23–48.

Török, L. J. & P. Röhlich, 1960. Elektronmikroszkopos vizsgalatok planariak eozinofil mirigysejtjein. Biol. Kozlemenyek 8: 117–124.

Tombes, A. S., 1970. An introduction to invertebrate endocrinology. Academic Press, New York, 217 pp.

Tulloch, G. S. & B. R. Hershenov, 1967. Fine structure of platyhelminth sperm tails. Nature 213: 299–300.

Tyler, S., 1973. An adhesive function for modified cilia in an interstitial turbellarian. Acta zool. 54: 139–151.

Tyler, S., 1975. Comparative ultrastructure of adhesive systems in the Turbellaria and other interstitial animals. Ph.D. Dissertation. University of North Carolina, Chapel Hill.

Tyler, S., 1976a. Comparative ultrastructure of adhesive systems in the Turbellaria. Zoomorphologie 84: 1–76.

Tyler. S., 1976b. Comparative morphology at the ultrastructural level: a unique type of cilium in lower invertebrates. Canadian Fed. Biol. Soc. 19 (1976): 519.

Tyler, S., 1977. Ultrastructure and systematics: an example from turbellarian adhesive organs. Mikrofauna Meeresboden 61: 271–286.

Tyler, S., 1979. Distinctive features of cilia in metazoans and their significance for systematics. Tissue Cell 11: 385–400.

Tyler, S., 1980. Development of cilia in embryos of turbellarians. Am. Zool. 20: 893.

Tyler, S., 1981. Development of cilia in embryos of the turbellarian Macrostomum. Hydrobiologia 84: 231–239.

Tyler, S., 1982. Functional anatomy of turbellarian epidermis at the ultrastructural level. Trans. am. microsc. Soc. 102: 87–88.

Tyler, S., 1984a. Ciliogenesis in embryos of the acoel turbellarian Archaphanostoma. Trans. am. microsc. Soc. 103: 1–15.

Tyler, S., 1984b. Turbellarian platyhelminths. In J. Bereiter-Hahn, A. G. Matoltsy, K. S. Richards (eds), Biology of the integument. Vol. 1. Invertebrates. Springer Verlag, Berlin: 112–131.

Tyler, S. & L. A. Melanson, 1979. Cytochemistry of adhesive organs in lower metazoans. Am. Zool.

19: 985.

Tyler, S. & R. M. Rieger, 1975. Uniflagellate spermatozoa in Nemertoderma (Turbellaria) and their phylogenetic significance. Science 188: 730–732.

Tyler, S. & R. M. Rieger, 1977. Ultrastructural evidence for the systematic position of the Nemertodermatida (Turbellaria). Acta zool. fenn. 54: 193–207.

Vagnetti, D., R. M. Farnesi, S. Tei & M. Marinelli, 1975. Rilievi ultrastrutturali su alcuni aspetti funzionale della borsa copulatrice in Dugesia lugubris s.l. Boll. Soc. ital. Biol. Sperim. 51: 581–587.

Vanfleteren, J. R., 1982. A monophyletic line of evolution? Ciliary induced photoreceptor membranes. In J. A. Westfall (ed.), Visual cells in evolution. Raven Press, New York: 107–136.

Vanfletern, J. R. & A. Coomans, 1976. Photoreceptor evolution and phylogeny. Z. zool. Syst. Evolutionsforsch. 14: 157–169.

Venturini, G., G. Palladini, V. Margotta, L. Medolago-Albani, A. Carolei & M. C. Hernandez, 1981. Ouabain-insensitive ATPase in planaria. Comp. Bioch. Physiol. 70B: 775–778.

Webb, R. I., 1979. Light and electron microscopic studies of spermatogenesis and spermatozoal morphology in temnocephalids (Platyhelminthes, Turbellaria). Thesis, University of Queensland, Brisbane, Australia, 129 pp.

Welsch, U. & V. Storch, 1976. Comparative animal histology and cytology. University of Washington Press, 343 pp.

Welsh, H. J. & L. D. Williams, 1970. Monoamine-containing neurons in planaria. J. comp. Neurol. 138: 103–116.

Werner, H. J., 1969. Electron microscope observations of the pharynx of the turbellarian Dugesia tigrina. Proc. La Acad. Sci. 32: 7–17.

Wetzel, B. K., 1961. Sodium permanganate fixation for electron microscopy. J. Biophys. Bioch. Cytol. 9: 711–716.

Wetzel, B. K., 1962. Contributions to the cytology of Dugesia tigrina (Turbellaria) protonophridia. In S. S. Breese (ed.), Proceedings Fifth International Congress of Electron Microscopy, Philadelphia, Vol. 2, Academic Press, New York: Q–10.

Wetzel, B. K., 1963. Studies on the fine structure of regenerating Dugesia tigrina (Girard). Thesis, Harvard University.

Williams, J. B., 1975. Studies on the epidermis of Temnocephala. I. Ultrastructure of the epidermis of Temnocephala novae-zealandiae. Aust. J. Zool. 23: 321–331.

Williams, J. B., 1977. Studies on the epidermis of Temnocephala. II. Epidermal sensory receptors of Temnocephala novae-zealandiae. Aust. J. Zool. 25: 187–191.

Williams, J. B., 1978. Studies on the epidermis of Temnocephala. III. Scanning electron microscope study of the epidermal surface of Temnocephala dendyi. Aust. J. Zool. 26: 217–224.

Williams, J. B., 1979. Studies on the epidermis of Temnocephala. IV. Observations on the ultrastructure of the epidermis of Temnocephala dendyi, with notes on the role of the Golgi apparatus in autolysis. Aust. J. Zool. 27: 483–499.

Williams, J. B., 1980. Studies on the epidermis of Temnocephala. V. Further observations on the ultrastructure of the epidermis of Temnocephala novae-zealandiae, including notes on the glycocalyx. Aust. J. Zool. 28: 43–57.

Williams, J. B., 1981a. Classification of the Temnocephaloidea (Platyhelminthes). J. nat. Hist., Lond. 15: 277–299.

Williams, J. B., 1981b. Protonephridial system of Temnocephala novae-zealandiae: structure of the flame cells and main vessels. Aust. J. Zool. 29: 131–146.

Williams, J. B., 1982. Studies on the epidermis of Temnocephala. VI. Epidermal topography of T. novae-zealandiae and other Australasian temnocephalids, with notes on microvillar and coated vesicle function and the evolution of a cuticle. Aust. J. Zool. 30: 375–390.

Williams, J. B., 1983. The genital system of Temnocephala. I. Ultrastructural features of the differentiating spermatid of Temnocephala novaezealandiae, including notes on a possible correlation between cellular autophagy and mitochondrial function. Aust. J. Zool. 31: 317–332.

Williams, J. B., 1984. Cells in the parenchyma of Temnocephala. I. A large secretory cell, with a 'honeycomb' nucleus surrounded by juxtanuclear desmosomes: a new cell type in Temnocephala novaezealandiae (Temnocephalidae: Platyhelminthes). Aust. J. Zool. 32: 207–218.

Williams, J. B., 1984b. The genital system of Temnocephala II. Further observations on the spermatogenesis of Temnocephala novaezealandiae, with particular reference to the mitochondria. Aust. J. Zool. 32: 447–461.

Williams, J. B., 1985. Phylogenetic relationships of the Temnocephaloidea (Platyhelminthes). This volume.

Wilson, R. A. & L. A. Webster, 1984. Protonephridia. Biol. Rev. 49: 127–160.

Wolken, J., 1958. Studies of photoreceptor structures. Ann. N.Y. Acad. Sci. 74: 164–181.

Yamasu, T., 1982. Five new species of acoel flat worms from Japan. Galaxea 1: 29–43.

Yamasu, T., Y. Tchimiya & T. Kanda, 1979. The structure and function of the ocelli of an acoel flatworm. Photomed. Photobiol. 1: 141–142.

Yanagita, Y. & T. S. Yamamoto, 1981. Formation and chemical property of the cocoon in a freshwater planarian, Dendrocoelopsis ezensis Ichikawa et Okugawa (Turbellaria, Tricladida). Zool. Anz. 207: 130–142.

Yoshida, M., 1979. Extraocular photoreception. In H. Autrum (ed.), Handbook of sensory physiology, Vol. VII/6A. Comparative physiology and evolution of invertebrate photoreceptors. Springer Verlag, New York: 581–640.

A computer-readable version of this bibliography is available from Seth Tyler.

Index to authors

Index to subjects

www.ingramcontent.com/pod-product-compliance
Ingram Content Group UK Ltd.
Pitfield, Milton Keynes, MK11 3LW, UK
UKHW051117190926
475688UK00005B/207

* 9 7 8 9 0 6 1 9 3 5 4 2 1 *